Aquatic Ecosystems Monitoring

This book collates traditional and modern applications of remote sensing in aquatic ecosystem monitoring. It covers conventional assessment methods like sampling, surveying, macroinvertebrates, and chlorophyll estimation for aquatic ecosystem health assessment. Advanced remote sensing technology provides timely spectral information for quantitative and qualitative assessment of water quality, shoreline changes, coral bleaching, and vegetation monitoring. The book covers different types of aquatic ecosystems like wetlands, rivers, lakes, saline, and the brackish lake. It also:

- Reviews the latest applications of remote sensing in the monitoring and assessment of aquatic ecosystems.
- Includes traditional methods like cartography, sampling, surveying, phytoplankton assessment, river interlinking, and chlorophyll estimation.
- Discusses the application of multi-source data and machine learning in monitoring aquatic ecosystems.
- Discusses aquatic ecosystem management, services, threats, and sustainability.
- Explores challenges, opportunities, and prospects of future Earth observation applications for aquatic ecosystem monitoring.

The book discusses space-borne, airborne, and drone geospatial data. The parts broadly cover aquatic ecosystem monitoring, vegetation management, advanced modeling practices, and challenges. It is meant for scientists, professionals, and policymakers working in environmental sciences, remote sensing, and geology.

Aquatic Ecosystems Monitoring

Conventional Assessment to Advanced Remote Sensing

Edited by
Prem Chandra Pandey, Prashant K. Srivastava, and
Sanjeev Kumar Srivastava

CRC Press
Taylor & Francis Group
Boca Raton London New York

CRC Press is an imprint of the
Taylor & Francis Group, an **informa** business

Designed cover image: Shutterstock

First edition published 2025
by CRC Press
2385 NW Executive Center Drive, Suite 320, Boca Raton FL 33431

and by CRC Press
4 Park Square, Milton Park, Abingdon, Oxon, OX14 4RN

CRC Press is an imprint of Taylor & Francis Group, LLC

ISBN: 978-1-032-39112-0 (hbk)
ISBN: 978-1-032-40623-7 (pbk)
ISBN: 978-1-003-35400-0 (ebk)

DOI: 10.1201/9781003354000

Typeset in Times
by Deanta Global Publishing Services, Chennai, India

Contents

Foreword

In recent decades, the rapid advancement of remote-sensing technologies has revolutionized the way we monitor and manage aquatic ecosystems. From the crystal-clear waters of pristine lakes to the sprawling expanses of coastal wetlands, the ability to observe and analyse these vital ecosystems from space, air, and ground-based platforms has ushered in a new era of environmental science. This book, *Aquatic Ecosystems Monitoring: Conventional Assessment to Advanced Remote Sensing*, stands as a testament to the evolution of these techniques and their crucial role in safeguarding our planet's water resources. This comprehensive volume brings together the collective expertise of scientists, researchers, and practitioners who are at the forefront of this field. It bridges the gap between traditional assessment methods and cutting-edge remote-sensing technologies, offering a holistic view of aquatic ecosystem monitoring. The book is meticulously structured into several parts, each dedicated to a specific aspect of the field, making it an invaluable resource for those involved in environmental sciences, remote sensing, and geology.

The opening sections lay a strong foundation by revisiting conventional methods of aquatic ecosystem assessment, such as sampling, surveying, and the use of bioindicators like macroinvertebrates and diatoms. These time-tested techniques continue to provide essential insights into ecosystem health, particularly in freshwater environments. However, as our understanding of aquatic systems has grown, so too has the need for more advanced and timely data. The subsequent chapters delve into the remarkable capabilities of remote-sensing technologies, including space-borne, airborne, and drone-based platforms. These tools offer unparalleled opportunities for monitoring water quality, detecting shoreline changes, assessing coral reef health, and mapping aquatic vegetation. By harnessing multi-source data and integrating machine learning algorithms, researchers are now able to generate more precise and comprehensive assessments of aquatic ecosystems, providing critical information for decision-makers.

In addition to monitoring, the book explores the complex challenges of managing aquatic ecosystems in the face of climate change, pollution, and other anthropogenic pressures. It discusses the sustainability of these ecosystems and offers insights into the management practices that can help mitigate the adverse impacts of human activities. The discussion extends to the broader implications of ecosystem services and the threats posed by phenomena such as plastic pollution. As we look to the future, the book concludes with an exploration of the challenges and opportunities that lie ahead in the field of aquatic ecosystem monitoring. It is a call to action for continued innovation and collaboration among scientists, professionals, and policymakers to ensure the long-term sustainability of our planet's water resources.

Aquatic Ecosystems Monitoring: Conventional Assessment to Advanced Remote Sensing is not just a compilation of research; it is a beacon for those committed to the protection and preservation of aquatic ecosystems. It serves as both a reference and a guide, inspiring future generations of environmental scientists to explore the possibilities that remote sensing offers in this ever-evolving field. As we embark on this journey into the depths of aquatic ecosystems, let us embrace the challenges with curiosity. May this volume inspire and empower the next generation of researchers and practitioners to harness the power of geospatial technology in service of the world's aquatic environments, ensuring their conservation and resilience for generations to come.

I am confident that this book will be an excellent resource for the students and researchers alike. I congratulate Dr Pandey and team for undertaking an important task of editing this book, which will serve a wide range of research and professional communities in upcoming years.

August 2024

Prof Timothy Dube
Director, Institute for Water Studies,
Professor, Department of Earth Science
University of the Western Cape, South Africa

Preface

As we stand at the threshold of the Anthropocene, our understanding of aquatic ecosystems has never been more crucial. With the rapid pace of environmental change and the myriad of stressors facing these delicate ecosystems, it is imperative that we employ a multidisciplinary approach to monitor, assess, and manage ecosystem health and resilience. In this volume, *Aquatic Ecosystems Monitoring: Conventional Assessment to Advanced Remote Sensing*, we embark on a journey through the realms of freshwater bodies to marine environments including riverine ecosystems, terrestrial or coastal wetlands, marsh, swamp, mangroves, delta regions, tidal flats, inter-tidal zones, exploring the intricacies of their monitoring using both traditional ecological knowledge and cutting-edge geospatial technologies.

The chapters within this volume represent a diverse array of perspectives, methodologies, and case studies, each shedding light on different aspects of aquatic ecosystem monitoring. From the conventional assessment of macroinvertebrates in freshwater ecosystems to the application of remote sensing for monitoring coastal chlorophyll-a concentration, each chapter offers valuable insights into the challenges and opportunities inherent in the monitoring of aquatic environments.

In Part I, the editorial message provides insightful to the overall book structure and contents about aquatic ecosystems and their monitoring using conventional and advanced techniques.

In Part II, authors delve into the traditional methods employed in monitoring aquatic ecosystems, from the evaluation of aquatic ecosystem health using diatoms and macroinvertebrates as biomonitoring tools to understanding the impact of river interlinkage. How macroinvertebrates are used to monitor aquatic ecosystems and assess their health in presented in a very articulated manner.

In Part III, authors demonstrate the advanced tools and techniques employed in monitoring aquatic ecosystems. From the evaluation of aquatic ecosystems using remote sensing and GIS to the mapping of shoreline changes in river deltas using geo-information technologies, the chapters in this part showcase the breadth and depth of approaches utilized in aquatic ecosystem monitoring. Moreover, this part also delves into a thorough analysis of stream channel dynamics within alluvial fan systems.

Part IV focuses on aquatic vegetation monitoring and assessment, highlighting the importance of remote sensing and machine learning in mapping and identifying aquatic vegetation species. Additionally, water quality analysis using chlorophyll concentration and its implications for aquatic ecosystem health are examined in detail.

In Part V, the authors explore the utilization of resources with sustainability, threats to these sensitive ecosystems, and challenges faced for the aquatic ecosystems. From the impact of climate change, land surface temperature, rainfall trends to land use/land cover dynamics on wetland ecosystems, along with the assessment of spatial distribution methods for water quality mapping, the chapters in this part provide valuable insights into the complex interactions between human activities and aquatic ecosystems such as fishing potential zonation in the ocean.

Finally, Part VI examines the challenges and future opportunities in aquatic ecosystem monitoring. From the management of aquatic ecosystems and vegetation for environmental sustainability to the estimation of potential fishing zones using geospatial technology, the chapter in this part offers a forward-looking perspective on the future of aquatic ecosystem monitoring.

Moreover, the editors wish to convey their sincere thanks to the esteemed publisher, CRC Press, Taylor & Francis Group, for providing the opportunity to bring together the valuable contributions of several contributors in this book. Special appreciation goes to all the authors for their enthusiastic efforts in delivering high-quality chapters and their cooperation during the screening, review, and copy-editing process. The editors are deeply indebted and grateful to the reviewers who dedicated their time and efforts to review the chapter manuscripts, providing invaluable voluntary support.

As we learn the complexities of aquatic ecosystems, let us heed the lessons learned from both traditional knowledge and advanced technology. May this volume serve as a catalyst for collaboration, research innovation, and stewardship in the ongoing effort to utilize and safeguard our planet's aquatic treasures for future generations.

We hope that this Preface has successfully provided insights into aquatic ecosystems, types, and conventional methods as well as advanced technology for sampling, assessment, and monitoring of aquatic ecosystems topics covered in this book. We encourage readers to adapt and utilize the content in a manner that best suits their needs, helping them understand the capabilities and potentials of emerging technologies and their applications for aquatic ecosystem monitoring, which is the core focus of this book. Users are welcome to inform the editors of any errors, suggestions, or comments to further improve and enhance the book's content.

July 2024
Dr. Prem C. Pandey, India

Dr. Prashant Srivastava, India

Dr Sanjeev Srivastava, Australia

Contributors

Prasannajit Acharya
Institute of Technical Education and Research, Department of Chemistry, Siksha 'O' Anusandhan (Deemed to be University), Bhubaneswar, India

Mubaraka S. Alghariani
Libyan Authority for Scientific Research, Tripoli, Libya

Mohit Arora
Punjab Remote Sensing Centre, PAU Campus, Ludhiana, Punjab, India

Heba Bedair
Botany Department, Faculty of Science, Tanta University, Tanta, Egypt
SILVIS Lab, Department of Forest and Wildlife Ecology, University of Wisconsin-Madison, Madison, WI, USA

Pinkey Bisht
Wadia Institute of Himalayan Geology, Dehradun, Uttarakhand, India

Darshanaa Chellaiah
School of Science, Technology & Engineering, University of the Sunshine Coast, Queensland, Australia

Anannya Dahiya
School of Agricultural Science, Sharda University, Noida, Uttar Pradesh, India

Spyridon E. Detsikas
Department of Geography, Harokopio University of Athens, Kallithea, Athens, Greece

Rakesh Kumar Dwivedi
Department of Botany, Bhakt Darshan Govt. P.G. College, Jaiharikhal, Uttarakhand, India

Suneet Dwivedi
K Banerjee Centre of Atmospheric and Ocean Studies, University of Allahabad, Prayagraj, Uttar Pradesh, India

Asmaa M. El-Makawy
Botany and Microbiology Department, Faculty of Science, Alexandria University, Alexandria, Egypt

Azzaoui Mohamed Essalah
Nature and Life Sciences Faculty, Ibn Khaldoun University of Tiaret, Algeria

Marwa Waseem A. Halmy
Department of Environmental Sciences, Faculty of Science, Alexandria University, Alexandria, Egypt

Bouacha Mohamed Islem
Nature and Life Sciences Faculty, Ibn Khaldoun University of Tiaret, Algeria

Efthimios Karymbalis
Department of Geography, Harokopio University of Athens, Kallithea, Athens, Greece

Basheer Ahammed KK
Universitas Udayana (UNUD), Denpasar, Bali, Indonesia

Amit Kumar
Department of Geography, Sikkim University, Gangtok, India

Dasputre Yash Sanjay Kumar
School of Agricultural Science, Sharda University, Noida, Uttar Pradesh, India

Suthar Shubham Kumar
Sharda University, Greater Noida, Uttar Pradesh, India

Soudani Leila
Nature and Life Sciences Faculty, Ibn Khaldoun University of Tiaret, Algeria

Abhijit Mazumder
Birbal Sahni Institute of Paleosciences, Lucknow, India

Manoj MC
Birbal Sahni Institute of Paleosciences, Lucknow, India

Nezha Mejjad
National Center for Energy, Sciences and Nuclear Techniques, Rabat, Morocco

Smrutisikha Mohanty
Department of Life Sciences, School of Natural Sciences, Shiv Nadar Institution of Eminence (deemed to be University), India

Pradipta R. Muduli
Wetland Research and Training Centre, Chilika Development Authority, Balugaon, India

Haidy Nasser
Botany and Microbiology Department, Faculty of Science, Kafrelsheikh University, Kafrelsheikh, Egypt

Safa Omar
Nature and Life Sciences Faculty, Ibn Khaldoun University of Tiaret, Algeria

Arvind Chandra Pandey
Department of Geoinformatics, School of Natural Resource Management, Central University of Jharkhand, Brambe, Ranchi, Jharkhand, India

Ashutosh K. Pandey
School of Agricultural Science, Sharda University, Noida, Uttar Pradesh, India

Manish Pandey
University Center for Research & Development (UCRD), Chandigarh University, Mohali, India
Department of Civil Engineering, Chandigarh University, Mohali, India

Prem Chandra Pandey
Department of Life Sciences, School of Natural Sciences, Shiv Nadar Institution of Eminence (deemed to be University), India

Suchita Pandey
K Banerjee Centre of Atmospheric and Ocean Studies, University of Allahabad, Prayagraj, Uttar Pradesh, India

Brijendra Pateriya
Punjab Remote Sensing Centre, PAU Campus, Ludhiana, Punjab, India

Subhashree Patra
Department of Environmental Sciences, Central University of Jharkhand, Ranchi, India

George P. Petropoulos
Department of Geography, Harokopio University of Athens, Kallithea, Athens, Greece

Parag Roy
Central University of Karnataka, Kadaganchi, Karnataka, India

Ehab M. Sagar
Department of Soil and Water, Faculty of Agriculture, University of Tripoli, Tripoli, Libya

Sashikanta Sahoo
Punjab Remote Sensing Centre, PAU Campus, Ludhiana, Punjab, India

Purabi Saikia
Department of Botany, Banaras Hindu University, Varanasi, India

Shilky
Department of Environmental Sciences, Central University of Jharkhand, Ranchi, India

Abha Singh
Birbal Sahni Institute of Paleosciences, Lucknow, India

Divya Singh
School of Science, Technology & Engineering, University of the Sunshine Coast, Queensland, Australia

Sudhir Kumar Singh
K Banerjee Centre of Atmospheric and Ocean Studies, University of Allahabad, Prayagraj, Uttar Pradesh, India

Prashant K. Srivastava
Institute of Environment and Sustainable Development, Banaras Hindu University, Varanasi, India

Prateek Srivastava
Department of Botany, University of Allahabad, Prayagraj, India

Sanjeev Kumar Srivastava
School of Science, Technology & Engineering, University of the Sunshine Coast, Queensland, Australia

Isha Swati
Punjab Agricultural University, Ludhiana, Punjab, India

Ioanna Tzanavari
Department of Geography, Harokopio University of Athens, Kallithea, Athens, Greece

Prem Raj Uddandam
Birbal Sahni Institute of Paleosciences, Lucknow, India

Jyoti Verma
Department of Zoology, CMP Degree College, University of Allahabad, Prayagraj, India

Md. Wasim
Department of Life Sciences, School of Natural Sciences, Shiv Nadar Institution of Eminence, Greater Noida, Uttar Pradesh, India

Anuradha Yadav
Department of Zoology, CMP Degree College, University of Allahabad, Prayagraj, India

Catherine M. Yule
School of Science, Technology & Engineering, University of the Sunshine Coast, Queensland, Australia

Arabi Zohra
Nature and Life Sciences Faculty, Ibn Khaldoun University of Tiaret, Algeria

About the Editors

Dr Prem Chandra Pandey is an Assistant Professor at the Shiv Nadar Institution of Eminence (Deemed to be University), Greater Noida, India. He was associated with Tel Aviv University as a Research Scientist and with Banaras Hindu University as a Post-Doctoral Scientist. Pandey received his Ph.D. degree from the University of Leicester, UK, under the Commonwealth Scholarship and Fellowship Plan, Association of Commonwealth Universities. Pandey is the recipient of several international and national awards including Inspire, Malviya Gold medal, DST-SERB, UGC-MHRD, and Tel Aviv University fellowship. He is serving as an editorial member for the journals *Geocarto International* (Taylor & Francis) and *Sustainable Development* (Wiley), and is associated with *Remote Sensing* (MDPI) as topic editor. He has authored over 60 research articles and edited more than eight books and 25 book chapters.

Dr Prashant K. Srivastava is a faculty member at the Institute of Environment & Sustainable Development, Banaras Hindu University. He was previously affiliated with the Hydrological Sciences, NASA Goddard Space Flight Center as a research scientist on SMAP satellite soil moisture retrieval algorithm development, instrumentation, and simulation for various applications. He received his Ph.D. degree from the Department of Civil Engineering, University of Bristol, Bristol, UK. Prashant is the recipient of several awards including the NASA Fellowship, USA; University of Maryland Fellowship, USA; and Commonwealth Fellowship, UK. He has published over 140 research articles, seven books, and over 100 chapters.

Dr Sanjeev Kumar Srivastava is a Senior Lecturer, Geospatial Analysis, University of Sunshine, Australia. He teaches and coordinates spatial science courses which include geographical information systems (GIS) and remote sensing and surveying courses that are part of Geospatial Analysis Minor. He obtained his Ph.D. degree in Geographical Sciences from The Australian National University where he tested several spatial models for predicting fish species' distribution in the inland aquatic ecosystem. Srivastava has worked with different organizations including the Australian Defence Force Academy, Bureau of Rural Sciences, The World Fish Centre, National Bureau of Fish Genetic Resources, and several locations at the Australian National University.

Part I

General

1 Introduction to Aquatic Ecosystems – Editorial Message

*Prem C. Pandey**
Department of Life Sciences, School of Natural Sciences, Shiv Nadar Institution of Eminence, Greater Noida, Uttar Pradesh, India

Prashant K. Srivastava
Institute of Environment and Sustainable Development, Banaras Hindu University, Varanasi, Uttar Pradesh, India

Sanjeev Kumar Srivastava
School of Science, Technology & Engineering, University of the Sunshine Coast, Queensland, Australia

1.1 INTRODUCTION

In today's dynamic world, understanding and preserving aquatic ecosystems have become more critical than ever. The health of these ecosystems directly impacts the well-being of both the environment and human societies. Therefore, it is imperative to employ effective monitoring techniques to assess the status of aquatic environments accurately. As we step ahead on this journey through the contents of *Aquatic Ecosystems Monitoring: Conventional Assessment to Advanced Remote Sensing* we encourage readers to explore the basic concepts, types of aquatic ecosystems, their importance, and monitoring techniques. Readers will find interesting and eye-catching contents for aquatic ecosystems, ranging from wetlands, riverine, coastal zone to ocean, along with advanced tools in monitoring. Users will explore the evolution of monitoring techniques, from the traditional yet important practices rooted in centuries-old wisdom to the cutting-edge technologies revolutionizing the field today, including remote sensing and geographic information system (GIS) techniques. By bridging the gap between tradition and emerging technologies, this edited volume aims to provide undergraduate, master's, PhD, and early career researchers, along with policymakers and conservationists with the knowledge and tools necessary to address the complex challenges facing aquatic ecosystems worldwide, mainly spatial and temporal variability. We extend our gratitude to the authors for their invaluable contributions and dedication to advancing our understanding of aquatic ecosystems monitoring: from molecular level eDNA to advanced remote sensing and GIS techniques. We hope this collection will serve as a treasure of insight and inspiration for all researchers working in the aforementioned research theme and committed to the conservation and sustainable use of our planet's aquatic treasures, and promise to safeguard freshwater ecosystems for future generations. The entire book is divided into six parts: Part I, Editorial; Part II, Conventional aquatic ecosystems monitoring; Part III, Advanced aquatic ecosystems monitoring; Part IV, Aquatic vegetation monitoring and assessment; Part V, Aquatic ecosystem management, services, threats, and sustainability; and finally Part VI, Challenges and future opportunities with concluding remarks.

* Corresponding author.

DOI: 10.1201/9781003354000-2

and Libyan waters. The study demonstrates the effectiveness of the cloud computing platform for monitoring Chl-a concentration and SST in marine coastal areas at large spatial scales.

Chapter 11 reviews the importance of monitoring aquatic vegetation for the well-being of aquatic life and overall ecosystem health, highlighting their functions such as providing food, habitat protection, water quality improvement, and nutrient absorption (Mohanty et al. 2024a). Leveraging geospatial technologies, particularly satellite-based sensors, drones, and radar data, enables the characterization and mapping of the spatio-temporal distribution of aquatic flora. The chapter offers a comprehensive review of various types of aquatic plants, emphasizing their significance and the necessity of mapping these species. This chapter has a unique section about the spectral discrimination of eight different aquatic vegetations and provides the steps to identify them using the continuum removal method. Moreover, it demonstrates the importance of multisource data in identifying aquatic plants and conducts scientometric analysis using Scopus database keywords, including co-occurrence maps, thematic maps, trend topic analysis, and countries' scientific production trends. The study aims to inspire future researchers to delve deeper into this promising field of study by collating important information from different databases or collective research.

Chapter 12 outlines a study conducted on the Arabian Sea (AS), focusing on its prominent blooming regions influenced by atmospheric and oceanic conditions (Pandey et al. 2024a). The variability of chlorophyll-a (Chl-a) in the basin is explored, affected by various physical mechanisms. Employing the Regional Ocean Modelling System (ROMS) coupled with a biogeochemical model and satellite observations, the study aims to understand the role of meteorological and oceanographic parameters on Chl-a variability during winter and summer monsoons. It highlights factors such as Ekman pumping, wind stress curl, wind-induced mixing, ocean waves, gyres, and lateral advection in driving Chl-a variability. Furthermore, the study delves into the seasonality of physical parameters and various marine components like oxygen, nitrate, phytoplankton, zooplankton, and detritus across different regions and different depths (surface and subsurface) of the Arabian Sea. Notable results are seen in this particular study; for detailed information refer to the above-mentioned chapter.

Part V provides content on sustainable management of aquatic ecosystems and its services while maintaining ecological balance. It also demonstrates how humans can utilize the rich ecosystems and present threats to aquatic ecosystems due to both natural as well as human impact. These impacts are caused by several factors such as the land surface temperature and rainfall trends across the ecosystems and climate change, while human-induced impacts include plastic pollution, a great menace to the environment of all types, mostly algal blooms of water bodies resulting from increased nutrient loads coming from agricultural run-offs affecting the aquatic ecosystem adversely. This part also demonstrates the sustainable use of resources to support aquatic ecosystems, for harnessing their services. It contains six well-curated chapters on the above topics.

Chapter 13 highlights the significance of aquatic ecosystems in supporting human activities such as fishing, aquaculture, and recreation, while also serving as habitats for biodiversity and regulating water quality (Shilky et al. 2024). However, global water pollution, exacerbated by factors like insufficient awareness and lax enforcement of eco-friendly regulations, poses a serious threat. Additionally, water scarcity affects a significant portion of the global population due to various factors including climate change and urbanization. Despite these challenges, aquatic ecosystems offer opportunities for sustainable management, such as using aquatic plants to absorb pollutants and produce bioenergy. In this regard, authors briefly mentioned aquatic vegetation and their role in ecosystem management. Geospatial technologies play an important role in monitoring and understanding aquatic ecosystems. The authors stress the implementation of a robust legislative framework and integrated maritime policies, and consider them as essential in each country to achieve Sustainable Development Goal-14 (which focuses on conserving and sustainably using oceans, seas, and marine resources). The chapter also discusses these challenges, remote sensing tools for monitoring, and proposes policy interventions and sustainable solutions to achieve SDG-14.

Chapter 14 focuses on the significance of fish harvesting for coastal communities in Peninsular India and the underutilization of available marine resources by fishermen (Wasim et al. 2024a). The study employed MODIS-derived data to identify potential fishing zones (PFZs) along the Indian coastline by analyzing sea surface temperature and chlorophyll-a concentration variations. It revealed seasonal SST patterns, with high temperatures on the western coast during summer and on the eastern coast during the monsoon, and moderate variations in winter along the Indian peninsula region. The study demonstrated a notably high concentration of Chl-a in certain regions along the coastline of India. By correlating SST and Chl-a together, it identified optimal conditions for PFZs. PFZ maps were generated, providing valuable information for safe and productive fishing activities, contributing to sustainable coastal management and enhancing the livelihoods of fishing communities.

Chapter 15 focuses on the vital role of wetlands in maintaining ecological balance and sustainability, highlighting their functions such as water filtration, weather moderation, and habitat provision (Arora et al. 2024). The authors stress that the shrinkage of wetlands is due to both natural factors and human activities. Specifically, the research investigates the impact of climate change and land use and land cover dynamics on wetlands. The authors examined spatial and temporal variations in wetland characteristics using climatic parameters and Landsat imagery from 2000 to 2022, particularly before and after the monsoon seasons. They classified study sites into four categories: agricultural land, built-up areas, water bodies, and wastelands. Results revealed substantial changes over the two-decade period, with analyses of land surface temperature (LST) and rainfall patterns providing insights into temperature and precipitation trends and its impact on wetlands. The findings emphasize the need for effective monitoring and conservation efforts, stressing the importance of remote sensing data for resource management and policy making. Finally, the study stresses the urgent need to preserve highly biodiverse wetland areas for sustainable development and ecosystem conservation.

Chapter 16 examines the coral reefs of Andaman-Nicobar and Lakshadweep islands, highlighting their diversity and ecological significance along with the threats and bleaching actions (Wasim et al. 2024b). The authors employed spatial-temporal analysis utilizing Landsat satellite data from 2002 and 2020, and mapped coral reefs. They used advanced techniques like deglinting and Depth Invariant Bottom Index (DII) for accurate coral mapping and identification of different coral reef classes using digital image processing and supervised classification. Additionally, surface algal bloom, turbidity, and colored dissolved organic matter (CDOM) were taken as standard parameters to assess their impact on coral health. The study revealed adverse trends in environmental parameters like sea surface temperature (SST), photosynthetically active radiation (PAR), salinity, and sea level rise, indicating coral bleaching and degradation. The authors suggest the decrease in reef extent and an increase in sand/mud over reef areas using temporal coral mapping and emphasize the need for comprehensive assessment of ocean and climate impacts on coral reef ecosystems.

Chapter 17 briefly reviews the impact of plastics on the environment with special emphasis on aquatic, ocean, and freshwater ecosystems, along with their impact on human health (Dahiya et al. 2024).

Part VI concludes with Chapter 18 which provides a concise overview of aquatic ecosystems, detailing their types and discussing assessment methods (Mohanty et al. 2024b). It delves into both traditional assessment and advanced remote sensing technologies, highlighting their significance in monitoring aquatic ecosystems. Additionally, the chapter explores models and software tools, including plugins in ArcGIS and QGIS commonly employed in this field. Furthermore, challenges associated with traditional methods and the advantages of advanced remote sensing techniques are examined, offering insights into the future implications of remote sensing and large datasets in platforms like Google Earth Engine for the monitoring of aquatic ecosystems.

1.2 CONCLUSION

From the serene depths of freshwater lakes to the expansive realms of the ocean, aquatic ecosystems harbor an incredible array of life forms ranging from micro- to macroorganisms with several ecological functions and processes. However, these delicate ecosystems face myriad threats, including pollution, climate change impact, habitat destruction, overharvesting of resources, and invasive species. Effective monitoring serves as a cornerstone in our efforts to safeguard these invaluable resources for future generations. Overall, the book highlights the crucial role of aquatic ecosystems in maintaining ecological balance, filtering water, and moderating local weather conditions. It demonstrates how wetlands are facing threats from natural impacts and human activities, leading to their shrinkage. Several chapters focus on the impact of climate change, land surface temperature, and land use dynamics on wetlands. The findings of all studies emphasize the importance of traditional as well as advanced technologies such as remote sensing, which are essential for monitoring purposes. Continuous monitoring of aquatic ecosystems is imperative for sustainable management of resources and to avoid the impact of climate change. Therefore, traditional methods, conventional techniques, and advanced emerging technologies such as remote sensing can provide valuable information on aquatic ecosystems, pertaining to lakes, rivers, wetlands, coastal areas, or ocean. These aquatic ecosystems monitoring and management are crucial for the achievement of Sustainable Development Goals, directly or indirectly. for example, wetlands are connected with few SDGs, therefore their monitoring, conservation and management efforts will assist in the achievement of SDGs (Mohanty et al., 2024c). Aquatic ecosystem monitoring and assessment is linked, directly and indirectly, and has common thread to achieving several Sustainable Development Goals (SDGs), particularly SDG 6 (Clean Water and Sanitation), SDG 14 (Life Below Water), and SDG 15 (Life on Land) i.e. wetlands contribution to achieving different SDGs (Mohanty et al., 2024c). Aquatic ecosystems such as marine, riverine, wetlands, mangroves, lakes are crucial to the sustenance of life on the earth while maintaining the integrity and balance of life. Aquatic ecosystems provide several ecosystem services which are essential to organisms, such as water purification, biodiversity support, and climate regulation, all of which are critical for sustainable development activities. Advanced remote sensing technology plays a crucial role in identifying, mapping, monitoring and managing by providing accurate, real-time data on location, water quality parameters, habitat conditions, and diversity, along with change assessment. This technology enhances our ability to track progress towards different SDGs, how they are connected and linked to it, and enabling timely interventions to protect and restore aquatic environments, thereby ensuring natural resources for future generations. Different methods and advanced techniques, also provide information on aquatic macrophyte and aquatic vegetation traits and aids in the detection and monitoring of various types of aquatic vegetation such as floating, emergent, and submerged (Mohanty et al. 2024a). Overall, this volume collectively highlights the importance of traditional data along emerging technologies in providing possibilities for the scientific community to understand and quantify the different processes affecting these ecosystems. It addresses challenges and research opportunities for achieving a better understanding of aquatic ecosystems. Furthermore, it discusses traditional methods alongside the most recent discoveries, encompassing pragmatic and realistic strategies. These strategies aim to allow real-time detection and response to threats facing aquatic ecosystems. The contents emphasize the importance of consolidating new approaches for controlling, monitoring, and alerting against constraints that could potentially harm these ecosystems.

ACKNOWLEDGEMENTS

The authors thank the Shiv Nadar Institution of Eminence (Deemed to be University) for continuous help and support. Dr. Pandey is grateful to the National Geospatial Program, DST Government of India (NGP/TPN/30705–2019 (G)) for financial support and assistance.

REFERENCES

Algharianі, MS., Mejjad, N., Sagar, EM., and Bedair, H. (2024). Monitoring coastal Chlorophyll-a concentration in The Coast of North Africa using remote sensing and machine learning. *In* Pandey, PC., Srivastava, PK., Srivastava, S. (Eds.) *Aquatic Ecosystem Monitoring: Conventional Assessment to Advanced Remote Sensing*, pp. 154–165. CRC Press.

Arora, M., Sahoo, S., Swati, I., Roy, P., and Pateriya, B. (2024). Impact of climate change and land use land cover dynamics on wetland ecosystem in Punjab using geospatial technology. *In* Pandey, PC., Srivastava, PK., Srivastava, S. (Eds.) *Aquatic Ecosystem Monitoring: Conventional Assessment to Advanced Remote Sensing*, pp. 233–251. CRC Press.

Bedair, H., El-Makawy, AM., Nasser, H., and Halmy, MWA. (2024). Application of remote sensing for monitoring aquatic vegetation: A case study of Egyptian coastal wetlands using MODIS/Terra data. *In* Pandey, PC., Srivastava, PK., Srivastava, S. (Eds.) *Aquatic Ecosystem Monitoring: Conventional Assessment to Advanced Remote Sensing*, pp. 133–153. CRC Press.

Dahiya, A., Kumar, DYS., Kumar, SS., Devi, and Pandey, AK. (2024). Plastic pollution is a serious menace to ecosystem health with special reference to aquatic ecosystems and its associated challenges, opportunities and mitigations. *In* Pandey, PC., Srivastava, PK., Srivastava, S. (Eds.) *Aquatic Ecosystem Monitoring: Conventional Assessment to Advanced Remote Sensing*, pp. 279–293. CRC Press.

Leila, S., Zohra, A., Mohamed Islem, B., Omar, S., and Mohamed Essalah, A. (2024). Aquatic ecosystems monitoring with remote sensing perspective. *In* Pandey, PC., Srivastava, PK., Srivastava, S. (Eds.) *Aquatic Ecosystem Monitoring: Conventional Assessment to Advanced Remote Sensing*, pp. 45–64. CRC Press.

Mohanty, S., Pandey, PC., and Srivastava, PK. (2024a). Aquatic vegetation species identification and mapping using multisource Data-Review. *In* Pandey, PC., Srivastava, PK., Srivastava, S. (Eds.) *Aquatic Ecosystem Monitoring: Conventional Assessment to Advanced Remote Sensing*, pp. 166–181. CRC Press.

Mohanty, S., Pandey, PC., Srivastava, PK., and Srivastava, SK. (2024b). Challenges and future implications in aquatic ecosystem monitoring and assessment. *In* Pandey, PC., Srivastava, PK., Srivastava, S. (Eds.) *Aquatic Ecosystem Monitoring: Conventional Assessment to Advanced Remote Sensing*, pp. 297–318. CRC Press.

Mohany, S., Pandey, P. C., Pandey, M., Srivastava, P. K., & Dwivedi, C. S. (2024c). Wetlands contribution andlinkage to support SDGs, its indicators and targets- A critical review. *Sustainable Development.* Online https://onlinelibrary.wiley.com/doi/full/10.1002/sd.2948

Muduli, PR., and Prasannajit Acharya, P. (2024). Monitoring of water quality of Asia's largest brackish water system, Chilika lagoon. *In* Pandey, PC., Srivastava, PK., Srivastava, S. (Eds.) *Aquatic Ecosystem Monitoring: Conventional Assessment to Advanced Remote Sensing*, pp. 65–87. CRC Press.

Pandey, M., and Bisht, P. (2024). Stream channel dynamics for an understanding of the evolution of forms and processes of alluvial fan systems: An overview. *In* Pandey, PC., Srivastava, PK., Srivastava, S. (Eds.) *Aquatic Ecosystem Monitoring: Conventional Assessment to Advanced Remote Sensing*, pp. 88–117. CRC Press.

Pandey, S., Dwivedi, S., and Singh, SK. (2024a). Investigation of seasonal variability of Chlorophyll-a in the Arabian Sea. *In* Pandey, PC., Srivastava, PK., Srivastava, S. (Eds.) *Aquatic Ecosystem Monitoring: Conventional Assessment to Advanced Remote Sensing*, pp. 182–201. CRC Press.

Petropoulos, GP., Detsikas, SE., Tzanavari, I., and Karymbalis, E. (2024). Mapping shoreline changes in two river deltas of Greece exploiting Geoinformation technologies. *In* Pandey, PC., Srivastava, PK., Srivastava, S. (Eds.) *Aquatic Ecosystem Monitoring: Conventional Assessment to Advanced Remote Sensing*, pp. 118–130. CRC Press.

Shilky, Patra, S., Kumar, A., and Saikia, P. (2024). Management of aquatic ecosystems and aquatic vegetation for environmental sustainability. *In* Pandey, PC., Srivastava, PK., Srivastava, S. (Eds.) *Aquatic Ecosystem Monitoring: Conventional Assessment to Advanced Remote Sensing*, pp. 205–217. CRC Press.

Singh, A., Uddandam, PR., Mazumder, A., and Manoj, MC. (2024a). Multiproxy (Calcareous nannofossil, benthic foraminiferal and TOC) records from the eastern Arabian Sea: Implications for monsoon induced nutrients and primary productivity changes during the Holocene. *In* Pandey, PC., Srivastava, PK., Srivastava, S. (Eds.) *Aquatic Ecosystem Monitoring: Conventional Assessment to Advanced Remote Sensing*, pp. 32–41. CRC Press.

Verma, J., Srivastava, P., Dwivedi, RK., and Yadav, A. (2024). Evaluating aquatic ecosystem health in the Indian River-Linking Program: Diatoms and macroinvertebrates as biomonitoring tools. *In* Pandey, PC., Srivastava, PK., Srivastava, S. (Eds.) *Aquatic Ecosystem Monitoring: Conventional Assessment to Advanced Remote Sensing*, pp. 24–31. CRC Press.

Wasim, Md., Mohanty, S., Ahammed, KK Basheer., Pandey, AC., and Pandey, PC. (2024b). Coral reefs threat assessment and comparative analysis of Neil and Agatti Island, India using multispectral approach. *In* Pandey, PC., Srivastava, PK., Srivastava, S. (Eds.) *Aquatic Ecosystem Monitoring: Conventional Assessment to Advanced Remote Sensing*, pp. 252–278. CRC Press.

Wasim, Md., Mohanty, S., Pandey, AC., and Pandey, PC. (2024a). Potential fishing zones estimation along the Coast of India using geospatial technology. *In* Pandey, PC., Srivastava, PK., Srivastava, S. (Eds.) *Aquatic Ecosystem Monitoring: Conventional Assessment to Advanced Remote Sensing*, pp. 218–232. CRC Press.

Yule, CM., and Chellaiah, D. (2024). Freshwater ecosystem monitoring: Conventional assessment of macroinvertebrates. *In* Pandey, PC., Srivastava, PK., Srivastava, S. (Eds.) *Aquatic Ecosystem Monitoring: Conventional Assessment to Advanced Remote Sensing*, pp. 13–23. CRC Press.

Part II

Conventional Aquatic Ecosystems Monitoring

2 Freshwater Ecosystem Monitoring
Conventional Assessment of Macroinvertebrates

Catherine M. Yule and Darshanaa Chellaiah
School of Science, Technology & Engineering,
University of the Sunshine Coast,
Queensland, Australia

2.1 INTRODUCTION

2.1.1 Why Sample Macroinvertebrates?

Fresh water is of fundamental importance to life on Earth, yet fresh waters contained in rivers, lakes, and swamps comprise a very small proportion of the waters of the world (approximately 0.01%) in comparison to the oceans (>97%) and water contained in ice and snow (2%) (Carpenter et al. 1992). Increasingly the world's fresh waters are adversely impacted by industries, urbanization, landfills, deforestation, agriculture, mining, oil spills, wildfires, salinization, thermal pollution, sea level rise, droughts, floods, channelization, and so on. Freshwater macroinvertebrates (invertebrates visible to the naked eye as opposed to microscopic forms) are immensely diverse, and the different species vary in their tolerance to physical conditions such as substrate, temperature, and flow rate, and water chemistry, including pH, conductivity, nutrients, dissolved oxygen, suspended solids, and pollutants. Consequently, they vary in their responses to adverse events and conditions. They are relatively sedentary in comparison to aquatic vertebrates, and many species have long life cycles in comparison to microscopic invertebrates. Whereas fish can often escape polluting events (e.g. by swimming to unaffected river tributaries) and water chemistry sampling may miss a polluting event, the presence and absence of tolerant and intolerant macroinvertebrate taxa can reveal a diverse array of stressors over a range of time periods. Consequently, macroinvertebrates are frequently used, together with water quality data, to monitor the health of fresh waters.

Freshwater macroinvertebrates include various kinds of snails, worms, crustaceans (e.g. shrimps and crabs), insects (e.g. beetles, bugs, mosquitoes, mayflies, dragonflies, and caddisflies), and a range of other arthropods. Many of these spend their entire lives in water (e.g. snails, worms, many beetles, and bugs), whereas others, such as dragonflies and mayflies, only have aquatic juvenile stages.

Macroinvertebrates are usually extremely abundant and diverse in fresh waters (e.g. 182 species were collected from one site in a small rainforest stream in the mountains of Bougainville Island, Papua New Guinea: Yule 1995), so sampling programs can provide enormous amounts of data to enable robust statistical analyses, for example, comparing communities subject to different physical, chemical, or biological conditions. This contrasts with sampling vertebrates such as fish, otters, or crocodiles, where numbers and diversity may be quite low and their presence may not necessarily be indicative of ecosystem health.

This chapter provides a description of simple qualitative and quantitative sampling methods which are commonly used to survey macroinvertebrates in a variety of freshwater habitats.

DOI: 10.1201/9781003354000-4

2.1.2 Diversity of Freshwater Habitats

Fresh waters are hotspots for biodiversity and exhibit high levels of endemism due to their isolation and physical diversity. Freshwater habitats include permanent and temporary rivers and streams, lakes, swamps, acidic peat swamps, mires and bogs, artificial reservoirs, pools, geothermal mineral springs, and phytotelmata such as tree holes, leaf axils, bamboo internodes, and pitcher plants. Small, rapidly developing species such as mosquitoes can colonize water contained in discarded rubbish such as tin cans, coconut husks, large fallen leaves, or wheel ruts on a muddy track.

At high latitudes or on mountains, streams and pools may freeze, while the temperatures of hot springs can be over 100°C. Underground waters may be in permanent darkness and can be completely isolated from the outside world. The deepest freshwater habitat is Lake Baikal (1620 m), and the deepest river is the Congo (220 m deep), while water flowing over bedrock or in a puddle may be only a millimeter or so deep. Some waters are highly acidic (e.g. peat swamps and mires) while others are alkaline (e.g. some hot springs and soda lakes).

2.2 SAMPLING METHODS

2.2.1 Qualitative vs Quantitative Sampling

Sampling the flora and fauna of the great variety of freshwater habitats presents various challenges, particularly if the aim is to collect quantitative rather than qualitative data.

Qualitative sampling is typically used for biomonitoring to collect information on biodiversity and relative abundance. It provides an indication of the health or pollution status of an ecosystem and a general description of the habitat. For qualitative sampling, the researcher will usually try to collect samples from the different microhabitats within an ecosystem (e.g., a lake's banks, substrate, and open water) or riffles, runs, pools, and waterfalls in a stream. Qualitative data will not be as useful as quantitative data for statistical analyses and robust comparisons.

Sampling can be as simple as picking invertebrates from rocks in a flowing stream or using complicated devices such as suction samplers by means of a water pump or freeze coring the substrate to collect hyporheic fauna. An introduction to traditional techniques in freshwater biomonitoring using benthic macroinvertebrates is provided by Rosenberg and Resh (1993). Field and laboratory techniques for sampling stream biota from bacteria to fungi, algae, macrophytes, and invertebrates are described in depth by Hauer and Lamberti (2017). A multitude of quantitative sampling devices have been devised to sample different habitats within fresh waters, such as lakes, rivers, riffles, waterfalls, mud, sand, pebbles, bedrock, peat, hyporheic zones (subsurface sediments), aquatic plants, leaf litter, wood, and drift (downstream transport of invertebrates in streams and rivers). A comprehensive list of quantitative aquatic invertebrate sampling devices and collecting techniques is given in Elliott et al. (1993) and Jackson et al. (2019), and various commonly used sampling methods are more fully described by Marchant and Yule (2019) and Callisto et al. (2021).

Different sampling techniques and devices will enable different levels of precision and will determine whether data will relate to density per unit area, abundance with respect to the amount of effort applied (e.g. with timed samples or samples taken over a certain distance), or relative numbers (e.g. percentage of taxa in different groups).

Quantitative sampling requires consideration not only of the sampling device but also of the question being answered, the number and size of samples, their location, the degree of aggregation of the populations being sampled, the size of the animals being sampled, and the timing of sampling (e.g. only once, monthly, seasonally, yearly, or before and after certain events like floods, fires, or dam construction). Timing may also relate to the life cycles of the organisms under study.

There may also be consideration of using random or haphazard sampling, sampling along transects, or stratified sampling. An example of stratified sampling of a stream could be:

1. Determine the proportion of a stream taken up by pools, riffles, runs, and waterfalls (e.g. 20% pools, 20% riffles, 50% runs, 10% waterfalls).
2. Collect random samples from each habitat (stratum) with the number of samples from each stratum depending on the proportion of the habitat in the stream.

Statistical considerations of quantitative sampling are thoroughly described elsewhere (e.g. Elliott 1977; Jackson et al. 2019; Rosenberg and Resh 1993).

2.2.1.1 How Many Samples to Take?

The number, size, location, and timing of samples to collect depend on the question/s being answered. For example, do you need to ascertain seasonal variation? Do you need to determine population densities? Are you undertaking a biodiversity survey? If your aim is to ensure that you collect representatives of the majority of species present at your site (a biodiversity survey), you will want to ensure the adequacy of your sampling method and/or adequacy of the number of replicates or size of samples taken. A simple way to do this is by creating a 'species accumulation curve' (species richness curve) using samples from a pilot study. This can also be used to estimate the number of species in the area sampled. Simply explained, it is a graph of the cumulative number of species collected with each increasing sample. Initially, the curve will typically rise quickly and then eventually level off as every species in the habitat has been collected.

2.2.1.2 Mesh Size

Whatever sampling device is used, mesh size is an important consideration. The finer the mesh, the greater the likelihood of collecting early instars of insect larvae and small invertebrates such as zooplankton. However, the finer the mesh size, the more time will be required to sort the samples as the net will collect more fine detritus and sediment. Typically nets of 100–500 μm are used for freshwater macroinvertebrate sampling; 250 μm is usually good for capturing most early instar specimens without too much unwanted detritus.

2.2.2 Sampling Still Waters: Lakes, Swamps, Mires, and Pools

2.2.2.1 Qualitative Sampling of Still Waters

The most commonly used method to sample still waters is dip netting (Figure 2.1). This is so simple that little children can do it. The best net to use is either D-shaped or triangular in shape, rather than round. Round butterfly nets are preferred for sampling terrestrial, flying insects. The net is simply swept through the water, paying particular attention to aquatic plants and the substrate, as well as the open water. The contents of the net are then deposited into a white plastic tray with a layer of water. Debris such as plant matter and sediment are carefully removed (ensuring there are no adhering fauna or fauna living inside plant material such as caddisfly larvae), and the material is then poured into a jar. The invertebrates are either brought back to the lab alive or the sample is preserved in ethanol (typically about 70%). The exact percentage of ethanol usually does not matter greatly as long as the sample smells quite strongly of alcohol so that the plant material and invertebrates do not decompose. However, if the sample is to be used to extract DNA from the fauna to submit sequence data for eDNA surveys (e.g. Benson et al. 2003; see Section 2.3), then 95% ethanol is required.

It is important to label the sample carefully with the location, the date, and any other details that may be required. Typically sampling of lakes and swamps is done around the edges where access is easy, particularly among the vegetation. This is usually where most of the macroinvertebrate fauna live, although the open water may have abundant microscopic zooplankton and phytoplankton. Deep lakes, such as Lake Baikal, and also deep rivers, pose particular sampling problems. The surface and edges of deep waters can be sampled with a dip net, while deeper waters can be sampled

FIGURE 2.1 Using a dip net to sample swamps. (a) Minjerribah (Stradbroke Island), Australia – note that round nets are not as efficient as triangular or D-shaped nets for sampling aquatic invertebrates. (b) K'gari (Fraser Island), Australia – note the triangular net.

with a net dropped from a boat and towed at different depths. The bottom sediments can be sampled with a grab sampler.

2.2.2.2 Quantitative Sampling of Still Waters

A variety of samplers have been devised to collect quantitative samples from lakes and pools (e.g., see Jackson et al. 2019). Some habitats are quite difficult to sample quantitatively, for example, heavily vegetated swamps, fens, and mires. In these situations, multiple timed dip net sweeps can provide a reasonable level of quantification, particularly if the same person collects each sample and uses the same technique, for example, four sweeps each of 20 seconds at each site.

2.2.3 Sampling Flowing Waters: Rivers and Streams

2.2.3.1 Shallow Streams and Rivers

One of the more commonly used quantitative samplers for shallow, flowing streams and rivers with a substrate of coarse sand, gravel, or cobbles is a Surber sampler (Figure 2.2). The operator disturbs the substrate in an area within a metal frame (e.g. 20 × 20 cm), and detritus, sediment, and fauna are washed into a net by the flow. The depth of the water must not exceed the height of the sampler, but it must be deep and fast enough for the invertebrates to be washed into the net. The base of the frame must be firmly embedded in the substrate so that all the animals within it are collected in the net. Any large cobbles within the frame should be washed carefully into the net, and the substrate beneath the rocks must be disturbed down a few centimeters to dislodge any fauna.

A dip net can also be used to sample flowing waters by holding the net downstream and kicking or disturbing the substrate upstream by hand so that fauna and detritus are washed into the net. A level of quantification can be achieved by trying to restrict the disturbance to a small, standard area, but it will not be as effective as a Surber sampler.

2.2.3.2 Large Rivers

Non-wadable rivers pose a challenge, particularly as sampling can be dangerous. Often just the banks are sampled, which can be done using a dip net, but clearly this is inadequate to fully survey the fauna of the river. A variety of methods are used to sample the macroinvertebrates from deep rivers, including dredges, grab samplers, corers, and suction samplers (air-lift or pump) as well as artificial substrates (Freshwater Biological Association 2014; Jackson et al. 2019).

FIGURE 2.2 Using a Surber sampler. (a) Queensland, Australia. (b) Kelian river, Kalimantan, Indonesia.

FIGURE 2.3 Typical container habitats: (a) Testing water quality in a bamboo drinking trough. Maganutu Village, Timor-Leste. (b) Sampling for mosquito larvae in a stump of a palm tree. Maganutu Village, Timor-Leste. (c) Pitcher plants, *Nepenthes ampullaria* – common phytotelmata. Home to specialized aquatic fauna. Bako National Park, Malaysia.

2.2.4 Sampling Phytotelmata and Other Small Container Habitats

The entire contents of small moveable containers, such as coconut husks, tin cans, or dead leaves, as well as phytotelmata such as *Nepenthes* pitcher plants (Figure 2.3), can simply be emptied into a jar. Bamboo internodes, leaf axils, and tree holes are usually more readily sampled by extracting the liquid and associated fauna and detritus using a plastic pipette (Figure 2.3B) or a tube attached to a syringe (e.g. 20 ml syringe). The fauna can be examined live or preserved in ethanol.

2.2.5 Collecting Field Data – Physical, Chemical, and Biological Variables

Collecting field data is a vital part of biomonitoring. The types of data collected will vary depending on the purpose of the study. The exact location of the study sites, size of the water body, flow

TABLE 2.1
Typical Field Data Sheet for Freshwater Biomonitoring

	Stream Riffle	Stream Pool	Pitcher plant: *Nepenthes gracilis*	Lake Banks	Lake Open Water
Coordinates N	03°10.0	03°10.095	03°09.910	03°09.897	03°09.964
E	101°46.840	101°46.833	101°46.631	101°46.588	101°46.647
Elevation (m)	231	228	153	156	151
Temperature – Air	28.3°C	30.7°C	32.8°C	35°C	29.9°C
Temperature – Water	25°C	25.2°C	31.8°C	31°C	29.9°C
Size – Depth	See below	See below	12 cm	0–0.5 m	14 m
Size – Width	See below	See below	3 cm	450 m × 110 m	450 m × 110 m
Average stream channel width (mean of five randomly selected widths). Mean width (cm)	1.16	4.88	—	—	—
Mean water depth (five depth measurements, spread evenly across the stream channel, cm)	3.9	14.8	—	—	—
Nepenthes volume of liquid	—	—	3.5 ml	—	—
Flow rate (metres/second)	0.75–1.0 m/s	0.05 m/s	0	0	0
Secchi depth	N/A	N/A	N/A	N/A	1.64 m
pH	6.55	6.95	2.19	7.22	7.23
Conductivity mS	197.5	40.5	3.13	46.7	3.45
TDS mg/l (conductivity × 0.67)	132.33	27.14	N/A	31.3	2.31
Oxygen ppm (%)	92.6	92.7	85.6	101.8	117.1
Substrate % Composition Bedrock, boulders, cobbles, gravel, sand, mud	Sand 10%, cobbles 20%, boulders 70%	Sand 90%, boulders 10%	N/A	70% cobbles, 30% mud	100% fine mud
Rock type	Granite	Granite	N/A	Granite	N/A
Leaf litter cover (%)	20	35	N/A	70	80
Canopy cover (%)	80	50	25	30	25
Light (Lux)	45,000	52,000	48,000	90,000	130,000

(Continued)

TABLE 2.1 (CONTINUED)

Typical Field Data Sheet for Freshwater Biomonitoring

	Stream Riffle	Stream Pool	Pitcher plant: *Nepenthes gracilis*	Lake Banks	Lake Open Water
Riparian vegetation	Dipterocarp tropical rainforest	Dipterocarp tropical rainforest	Dipterocarp tropical rainforest – pitcher plants in cleared area by lake banks	Dipterocarp tropical rainforest, but some clearing by lake banks	Dipterocarp tropical rainforest, but some clearing by lake banks
Riparian tree species richness estimate: <15, 15–50, >50 spp	>50 species	>50 species	15–50 species	15–50 species	15–50 species
Aquatic vegetation	Moss on rocks, biofilm (with diatoms)	Moss on rocks, biofilm (with diatoms)	None	Algae, *Utricularia* sp. (bladderworts)	Chlorophytes, Cyanobacteria, diatoms, dinoflagellates
Notes	Pristine condition	Pristine condition	*Nepenthes gracilis* abundant by lake banks	Popular walking trail around lake. Some litter observed in water along banks of the lake	Popular walking trail around lake. Some litter observed in water along banks of the lake

This example is from Malaysia with typical data for a lake, a forested stream, and a container habitat (*Nepenthes* pitcher plant). Note that multiple readings should be collected from each site.

environments. It is recommended that a pilot study be undertaken for the targeted species to ensure the appropriate design of sampling protocols (Shogren et al. 2017).

In many regions, particularly in the tropics, the aquatic invertebrate fauna remains poorly known, and there are probably hundreds of thousands of species globally that are yet to be described, particularly among the aquatic Diptera (larvae), Annelida, and Nematoda. In Yule's 1995 study of the ecology of a tropical mountain stream on Bougainville Island, of 182 morpho-species collected, only five had already been described and so could be identified to species (Yule 1995), and no further freshwater research has been undertaken on the island since then. The use of eDNA sampling in such unstudied environments can be disappointing. Taxonomic experts are becoming increasingly rare, which hinders both identification and descriptions of novel taxa. Even experienced freshwater ecologists may not be able to identify specimens past genus or even family or higher (e.g. for taxa in the phylum Nematoda). Until a species has been officially described and its genome deposited in a database, it cannot be identified using eDNA. However, the information provided can aid in determining what relationships the target specimen has with other organisms and what it might be.

There are also many practical considerations when using eDNA sampling. How much water should be collected? How many samples should be taken? Where in the water body should the samples be collected – the surface, the substrate, mid-water, the banks, under or on rocks, or vegetation? Over what time frame? How often? How should one manage clogging filter paper in turbid waters, particularly swamps or sediment-polluted sites? And then there are the questions regarding the laboratory methods: how best to amplify the extracted DNA – PCR, ddPCR (Droplet Digital PCR), or qPCR (Quantitative PCR)? Further, results obtained can be limited by false positive or negative readings, the rate of eDNA degradation, as well as bioinformatic pipelines such as incomplete reference databases for aquatic biodiversity (Deiner et al. 2015; Roussel et al. 2015). All of these issues are yet to be resolved.

Despite these issues, ecosystem-wide applications of eDNA provide enormous potential to improve conservation, biological monitoring, and ecosystem assessments, as they can also be used to detect interactions between species and ecological patterns and processes across large temporal and spatial scales (Deiner et al. 2017). As reference databases improve and sequencing costs decrease, eDNA will become a common feature of aquatic macroinvertebrate surveys, either complementing conventional sampling methods or increasingly by itself.

REFERENCES

Benson, D.A., Karsch-Mizrachi, I., Lipman, D.J., Ostell, J., Wheeler, D.L. (2003). GenBank. *Nucleic Acids Research,* 31(1): 23–27. https://doi.org/10.1093/nar/gkg057. PMID: 12519940; PMCID: PMC165504.

Brantschen, J., Blackman, R.C., Walser, J.-C., Altermatt, F. (2021). Environmental DNA gives comparable results to morphology-based indices of macroinvertebrates in a large-scale ecological assessment. *PLoS One,* 16(9): e0257510. https://doi.org/10.1371/journal.pone.0257510

Callisto, M., Mugnai, R., Castro, D.M.P., Linares, M.S. (2021). Sampling methods for aquatic insects. In: Santos, J.C., Fernandes, G.W. (eds) *Measuring Arthropod Biodiversity.* Springer, Cham. https://doi.org/10.1007/978-3-030-53226-0_20

Carpenter, S.R., Fisher, S.G., Grimm, N.B., Kitchell, J.F. (1992). Global change and freshwater ecosystems. *Annual Review of Ecology and Systematics,* 23: 119–139. http://www.jstor.org/stable/2097284

Damborenea, C., Rogers, D.C., Thorp J.H. (Eds.). (2020). *Keys to Neotropical and Antarctic Fauna 2020 Volume V in Thorp and Covich's Freshwater Invertebrates.* Academic Press.

Deiner, K., Bik, H.M., Mächler, E., Seymour, M., Lacoursière-Roussel, A., Altermatt, F., Creer, S., et al. (2017). Environmental DNA metabarcoding: Transforming how we survey animal and plant communities. *Molecular Ecology,* 26(21): 5872–5895.

Deiner, K., Walser, J.-C., Mächler, E., Altermatt, F. (2015). Choice of capture and extraction methods affect detection of freshwater biodiversity from environmental DNA. *Biological Conservation,* 183: 53–63.

Elliott, J.M. (1977). *Some Methods for the Statistical Analysis of Samples of Benthic Invertebrates.* 2nd Edition. Scientific Publication No. 25. Freshwater Biological Association, Ambleside.

Elliott, J.M., Tullett, P.A., Elliott, J.A. (1993). *A New Bibliography of Samplers for Freshwater Benthic Invertebrates.* Occasional Publication No. 30. Freshwater Biological Association, Ambleside.

Fernandez, S., Rodriguez, S., Martinez, J.L., Garcia-Vazquez, E., Ardura, A. (2019). How can eDNA contribute in riverine macroinvertebrate assessment? A metabarcoding approach in the Nalón River (Asturias, Northern Spain). *Environmental DNA,* 1(4): 385–401.

Freshwater Biological Association (FBA). (2014). *Review of Techniques for Sampling Benthic Macroinvertebrates in Deep Rivers.* Science Report, Environment Agency.

Gleason, J.E., Elbrecht, V., Braukmann, T.W., Hanner, R.H., Cottenie, K. (2021). Assessment of stream macroinvertebrate communities with eDNA is not congruent with tissue-based metabarcoding. *Molecular Ecology*, 30(13): 3239–3251.

Hamada, N., Thorp, J.H., Rogers, D.C. (Eds.). (2018). *Keys to Neotropical Hexapoda Volume III in Thorp and Covich's Freshwater Invertebrates.* Academic Press, London.

Harper, L.R., Buxton, A.S., Rees, H.C., et al. (2019). Prospects and challenges of environmental DNA (eDNA) monitoring in freshwater ponds. *Hydrobiologia,* 826: 25–41. https://doi.org/10.1007/s10750-018-3750-5

Hauer, F.R., Lamberti, G.A. (Eds.). (2017). *Methods in Stream Ecology.* 3rd Edition. Academic Press, London.

Jackson, J.K., Resh, V.H., Batzer, D.P., Merritt R.W., Cummins, K.W. (2019). Sampling aquatic insects: Collection devices, statistical considerations, and rearing procedures. Chapter 3. In: Merritt, R., Cummins, K., Berg, M.B. (eds) *An Introduction to the Aquatic Insects of North America.* 5th Edition. Kendall Hunt.

Marchant, R., Yule, C.M. (2019). Aquatic macroinvertebrates. Chapter 11 In: Hughes, J. (ed) *Freshwater Ecology and Conservation.* Oxford University Press.

Merritt, R., Cummins, K., Berg, M.B. (Eds.). (2019). *An Introduction to the Aquatic Insects of North America.* 5th Edition. Kendall Hunt.

Rosenberg, D.M., Resh, V.H. (Eds.). (1993). *Freshwater Biomonitoring and Benthic Macroinvertebrates.* Chapman & Hall, New York.

Roussel, J.M., Paillisson, J.M., Treguier, A., Petit, E. (2015). The downside of eDNA as a survey tool in water bodies. *Journal of Applied Ecology*, 52: 823–826. https://doi.org/10.1111/1365-2664.12428

Ruppert, K.M., Kline, R.J., Rahman, M.S. (2019). Past, present, and future perspectives of environmental DNA (eDNA) metabarcoding: A systematic review in methods, monitoring, and applications of global eDNA. *Global Ecology and Conservation,* 17: e00547.

Shogren, A.J., Tank, J.L., Andruszkiewicz, E., Olds, B., Mahon, A.R., Jerde, C.L., Bolster, D. (2017). Controls on eDNA movement in streams: Transport, retention, and resuspension. *Scientific Reports*, 7(1): 5065.

Thorp, J.H., Rogers, D.C. (Eds.). (2015). *Ecology and General Biology. Volume I in Thorp and Covich's Freshwater Invertebrates.* 4th Edition. Elsevier.

Thorp, J.H., Rogers, D.C. (2016a). *Thorp and Covich's Freshwater Invertebrates.* 4th Edition. Academic Press.

Thorp, J.H., Rogers, D.C. (Eds.). (2016b). *Keys to Nearctic Fauna Volume II in Thorp and Covich's Freshwater Invertebrates.* 4th Edition. Elsevier.

Thorp, J.H., Rogers, D.C. (Eds.). (2019). *Keys to Palaearctic Fauna 2019 Volume IV in Thorp and Covich's Freshwater Invertebrates.* 4th Edition. Academic Press.

Tsuji, S., Takahara, T., Doi, H., Shibata, N., Yamanaka, H. (2019). The detection of aquatic macroorganisms using environmental DNA analysis—A review of methods for collection, extraction, and detection. *Environmental DNA*, 1(2): 99–108.

Wu, F., Zou, Y., Qin, S., Li, F., Zhang, Y. (2023). eDNA biomonitoring of macroinvertebrate communities for the bioassessment of a river's ecological status. *Water,* 15: 308. https://doi.org/10.3390/w15020308

Yule, C.M. (1995). The benthic invertebrate fauna of an aseasonal tropical stream on Bougainville Island, Papua New Guinea. *Marine and Freshwater Research*, 46: 507–518.

Yule, C.M., Yong, H.S. (Eds.). (2004). *Freshwater Invertebrates of the Malaysian Region.* Academy of Sciences Malaysia.

3 Evaluating Aquatic Ecosystem Health in the Indian River Linking Program

Diatoms and Macroinvertebrates as Biomonitoring Tools

Jyoti Verma
Department of Zoology, CMP Degree College, University of Allahabad, Prayagraj, India

Prateek Srivastava and Rakesh Kumar Dwivedi
Department of Botany, University of Allahabad, Prayagraj, India

Anuradha Yadav
Department of Botany, Bhakt Darshan Govt. P.G. College, Jaiharikhal, Uttarakhand, India

3.1 INTRODUCTION

Endowed with vast land and water resources, India is facing rapid depletion of its substantial water reservoirs. Despite significant advancements in water conservation through reservoirs and dams, the challenges of a burgeoning population, urbanization, and industrialization have intensified the issue of water scarcity. Addressing this future water crisis requires efficient water usage measures and harnessing surplus monsoon water. The Indian River Linking Program (IRLP) is a major initiative to augment storage capacity and facilitate water transfer from water-abundant regions to drought-prone areas through intra- and interbasin transfers (NWDA 2005; Lakra et al. 2011).

Macroinvertebrates and diatoms have emerged as prominent biomonitoring tools because of their sensitivity to environmental shifts and their ability to mirror the ecological health of aquatic systems. Diatoms, the single-celled algae characterized by intricate silica shells, and macroinvertebrates visible to the naked eye, which include insects, crustaceans, and mollusks, play pivotal roles in freshwater ecosystems. They offer valuable insights into water quality, pollution levels, and overall ecosystem integrity. Diatoms, with their high diversity and rapid response to environmental variations, are particularly valuable as bioindicators for changes in water chemistry, temperature, nutrient levels, and habitat conditions. The assessment of diatom composition, abundance, and community structure enables the evaluation of the ecological conditions and impacts of river interlinking activities within the IRLP. Moreover, this study supports the applicability of diatom-based matrices for assessing the effect of hydroelectric projects (HEP) in Himalayan rivers (Kumar & Nautiyal 2023).

Biodiversity in freshwater systems is primarily dispersed differently from that of aquatic or terrestrial environments. Living in nearly continuous media across vast regions, land and marine

 DOI: 10.1201/9781003354000-5

organisms modify their ranges to some extent in response to changes in climate or biological conditions. Nevertheless, many freshwater species find it difficult to spread over the geographical barriers that divide river drainages into distinct units because freshwater habitats are generally discontinuous. Freshwater species diversity is high even in regions where the number of species at any given site is low because species differ between one site and the next. Freshwater biodiversity is typically highly localized, and even small lakes or stream systems frequently harbor unique, locally evolved forms of life (Scuri et al. 2006; Carlisle et al. 2008).

According to what is now known, major dams, interbasin transfers, and river water withdrawals affect freshwater aquatic ecosystems in both positive and negative ways (Korse 2004; Zhang 2009; Vörösmarty et al. 2010). Adaptation and diversification have meant that there has been an increase in species numbers in the last billion years. However, a loss of biodiversity has begun to be observed with the growth in human populations, and Bradley et al. (2010) show that this situation is worsening. Future habitat loss, canal obstructions, irresponsible species introductions, and widespread fish disease may make the process of species extinction and genetic loss in tropical developing nations more severe (Millennium Ecosystem Assessment 2005). However, before linking these rivers, it is important to understand, through scientific studies, the effects that may occur on the health of these rivers.

Water transfer between distinct river basins, or river interlinking, has been discussed and put into practice in a number of nations. Notably, in an effort to correct regional water imbalances, China has launched significant water diversion projects, such as the South-North Water Transfer Project. Water must be moved across different river basins as part of US water management initiatives, and see also the South-African–Lesotho Highlands Water Projects. In order to tackle water scarcity in arid regions, Russia has investigated water diversion efforts, while Australia has established water transfer projects, such as the Snowy Mountains Scheme. River interlinking projects come in a variety of shapes and sizes, but their practicality and effects vary as well, frequently provoking discussions on social, political, and economic issues (Jeet et al. 2020).

The stabilization of groundwater has been demonstrated by river interlinking experiences worldwide, although the ecology of the aquatic ecosystem and fish species may be greatly impacted by such an ambitious strategy (Long et al. 2020; Chen & Chen 2004; Lakra et al. 2011). Additionally, research indicates that tighter river regulations may result in a larger water footprint (Jaramillo & Destouni 2015). In order to reconcile the competing goals of ecological sustainability and satisfying water demands, interlinking must be carefully planned (Chen & Olden 2017).

This study explores the potential of macroinvertebrates and diatoms as biomonitoring tools within India's Inter-river Linking Project. Moreover, it discusses these organisms' applicability, advantages, and disadvantages in assessing river management strategies. Examining the ecological impacts on the aquatic habitats associated with rivers is imperative, and biomonitoring, which uses biological indicators, serves as a vital tool for assessing aquatic habitat health and quality.

3.2 TOOLS FOR BIOMONITORING IN RIVER INTERLINKING

To evaluate the health of aquatic ecosystems within the context of river interlinking projects, bioindicators, which are essential for environmental monitoring, must be used. Effective bioindicators must be able to adapt quickly to changes in the environment and offer information on long-term ecological interactions. Benthic macroinvertebrates, periphytons, and fish are the top candidates among the potential bioindicators for determining the health of streams and rivers (Li et al. 2012). The dependability and efficacy of these bioindicators have been confirmed by several independent academic studies.

Diatom indices play an important role in the biomonitoring of waterway interlinking projects because they can provide important information about the quantity and composition of aquatic ecosystems. To evaluate various ecological features, including water quality, various diatom indices have been established. The Siltation Diatom Index (SDI), Specific Pollution Sensitivity Index

(IPS), Trophic Diatom Index (TDI), Diatom Assemblage Composition Index (DACI), Diatom-based Habitat Quality Assessment (DHQA), Biotic Index for Diatoms (BID), Periphyton-based Diatom Index (PDI), and Diatom-based Nutrient Assessment (DNA) are some notable indices (Whitton & Rott 1996; Coste et al. 2008; Carlisle et al. 2008; Kumar & Nautiyal 2023). These indices collectively provide comprehensive information on water quality, pollution sensitivity, habitat integrity, nutrient enrichment, and overall ecological health of the water body. Researchers and environmental managers involved in river interlinking projects can leverage these indices in combination to make up-to-date decisions, ensuring the sustainable management of river resources and the preservation of aquatic biodiversity.

Macroinvertebrate indices are of utmost importance when assessing the health of aquatic ecosystems for river interlinking initiatives. To assess habitat conditions, water quality, and the system's overall ecological integrity, these indices depend on the variety and quantity of macroinvertebrate species. The Trophic Level Index (TSL), which concentrates on trophic composition, the Hilsenhoff Biotic Index (HBI), which ranks macroinvertebrates based on pollution tolerance, and the Ephemeroptera-Plecoptera-Trichoptera (EPT) Index, which focuses on sensitive taxa, are some notable macroinvertebrate indices (Lenat & Barbour 1994; Buffagni et al. 2004; Torrisi et al. 2010). Some biomonitoring initiatives use multimeric indices, combining many metrics to assess ecological health holistically. These macroinvertebrate indicators are crucial for understanding how river interlinking projects affect aquatic ecosystems, enabling us to make well-informed decisions.

3.3 IMPACT OF DAMS, DIVERSIONS, AND INTER-RIVER LINKING WATER TRANSFERS

3.3.1 Biological Impacts

The populations of biological species in aquatic habitats are significantly impacted by the construction of dams, diversions, and interbasin water transfers (IBWTs). Both direct and indirect effects on the creatures that live in these settings are caused by these water management efforts, which frequently result in considerable alterations to the natural flow of rivers and streams. For instance, dams can alter rivers' normal course and result in the submergence of ecosystems such as floodplains and riparian areas, which are essential for the survival of a variety of species. The availability of suitable habitats for aquatic species, as well as fish migration and spawning, may be impacted by these changing flow patterns (Parveen & Ilyas 2021). Additionally, dams change water temperature, sediment transport, and nutrient concentrations, all of which have an impact on the composition of aquatic ecosystems and water quality (Grant et al. 2012).

In addition to dams, water diversions frequently alter the course of rivers and streams, which has a direct impact on the distribution of aquatic creatures. Aquatic habitats may become more visible because of reduced water levels caused by diversions, which could damage species that depend on these places for life (Nath & Srivastava 1999; Pathak & Tyagi 2010). Additionally, diversions can alter flow, which can impact sediment transport and benthic creatures as well as overall water quality. IBWTs move water between basins, altering flow patterns and affecting ecological connectivity. The dispersion and movement of aquatic species are also hampered by this modification. Changes in water quality caused by water transfer may also alter the suitability of a species' habitat (Li et al. 2010). Transfer of water between basins may unintentionally introduce invasive species, further upsetting the balance of local ecosystems.

Habitat fragmentation and population isolation are common side effects of these water management measures and occur in all these scenarios. Such fragmentation can lower genetic diversity and impede the ability of aquatic species to spread. Hydropower activities sometimes go hand in hand with dam construction and have downstream consequences on aquatic species' distribution and behavior, such as changing flow patterns and temperature. Fish ladders, wildlife tunnels, and habitat restoration are measures taken to reduce the effects of dams, diversions, and IBWTs on aquatic fauna (Bartwal & Nautiyal 2023; Tyagi et al. 1999; Lakra et al. 2011). These steps are essential for reducing negative effects and maintaining the ecological well-being of aquatic environments. In addition, achieving a balance between human water resource needs and the preservation of aquatic biodiversity depends on thoroughly thought-out and well-regulated water management policies.

Interlinking rivers can have severe ecological ramifications by altering river flows, habitats, and ecosystems, as well as leading to loss of biodiversity and endangering aquatic species' existence.

3.3.2 Ecological Impact

Growing concerns have been expressed about the ecological effects of dams, diversions, and inter-river water transfers. Although these water management techniques can be essential for providing water resources, flood control, and hydroelectric energy, they frequently have high environmental costs. Dams can damage natural river ecosystems, change water flow patterns, and displace both wildlife and human populations. Inter-river connecting projects can change the natural flow of several rivers, impacting biodiversity and water quality, while diversion projects can deplete upstream water bodies and adversely affect aquatic life (Gopal & Marothia 2015, 2016). Since reconciling the requirement for water resource management with the preservation of our natural environment is a significant challenge considering rising water demands and climate change, careful planning and mitigation strategies are required to avoid these ecological effects.

Because of their substantial effects on the hydrologic regime, which is the main driver of freshwater ecosystems, dams and diversions profoundly change the ecology of freshwater systems. Because it impacts the hydrologic regime of both the donor and the receiving ecosystems, the interlinking of rivers project, a particular type of water diversion, has a substantial impact on freshwater ecosystems (Bandyopadhyay & Perveen 2003). Dams and water diversions alter the hydrologic regime, which in turn alters the sediment regime, habitat structure, and water chemistry. Together, the variations in the freshwater ecosystems' habitat layout, hydrologic regime, sediment regime, and water chemistry have a profound, and frequently irreversible, impact on the structure and function of the biological community (Bandyopadhyay & Perveen 2008).

Through feedback loops, changes in the biological community lead to additional changes in the sediment regime, habitat structure, hydrologic regime, and water chemistry. Thus, a once stable freshwater ecosystem becomes unstable because of dams and diversions. When the newly formed freshwater ecosystem reaches equilibrium, in years to millennia, it will resemble something very different from the original ecosystem. The Ganges-Brahmaputra river would be significantly impacted by the proposed ILRP because it will include several run-of-the-river dams (for example, barrages) with numerous functions (for example, flood control, water diversion, and hydropower generation, among others).

3.3.3 Economic Effects

From India's perspective, the interlinking of rivers represents a substantial national endeavor with the potential to profoundly impact the country's economy. This large-scale initiative addresses the persistent issue of water scarcity in specific regions while simultaneously optimizing the economic use of water resources. Notably, the project stands to significantly enhance agricultural productivity, a linchpin of the Indian economy, by enabling improved water management and irrigation practices (Gopal & Marothia 2015, 2016). Additionally, this may lead to a rise in food production and farmer earnings. Moreover, constructing hydropower-generating facilities on connected rivers can guarantee the energy security of an entire nation. It can lessen India's reliance on fossil fuels, which would have positive effects on the environment and the economy. This strategic shift toward cleaner energy sources also stimulates economic growth (Bandyopadhyay & Perveen 2008).

The project's economic impact extends to various sectors, notably through the enhancement of navigation and transportation. This can expedite trade and commerce, benefiting various industries. However, it is imperative to acknowledge that the economic ramifications of the project are a subject of intensive scrutiny and debate in the Indian context (Bandyopadhyay & Perveen 2003). This is because of the substantial financial, environmental, and social considerations associated with the river interlinking initiative. A thorough examination and careful planning are essential to ensure a balanced approach that maximizes economic benefits while mitigating potential adverse effects.

3.4 CAN THE INTERLINKING OF RIVER PROJECTS BALANCE BIODIVERSITY?

The question of whether inter-river linking projects can effectively reconcile the growing global demand for water resources with the imperative to preserve biodiversity is complex and multifaceted. It necessitates a detailed analysis of the ecological effects of the IBWT plan, which is projected to result in a reduction in biodiversity and substantial alterations to the biogeography of peninsular rivers through the disruption of connectivity within the riverine network. Moreover, the specific sequencing of rivers in the interlinking process plays a pivotal role in shaping continental-scale biodiversity patterns. It is evident that adopting an ecological approach, rather than purely an economic one, can mitigate these ecological consequences, resulting in more modest spatial alterations, slower shifts in species distribution, and a gradual decline in inter-community diversity as network connectivity is enhanced.

When evaluating the advantages of each canal while connecting regions with varying levels of endemicity and species distribution, decision-makers are confronted with a critical task. It is essential to recognize that the results presented here represent the minimal impacts of an IBWT project on biodiversity and biogeography, with additional ecological factors such as water transfer rates, species interactions, and habitat use potentially amplifying these effects. This underscores the pressing need for ethical and sustainable approaches to water resource management to strike a harmonious balance between human water resource needs and the preservation of aquatic biodiversity.

3.5 ADVANTAGES, CHALLENGES, AND OPPORTUNITIES IN INTERLINKING OF RIVERS

As we discussed earlier, the river interlinking program is basically meant for two important purposes. On one hand, it helps transfer the excess water in the rivers during the monsoon season to the rivers that receive less water during the season; on the other hand, it helps to provide much-needed water in parched regions of the country for drinking and agriculture. Thus, such a project helps to improve water management during periods of flood and drought, which supports the income and living standards of the people by protecting agriculture from flooding and increases the sowing area in drought-prone regions by providing water for agriculture. In both ways, the river interlinking program helps the exchequer every year by preventing losses and offering more revenue options to the country. The formation of interlinking channels of rivers would offer new waterways for transportation and trade and boost the economy.

The creation of the perennial water channel for the interlinking of rivers helps to improve the underground water table and submerged aquifers in the area. It will boost the rejuvenation of wetlands and promote the forest health of the area. In many regions where the river interlinking projects have been envisioned in India, the project will help to stop the spread of desertification and prevent soil erosion by improving the soil moisture content.

But such mega projects are looked at with great apprehension by ecologists, and deep concerns are expressed over their viability over a long period. It is generally believed that the interlinking of rivers will mix the biota of two distinct water bodies, each having their own physicochemical characteristics. This would lead to the invasion of new taxa into new habitats and start competition among species by changing the biochemistry of the habitat. No one can predict which species will prevail and which will perish in such a situation, given the emerging competition and altered physicochemical properties of the river waters arising due to river interlinking projects. Another concern is the possibility of increased salinization of the soil and water due to the mixing of the waters from two different watershed areas, particularly if one of them flows from the desert or run off from agriculture.

River interlinking projects require the digging of linking canals and channels connecting the two separate rivers, which is different in hydro-ecological nature from the natural confluence of rivers that was achieved over long geological periods and reached homeostasis long ago. It is understood that the global warming effects on the Himalayan glaciers and the gradual reduction in the forest cover in central highland areas would reduce the amount of water in so-called perennial rivers flowing in the Indo-Gangetic plains and central highland areas, which would question the availability of enough water to channel the excess water to parched rivers. Such projects require the acquisition of huge areas of land alongside the channel, which would require the displacement of the people and the loss of forests and agricultural lands that are already dwindling every year due to many other development projects such as the construction of highways and railway tracks. Huge inputs of money and manpower would be needed for the interlinking of the rivers, and the viability of the project over the long run is uncertain.

3.6 CONCLUSIONS AND RELEVANCE

A paradigm shift in India's water resource management is urgently required, as demonstrated by the assessment of aquatic environmental health within the context of the River Linking Program in India. The implementation of large-scale interlinking projects and traditional methods of centralized water transfer represents a significant departure from international trends that support decentralized, transparent, participatory, and environmentally and socially integrated water resource management. Analysis of the available data and literature reveals a glaring discrepancy between the benefits of these projects as advertised and their viability. A plea for more focus on demand management and effective use of water resources questions the emphasis on supply-side hydrology through dams and canals.

It is clear that the proposed river linking project may benefit multinational firms and the building sector more than it will assist disadvantaged communities. Concerns are also raised by the potential harm to the ecology and the ecosystem, as well as by the financial obstacles and more affordable, sustainable options. It is crucial to conduct accurate environmental and social impact assessments and address problems such as global warming and its effects on glacial retreat. The purported benefits and potential environmental and ecological dangers, such as the ambiguity surrounding water changes, effects on river courses, and socioeconomic effects, including relocation, water quality problems, and land submersion, are both subject to significant knowledge gaps. Finally, to address these complex challenges and ensure equitable and sustainable management of aquatic biodiversity in the region, a thorough review of the Indian River Connectivity Program needs to be conducted.

This study contains new findings from the study of the diversity of benthic communities, assemblages, and habitat conditions in the Indian river system that are likely to change significantly over time during the connection project. This work provides information for the first time. The most ambitious initiative to date is the assessment of benthic communities, specifically with regard to the projected river connecting project in India. The sustainable aquatic biodiversity conservation of these rivers, which are part of the same ecoregional rivers of India, requires the visualization of numerous pertinent issues. In the current study, baseline data on the state of the river's benthic communities that comprise its priority habitat characteristics will aid in the incorporation of appropriate control measures to mitigate adverse effects from project planning to various stages of development. Additionally, the data may be useful in quantifying the degree of species and habitat change or loss following river interlinking. This study is anticipated to be useful as a tool for decision-making in the evaluation of variables associated with suggested connections, fish conservation, and fish, ecology, and biodiversity management.

REFERENCES

Bandyopadhyay, J., and Perveen, S. (2003). The interlinking of Indian rivers: Some questions on the scientific, economic and environmental dimension of the proposal. School of Oriental and African Studies (SOAS)/King's College London Occasional Paper 60, 35.

Bandyopadhyay, J., and Perveen, S. (2008). Chapter 4: The interlinking of Indian Rivers: Questions on the scientific, economic and environmental dimensions of the proposal. In: Mizra, M.M.Q., Ahmed, A.U., and Ahmad, Q.K. (eds.), *Interlinking of Rivers in India: Issues and Concerns*, 53–76. London: Taylor and Francis.

Bartwal, T., and Nautiyal, P. (2023). Hydropower development in the Himalaya: Identifying critical river stretches. *Environmental Science and Pollution Research*, 30, 46741–46747.

Bradley, D.N., Tucker, G.E., and Benson, D.A. (2010). Fractional dispersion in a sand bed river. *Journal of Geophysical Research*, 115, F00A09. https://doi.org/10.1029/2009JF001268.

Buffagni, A., Erba, S., Cazzola, M., and Kemp, J.L. (2004). The AQEM multimetric system for the southern Italian Appennines: Assessing the impact of water quality and habitat degradation on pool macroinvertebrates in Mediterranean rivers. *Hydrobiologia*, 516, 313–329.

Carlisle, D.M., Hawkins, C.P., Meador, M.R., Potapova, M., and Falcone, J. (2008). Biological assessments of Appalachian streams based on predictive models for fish, macroinvertebrate, and diatom assemblages. *Journal of the North American Benthological Society*, 27(1), 16–37.

Chauhan, T., Devanand, A., Roxy, M.K., Ashok, K., and Ghosh, S. (2023). River interlinking alters land-atmosphere feedback and changes the Indian summer monsoon. *Nature Communications,* 14, Article number: 5928. https://doi.org/10.1038/s41467-023-41668-x.

Chen, A., and Chen, C. (2004). Ecological and political costs of river diversion. *Nature,* 429, 501–501.

Chen, W., and Olden, J.D. (2017). Designing flows to resolve human and environmental water needs in a dam-regulated river. *Nature Communications*, 8, 2158.

Coste, M., Botry, S., Tison-Rosebery, J., and Delmas, F. (2008). Improvements of the Biological Diatom Index (BDI): Description and efficiency of the new version (BDI-2006). *Ecological Indicators*, 9, 621–650.

Gopal, B., and Marothia, D.K. (2015). Integrating the economics of wetland biodiversity and ecosystem services in management of water resources of River Ken: A study, economics of ecosystems and biodiversity (TEEB)–India initiative, GIZ, and Ministry of Environment, Forest and Climate Change, Government of India, New Delhi.

Gopal, B., and Marothia, D.K. (2016). Seeking viable solutions to water security in Bundelkhand *Economic & Political Weekly EPW* November 5, 2016 vol lI nos 44 & 45.

Grant, E.H., Lynch, H.J., Muneepeerakul, R., Arunachalam, M., Rodriguez-Iturbe, I., and Fagan, W.F. (2012). Interbasin water transfer, riverine connectivity, and spatial controls on fish biodiversity. *Plos One,* 7(3), e34170. https://doi.org/10.1371/journal.pone.0034170.

Jaramillo, F., and Destouni, G. (2015). Local flow regulation and irrigation raise global human water consumption and footprint. *Science,* 350, 1248–1251.

Jeet, P., Kumar, A., and Sundaram, P.K. (2020). Interlinking of river: Issues and challenges. https://doi.org/10.5772/intechopen.93594.

Korse, K.H. (2004). River linking project: Case-specific approach is needed. *Current Science*, 87(12), 1643.

Kumar, S., and Nautiyal, P. (2023). Assessment of the ecological state of serially impounded Himalayan River using diatoms-based matrices: Identification of critical stretches for management. *Acta Ecologica Sinica.* https://doi.org/10.1016/j.chnaes.2023.07.009.

Lakra, W.S., Sarkar, U.K., Dubey, V.K., Sani, R., and Pandey, A. (2011). River interlinking in India: Status, issues, prospects and implications on aquatic ecosystems and freshwater fish diversity. *Reviews in Fish Biology and Fisheries,* 21, 463–479.

Lenat, D.R., and Barbour, M.T. (1994). Using benthic macroinvertebrate community structure for rapid, cost-effective, water quality monitoring: rapid bioassessment. In: Loeb, S.L., and Spacie, A. (eds.), *Biological Monitoring of Aquatic Systems.* Boca Raton, FL: Lewis Publishers, pp. 187–215.

Li, G.C., Liu, L.S., Wang, X., and Li, L. (2012). Applications of diatom in river health assessment: A review. *Ying Yong Sheng Tai Xue Bao*, 23(9), 2617–2624.

Li, L., Zheng, B., and Liu, L. (2010). Biomonitoring and bioindicators used for river ecosystems: Definitions, approaches, and trends*. *Procedia Environmental Sciences,* 2, 1510–1524.

Long, D., et al. (2020). South-to-North water diversion stabilizing Beijing's groundwater levels. *Nature Communications,* 11, 3665.

M.E.A. (2005). *A Report of the Millennium Ecosystem Assessment. Ecosystems and Human Well-Being.* Island Press, Washington DC.

Nath, D., and Srivastava, N.P. (1999). The decline in carp fishery in Narmada in the context of the construction of a dam on the river and its tributaries. *Journal of the Inland Fisheries Society of India*, 31(2), 25–27.

NWDA. (2005). Water for life. 11th National Water Convention, org. by National Water Development Agency, Govt. of India Delhi.

Parveen, T., and Ilyas, O. (2021). The inter-linking of rivers and biodiversity conservation: A study of Panna Tiger Reserve, Madhya Pradesh India. *Current Science*, 121(12), 1572–1583.

Pathak, V., and Tyagi, R.K. (2010). Riverine ecology and fisheries visa'-vis hydrodynamic alterations: Impacts and remedial measures. Bull. No.-161, Central Inland Fisheries Research Institute, Barrackpore, Kolkata, West Bengal.

Scuri, S., Torrisi, M., Cocchioni, M., and Dell'Uomo, A. (2006). The European Water Framework Directive 2000/60/EC in the evaluation of the ecological status of watercourses. Case study: The river Chienti (central Apennines, Italy). *Acta Hydrochimica et Hydrobiologica*, 34, 498–505.

Torrisi, M., Scuri, S., Dell'Uomo, A., and Cocchioni, M. (2010). Comparative monitoring by means of diatoms, macroinvertebrates and chemical parameters of an Apennine watercourse of central Italy: The river Tenna. *Ecological Indicators*, 10, 910–913.

Tyagi, R.K., Panwar, R.C., and Chandra, R. (1999). Impact of dams and barrages on riverine fisheries National seminar on Ecofriendly management of resource for doubling fish production-strategies for 21 st century. Inland fisheries society of India and CIFRI, Barrackpore. 22–23 December 1999.

Vörösmarty, C.J., Gessner, M.O., Dudgeon, D., Prusevich, A., Green, P., Glidden, S., Bunn, S.E., Sullivan, C.A., Liermann, C.R., and Davies, P.M. (2010). Global threats to human water security and river biodiversity. *Nature,* 467, 555–561.

Whitton, B.A., and Rott, E. (Eds.). (1996). *Use of Algae for Monitoring Rivers II.* Institut für Botanik. Universität Innsbruck, Innsbruck, Austria, 196 pp.

Zhang, Q. (2009). The South-to-North water transfer project of China: Environmental implications and monitoring strategy. *The Journal of the American Water Resources Association,* 45(5), 1238–1247. https://doi. org/10.1111/j.1752-1688.2009.00357.x.

4 Multiproxy (Calcareous Nannofossil, Benthic Foraminiferal, and Total Organic Carbon) Records from the Eastern Arabian Sea

Implications for Monsoon-Induced Nutrients and Primary Productivity Changes during the Holocene

Abha Singh, Prem Raj Uddandam, Abhijit Mazumder and MC Manoj*
Birbal Sahni Institute of Paleosciences, Lucknow, India

4.1 INTRODUCTION

The Arabian Sea (AS) is one of the most productive oceans in the world (Bauer et al., 1991). The biological productivity is strongly governed by the nutrient input in the Arabian Sea, which shows spatial and temporal heterogeny. Two major processes that enhance nutrient availability in the AS are upwelling, which occurs in the western and southeastern regions during summers, and vertical mixing, which occurs in the northeastern region during the winters. In the eastern Arabian Sea, although runoff may benefit the productivity locally, the lack of intense upwelling and stratification due to high precipitation and runoff produces little productivity during the summers (Cabarcos et al., 2014). The sporadic moderate upwelling/mixing during the northeast (NE) monsoon promotes biological productivity in the eastern Arabian Sea.

The biological productivity has been shown to vary on temporal scales in the Arabian Sea. Several studies reported strong contrasts in productivity between the western, northern, and eastern regions during glacial–interglacial periods of the last glacial cycle. During the glacial period, the western and northern regions experienced diminished biotic productivity due to decreased summer monsoon strength. On the other hand, the eastern Arabian Sea experienced enhanced biotic productivity attributed to upwelling and/or vertical mixing in the water column due to the cold and strong NE monsoon (Singh et al., 2011, 2018; Cabarcos et al., 2014; López-Otálvaro et al., 2008). However, most of the studies are based on planktic foraminifera, % Total Organic Carbon (TOC), and other geochemical proxies, with less emphasis on other biotic organisms such as coccolithophores (Agnihotri et al., 2003; Bassinot et al., 2011).

* Corresponding author.

 DOI: 10.1201/9781003354000-6

Coccolithophores are unicellular flagellate nannoplankton and major primary producers in the ocean (Bown, 1998). They possess characteristic fossilizable calcified scales called 'coccoliths'. Coccolith distribution in the modern ocean is controlled by environmental parameters, such as temperature, light, nutrient availability, salinity, and trace elements (Winter et al., 1994; Bown, 1998). They have been utilized as a potential tool to reconstruct paleoclimatic and paleoceanographic changes from different oceanic regimes (Flores and Marino, 2002; Flores et al., 2003; Jordon et al.Jordan & Winter, 2000; Baumann and Freitag, 2004; López-Otálvaro et al., 2009; Saavedra-Pellitero et al., 2011; Amore et al., 2012; Cabarcos et al., 2014, Zhou et al., 2020). Modern distribution and sediment trap studies of coccolithophores from the Arabian Sea revealed their sensitivity to nutrient availability in response to the monsoon winds and runoff (Mergulho et al., 2006).

Benthic foraminifera are eukaryotic shelled, single-celled protists that are found in all marine environments. They have been widely used to reconstruct the records of past changes in deep-water environments, oxygen levels, and organic matter flux (Nigam et al., 2007; Singh et al., 2015; Ma et al., 2019; Verma et al., 2018, 2021). Benthic foraminifera from the eastern Arabian Sea have been utilized to document variation in the oxygen minimum zone (Nigam et al., 2007, 2009; Mazumder and Nigam, 2014; Verma et al., 2018, 2021). The planktic and benthic foraminifera ratio is useful to record changes in the sea level.

This study provides coccolith and benthic foraminiferal assemblage records during the last 12 ka BP from the eastern Arabian Sea. The combined results of coccoliths, benthic foraminifera, and TOC % are applied to decipher the changes in productivity variation in the eastern Arabian Sea during the Holocene. The productivity variation is crucial to understanding monsoon and climate change during the Holocene.

4.2 STUDY AREA

The Arabian Sea is a semi-enclosed basin profoundly influenced by semi-annually reversing monsoon winds, that is, southwest (SW0 summer monsoon winds during June to September and NE winter monsoon winds during October–January (Qasim, 1982). During the summer monsoon the coastal region of the eastern Arabian Sea receives high precipitation and runoff from the Western Ghats, forming a low salinity tongue (Sarkar et al., 2000). During winters, surface cooling north of 15° N leads to vertical mixing (Madhupratap et al., 1996). During the summer, surface currents flow clockwise and West India coastal currents (WICC) flow toward the south, bringing Arabian Sea waters into the Bay of Bengal (Colborn, 1975). At the same time, the West Indian undercurrent flows toward the north (up to 17° N) at 200 m water depth (Shanker et al., 2002; Schott and McCreary, 2001). During the winter, surface currents move anticlockwise and WICC flows northward, bringing low salinity waters from the southern region (Bay of Bengal). The oxygen minimum zone present in the Arabian Sea is perennial and exists between the depth of 100–120 m to 1200 m with intense denitrification in the upper water column.

4.3 MATERIAL AND METHODS

The presently studied core SC-26 (25 cm long) was retrieved onboard the ORV Sagar Kanya 117 cruise from the eastern AS (off Goa, India: 14.47° N, 73.23° E), at a water depth of 240 m (Figure 4.1). Core lithology is silty-clay throughout the core.

4.3.1 Calcareous Nannofossils Slide Preparation

For the calcareous nannofossil study, permanent slides were prepared in duplicate, one containing a coarser fraction and the other having a comparatively finer fraction (Bown, 1998). These slides were examined under a Leitz Laborlux 12 POL microscope. For additional information on the morphology, scanning electron microscope (SEM) analysis of some samples was carried out at the

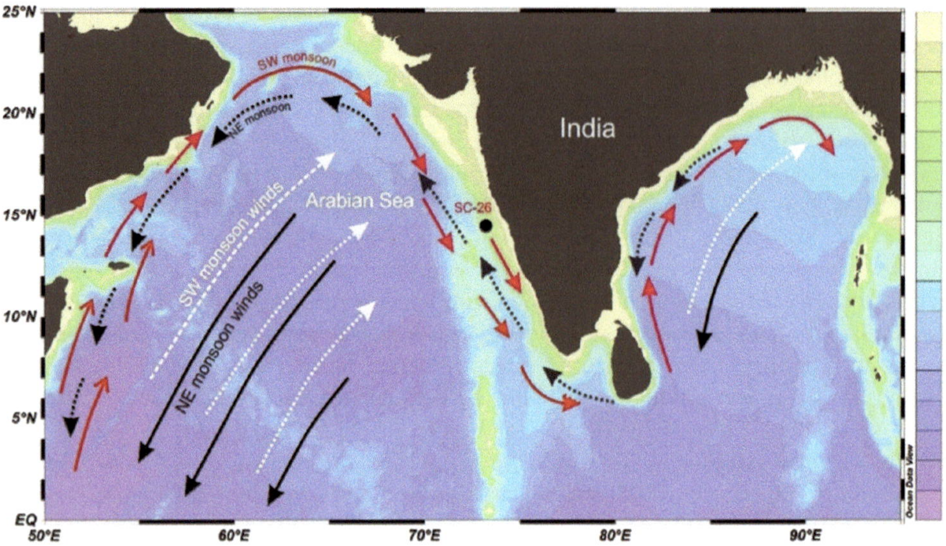

FIGURE 4.1 Schematic map with core site SC-26 (black dot). Summer monsoon currents are represented by red arrows, winter monsoon currents by black dashed arrows.

Birbal Sahni Institute of Paleosciences, India. For SEM, these samples were prepared following the technique of dilution and filtering of Boeckel et al. (2006).

Coccolithophore preservation was estimated with the scale reported by Roth and Thierstein (1972) and Flores and Marino (2002), which reveals several degrees of preservation according to visual estimations. The CEX preservation index (Dittert et al., 1999), which represents the relationship between small and delicate placoliths and strongly calcified coccoliths, is calculated for all the samples: [CEX = (% *Emiliania huxleyi*)/(% *Emiliania huxleyi* + % *Calcidiscus leptoporus*)] is calculated for all the samples.

4.3.2 Benthic Foraminiferal Sample Preparation

The samples were processed following the standard procedure, including drying, soaking, and treatment with sodium hexametaphosphate and hydrogen peroxide. From the >63 μm fraction, relict and fresh foraminifera were picked up for analysis (Mazumder et al., 2012). The taxonomic classification of benthic foraminiferal species were based on Loeblich and Tappan (1988, 1990).

Chronological control was obtained by accelerator mass spectrometry (AMS) radiocarbon (^{14}C) dating on total organic carbon in the sediment selected intervals at the Birbal Sahni Institute of Palaeosciences, Lucknow (Uttar Pradesh, India). The age model is based on linear interpolation between the dated horizons. All radiocarbon ages were calibrated to calendar ages using the Oxcal 4.3 program (Bayliss et al., 1995), and all dates cited in the text thus refer to calendar years (Table 4.1). Assuming the sedimentation rate was constant for the whole sediment profile, they have been

TABLE 4.1
Radiocarbon Ages of the Studied Core

Sample depth (cm)	^{14}C age	Calibrated age (cal year)
10–12	7130 ± 110	7317
20–22	9830 ± 30	10464
25	11235 ± 30	12351

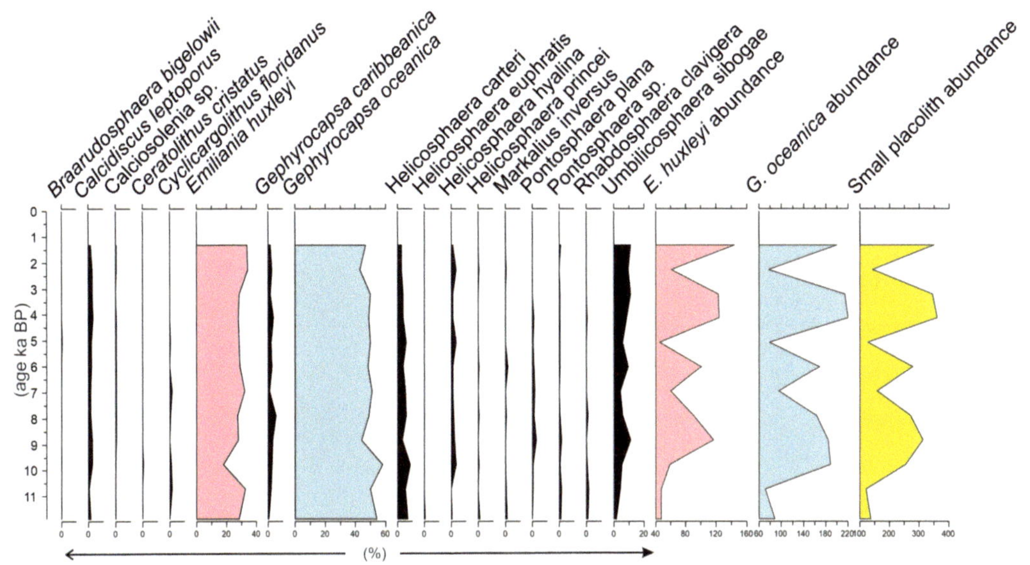

FIGURE 4.2 Relative and absolute abundance of coccoliths.

extrapolated to decipher the environmental changes and coeval climatic events in the region within a definite time frame.

4.4 RESULTS

4.4.1 Calcareous Nannofossils Assemblage Composition

The nannofossil assemblages in the SC-26 core are composed of 17 species (Figure 4.2). The assemblage is typically dominated by placolith-bearing coccoliths such as *Emiliania huxleyi* and *Gephyrocapsa oceanica* that dominate the assemblage up to ~72–80%. Other major species are *Umbilicosphaera sibogae* (7.2%), *Helicosphaera carteri* (4.7%), *Gephyrocapsa caribbeanica* (2.3%), *Calcidiscus leptoporus* (2.10%), *Helicosphaera hyalina* (1.6%), *Pontosphaera plana* (0.8%), *Pontosphaera* sp. (0.6%), *Cyclicargolithus floridanus* (0.42%), *Rhabdosphaera clavigera* (0.4%), *Markalius inversus* (0.3%), *Calciosolenia* sp. (0.3%), *Helicosphaera princei* (0.2%), *Ceratolithus cristatus* (0.07%), *Braarudosphaera bigelowii* (0.05%), and *Helicosphaera euphratis* (0.05%). Dominant species are represented with a mean relative abundance of ~49% for *Gephyrocapsa oceanica* and ~29% for small placoliths.

The relative abundance of *E. huxleyi* shows a gradual increase from 11.8 to 10.8 ka BP and then decreases from 10.7 to 9.7 ka BP followed by an increase up to 7.9 ka BP. Thereafter, its relative abundance shows minor fluctuations. *G. oceanica*, the dominant species of *Gephyrocapsa* spp., reached its maximum relative abundance during 9.7 ka BP (~45%) (Figure 4.2). The CEX index of the core ranges from 0.88 to 0.95 (Figure 4.3).

4.4.2 Benthic Foraminifera Assemblage Composition

Benthic foraminiferal assemblages show high species diversity upto 352 species) (Mazumder, 2005) in the studied core. Benthic foraminifera are mainly represented by the genera *Bolivina*, *Uvigerina*, *Cassulidinids*, *Gyroidina*, *Gyroidinoides*, *Melonis*, *Nonion*, *Nonionoides*, *Osangularia*, *Pullenia*, *Quinqueloculina*, *Rotalidium*, and *Trifarina*. The relative abundance of important taxa – uvigerinids, bolivinids, bulliminids, cassidulinids, and cibicids – are presented in Figure 4.4. Benthic groups, which are related to high organic matter flux, do not reveal temporally similar changes but

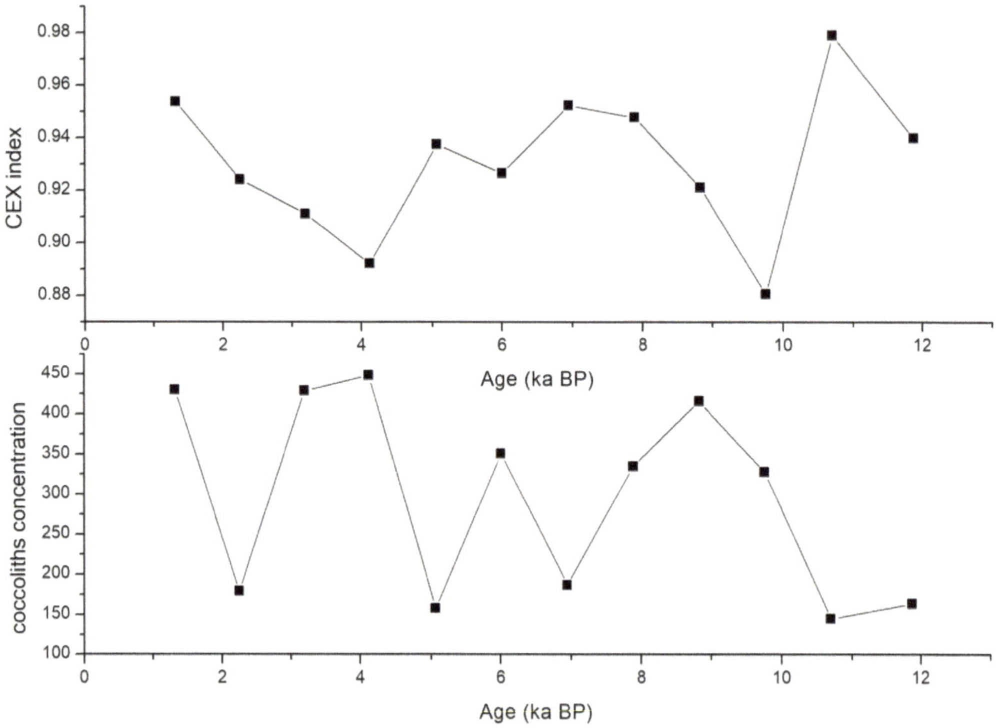

FIGURE 4.3 Coccolith absolute abundances and dissolution (CEX) index.

agree with a low abundance during the early Holocene period and a high abundance during the late Holocene period (Figure 4.4).

The TOC % ranges between 3 and 8 in the studied core. TOC % is high in the late Holocene compared to the early Holocene. It shows an overall gradual increasing trend during the last 12 ka BP (Figure 4.4).

4.5 DISCUSSION

4.5.1 Coccolith Indicators Used in the Present Study

E. huxleyi is the most abundant living coccolithophore in the majority of the world's oceans and has an extremely wide geographical distribution (Winter et al., 1994; Hattori et al., 2004). It thrives in broad ranges of salinity and temperatures (1–30°C) and lives in the entire photic zone (Winter, 1982). Its high abundances are documented from the tropical and subtropical ocean margins and continental shelves (Okada and Honjo, 1973; Broerse et al., 2000; Guptha et al., 2005; Mergulhao et al., 2006) with nutrient-rich environments (Smayda, 1966; McIntyre and Be, 1967; Berger, 1976; Stoll et al., 2007), and are rarely observed in oligotrophic regions. Lower fluxes of *E. huxleyi* in the central-eastern Arabian Sea are documented during the early SW monsoon, attributed to shallow mixed layer depth associated with the thermal stratification of the upper ocean layer (Mergulhao et al., 2006). *E. huxleyi* is used as a moderate saline waters taxon (Smayda, 1966; Jordan and Winter, 2000) and also as an indicator of warming and runoff (Weaver and Pujol, 1988).

G. oceanica is predominantly a low-latitude species (Okada and McIntyre, 1979). A positive relationship between nutrients and *G. oceanica* has been reported from the Arabian Sea (Andruleit and Rogalla, 2002). Sediment trap studies in the upwelling area off Somalia revealed that *G. oceanica* is an indicator of maximum upwelling conditions (Broerse et al., 2000). In the central-eastern

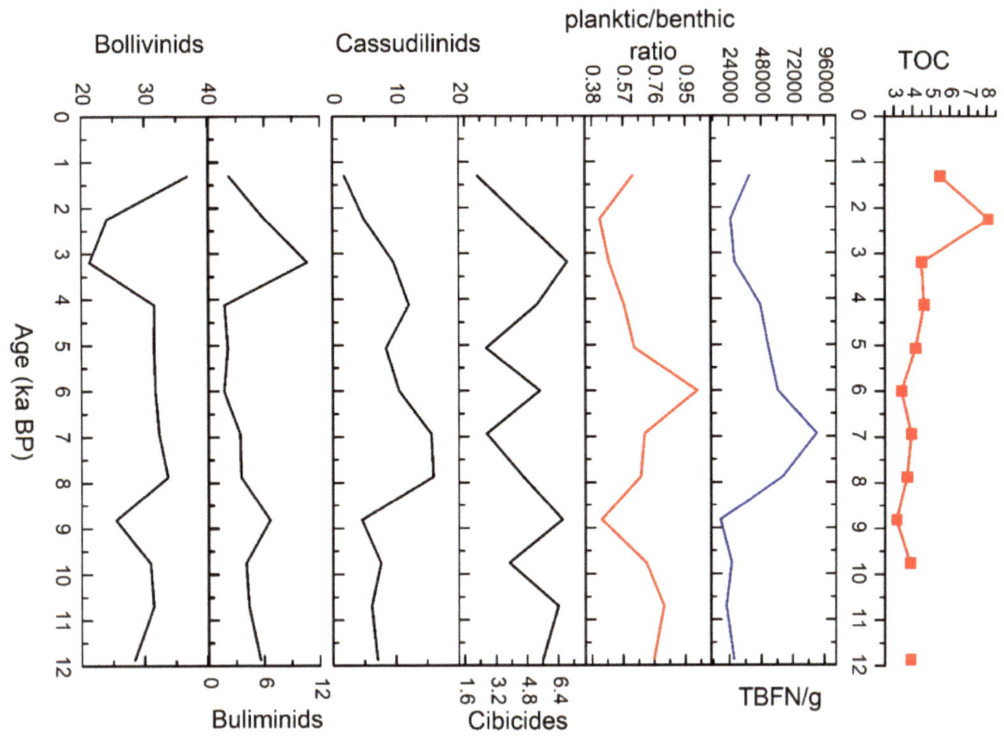

FIGURE 4.4 Benthic foraminiferal abundance and TOC % in the studied core.

Arabian Sea, its high fluxes are documented during the NE monsoon due to upwelling and nutrient entrainment to the upper photic zone (Mergulho et al., 2006). Its high abundance over *E. huxleyi* is generally related to its preference for nutrient enrichment.

C. leptoporus is one of the eurythermal species and widely distributed from equatorial to polar waters (McIntyre and Bé, 1967). Higher abundance of *C. leptoporus* is reported during the NE monsoon compared to the SW monsoon in the central Arabian Sea (Broerse et al., 2000; Andruleit and Rogalla, 2002; Merghulaho et al., 2006).

U. sibogae is normally associated with temperate and oligotrophic waters (Okada and McIntyre, 1979). However, it shows a positive relationship with the nutrient supply in the Arabian Sea (Andruleit and Rogalla, 2002). *U. sibogae* is also documented from the nutrient-rich upwelling regions (Mergulho et al., 2006). In the central Arabian Sea, the high flux of *U. sibogae* is observed during the NE monsoon, and decreased fluxes are related to the summer monsoon-induced stratification and low nutrients (Mergulho et al., 2006). The other documented species in the present study have ecologies that are not well established.

4.5.2 Benthic Foraminifera Used in the Present Study

The benthic foraminiferal groups Bollivinids, Bulliminids, and Cassidulinids, which are mainly related to the low oxygenated environments and high organic matter flux, which in turn are governed by the primary productivity in surface waters, are used in the present study (Figure 4.4). Cibicidids, which are related to high oxygenated environments and low organic matter flux, do not show any relationship with the productivity in the present study (Figure 4.4).

4.5.3 Effect of Dissolution on the Preservation in the Studied Core

The presently studied proxies are composed of calcium carbonate, which is a sensitive substance to the dissolution process that can alter the original signals. If dissolution removed the susceptible species of coccoliths and benthic foraminifera, recovered assemblages then would not depict the actual sea surface and bottom conditions and primary productivity. However, selective preservation can be excluded from influencing the studied core assemblages. The support for this statement comes from the coccolith records. The degree of preservation of coccoliths is found to be good, as per the visual scale given by Roth and Thierstein (1972) and Flores and Marino (2002). All the specimens of coccoliths are well preserved and can be identified at the species level with no or negligible signs of dissolution or fragmentation. This is also supported by the CEX index, which is high throughout the studied core (>0.88) providing the support for no dissolution effect on the coccoliths' signal (Figure 4.3). No dissolution is also observed in the benthic foraminifera.

4.5.4 Paleoproductivity Changes during the Holocene

Coccolithophore assemblages indicate primary productivity and nutrient availability, which is mainly influenced by vertical mixing or upwelling and runoff in the eastern Arabian Sea. The lowest concentrations of total cocooliths are observed during the earliest Holocene (12–10.7 ka BP) (Figures 4.2, 4.3), reflecting low productivity during this period. The absolute abundance of eutrophic species *Emiliania huxleyi* and *Gephyrocapsa oceanica* also shows low values during this period, indicating a low productivity. The relative abundance of *U. sibogae* is low during this period, which also reveals the low productivity and oligotrophic conditions in the study area. This clearly indicates the reduction in the nutrient source, that is, summer monsoon runoff or upwelling in the study area during the earliest Holocene. This time period occurs just after the Younger Dryas cold period in the North Atlantic Ocean, which, through teleconnections, affected the primary productivity in the Arabian Sea (Singh et al., 2018). The summer monsoon has been documented to show decreased intensity during this period, resulting in low nutrient availability and low primary productivity, which is also clearly reflected in the present core. Besides, the sea level was lower than present day during this period (~120 m during the last glacial period), which could have played a role in the low abundance of coccoliths during this period. The benthic foraminiferal total abundance also shows low values during this period, revealing low benthic foraminiferal productivity. The foraminiferal groups Bollivinids, Bulliminids, and Basusilids all show comparatively low abundances during this period, revealing low organic matter flux to the sea floor as primary productivity decreased. It is also interesting to note that primary productivity records show contrasting evidence during this period in the eastern Arabian Sea. Planktic foraminiferal records show low primary productivity, whereas coccoliths from the same region show high primary productivity (Cabarcos et al., 2014; Singh et al., 2018). The present study, however, supports the planktic foraminiferal records, indicating that productivity is comparatively weak during this period.

The coccolith concentration shows a gradual increase from ~10.5 to 9 ka BP, indicating the increasing trend in productivity, which suggests the strengthening of the summer monsoon and nutrient availability. Increased relative abundances of *U. sibogae* and *G. oceanica* are also in agreement with the increased nutrients at the core site. A slight increase in *C. leptoporus* during this period suggests increased upwelling that could have occurred due to the increased summer monsoon winds. This period is known as the early Holocene humid phase. Several studies have previously documented the intensification of summer monsoons during this period across the Arabian Sea, Bay of Bengal, and Indian continent (Swain et al., 1983; Rashid et al., 2011). The increased insolation during this period resulted in warm and humid conditions that enhanced runoff nutrients to the Arabian Sea. Coccolith productivity shows a drop again at 7 ka BP and 5 ka BP, which could be related to the middle Holocene transition period that experienced arid conditions and decreased summer monsoon rainfall and nutrients (Saraswat et al., 2016). During the late Holocene, coccolith abundances and TOC % show very high values, indicating high primary productivity. Increased primary productivity during the late Holocene is also earlier observed from the eastern Arabian Sea (Singh et al., 2018), which is related to enhanced wind mixing during winters. *U. sibogae* abundance shows an increase during the late Holocene, which has been previously documented to show high abundances during NE monsoon mixing and a decrease during the summer monsoon period, suggesting that wind mixing during the late Holocene enhanced productivity.

4.6 CONCLUSIONS

A combined study of phytoplankton coccoliths and benthic foraminiferal proxies reveals changes in the monsoon and thus the nutrient and primary productivity during the past 12 Ka. *G. oceanica* and *E. huxleyi* are the two dominant species in the coccoliths throughout the Holocene, suggesting eutrophic conditions, which is also reflected by the high abundances of benthic foraminiferal groups Bulliminids, Bollivinids, and Cassudilinids that are related to the high organic matter flux. Both proxies show low abundances prior to 10.7 ka BP revealing comparatively low productivity due to the weak monsoon. The core site shows high productivity reflected in coccoliths and benthic foraminifera during the early Holocene (10–9 Ka BP) related to the intensified summer monsoon and nutrients. Coccoliths, benthic foraminifera, and TOC % also show high productivity during the late Holocene in comparison to the early and middle Holocene, indicating increased wind mixing in the eastern Arabian Sea. The present study shows the potential of coccoliths to reconstruct primary productivity changes related to monsoon changes in the eastern Arabian Sea.

ACKNOWLEDGEMENTS

The authors thank the Director BSIP for the facilities provided to carry out the present study. Dr. Rajiv Nigam and Dr. Vandana Prasad are duly acknowledged for their support during the study. The present work is funded by the SERB project SRG/2022/002129.

REFERENCES

Agnihotri, R., Bhattacharya, S.K., Sarin, M.M., Somayajulu, B.L.K., 2003. Changes in surface productivity and subsurface denitrification during the Holocene: A multiproxy study from the eastern Arabian Sea. *The Holocene* 13(5), 701–713.

Amore, F.O., Flores, J.A., Voelker, A.H.L., Lebreiro, S.M., Palumbo, E., Sierro, F.J., 2012. A middle Pleistocene Northeast Atlantic coccolithophore record: Paleoclimatology and paleoproductivity aspects. *Marine Micropaleontology* 90–91, 44–59.

Andruleit, H., Rogalla, U., 2002. Coccolithophores in surface sediments of the Arabian Sea in relation to environmental gradients in surface waters. *Marine Geology* 186(3–4), 505–526.

Bassinot, F.C., Marzin, C., Braconnot, P., Marti, O., Mathien-Blard, E., Lombard, F., Bopp, L., 2011. Holocene evolution of summer winds and marine productivity in the tropical Indian Ocean in response to insolation forcing: Data-model comparison. *Climate of the Past Discussions* 7(3), 815–829.

Bauer, S., Hitchcock, G.L., Olson, D.B., 1991. Influence of monsoonally forced Ekman dynamics upon surface-layer depth and plankton biomass distribution in the Arabian Sea. *Deep-Sea Research Part A* 38(5), 531–553.

Baumann, K.-H., Freitag, T., 2004. Pleistocene fluctuations in the northern Benguela Current system as revealed by coccolith assemblages. *Marine Micropaleontology* 52(1–4), 195–215.

Bayliss, A., Ramsey, C. B., Cook, G., McCormac, G., Otlet, R., & Walker, J. (1995). Radiocarbon dates. *Ratcliffe J. Duckpool, Morwenstow: a Romano-British and early medieval industrial site and harbour.* Cornish Archaeology, 34, 81–171.

Berger, W.H., 1976. Biogenous deep sea sediments: Production, preservation and interpretation. In: Riley, J.P., Chester, R. (Eds.), *Chemical Oceanography 5*. London: Academic Press, 265–388.

Boeckel, B., Baumann, K.-H., Henrich, R., Kinkel, H., 2006. Coccolith distribution patterns in South Atlantic and Southern Ocean surface sediments in relation to environmental gradients. *Deep-Sea Research I Oceanography Research Papers* 53(6), 1073–1099.

Bown, P.R., 1998. *Calcareous Nannoplankton Biostratigraphy.* London: Chapman Hall, 1–324.

Broerse, A.T.C., Brummer, G.-J.A., van Hinte, J.E., 2000. Coccolithophore export production in response to monsoonal upwelling off Somalia (north western Indian Ocean). *Deep-Sea Research II* 47, 2179–2205.

Cabarcos, E., Flores, J.A., Sierro, F.J., 2014. High-resolution productivity record and reconstruction of ENSO dynamics during the Holocene in the Eastern Equatorial Pacific using coccolithophores. *The Holocene* 24(2), 176–187.

Colborn, J.G., 1975. *The Thermal Structure of the Indian Ocean.* Int. Indian Ocean Expedition Monograph. Honolulu: University of Hawaii Press, 1–173.

Dittert, N., Baumann, K.H., Bickert, T., Henrich, R., Kinkel, H., Meggers, H., 1999. Carbonate dissolution in the deep-sea: Methods, quantification and paleoceanographic application. In: Fischer, G., Wefer, G. (Eds.), *Proxies in Paleoceanography: Examples From the South Atlantic.* Berlin: Springer, 255–284.

Flores, J.A., Marino, M., 2002. Pleistocene calcareous nannofossil stratigraphy for ODP Leg 177 (Atlantic sector of the Southern Ocean). *Marine Micropaleontology* 45(3–4), 191–224.

Flores, J. A., Marino, M., Sierro, F. J., Hodell, D. A., & Charles, C. D. (2003). Calcareous plankton dissolution pattern and coccolithophore assemblages during the last 600 kyr at ODP Site 1089 (Cape Basin, South Atlantic): paleoceanographic implications. 196(3-4), 409–426.*Palaeogeography, Palaeoclimatology, Palaeoecology,*

Guptha, M.V.S., Mergulhao, L.P., Murty, V.S.N., Shenoy, D.M., 2005. Living coccolithophores during the northeast monsoon from the Equatorial Indian Ocean: Implications on hydrography. *Deep-Sea Research II* 52, 2048–2060.

Hattori, H., Koike, M., Tachikawa, K., Saito, H., Nagasawa, K., 2004. Spatial variability of living coccolithophore distribution in the western subarctic Pacific and western Bering Sea. *Journal of Oceanography* 60, 505–515.

Jordan, R.W., Winter, A., 2000. Assemblages of coccolithophorids and other living microplankton off the coast of Puerto Rico during January-May 1995. *Marine Micropaleontology* 39, 113–130.

López-Otálvaro, G.E., Flores, J.A., Sierro, F.J., Cacho, I., 2008. Variations in coccolithophorid production in the Eastern Equatorial Pacific at ODP Site 1240 over the last seven glacial–interglacial cycles. *Marine Micropaleontology* 69(1), 52–69.

López-Otálvaro, G.E., Flores, J.A., Sierro, F.J., Cacho, I., Grimalt, J.O., Michel, E., Cortijo, E., Labeyrie, L., 2009. Late Pleistocene paleoproductivity patterns turing the last climatic cycle in the Guyana Basin as revealed by calcareous nannoplankton. *eEarth* 4, 1–13.

Ma, R., Sépulcre, S., Licari, L., Bassinot, F., Liu, Z., Tisnérat-Laborde, N., 2019. Changes in intermediate circulation in the Bay of Bengal since the last glacial maximum as inferred from benthic foraminifera assemblages and geochemical proxies. *Geochemistry Geophysics Geosystems* 20, 1592–1608.

Madhupratap, M., Prasanna Kumar, M.S., Bhattathiri, P.M.A., Dileep Kumar, M., Raghukumar, S., Nair, K.K.C., Ramaiah, N., 1996. Mechanism of the biological response to winter cooling in the northeastern Arabian Sea. *Nature* 384, 549–552.

Mazumder, A., 2005. Paleoclimatic reconstruction through the study of foraminifera in marine sediments off central west coast of India (Doctoral dissertation, Goa University).

Mazumder, A., Nigam, R., & Henriques, P. J. 2012. Deterioration of Early Holocene coral reef due to sea level rise along west coast of India: Benthic foraminiferal testimony. *Geoscience Frontiers, 3(5),* 697–705.

Mazumder, A., & Nigam, R. (2014). Bathymetric preference of four major genera of rectilinear benthic foraminifera within oxygen minimum zone in Arabian Sea off central west coast of India. *Journal of Earth System Science,* 123, 633-639.

McIntyre, A., Be, A.W.H., 1967. Modern coccolithophores of the Atlantic Ocean. I: Placoliths and cyrtoliths. *Deep Sea Research* 14, 561–597.

Mergulhao, L.P., Mohan, R., Murty, V.S.N., Guptha, M.V.S., Sinha, D.K., 2006. Coccolithophores from the central Arabian Sea: Sediment trap results. *Journal of Earth System Science* 115(4), 415–428.

Nigam, R., Prasad, V., Mazumder, A., Garg, R., Saraswat, R., Henriques, P.J., 2009. Late Holocene changes in hypoxia off the west coast of India: Micropalaeontological evidences. *Current Science* 96(5), 708–713.

Nigam, R., Mazumder, A., Henriques, P. J., & Saraswat, R. (2007). Benthic foraminifera as proxy for oxygen-depleted conditions off the central west coast of India. Journal of Geological Society of India. 70, 1047-1054. http://drs.nio.org/drs/handle/2264/75

Okada, H., Honjo, S., 1973. The distribution of oceanic coccolithophorids in the Pacific. *Deep-Sea Research* 20, 355–374.

Okada, H., McIntyre, A., 1979. Seasonal distribution of modern coccolithophores in the western North Atlantic Ocean. *Marine Biology* 54(4), 319–328.

Qasim, S.Z., 1982. Oceanography of the northern Arabian Sea. Deep Sea Research Part A. *Oceanographic Research Papers* 29(9), 1041–1068.

Ramsey, C.B., 2009. Bayesian analysis of radiocarbon dates. *Radiocarbon* 51(1), 337–360.

Rashid, H., England, E., Thompson, L., and Polyak, L., 2011. Late glacial to Holocene Indian summer monsoon variability based upon sediment records taken from the Bay of Bengal. *TAO: Terrestrial, Atmospheric and Oceanic Sciences* 22(2), 2.

Roth, P., Thierstein, H., 1972. Calcareous nannoplankton: Leg 14 of the deep sea drilling project. In: Hayes, D.E., Pimm, A.C., et al. (Eds.), *Initial Reports DSDP 14.* U.S. Govt. Printing.

Saavedra-Pellitero, M., J. A. Flores, F. Lamy, F. J. Sierro, and A. Cortina (2011), Coccolithophore estimates of paleotemperature and paleoproductivity changes in the southeast Pacific over the past ~27 kyr, 26, PA1201, 1–16. doi:10.1029/2009PA001824.*Paleoceanography,*

Saraswat, R., Naik, D.K., Nigam, R., Gaur, A.S., 2016. Timing, cause and consequences of mid-Holocene climate transition in the Arabian Sea. *Quaternary Research* 86(2), 162–169.

Sarkar, A., Ramesh, R., Somayajulu, B.L.K., Agnihotri, R., Jull, A.J.T., Burr, G.S., 2000. High-resolution Holocene monsoon record from the eastern Arabian Sea. *Earth and Planetary Science Letters* 177, 209–218.

Schott, F.A., McCreary, J.P., 2001. The monsoon circulation of the Indian Ocean. *Progress in Oceanography* 51(1), 1–123.

Shankar, D., Vinayachandran, P.N., Unnikrishnan, A.S., 2002. The monsoon currents in the north Indian Ocean. *Progress in Oceanography* 52, 63–120.

Singh, A.D., Rai, A.K., Verma, K., Das, S., Bharti, S.K., 2015. Benthic foraminiferal diversity response to the climate-induced changes in the eastern Arabian Sea oxygen minimum zone during the last 30 ka BP. *Quaternary International* 374, 118–125.

Singh, A.D., Jung, S.J., Anand, P., Kroon, D., Ganeshram, R.S., 2018. Rapid switch in monsoon-wind induced surface hydrographic conditions of the eastern Arabian Sea during the last deglaciation. *Quaternary International* 479, 3–11.

Singh, A.D., Jung, S.J.A., Darling, K., Ganeshram, R., Ivanochko, T., Kroon, D., 2011. Productivity collapses in the Arabian Sea during glacial cold phases. *Paleoceanography*, PA3210, 26 (3), 1–10, doi:10.1029/2009PA001923

Smayda, T.J., 1966. A quantitative analysis of the phytoplankton of the Gulf of Panama. III. General ecological conditions, and the phytoplankton dynamics at 8,45°N, 79,23°W from November 1954 to May 1957. *Bulletin of Inter-American Tropical Tuna Commission* 11, 355–612.

Stoll, H. M., Ziveri, P., Shimizu, N., Conte, M., & Theroux, S. (2007). Relationship between coccolith Sr/Ca ratios and coccolithophore production and export in the Arabian Sea and Sargasso Sea. *Deep Sea Research Part II: Topical Studies in Oceanography,* 54(5-7), 581–600.

Swain, A.M., Kutzbach, J.E., Hastenrath, S., 1983. Estimates of Holocene precipitation for Rajasthan, India, based on pollen and lake-level data. *Quaternary Research* 19, 1–17. https://doi.org/10.1016/0033-5894(83)90024-8.

Verma, K., Bharti, S.K., Singh, A.D., 2018. Late Glacial–Holocene record of benthic foraminiferal morphogroups from the eastern Arabian Sea OMZ: Paleoenvironmental implications. *Journal of Earth System Sciences* 127, 21.

Verma, K., Singh, H., Singh, A.D., Singh, P., Satpathy, R.K., Naidu, P.D., 2021. Benthic foraminiferal response to the millennial-scale variations in monsoon-driven productivity and deep-water oxygenation in the Western Bay of Bengal during the last 45 ka. *Frontiers in Marine Science* 8, 733365.

Weaver, P.P.E., Pujol, C., 1988. History of the last deglaciation in the Alborán Sea (western Mediterranean) and adjacent north Atlantic as revealed by coccolith floras. *Palaeogeography Palaeoclimatology Palaeoecology* 64, 35–42.

Winter, A., 1982. Paleoenvironmental interpretation of Quaternary coccolith assemblages from the Gulf of Aqaba (Elat), Red Sea. *Revistaespañola de micropaleontología* 14, 291–314.

Winter, A., Jordan, R.W., Roth, P.H., 1994. Biogeography of living coccolithophores in ocean waters. In: Winter, A., Siesser, W.G. (Eds.), *Coccolithophores.* Cambridge University.

Zhou, X., Duchamp-Alphonse, S., Kageyama, M., Bassinot, F., Beaufort, L., Colin, C., 2020. Dynamics of primary productivity in the northeastern Bay of Bengal over the last 26 000 years. *Climate of the Past* 16(5), 1969–1986.

Part III

Advanced Aquatic Ecosystems Monitoring

5 Aquatic Ecosystems Monitoring with Remote Sensing Perspective

Soudani Leila, Arabi Zohra, Bouacha Mohamed Islem, Safa Omar and Azzaoui Mohamed Essalah*
Nature and Life Sciences Faculty, Ibn Khaldoun University of Tiaret, Algeria

5.1 INTRODUCTION

Aquatic ecosystems constitute a crucial part of life as we recognize it on Earth, playing a pivotal role in the global water cycle and providing habitats for a variety of species. Despite their significance, these ecosystems face multiple challenges, including pollution, climate change, and human activities, placing their well-being at risk (OFEV, 2020). Consequently, the decline in aquatic biodiversity exceeds that observed in terrestrial ecosystems, which are also significantly affected (Dudgeon et al., 2005). Changes in biodiversity have major consequences for the services humans derive from ecosystems, given their connection to ecosystem processes and resilience in the face of environmental changes (Grizzetti et al., 2016; Chapin et al., 2000).

To ensure the protection and conservation of aquatic ecosystems, it is imperative to engage in monitoring and assessment efforts. Monitoring offers insights into long-term changes in the aquatic ecosystem, while assessment shows the ecosystems' current state (Karr, 1991; Karr et al., 1986). They play a crucial role in supporting policymakers in their policy development, helping water managers make decisions, and informing stakeholders so that they might be inspired to get involved in the process of making and putting policies into action (Goethals and Forio, 2018). Traditional methods for monitoring and assessment involve physical–chemical and biological approaches, relying on experimental measurements and established sampling protocols. However, they are limited in their ability to understand the entirety of water systems, especially in terms of the integrity and health of the ecosystems (Brachet et al., 2015; Tockner and Stanford, 2002). Using geographic information system (GIS) technology for aquatic ecosystem monitoring and assessment is a creative way to identify and evaluate the problems influencing the quality of the water. Thus, the sophisticated use of GIS in today's world has greatly facilitated the identification of water quality (Nazzal, 2019). This chapter offers a comprehensive review of various aquatic ecosystems, exploring the utilization and potential of remote sensing and GIS in monitoring, as well as understanding the spatiotemporal evolution of these ecosystems. In addition, traditional methods and the most recent discoveries encompassing pragmatic and realistic strategies that allow real-time evocation and consolidation of new approaches in order to control, monitor, and alert all the constraints that can threaten these ecosystems are discussed.

5.2 TYPES OF AQUATIC ECOSYSTEMS

More than two-thirds of our planet is made up of aquatic ecosystems, which are crucial for maintaining climate stability and providing a diverse array of services to a rapidly growing human population (Häder et al., 2020). Nearly 70% of people on Earth reside in regions that are adjacent to lakes, rivers,

* Corresponding author.

DOI: 10.1201/9781003354000-8

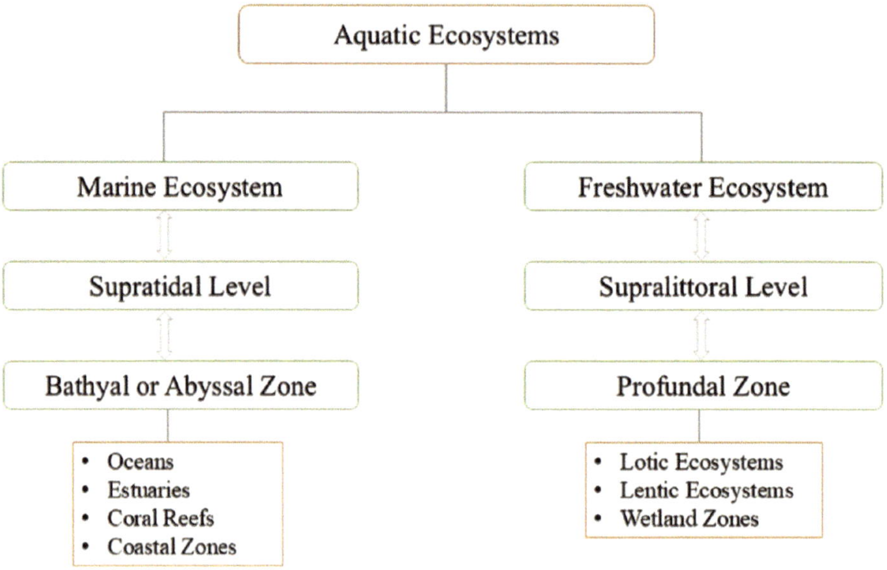

FIGURE 5.1 Aquatic ecosystems types.

and other bodies of water, which is where civilizations have historically developed. Water shortages already impact approximately 40% of the global population, residing in 80 different countries (Dugan, 1993; Lewis, 1996). Aquatic ecosystems constitute biological communities primarily existing in water environments and are categorized into two types: the marine ecosystem and the freshwater ecosystem (Figure 5.1), based on the type of water with which organisms interact (Barange et al., 2010). Encompassing over 70% of the Earth's surface, the marine ecosystem is the largest water ecosystem, including oceans, estuaries, coral reefs, and coastal ecosystems as distinct categories. In contrast, freshwater ecosystems cover less than 1% of the Earth's surface and comprise the lotic ecosystem, lentic ecosystem, and wetland ecosystem as different types (Bashir et al, 2020).

5.2.1 Marine Ecosystems

The marine ecosystem stands as one of Earth's largest ecosystems, playing a primordial role in maintaining the planet's health and biodiversity. Generally encompassing aquatic ecosystems with saline water, marine ecosystems exhibit varying salt content, ranging from low to high and showing significant variability or relative constancy over time, depending on the specific type of marine ecosystem (Nybakken and Bertness, 2005). Subdividing the marine environment takes into account vertical stratification into a benthic region (sea bottom) and a pelagic region (water column). On a horizontal scale, it further stratifies into open oceanic, neritic, and coastal waters (Castro and Huber, 2012). These waters harbor a diverse array of frequently interconnected ecosystems, such as seagrass beds, coral reefs, estuaries, coastal wetlands, sand beaches, and the largely uncharted deep sea. From the coastline to the deep seabed, the marine ecosystem provides a wide range of habitats for aquatic flora and fauna. Approximately 90% of the marine ecosystem is composed of extended sea and ocean water, hosting about 10% of the total marine animal population (Postel and Richter, 2012).

5.2.1.1 Ocean Ecosystem

The ocean ecosystem, also known as the marine ecosystem, is a vast and complex ecological system that encompasses all living organisms and their physical and chemical environment in the earth's oceans and seas. Encompassing over 70% of the Earth's surface, equivalent to 140 million square miles, this ecosystem is among the largest and most diverse on the planet. Ninety-seven percent of the planet's free water is found in the ocean (Stewart and Jamieson, 2019). The world's oceans range in depth from 2.4 miles on average to over 6 miles at their deepest points. The Pacific, Indian, Arctic, Atlantic, and Southern (Antarctic) oceans are the five great oceans on Earth. Despite their interconnectedness, the oceans each have distinct species and characteristics. Ranked as the second largest and deepest ocean globally, the Atlantic Ocean follows the Pacific Ocean in size and depth. Diverse life forms can be found in the oceans. Despite being extremely cold, the waters of the Southern and Arctic Oceans are teeming with life. The Southern Ocean's ice cover is home to the world's largest population of krill, which are tiny marine animals that resemble shrimp. Oceans are home to a wealth of essential natural resources. They supply minerals, food, and energy. Ships use the oceans as waterways to move cargo and passengers between continents. Seas and oceans play a major role in preserving the global climate by controlling air temperature and providing moisture for precipitation. Life on Earth would not have begun if there were no oceans (Fox-Kemper et al., 2021; Bar-On et al., 2018; Appeltans et al., 2012).

5.2.1.2 Estuarine Ecosystem

Estuaries are areas of transition that are different from both the sea and rivers in terms of both biotic and abiotic factors. Estuarine waters experience greater daily fluctuations in temperature, salinity, and turbidity than do rivers or the ocean. Estuaries are extremely productive ecosystems that are comparable in productivity to mangrove swamps and coral reefs from a biological perspective. Both the sediment and the water column have high nutrient levels, which maintain an elevated productivity. Estuaries have the capacity to sustain euryhaline biological species throughout part or the entirety of their life cycle, as noted by McLusky and Elliott (2004). These marine environments exhibit fluctuations in salinity, pH, and water levels, influenced by the combined factors of ocean salinity and river inflow into the estuary. Ocean and land combine to create a special environment to which plants and animals have evolved (organisms have specialized capacity for adaptation). Reefs, barrier islands, headlands, and deltas shield estuaries from the forces of the ocean. Estuarine circulation, a significant process in estuaries, is characterized by the movement of fresh or brackish water flowing outward near the surface, while denser saline water moves inward near the bottom. This represents one of the two main circulation patterns observed in estuaries. Conversely, the contrasting phenomenon, termed "anti-estuarine flow", occurs when less dense water circulates inward at the surface, and denser water flows outward near the bottom (McLusky and Elliott, 2004; Kennish, 2005; Day et al., 1989).

Millions of plants and animals find a tranquil haven in estuaries away from the open ocean, supporting a vast diversity and abundance of species. Estuaries support a diverse array of marine life, including fish, shellfish, lobsters, marine worms, seagrasses, algae, and phytoplankton. Several fish species use estuaries for breeding, providing a conducive environment for their young to develop and mature before migrating to the open ocean. Many commercially valuable fish species, including snapper, flounder, mullet, rockfish, sole, kahawai, trevally, parore, red cod, koral, lakhoua, gurnard, and others, incorporate estuaries into various stages of their life cycles (McLusky and Elliott, 2004; Dyer, 1997).

5.2.1.3 Coral Reefs Ecosystem

According to Moberg and Folke (1999), coral reefs rank among the most productive and diverse ecosystems. They offer a wide range of goods and services to substantial populations, encompassing seafood, recreational opportunities, coastal protection, as well as aesthetic and cultural benefits (Burke et al., 2011; Burke et al., 2002; Speers et al., 2016; Wilkinson, 2008). The estimated coverage of coral reefs on the ocean floor ranges from 0.1% to 0.5%. Despite this relatively small footprint, coral reefs support remarkable biodiversity, hosting nearly 30% of marine fish species globally. Moreover, approximately 10% of the fish consumed by humans is sourced from reef areas through fishing activities (Reaka-Kudla and Wilson, 1997; Smith, 1987). There are coral reefs along the coastlines of more than 100 countries. Tens of millions of people in those nations rely on coral reefs for a portion of their protein intake or way of life (Fisher et al., 2015; Salvat, 1992). Atolls, platform reefs, barrier reefs, and fringing reefs are the four primary types of coral reefs. These reef types differ greatly in their functions, and they are connected to other systems, like seagrass beds, mangrove forests, and the open ocean, to differing degrees. The ecosystem of coral reefs is open and intricate, with human modifications to water quality, biogeochemical processes, and hydrological flows (locally or remotely) having the potential to impact its structure, function, biodiversity, and resilience. Reef ecosystems provide most of their ecological goods and services through a wide range of intricate and dynamic interactions between species networks both within and between ecosystems (Moberg and Folke, 1999).

5.2.1.4 Coastal Ecosystem

Many of the services and benefits that coastal ecosystems provide are tangible, like storm protection, food supply, water quality control, and carbon storage. The coast is a very dynamic interface between the sea and the land that is shaped by both marine and terrestrial processes. However, non-material benefits that impact on people's spiritual, social, and cultural aspects make up a significant portion of the services that coastal ecosystems provide to humans. Coastal ecosystems encompass diverse habitat types, a plethora of species, and genetic diversity. These ecosystems serve vital functions, including nutrient storage and cycling, filtration of pollutants from inland freshwater systems, and providing protection to shorelines against storms and erosion. Conversely, oceans on the other side of shorelines are essential sources of oxygen and significant carbon dioxide sinks, largely due to the high productivity of phytoplankton. Oceans also play a pivotal role in global hydrology and climate regulation. Beyond their ecological significance, coastal ecosystems attract the world's population due to their natural beauty, making coastal areas popular places for residence, tourism, and leisure activities (Barbier et al., 2011; Plag and Jules-Plag 2013; Kusky, 2008).

5.2.2 Freshwater Ecosystems

Within the aquatic ecosystems of Earth, freshwater ecosystems make up a subset. Wetlands, rivers, streams, and springs are among them, as are lakes and ponds. Freshwater ecosystems are characterized by being both disproportionately rich and disproportionately threatened in terms of biological assets. Freshwaters, especially for vertebrates, have a disproportionately high species richness for their surface area. Though they vary widely in type, location, and climate, freshwater ecosystems nevertheless have many essential characteristics in common. A vast number of species on Earth's surface live in freshwater habitats. Much of the biological diversity of Earth is comprised of these aquatic species and the environments in which they coexist (Baron et al., 2003; Strayer and Dudgeon, 2010).

5.2.2.1 Lotic Ecosystem

Systems with swiftly moving waters are referred to as lotic ecosystems. These landscape features are linear. Among aquatic ecosystems, running waters are distinguishable by their unidirectional flow, adding to their distinctive linear form (Malmqvist and Rundle, 2002).

The linear form characteristic of running waters not only shapes river channel morphology but also contributes to the intrinsic variability of these environments in space and time (Giller and Malmqvist, 1998). In terms of energy, particularly in the form of organic matter flowing consistently in one direction, the lotic ecosystem operates essentially as an open system. However, the energetic viability of running waters is sustained by the input of solar radiation and the close connection between streams and nearby terrestrial ecosystems. It's important to note that, despite their significance, streams and rivers only account for a tiny fraction (0.006%) of the Earth's freshwater at any given time (Shiklomanov, 1993). Lotic ecosystems offer humanity a multitude of benefits, such as a supply of water for residential, commercial, and agricultural use; a way to generate electricity and dispose of waste; pathways for travel; and places to engage in recreational activities. Because flowing waters have drawn people to them, their immense value to humanity has also proved to be their downfall. Lotic ecosystems include a broad range of habitats, from massive lowland rivers to tiny mountain springs. Many different species, including some of the most endangered, can be found in lotic systems (Allan and Flecker, 1993).

5.2.2.2 Lentic Ecosystem

Characterized by their still waters, lentic ecosystems represent a crucial component of the Earth's aquatic landscape. Over vast geological timescales, these bodies of water undergo a gradual transformation, transitioning from expansive lakes to smaller ponds and ultimately evolving into wetland habitats. Lentic ecosystems harbor intricate networks of biotic and abiotic interactions, playing a central role as integrators of climate signals from both the atmosphere and surrounding catchments. Situated at low points in the landscape, these ecosystems serve as sensitive receptors of environmental change (Williamson et al., 2009). Lentic ecosystems encompass areas of standing water that serve as habitats for a diverse array of organisms, including fish, birds, frogs, and plants featuring both rooted and floating leaves. In lentic waters, a large number of tiny invertebrates known as zooplankton reside. Zoobenthos play a more significant role in lentic ecosystems than do pelagic animal communities, especially in shallow systems. They supply food for other organisms like snails and water strider insects by feeding on algae and plants. Numerous variables, including the water's depth, salinity, and amount of light, affect the types of fish and other vertebrates (Reddy and Campus, 2018).

5.2.2.3 Wetland Ecosystem

Wetland ecosystems are characterized by their dependence on flooding or shallow saturation at or near the surface of the substrate, either continuously or periodically. They share wide boundaries with aquatic and terrestrial ecosystems. A preponderance of hydrophytes and growing conditions that are at least occasionally moist enough to create anaerobic soil conditions during the growing season are characteristics of wetlands that impact plant growth. Hence, bodies of water that exhibit primarily permanent characteristics, such as lakes, ponds, rivers, and lagoons located below the level of rooted vegetation (littoral zone), are excluded from the classification of wetland ecosystems (Reddy and Campus, 2018).

Wetlands derive their distinctive characteristics from many interdependent elements, including the volume, velocity, salinity, and permanent or transient nature of the water. The characteristics of wetlands are strongly influenced by underlying factors such as geology, soil, climate, groundwater levels, water chemistry and fertility, and the flora and fauna that inhabit them. Freshwater wetlands are classified into two main categories according to the duration of waterlogging: perennial wetlands, characterized by waterlogging throughout the year and including habitats such as riverbanks and the shores of large lakes or reservoirs, and seasonal wetlands, which dry out completely for variable periods depending on the monsoon season. Among freshwater ecosystems, wetlands stand out for their high biological productivity. They are essential habitats for migratory species and serve as vital breeding grounds, particularly for specialist vascular plants, mosses, invertebrates, and birds. As well as a variety of fish species, freshwater wetlands are home to butterflies, dragonflies, mosquitoes, frogs, turtles, and ducks. Unfortunately, man's influence has led to a significant overall decline in these essential ecosystems (Figure 5.2) (Reddy and Campus, 2018; Solimini et al., 2008; Stenert and Maltchik, 2007; Basset and Abbiati, 2004; Shine and Klemm, 1999).

5.3 TRADITIONAL METHODS MONITORING

5.3.1 Physical–Chemical Monitoring

In the last century, human activities have altered the chemical composition and physical attributes of water bodies. Rivers, in particular, have experienced an increase in concentrations of various chemical pollutants, encompassing pesticides, organic pollutants, trace metals, pathogens, nutrients, and endocrine disruptors. Beyond disrupting naturally occurring hormones, these pollutants have the potential to induce direct physiological effects, indirect impacts on the food chain, heightened

FIGURE 5.2 A wetland in the Tiaret region (Algeria).

primary production, algal blooms, toxic effects on both humans and biota, and toxicity through biomagnification (Deknock et al., 2019; Malmqvist and Rundle, 2002). The physical and chemical qualities of water play a crucial role in supporting ecosystem functioning, the products and services provided by ecosystems, and the ecological processes vital for organism survival. Consequently, alterations to the chemical and physical properties of water may significantly impact how an ecosystem operates and the services it delivers (Keeler et al., 2012).

The chemical quality of water is commonly assessed through analyses of nutrient concentrations, dissolved oxygen concentration, pH, chemical oxygen demand, biological oxygen demand, pesticide concentrations, and trace metal concentrations. Concurrently, the physical characteristics of water are typically monitored, including temperature, particle size of suspended and deposited material, conductivity, light penetration, flow velocity, dimensions of the water body, and hydrological balance. Some of these analyses can be conducted directly in the field, particularly for parameters like dissolved oxygen and water temperature, utilizing electronic devices such as electronic thermometers and dissolved oxygen probes. However, for most chemical parameters, water samples are collected and sent to the laboratory for analysis. It is crucial to follow established protocols for water sample collection, preservation, and analysis. It's important to note that the chemical and physical properties of water can exhibit significant variations over time (Jerves-Cobo et al., 2020; Mercado-Garcia et al., 2018; Bartram and Ballance, 1996).

5.3.2 Biological Monitoring

One indicator of the health and integrity of an ecosystem is its organisms (Lu et al., 2015). Biological monitoring involves assessing environmental changes through biological responses in order to use the data for ecosystem quality control. The two primary elements of the biomonitoring process hierarchy are biomarkers and bioindicators. Bioindicators are a useful tool for tracking ecosystem changes and differentiating between natural variability and changes caused by human activity. Many aquatic taxa, including macroinvertebrates, fish, periphyton, and planktons, are effectively employed in the biomonitoring process (Van Echelpoel et al., 2019; Parmar et al., 2016; Hering et al., 2006).

Macroinvertebrates, an important class of aquatic organisms, use the trophic levels of the aquatic food chain for the transfer of material and energy, and their susceptibilities to environmental changes can vary. Assessing the health of aquatic ecosystems involving macroinvertebrates often relies on methods such as diversity indices, biotic indices, multimetric approaches, multivariate approaches, Indices of Biological Integrity (IBI), and trait-based approaches. Among these, multimetric techniques and biotic indices are commonly used to assess the impacts of pesticides on aquatic systems. However, the non-random distribution of macroinvertebrates in the riverbed introduces certain limitations. Firstly, their seasonal nature may result in their absence during certain times of the year. Secondly, the geographic limitations may restrict the incidence and frequency of occurrence of certain taxa (Sumudumali and Jayawardana, 2021; Damanik-Ambarita et al., 2016; Van den Brink et al., 2011; De Pauw et al., 2006; Rosenberg and Resh, 1996).

As fish communities hold a prominent position in the aquatic trophic chain, their distribution and content are employed as valuable indicators of the ecological state of aquatic ecosystems (Zou et al., 2020; Gao et al., 2015; Abell et al., 2008). Many fish species are easily identifiable, and once sampled and identified, they can be promptly returned to the water body. The suitability of numerous fish species for ecotoxicological testing, coupled with public awareness of their economic importance and role in aquatic ecosystems, facilitates the enforcement of monitoring activities. However, the mobility of fish based on seasonal, daily, or sporadic patterns can introduce biases in sampling, and certain fish species may actively avoid environmental disturbances. These considerations highlight the limitations of using fish for monitoring and assessment, with unbiased sampling proving challenging in specific situations, such as in large rivers (Figure 5.3) (Scardi et al., 2006).

Even though it hasn't been extensively used in monitoring programs, periphyton is helpful for tracking water quality. They show a range of sensitivities, represent the primary producer trophic level, and frequently point to effects that are only marginally noticeable in the fish and benthic communities. The most effective way to evaluate impairment is to integrate structural/compositional and functional characteristics, just as in the case of fish and benthic macroinvertebrate assemblages (Rodgers et al., 1979). Periphyton communities are better suited for monitoring than other organisms because they are fixed in their habitats, making pollution unavoidable; they recolonize quickly

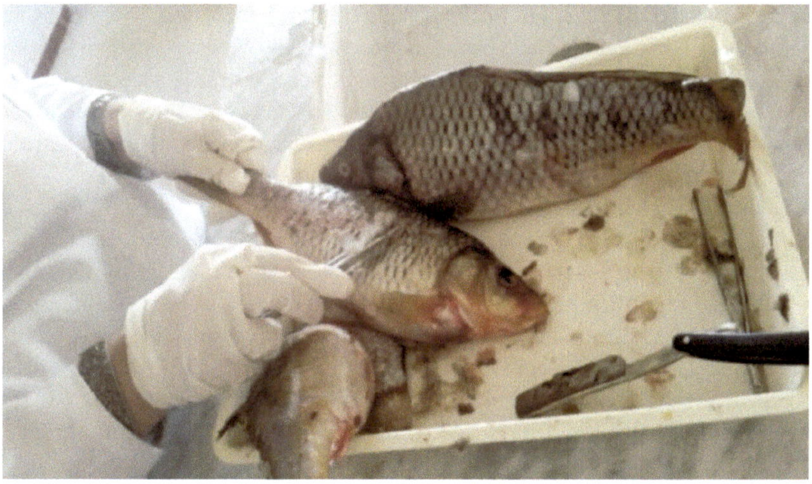

FIGURE 5.3 Determination of water quality using fish from the dam in the Tiaret region (Algeria).

following changes in water quality or flow, making monitoring easier to resume; sample preparation for analysis is simple; and they are widespread and common enough to have well-known pollution tolerances (Dalu and Froneman, 2016; Mancini, 2006; Biggs, 1985).

Plankton, which includes microalgae, copepods, and other small crustaceans, is a type of bio-indicator species that can be observed in water bodies to detect changes in biochemistry, physiology, and behavior. Additionally, these species have the capability to offer valuable insights into the cumulative effects of pollutants in the aquatic ecosystem (Cuadro et al., 2022). In particular, they show signs of eutrophication and predation, shifts in the climate, and advance knowledge of the composition and dynamics of ecosystems (Jeppesen et al., 2011).

5.4 UTILIZING REMOTE SENSING AND GIS FOR EFFECTIVE MONITORING OF AQUATIC ECOSYSTEMS

In recent times, remote sensing and geographic information systems have emerged as tangible tools, demonstrating high suitability for the comprehensive monitoring of ecosystems (Parra, 2022) and aquatic ecosystems in particular. The particularity of these digital solutions lies in their ability to manage, provide, and share geolocated data in time and space. Their use in the fields of water (Chawla et al., 2020) and marine ecosystems (Pérez-Castro et al., 2022; Lechner et al., 2020) continues to increase (Parra, 2022). They have become intelligent and essential tools for the development and monitoring of aquatic ecosystems.

Remote sensing is defined as

> the set of techniques that make it possible, through the acquisition of images, to obtain information on the surface of the earth (including the atmosphere and the oceans) without direct contact with this one. Remote sensing encompasses the entire process of capturing and recording the energy of emitted or reflected electromagnetic radiation, processing and analysing the information it represents, and then applying this information.
>
> **(CCCOT, 2015)**

Remote sensing is based on the flow between the target and the sensor, which is the electromagnetic radiation emitted or reflected by the object studied, and is necessary for the acquisition of information remotely (Ranaivoson, 2022). Visible and infrared wavelengths from 400 nm to 900 nm will mainly be used to observe water (Dörnhöfer and Oppelt, 2016). The temperature of the target and that reflected, the source of emission of the radiation, and the nature of the target determine the quantity and type of radiation emitted (Bonn and Rochon, 1992), whose variation in energy is estimated and detected by sensors. These can measure a variety of wavelengths, ranging from gamma rays to radio waves, providing information on various aspects of aquatic ecosystems, such as water surface temperature, turbidity, chlorophyll concentration, and even the topographical characteristics of the seabed.

A varied set of sensors is used to study water quality by remote sensing (Table 5.1), attached to satellites, drones, or planes. The advantage of their use lies in the possibility of observing in detail the variations in reflectance depending on the wavelength used (Vinel, 2022). This enables the collection of high-resolution spectral and spatial data on aquatic ecosystems. For example, airborne sensors provide access to a resolution of between 4 cm^2 (visible) and 100 cm^2 (thermal), which is much more suitable than satellite sensors (from approximately 1 to 60 m^2) (Vincourt, 2016).

GIS is defined according to the French Society of Photogrammetry and Remote Sensing (1989) as "a computer system allowing, from various sources, to gather and organise, manage, analyse, combine, develop, and present geographically located information, contributing in particular to space management". GIS makes it possible to clearly convey and understand the real aspects of a problem. These digital solutions attempt to model reality on the ground as faithfully as possible in

TABLE 5.1
Sensors Used in Remote Sensing for Monitoring Aquatic Ecosystems

Types	Use	Examples	References
Satellite Sensors			
Multispectral imaging	Turbidity, water color, chlorophyll-a concentration, and other parameters	CZCS, OCTS, SeaWiFS	Sylla, 2014; Hovis, 1981; Kawamura, 1998; Hooker and McClain, 2000
Hyperspectral imaging	More detailed information on the chemical and biological composition of water (state of health). Characterization of the distribution of coastal benthic habitats. Management support	DESIS, PRISMA, EnMap, PACE, AquaWatch	
Radar altimeter	Measuring the height of the water surface. Fluctuations over time in sea level, tides, and ocean waves. Thickness of sea ice and snow. Measuring ocean surface topography	RA-2 and Poséidon-2, Jason-2	
UAV			
Multispectral cameras	Detailed observations at lower altitudes (lakes, rivers, etc.). General delimitation of a wetland. Aquatic herbarium characterization. Accurately measures reflectance, temperature, and surface topography	MicaSense RedEdge 3	Madden et al., 2015; Caron, 2000
LiDAR sensors	Creation of precise topographical models (seabed, coastal zones, etc.). Bottom topography. Mapping wetland. Vegetation mapping. Microtopography. Monitoring changes (bottom topography, water level). Accurate measurement of reflectance (temperature, surface topography)		Wu et al., 2019; Rapinel et al., 2016
Thermal cameras	Detection of temperature variations on the water surface. Detection and monitoring of aquatic fauna. Fisheries management. Delimitation of water surfaces. Accurate measurement of reflectance (temperature, surface topography)		
Aircraft Mounted Sensors			
Airborne spectrometers	Multispectral or hyperspectral observations at high spatial and spectral resolutions	AVIRIS CASI ROSIS	RapInel, 2012
Fluorimeters	Information on water quality (presence of algae, dissolved organic matter) eutrophication	IfW-fluorometer	Schmitz et al., 1993 ; Serre-Fredj, 2022

a way that users can understand. GIS is based on the integration and storage of spatial and alphanumeric data in a spatially referenced database, thus allowing their management, updating, and analysis. Spatial data are represented as a point, line, or polygon, while alphanumeric data can include information such as the number of individuals in a population, biodiversity, environmental variables, pollution levels, etc.

GIS enables complete mapping of aquatic ecosystems because it offers the possibility of integrating multi-source data into a georeferenced database. These data can be of all types (aerial photos, satellite images, alphanumeric data, etc.) and relate to different sources (topographic characteristics, aquatic vegetation, wildlife habitats, hydrological characteristics, etc.).

GIS makes it possible to monitor environmental changes in real time, such as water quality, plant cover, and water level, which allows for the identification of areas at risk, the monitoring of phenomena, and the prediction of disasters and scenarios. It enables the planning of relief operations, the minimization of environmental damage, and the direction of restoration efforts. GIS is used to assess the impact on aquatic ecosystems and plan relief and restoration efforts. Geographic information systems provide essential functionality for understanding, monitoring, and managing these complex ecosystems. Integrating these attributes into GIS technology enables a comprehensive understanding of aquatic ecosystems, which helps researchers, managers, and policymakers make informed decisions regarding the preservation and sustainable administration of these environments.

5.4.1 Some Applications of Remote Sensing in Monitoring Aquatic Ecosystems

5.4.1.1 Mapping Aquatic Habitats

Mapping aquatic habitats requires expertise in remote sensing, GIS, and aquatic ecology. Information from multispectral and hyperspectral images contributes to the characterization of aquatic habitats. Characterizing aquatic habitats serves as a valuable tool not only for understanding the features of these environments but also for detecting indicators of specific species and comprehending the migration patterns of certain groups, notably avian species. This approach enables researchers to identify key environmental factors influencing the presence and behavior of wildlife, contributing to more informed conservation and management strategies for aquatic ecosystems.

5.4.1.2 Water Quality Detection

Remote sensing plays a crucial role in water quality monitoring, essential for characterizing aquatic ecosystems (Uudeberg et al., 2020; Yadav et al., 2019; Ansper and Alikas, 2018). Water quality is also a concern highlighted by the Sustainable Development Goals (SDGs) (UNESCO, 2022). This is crucial for assessing water quality and monitoring the effects of pollution or natural phenomena such as erosion.

5.4.1.3 Turbidity Measurement

The spatiotemporal evolution of turbidity in aquatic ecosystems can be detected and measured effectively by remote sensing. This is possible either by using infrared light or by using attenuated ultrasound frequencies (Fajerwerg et al., 2015).

Drones, or drones equipped with satellites, offer high-resolution imaging capabilities to measure the spectral reflectance of water in the visible and near-infrared bands. These data are used with regression algorithms to predict and estimate water turbidity in aquatic ecosystems such as coastal areas, estuaries, deltas, lakes, and rivers.

Several sensors have been used to measure water turbidity, such as Landsat TM (Fraser, 1998), SPOT and HVR (Shafique et al., 2003), Ikonos (Ustun, 2011), Quickbird (Somvanshi et al., 2011), and Sentinel 2 (Rouzeau, 2017). Some are classified as empirical indicators, which, according to Dekker in 1993, were developed to quantify turbidity from images from airborne hyperspectral sensors, and others are classified as theoretical indicators that reveal the relative turbidity of each pixel. Some are categorized based on the utilization of a single band or a combination of bands (Chamaillé, 2008).

5.4.1.4 Measurement of the Concentration of Suspended Solids

Satellites and drones can measure the concentration of suspended particles in water using multispectral and hyperspectrally modified light responses. Image processing algorithms, such as linear regression or classification, compare detected data to those measured in the field.

5.4.1.5 Characterization of Water Color

The color of the water corresponds to the spectral variation of the absorption and diffusion processes of incident solar light by the water column. Temporal trends are characterized by remote sensing and are reconverted by spectrometry algorithms and water color indices (WCI, ACI).

5.4.1.6 Estimation of the Concentration of Chlorophyll A

Estimating temporal variations in chlorophyll concentration in aquatic ecosystems is achievable by measuring spectral reflectance in the visible and near-infrared bands using specific sensors like MODIS, MERIS, and SeaWiFS.

5.4.1.7 Temperature Estimation: Surface Temperature (Skin Temperature)

Satellites currently measure water surface temperature in °C, called "skin temperature". These algorithms are used to increase precision while minimizing the error resulting from stray light effects on these bands on the one hand (Jimenez et al., 2018). Conversely, verifying the precision of water vapor measurements involves subjecting them to testing, assessing their sensitivity across various algorithms, and allowing a better estimation of surface water temperature while considering other water quality parameters. Infrared sensors from satellites, drones, or manned aircraft allow temperature measurements based on the principle that higher water temperatures increase infrared radiation.

5.5 MEASUREMENT OF SUSPENDED SEDIMENT CONCENTRATION

Numerous researchers have affirmed the significance of remote sensing in the assessment of suspended sediments (Condé et al., 2019; Gallay, 2017; Espinoza Villar et al., 2013; Martinez et al., 2009a; Yepez et al., 2018). The concentration of suspended sediments makes it possible to monitor the evolution of erosion processes in relation to natural processes and human activities. Spatial data are used to compensate for gaps in field data, enabling precise monitoring of the evolution of suspended sediments and the quantity of material transported by a river in poorly instrumented regions. Among the sensors used were MODIS, MERIS, Landsat, Sentinel-2, HICO, OLCI, and AVIRIS (Jhan Carlo Espinoza Villar et al., 2012).

5.5.1 Pollutant Detection

Oil spills or spills of other polluting substances in oceans, lakes, and rivers can be detected by specific sensors, such as Synthetic Aperture Radar (SAR) sensors. The technique involves using advanced sensors to detect emerging pollutants such as trace elements, pesticides, herbicides, polyaromatic hydrocarbons, antibiotics, toxins, xenobiotics, and radioelements, using innovative technologies. Other sensors are thermal (Landsat IRT satellite, airborne thermal cameras, thermal probes, infrared radiometers) to measure water temperature, affecting oxygen solubility.

5.5.2 Track and Monitor Aquatic Species (Movement of Aquatic Species)

Global positioning system (GPS) and satellite remote sensing enable real-time monitoring of environmental changes, contributing to sustainable conservation and early warning of threats to aquatic species and their habitats.

5.5.3 Monitoring Populations of Threatened or Endangered Species

High-resolution satellite images help count animal populations and monitor migrations. Examples include tracking large mammals such as whales, tracking seabirds (penguins in Antarctica and albatrosses in the oceans), tracking reptiles (sea turtles) and amphibians, tracking fish (sturgeons) and corals, and monitoring rare plants. By combining GPS tags and animal body collars, researchers can geolocate breeding, nesting, or wintering areas.

5.6 CHALLENGES AND OPPORTUNITIES IN AQUATIC ECOSYSTEMS MONITORING

Monitoring is crucial for observing and managing aquatic ecosystems. It facilitates the collection and transmission of data from distant locations, offering scientists a comprehensive insight into the health of aquatic ecosystems. Nevertheless, there are challenges and opportunities linked to the utilization of monitoring for these objectives (Chen et al., 2022; Buytaert et al., 2014; Biber, 2013; Li et al., 2014; Kröger et al., 2009).

5.6.1 Challenges

5.6.1.1 Aquatic Ecosystems Are Diverse and Complex

It is difficult to observe all the elements and species that interact in these ecosystems.

5.6.1.2 Data Collection

Underwater data collection is complex and requires special equipment and methods. Water quality, temperature, and chemical parameters have to be measured precisely.

5.6.1.3 Environmental Changes

Aquatic ecosystems are highly vulnerable to environmental changes such as pollution, climate change, and habitat loss, making it important to observe and react quickly.

5.6.1.4 Spatial and Temporal Variability

Aquatic ecosystems change a great deal in space and time, requiring long-term and large-scale observation efforts.

5.6.2 Opportunities

Opportunities for observing aquatic ecosystems include:

5.6.2.1 Technological Progress

Remote sensing, underwater robotics, and sensor technologies have improved our ability to observe aquatic ecosystems more efficiently and accurately.

5.6.2.2 Data Integration

Combining data from different sources provides a better understanding of aquatic ecosystems and their evolution.

5.6.2.3 Conservation and Restoration

Observation data can support conservation actions and ecosystem restoration projects, helping to preserve and restore these vital habitats.

5.6.2.4 Citizen Science

Involving the public in data collection through citizen science initiatives can broaden observation efforts and raise awareness of the importance of aquatic ecosystems.

5.6.2.5 Scientific Discovery

Continuous observation leads to new scientific discoveries, showing the complex relationships within aquatic ecosystems and helping us to understand their importance in the wider context of the environment.

REFERENCES

Abell, R., Thieme, M.L., Revenga, C., Bryer, M., Kottelat, M., Bogutskaya, N., Coad, B., Mandrak, N., Balderas, S.C., Bussing, W., Stiassny, M.L.J., Skelton, P., Allen, G.R., Unmack, P., Naseka, A., Ng, R., Sindorf, N., Robertson, J., Armijo, E., Higgins, J.V., Heibel, T.J., Wikramanayake, E., Olson, D., López, H.L., Reis, R.E., Lundberg, J.G., Sabaj Pérez, M.H., and Petry, P., 2008. Freshwater ecoregions of the world: A new map of biogeographic units for freshwater biodiversity conservation. *BioScience*, 58(5), 403–414.

Allan, J.D., and Flecker, A.S., 1993. Biodiversity conservation in running waters. *BioScience*, 43, 32–43. https://doi.org/10.2307/1312104.

Ansper, A., and Alikas, K., 2018. Retrieval of chlorophyll a from Sentinel-2 MSI data for the European Union water framework directive reporting purposes. *Remote Sensing*, 11(1), 64.

Appeltans, W., et al., 2012. The magnitude of global marine species diversity. *Current Biology*, 22(23), 2189–2202. https://doi.org/10.1016/j.cub.2012.09.036.

Barange, M., Field, J.G., Harris, R.P., Eileen, E., Hofmann, E.E., Perry, R.I., and Werner, F., 2010. *Écosystèmes marins et changement global.* Oxford: Presse universitaire d'Oxford, 464.

Barbier, E.B., Hacker, S.D., Kennedy, C., Koch, E.W., Stier, A.C., and Silliman, B.R., 2011. The value of estuarine and coastal ecosystem services. *Ecological Monographs,* 81, 169–193. https://doi.org/10.1890/10-1510.1.

Bar-On, Y.M., Phillips, R., and Milo, R., 2018. The biomass distribution on Earth. *Proceedings of the National Academy of Sciences of the United States of America*, 115(25), 6506–6511. https://doi.org/10.1073/pnas.1711842115.

Baron, J.S., Poff, N.L., Angermeier, P.L., Dahm, C.N., Gleick, P.H., Hairston, Jr. N.G., Jackson, R.B., Johnston, C.A., Richter, B.D., and Steinman, A.D., 2003. Sustaining healthy freshwater ecosystems. *Issues in Ecology*, 10, 1–16.

Bartram, J., Ballance, R., United Nations, and World Health Organization, eds., 1996. *Water quality monitoring: A practical guide to the design and implementation of freshwater quality studies and monitoring programmes.* 1st ed. London; New York: E & FN Spon.

Bashir, N., Saeed, R., Afzaal, M., Ahmad, A., Muhammad, N., Iqbal, J., … Hameed, S., 2020. Water quality assessment of lower Jhelum canal in Pakistan by using geographic information system (GIS). *Groundwater for Sustainable Development*, 10, 100357. https://doi.org/10.1016/j.gsd.2020.100357.

Basset, A., and Abbiati, M., 2004. Challenges to transitional water monitoring: Ecological descriptors and scales. *Aquatic Conservation: Marine and Freshwater Ecosystems*, 14, S1–S3.

Biber, E. 2013. The challenge of collecting and using environmental monitoring data. Ecology and Society18(4): 68. http://dx.doi.org/10.5751/ES-06117-180468

Biggs, B.J.F., 1983. Phytoplankton enumeration by the inverted microscope method. In Biggs, B.J.F., Gifford, J.S., and Smith, D.G. (Eds.), *Biological methods for water quality surveys.* Wellington: Ministry of Works and Development.

Biggs B.J.F., 1985. The use of periphyton in the monitoring of water quality. In: Biological Monitoring in Freshwaters - Proceedings of a Seminar (Eds R. D. Pridmore and A. B. Cooper). Water and Soil Miscellaneous Publication, 82, 117–142.

Bonn, F.J., and Rochon, G., 1992. *Précis de télédétection.* Sillery: Presses de l'Université du Québec.

Bouju, T., and Barthelemy, K., 2013. Télédétection amazonienne. *Latitude*, 5(99), 15–19.

Brachet, C., Magnier, J., and Fribourg-Blanc, B., 2015. Manuel sur la gestion et la restauration des écosystèmes aquatiques dans les bassins des fleuves et des lacs | RIOB (Réseau International des Organismes de Bassin) [online], 2023. Available from: https://www.inbo-news.org/fr/documents/manuel-sur-la-gestion-et-la-restauration-des-ecosystemes-aquatiques-dans-les-bassins-des [Accessed 25 Sep 2023].

Buma, W., Lee, S.-I., and Seo, J., 2018. Recent surface water extent of Lake Chad from multispectral sensors and GRACE. *Sensors*, 18(7), 2082.

Burke, L., Reytar, K., Spalding, M., and Perry, A., 2011. Reefs at risk revisited. World Resources Institute, Washington, DC. https://doi.org/10.1016/0022-0981(79)90136-9.

Burke, L., Selig, E., and Spalding, M., 2002. Reefs at risk in Southeast Asia, world fishing. World Resources Institute in Collaboration with the United Nations Environment Programme-World Conservation Monitoring Centre. The World Fish Center, and the International Coral Reef Action Network.

Buytaert, W., Zulkafli, Z., Grainger, S., et al. 2014. Citizen science in hydrology and water resources: Opportunities for knowledge generation, ecosystem service management, and sustainable development. *Frontiers in Earth Science*, 2, 26.

Caron, J., 2000. Potentiel de l'imagerie par drone pour la caractérisation d'herbiers aquatiques: Le cas de l'archipel des Îles-de-la-Paix, Québec. Université de Sherbrooke.

Castro P., & Huber M. E., 2012. Marine biology. McGraw-Hill Education, Edition 9, ISBN: 0073524204, 9780073524207. pp: 480.

Catteau, S., & Couderchet, L. (2022). Les Systèmes d'Information Géographique pour discuter l'articulation entre le fonctionnement des zones humides et l'aménagement du territoire. *Géocarrefour*, 96(96/2).

CCCOT, 2015. Imagerie satellitaire et photos aériennes. (Centre canadien de cartographie et d'observation de la Terre) Consulté le Janvier 15, 2023, sur Ressources naturelles Canada. https://www.rncan.gc.ca/sciences-terre/geomatique/imagerie-satellitaire-photosaeriennes/10800.

Chamaillé, L., 2008. Apport de la télédétection pour la détection de l'eau et la caractérisation des propriétés de l'eau. Application de la télédétection à l'étude des facteurs de risque liés à l'eau pour la survie du virus H5N1 dans l'environnement. Mémoire de master professionnel. Ecole nationale supérieure agronomique de Toulouse.

Chapin, F.S., Zavaleta, E.S., Eviner, V.T., Naylor, R.L., Vitousek, P.M., Reynolds, H.L., Hooper, D.U., Lavorel, S., Sala, O.E., Hobbie, S.E., et al., 2000. Consequences of changing biodiversity. *Nature*, 405, 234–242.

Chawla, I., Karthikeyan, L., and Mishra, A.K., 2020. A review of remote sensing applications for water security: Quantity, quality, and extremes. *Journal of Hydrology*, 585, 124826.

Condé, R., Martinez, J.-M., Pessotto, M., Villar, R., Cochonneau, G., Henry, R., Lopes, W., and Nogueira, M., 2019. Indirect assessment of sedimentation in hydropower dams using MODIS remote sensing images. *Remote Sensing*, 11(3), 314.

Chen, J., Chen, S., Fu, R., et al. 2022. Remote sensing big data for water environment monitoring: Current status, challenges, and future prospects. *Earth's Future*, 10(2), e2021EF002289.

Cordeiro, M.C.R., 2022. Use of data science tools for assessing inland water surface and quality on regional scales through high-resolution Sentinel-2 remote sensing images. phdthesis. Université Paul Sabatier - Toulouse III.

Cordeiro, M.C.R., Martinez, J.-M., and Peña-Luque, S., 2021. Automatic water detection from multidimensional hierarchical clustering for Sentinel-2 images and a comparison with Level 2A processors. *Remote Sensing of Environment*, 253, 112209.

Cuadro D.L.C., Berasategui A.A., and Menéndez M.C., 2022. Zooplankton: The Ocean Drifters. In Marine Biology A Functional Approach to the Oceans and their Organisms. CRC Press, 2022, 113–149.

Dalu, T., and Froneman, P.W., 2016. Diatom-based water quality monitoring in southern Africa: Challenges and future prospects. *Water SA*, 42(4), 551.

Damanik-Ambarita, M.N., Lock, K., Boets, P., Everaert, G., Nguyen, T.H.T., Forio, M.A.E., Musonge, P.L.S., Suhareva, N., Bennetsen, E., Landuyt, D., Dominguez-Granda, L., and Goethals, P.L.M., 2016. Ecological water quality analysis of the Guayas river basin (Ecuador) based on macroinvertebrates indices. *Limnologica*, 57, 27–59.

Day, J.W., Hall, C.A.S., Kemp, W.M., Yanez-Aracibia, A., editors, 1989. *Estuarine ecology*. 1st ed. New York: John Wiley and Sons, 558.

Dekker, A.G., 1993. *Detection of optical water quality parameters for entrophic waters by high resolution remote sensing*. Amsterdam: Free University.

Deknock, A., De Troyer, N., Houbraken, M., Dominguez-Granda, L., Nolivos, I., Van Echelpoel, W., Forio, M.A.E., Spanoghe, P., and Goethals, P., 2019. Distribution of agricultural pesticides in the freshwater environment of the Guayas river basin (Ecuador). *Science of The Total Environment*, 646, 996–1008.

De Pauw, N., Gabriels, W., and Goethals, P., 2006. River monitoring and assessment methods based on macroinvertebrates. In Ziglio, G., Siligardi, M., and Flaim, G. (Eds.), *Biological monitoring of rivers: Applications and perspectives*. Chichester; Hoboken, NJ: John Wiley & Sons, Ltd.

Dörnhöfer, K., and Oppelt, N., 2016. Remote sensing for lake research and monitoring – Recent advances. *Ecological Indicators*, 64, 105–122.

Dudgeon, D., Arthington, A.H., Gessner, M.O., Kawabata, Z.-I., Knowler, D.J., Lévêque, C., Naiman, R.J., Prieur-Richard, A.-H., Soto, D., Stiassny, M.L.J., et al., 2005. Freshwater biodiversity: Importance, threats, status and conservation challenges. *Biological Reviews*, 81, 163–182.

Dugan, P., 1993. *Review: Wetlands in danger: A world conservation atlas*. New York: Oxford University Press, 192.

Dyer, K.R., 1997. *Estuaries: A physical introduction*. Chichester: John Wiley and Sons.

Echelpoel, F., Heyden, B., Ho, M., Narea, G., Sanchez, D., and Goethals, P.L.M., 2019. Spatial characteristics and temporal evolution of chemical and biological freshwater status as baseline assessment on the Tropical Island San Cristóbal (Galapagos, Ecuador). *Water*, 11(5), 880.

Espinoza Villar, R., Martinez, J.-M., Guyot, J.-L., Fraizy, P., Armijos, E., Crave, A., Bazán, H., Vauchel, P., and Lavado, W., 2012. The integration of field measurements and satellite observations to determine river solid loads in poorly monitored basins. *Journal of Hydrology*, 444–445, 221–228.

Espinoza Villar, R., Martinez, J.-M., Le Texier, M., Guyot, J.-L., Fraizy, P., Meneses, P.R., and Oliveira, E.D., 2013. A study of sediment transport in the Madeira River, Brazil, using MODIS remote-sensing images. *Journal of South American Earth Sciences*, 44, 45–54.

Fajerwerg, K., Behra, P., et Gaillardet, J., 2015. 9. Capteurs de la qualité des eaux. In *L'eau à découvert [en ligne]*. Paris: CNRS Éditions (généré le 24 septembre 2023). Disponible sur Internet: <http://books.openedition.org/editionscnrs/9994>. ISBN: 9782271119117. https://doi.org/10.4000/books.editionscnrs.9994.

FAO and COOPI, 2014. *Systèmes de Gestion des Connaissances et des Informations: Pratiques clés pour les praticiens de la RRC*. COOPI. http://www.fao.org/3/a-i3772f.pdf

Feyisa, G.L., Meilby, H., Fensholt, R., and Proud, S.R., 2014. Automated water extraction index: A new technique for surface water mapping using Landsat imagery. *Remote Sensing of Environment*, 140, 23–35.

Fisher, R., O'Leary, R.A., Low-Choy, S., Mengersen, K., Knowlton, N., Brainard, R.E., et al. 2015. Species richness on coral reefs and the pursuit of convergent global estimates. *Current Biology*, 25, 500–505. https://doi.org/ 10.1016/j.cub.2014.12.022.

Fox-Kemper, B., et al., 2021. Ocean, cryosphere and sea level change. In *Climate change 2021: The physical science basis. Contribution of Working Group I to the Sixth Assessment Report of the Intergovernmental Panel on Climate Change*.

Fraser, R.N., 1998. Multispectral remote sensing of turbidity among Nebraska Sand Hills lakes. *International Journal of Remote Sensing*, 19(15), 3011–3016.

Gallay, M., 2017. Etude de la dynamique hydro-sédimentaire de fleuves tropicaux: le cas du Maroni, de l'Oyapock et de l'Orénoque. Thèse de Doctorat Thesis, Université de la Guyane, 316 pp.

Gao, X., Ding, S., Zhang, Y., Ma, S. Q., Liu, S. S., & Meng, W. (2015). Exploring the relationship among land-use, riparian habitat quality, and biological integrity of a fish community. *Acta Ecol. Sin*, *35*, 7198–7206.

Giller, P.S., and Malmqvist, B., 1998. *The biology of streams and rivers*. Oxford: Oxford University Press.

Goethals, P., and Forio, M., 2018. Advances in ecological water system modeling: Integration and leanification as a basis for application in environmental management. *Water*, 10, 1216.

Grizzetti, B., Lanzanova, D., Liquete, C., Reynaud, A., and Cardoso, A.C., 2016. Assessing water ecosystem services for water resource management. *Environmental Science & Policy*, 61, 194–203.

Häder D-P., Banaszak A. T., Villafañe V. E., Narvarte M. A., González R. A., E. Helbling W., 2020. Anthropogenic pollution of aquatic ecosystems: Emerging problems with global implications, Science of The Total Environment, Volume 713, 136586, ISSN 0048-9697, https://doi.org/10.1016/j.scitotenv.2020.136586.

Hering, D., Johnson, R.K., Kramm, S., Schmutz, S., Szoszkiewicz, K., and Verdonschot, P.F.M., 2006. Assessment of European streams with diatoms, macrophytes, macroinvertebrates and fish: A comparative metric-based analysis of organism response to stress. *Freshwater Biology*, 51(9), 1757–1785.

Hooker, S.B., and McClain, C.R., 2000. The calibration and validation of SeaWiFS data. *Progress in Oceanography*, 45(3–4), 427–465.

Hovis, W.A., 1981. The Nimbus-7 Coastal Zone Color Scanner (CZCS) program. In Gower, J.F.R. (Ed.), *Oceanography from space*. Marine Science. Springer US, 13, pp. 213–225.

Hubert-Moy, L., 2005. La télédétection: des outils multiples pour mieux connaître les zones humides. In *Caractérisation des zones humides*. MEDD, Agences de l'eau, BRGM, 54–62.

Karr, J.R., 1991. Biological integrity—A long-neglected aspect of water-resource management. *Ecological Applications*, 1, 66–84.

Karr, J.R., Fausch, J.D., Yant, P.R., and Schlosser, I.L., 1986. Assessing biological integrity in running waters: A method and its rationale. Champaign, IL: Illinois Natural History Survey Special Publication 5.

Kawamura, H., and The OCTS Team, 1998. OCTS mission overview. *Journal of Oceanography*, 54(5), 383–399.

Keeler, B.L., Polasky, S., Brauman, K.A., Johnson, K.A., Finlay, J.C., O'Neill, A., Kovacs, K., and Dalzell, B., 2012. Linking water quality and well-being for improved assessment and valuation of ecosystem services. *Proceedings of the National Academy of Sciences*, 109(45), 18619–18624.

Kennish, M.J., editor, 2005. *Estuarine research, monitoring, and resource protection*. Boca Raton, FL: CRC Press, 297.

Kordelas, G., Manakos, I., Aragonés, D., Díaz-Delgado, R., and Bustamante, J., 2018. Fast and automatic data-driven thresholding for inundation mapping with Sentinel-2 Data. *Remote Sensing*, 10(6), 910.

Kordelas, G.A., Manakos, I., Lefebvre, G., and Poulin, B., 2019. Automatic inundation mapping using Sentinel-2 data applicable to both Camargue and Doñana biosphere reserves. *Remote Sensing*, 11(19), 2251.

Kröger, S., Parker, E.R., Metcalfe, J.D., et al. 2009. Sensors for observing ecosystem status. *Ocean Science*, 5(4), 523–535.

Kusky, T., 2008. *The coast: Hazardous interactions within the coastal environment*. New York: Facts On File, Inc.

Lechner, A.M., Foody, G.M., and Boyd, D.S., 2020. Applications in remote sensing to forest ecology and management. *One Earth*, 2(5), 405–412.

Lewis, P., 1996. U.N. Report warns of problems over dwindling water supplies. *The New York Times*, January 20.

Li, Z., Xu, D., and Guo, X., 2014. Remote sensing of ecosystem health: Opportunities, challenges, and future perspectives. *Sensors,* 14(11), 21117–21139. https://doi.org/10.3390/s141121117.

Lu, Y., Wang, R., Zhang, Y., Su, H., Wang, P., Jenkins, A., Ferrier, R.C., Bailey, M., and Squire, G., 2015. Ecosystem health towards sustainability. *Ecosystem Health and Sustainability,* 1, 1–15.

Madden, M., Jordan, T., Bernardes, S., Cotten, D., O'Hare, N., and Pasqua, A., 2015. Unmanned Aerial Systems (UAS) and Structure from Motion (SfM) revolutionize wetlands mapping. In Tiner, R.W., Lang, M.W., and Klemas, V. (Eds.), *Remote sensing of wetlands: Applications and advances.* Boca Raton: CRC Press, Taylor & Francis Group.

Malmqvist, B., and Rundle, S., 2002. Threats to the running water ecosystems of the world. *Environmental Conservation*, 29(2), 134–153.

Mancini, L., 2006. Organization of biological monitoring in the European Union. In Ziglio, G., Siligardi, M., and Flaim, G. (Eds.), *Biological monitoring of rivers: Applications and perspectives.* Chichester; Hoboken, NJ: John Wiley & Sons, Ltd.

Markert, K.N., Griffin, R.E., Limaye, A.S., and McNider, R.T., 2018. Spatial modeling of land cover/land use change and its effects on hydrology within the lower Mekong Basin. In Vadrevu, K.P., Ohara, T., and Justice, C., (Eds.), *Land-Atmospheric research applications in South and Southeast Asia.* Cham: Springer International Publishing, 667–698.

Martinis, S., Twele, A., and Voigt, S., 2011. Unsupervised extraction of flood-induced backscatter changes in SAR data using Markov image modeling on irregular graphs. *IEEE Transactions on Geoscience and Remote Sensing*, 49(1), 251–263.

Martinez, E., 2006. Circulation des eaux de la thermocline en Polynésie française. Utilisation de données satellites, in situ, et d'une modèle de circulation océanique régional. *Thèse de doctorat d'Océanographie physique.* Université de la Polynésie française, avril 2006, 180p.

Martinez, E., Ganachaud, A., Lefevre, J., et Maamaatuaiahutapu, K., 2009a. Central South Pacific thermocline water circulation from a high-resolution ocean model validated against satellite data: Seasonal variability and El Niño 1997–1998 influence. *Journal of Geophysical Research,* 114, C05012. https://doi.org/10.1029/2008JC004824.

Martinez, E., Maamaatuaiahutapu, K., and Taillandier, V., 2009b. Floating marine debris surface drift: Convergence and accumulation toward the South Pacific subtropical gyre. *Marine Pollution Bulletin,* 58, 1347–1355. https://doi.org/10.1016/j.marpolbul.2009.04.022.

Martínez S. G., Andreani L., Gusano M.G., Geuna F., and Ruiz J.J. 2006. Evaluation of amplified fragment length polymorphism and simple sequence repeats for tomato germplasm fingerprinting: utility for grouping closely related traditional cultivars. Genome. 49(6): 648-656. https://doi.org/10.1139/g06-016.

McLusky, D.S., and Elliott, M., 2004. *The estuarine ecosystem: Ecology, threats, and management.* Oxford: Oxford Press.

Mercado-Garcia, D., Wyseure, G., and Goethals, P., 2018. Freshwater ecosystem services in mining regions: Modelling options for policy development support. *Water,* 10(4), 531.

Moberg, F., and Folke, C., 1999. Ecological goods and services of coral reef ecosystems. *Ecological Economics,* 29, 215–233. https://doi.org/10.1016/S0921-8009(99)00009-9.

Nazzal Y., Howari F. M., Iqbal J., Izrar A., Bou Orm N., Yousef A., 2019. Investigating aquifer vulnerability and pollution risk employing modified DRASTIC model and GIS techniques in Liwa area, United Arab Emirates. *Groundwater for Sustainable Development*, Volume 8, 567–578.

Nybakken JW. Bertness MD., 2005. Marine biology: an ecological approach. Pearson/Benjamin Cummings, San Francisco., Edition 9, ISBN: 0 8053 4582 5. pp: 579.

OFEV, 2020. Infrastructure écologique. Available at: https://www.bafu.admin.ch/bafu/fr/home/ themes/biodiversite/info-specialistes/mesures de-conservation-de-la-biodiversite/infrastructureecologique.html (accessed 15/04/2023).

Parmar, T.K., Rawtani, D., and Agrawal, Y.K., 2016. Bioindicators: The natural indicator of environmental pollution. *Frontiers in Life Science*, 9(2), 110–118.

Parra, L., 2022. Remote sensing and GIS in environmental monitoring. *Applied Sciences*, 12(16), 8045.

Peña-Luque, S., Ferrant, S., Cordeiro, M.C.R., Ledauphin, T., Maxant, J., and Martinez, J.-M., 2021. Sentinel-1&2 multitemporal water surface detection accuracies, evaluated at regional and reservoirs level. *Remote Sensing*, 13(16), 3279.

Pérez-Castro, M.Á., Schubert, N., Ang-Montes De Oca, G., Leyte-Morales, G.E., Eyal, G., and Hinojosa-Arango, G., 2022. Mesophotic Coral Ecosystems in the Eastern Tropical Pacific: The current state of knowledge and the spatial variability of their depth boundaries. *Science of The Total Environment*, 806, 150576.

Plag, H.P., & Jules-Plag, S. 2013. Sea-level rise and coastal ecosystems. In *Climate vulnerability. Understanding and addressing threats to essential resources.* Amsterdam: Elsevier. https://doi.org/10.1016/B978-0-12-384703-4.00419-6.

Postel S., and Richter B., 2012. Rivers for Life: Managing Water for People and Nature. Ed. Island Press, Washington. ISBN 1597267805, 9781597267809. pp: 220.

Ranaivoson, B.N.J., 2022. Monitoring des forêts de mangrove exploitées à Madagascar par télédétection: de l'observation du géosystème à l'estimation d'un potentiel de la ressource crabe de mangrove. These de doctorat. La Réunion.

Rapinel, S., 2012. Contribution de la télédétection à l'évaluation des fonctions des zones humides : de l'observation à la modélisation prospective. phdthesis. Université Rennes 2.

Rapinel, S., Hubert-Moy, L., Clément, B., and Maltby, E., 2016. Mapping wetland functions using Earth observation data and multi-criteria analysis. *Environmental Monitoring and Assessment*, 188(11), 641.

Reaka-Kudla M.L., and Wilson, D.E. 1997. Global biodiversity of coral reefs: A comparison with rainforests. *Biodiversity II: Understanding and protecting our biological resources.* Joseph Henry Press.

Reddy, M. T., Sivaraj, N., Kamala, V., Pandravada, S. R., Sunil, N., & Dikshit, N. (2018). Classification, characterization and comparison of aquatic ecosystems in the landscape of Adilabad District, Telangana, Deccan Region, India. *Open Access Library Journal*, *5*(4), 1–49.

Rodgers J.H., Dickson K.L., and Cairns J. Jr., 1979. A review and analysis of some methods used to measure functional aspects of periphyton. Methods and Measurements of Periphyton Communities: A Review, ASTM STP 690 (Ed. R. L. Weitzel). pp. 142–167, *American Society for Testing and Materials*, Philadelphia, Pennsylvania.

Rosenberg, D.M., Resh, V.H., 1996. Use of aquatic insects in biomonitoring. In Merritt, R.W., and Cummins, K.W. (Eds.), *An introduction to the aquatic insects of North America.* 3rd ed. Dubuque, Iowa: Kendall/Hunt Pub. Co.

Rouzeau, O., and Capar, L., 2017. Suivi de la turbidité des rivières guyanaises par télédétection multispectrale SENTINEL 2 - Acquisitions, algorithmes et masques des portions de rivières valides. Rapport final. BRGM/RP67008-FR. 91p.

Sakamoto, T., Van Nguyen, N., Kotera, A., Ohno, H., Ishitsuka, N., and Yokozawa, M., 2007. Detecting temporal changes in the extent of annual flooding within the Cambodia and the Vietnamese Mekong Delta from MODIS time-series imagery. *Remote Sensing of Environment*, 109(3), 295–313.

Salvat B., 1992. Coral reefs — a challenging ecosystem for human societies, Global Environmental Change, Volume 2, Issue 1, 1992, Pages 12-18, ISSN 0959-3780, https://doi.org/10.1016/0959-3780(92)90032-3.

Scardi M., Tancioni L., Cataudella S. (2006). Monitoring methods based on fish. In: Ziglio G., Siligardi M., Flaim G. (eds.), Biological monitoring of rivers, Wiley, pp. 135-153.

Schmitz, Jr., William J., Luyten, J.R., and Schmitt, R.W., 1993. On the Florida current T/S envelope. *Bulletin of Marine Science*, 53(3), 1048–1065.

Serre-Fredj, L., 2022. Production primaire et dynamique phytoplanctonique le long d'un gradient d'eutrophisation: Continuum estuaire-baie de Seine. Normandie Université.

S.F.P.T, 1989. Programme national télédétection spatiale. Colloque d'Orléans : la recherche et les satellites à haute résolution spatiale : SPOT, Landsat-TM, n° 114.

Shafique, N.A., Autrey, B.C., Fulk, F., and Cormier, S.M. 2003. The selection of narrow wavebands for optimizing water quality monitoring in the Great Miami River, Ohio using hyperspectral remote sensor data. *Journal of Spatial Hydrology*, 1, 1–22.

Shiklomanov, L.A., 1993. World freshwater resources. In Gleick, P.H. (Ed.), *Water in crisis: A guide to world's freshwater resources.* New York: Oxford University Press, 13–24.

Shine, C., and Klemm, C., 1999. *Wetlands, water and the law: Using law to advance wetland conservation and wise use.* IUCN, Gland.

Smith, L.C., 1997. Satellite remote sensing of river inundation area, stage, and discharge: A review. *Hydrological Processes*, 11(10), 1427–1439.

Smith S., 1987. Coral-reef area and the contributions of reefs to processes and resources of the world's oceans. *Nature* 273, 225–226 (1978). https://doi.org/10.1038/273225a0.

Solimini, A.G., Bazzanti, M., Ruggiero, A., and Carchini, G., 2008. Developing a multimetric index of ecological integrity based on macroinvertebrates of mountain ponds in Central Italy. *Hydrobiologia*, 597, 109–123.https://doi.org/10.1007/s10750-007-9226-7

Somvanshi, S., Kunwar, P., Singh, N.B., and Kachhwaha, T.S., 2011. Water turbidity assessment in part of Gomti River using high resolution Google Earth's Quickbird satellite data. In *Geospatial World Forum*, 18–21.

Souza, C., Kirchhoff, F., Oliveira, B., Ribeiro, J., and Sales, M., 2019. Long-term annual surface water change in the Brazilian Amazon Biome: Potential links with deforestation, infrastructure development and climate change. *Water*, 11(3), 566.

Speers, A.E., Besedin, E.Y., Palardy, J.E., Moore, C., 2016. Impacts of climate change and ocean acidification on coral reef fisheries: An integrated ecological economic model. *Ecological Economics*, 128(11), 33–43). http://dx.doi.org/10.1016/j.ecolecon.2016.04.012.

Stenert, C., and Maltchik, L., 2007. Influence of area, altitude and hydroperiod on macroinvertebrate communities in Southern Brazil wetlands. *Marine and Freshwater Research*, 58, 993–1001. https://doi.org/10.1071/MF07073.

Stewart, H.A., and Jamieson, A.J., 2019. The five deeps: The location and depth of the deepest place in each of the world's oceans. *Earth-Science Reviews*, 197, 102896. https://doi.org/10.1016/j.earscirev.2019.102896.

Strayer, D.L., and Dudgeon, D., 2010. Freshwater biodiversity conservation: Recent progress and future challenges. *Journal of the North American Benthological Society*, 29, 344–358.

Sumudumali, R.G.I., and Jayawardana, J.M.C.K., 2021. A review of biological monitoring of aquatic ecosystems approaches: With special reference to macroinvertebrates and pesticide pollution. *Environmental Management*, 67(2), 263–276.

Sylla, D., 2014. Fusion d'images provenant de différents capteurs satellitaires pour le suivi de la qualité de l'eau en zones côtières. Université de TOULON.

Tockner, K., and Stanford, J.A., 2002. Riverine flood plains: Present state and future trends. *Environmental Conservation*, 29, 308–330.

UNESCO, 2022. *La qualité de l'eau depuis l'espace grâce à l'observation de la Terre par satellite*, n.d.

Ustun, B., 2011. Mapping water quality by using satellite imagery. In *FiG working week 2011 bridging the gap between cultures Marrakech, morocco, 18,–22 May 2011,* 13p.

Uudeberg, K., Aavaste, A., Kõks, K.-L., Ansper, A., Uusõue, M., Kangro, K., Ansko, I., Ligi, M., Toming, K., and Reinart, A., 2020. Optical water type guided approach to estimate optical water quality parameters. *Remote Sensing*, 12(6), 931.

Van Den Brink, P.J., Alexander, A.C., Desrosiers, M., Goedkoop, W., Goethals, P.L., Liess, M., and Dyer, S.D., 2011. Traits-based approaches in bioassessment and ecological risk assessment: Strengths, weaknesses, opportunities and threats. *Integrated Environmental Assessment and Management*, 7(2), 198–208.

Vincourt, S., 2016. Contribution de l'imagerie multispectrale et infrarouge thermique aéroportée à l'étude de l'effet de la contrainte hydrique sur plusieurs variétés de pommier. other. France. Université Paris Diderot - Paris 7 (UPD7), FRA.

Vinel, A., 2022. Characterization of the thermomechanical behaviour of metals for high strain-rates, using ultrahigh speed imaging cameras. phdthesis. École centrale de Nantes.

Wieland, M., and Martinis, S., 2019. A modular processing chain for automated flood monitoring from multi-spectral satellite data. *Remote Sensing*, 11(19), 2330.

Wilkinson, C., 2008. Status of coral reefs of the world: 2008. GCRMN, Townsville.

Williamson, C.E., Saros, J.E., Vincent, W.F., and Smol, J.P., 2009. Lakes and res-ervoirs as sentinels, integrators, and regulators of climate change. *Limnology and Oceanography*, 54, 2273–2282.

Wu, B., Zhou, X., Zhao, S., Yue, X., and Keutzer, K., 2019. SqueezeSegV2: Improved model structure and unsupervised domain adaptation for road-object segmentation from a LiDAR point cloud. In *2019 International Conference on Robotics and Automation (ICRA).* Montreal, QC, Canada: IEEE, 4376–4382.

Yadav, S., Yamashiki, Y., Susaki, J., Yamashita, Y., and Ishikawa, K., 2019. Chlorophyll estimation of lake water and coastal water using landsat-8 and sentinel-2a satellite. *The International Archives of the Photogrammetry, Remote Sensing and Spatial Information Sciences*, XLII-3/W7, 77–82.

Yepez S., Laraque A., Martinez J. M., Sa J. D., Carrera J.M, Castellanos, B., Gallay M., and Lopez J.L. 2018. Retrieval of suspended sediment concentrations using Landsat-8 OLI satellite images in the Orinoco River (Venezuela). *Comptes Rendus Geoscience*, 350(1–2), 20–30. https://doi.org/10.1016/j.crte.2017.08.004.

Zhan, X., Miller, S., Chauhan, N., Di, L., Ardanuy, P., and Running, S., 2002, Soil moisture visible/infrared imager/radiometer suite algorithm theoretical basis document, Version 5..

Zou, K., Chen, J., Ruan, H., Li, Z., Guo, W., Li, M., and Liu, L., 2020. eDNA metabarcoding as a promising conservation tool for monitoring fish diversity in a coastal wetland of the Pearl River Estuary compared to bottom trawling. *Science of The Total Environment*, 702, 134704.

6 Monitoring of Water Quality of Asia's Largest Brackish Water System, Chilika Lagoon

*Pradipta R. Muduli**
Wetland Research and Training Centre, Chilika Development Authority, Balugaon, India

Prasannajit Acharya
Institute of Technical Education and Research, Department of Chemistry, Siksha 'O' Anusandhan, Bhubaneswar, India

6.1 INTRODUCTION

The water quality of aquatic ecosystems indicates environmental health and bioresources. Due to human-driven nearshore activities and climate change, which pose a significant risk to aquatic ecosystems (Gurlin et al., 2011), there has been a rise in the demand for sustainable management of water quality. For example, eutrophication, which is primarily caused by increased agricultural and industrial activities, is one of the primary factors that contribute to the development of significant challenges for the ecosystem's health, as well as for aquaculture and fisheries activities, and also for tourism and tourism-related activities (Anderson et al., 2002). In this regard, knowledge of water quality parameters is essential to the advancement of our understanding of ecosystem services and health, as well as the evaluation of the impact that humans have on the environment (Maeada et al., 2019).

Monitoring the health of the ecosystem periodically helps to recognize and prevent contamination issues. Water quality monitoring is an essential practice conducted worldwide, encompassing lakes, streams, coastal waterways, and estuaries (Muduli et al., 2011; Sarma et al., 2010). The fundamental method for understanding the hydrologic cycle in any aquatic ecosystem is the monitoring of water quality; hence, an effective monitoring system is key to the management of a water basin. Water quality is most frequently used as a reference to a set of standards against which compliance can be assessed and can be used as a fundamental tool in the management of aquatic ecosystems. A better understanding of the hydrologic cycle, as well as the ecosystem services, especially the use of water, helps to improve planning efforts. The long-term collection and availability of water quality data are crucial to understanding climate change, biodiversity, and the health of ecosystems. Since the year 2000, the implementation of the water framework directive in India has made the biological and chemical evaluations of water quality crucial for efficient global monitoring. Our local, state, and national governments use monitoring information to formulate a management action plan to control pollution levels. Ecological health monitoring in terms of water quality provides the objective evidence necessary to make sound decisions on managing lagoon ecosystems.

* Corresponding author.

DOI: 10.1201/9781003354000-9

A lagoon inhabited by various aquatic life forms is a large body of water surrounded by land and connected to the sea. The natural processes as well as anthropogenic activities such as unprecedented developmental activities significantly influence the lagoon's health. River water discharge, sewage, and industrial discharge into lagoons are some of the common causes that bring various nutrients and contaminants into the ecosystem, causing eutrophication and trace metal accumulation in the bioresources of the ecosystem (Parida et al., 2017). Chilika lagoon is the largest brackish water ecosystem in Asia, surrounded by 52 rivers and rivulets, and is hence vulnerable to uncontrolled nutrient (Mukherjee et al., 2018; Patra et al., 2017) as well as contaminant discharge through monsoonal discharge. Most of the freshwater discharge comes from the Mahanadi tributaries (Gupta et al., 2008; Muduli et al., 2012, 2013), and the rivers are surrounded by agricultural fields, which indicates the possibility of pesticide contamination as well as a high nutrient load due to the uncontrolled application of fertilizers. There are also chances of oil pollution due to motorized boats that operate for fishing and tourism purposes (Mohanty et al., 2017). Such environmental conditions can make Chilika vulnerable to pollutants. Hence, periodic monitoring of water quality is warranted for the Chilika lagoon to ensure the sustainable maintenance of good health and valued ecosystem services. In terms of basic water quality parameters, nutrient load, pesticides, organic load, and petroleum hydrocarbons are very important to develop the management action plan. The following sections provide information on the importance of each parameter that could be considered for a monitoring program by conventional methods as well as advanced methods.

The conventional methods for determining the quality of water involve on-site sampling and data collection, which can be time-consuming, expensive, and labor-intensive. Following this, the samples must be transported to laboratories, either on land or at sea, for examination. While some studies have been comprehensive, the majority of them have not been carried out at an appropriate temporal and spatial scale to properly analyze the variables that impact the occurrence of events like harmful algal blooms, oxygen depletion, fish kills, and so on. Inadequate quality control and quality assurance procedures, such as using non-standard methodologies and storing data for extended periods before analysis, can also make the data unusable. Moreover, the data often exhibit vulnerability to errors in recording and geo-referencing during interpretation.

To resolve these issues, an advanced method, remote sensing (RS), provides value to the earth science community as a source of primary data. The electromagnetic radiation data received by sensors must be calibrated properly for transformation into usable scientific signals. It provides high-quality spatiotemporal data. Many different methods have been created for the spectrally based retrieval of components, and these methods can generally be broken down into three primary categories (Niroumand-Jadidi et al., 2019). The initial method involves evaluating and conducting tests to train and fine-tune a regression model, such as a polynomial, using image-derived properties (such as band ratios). This model is then used to estimate the concentrations of the desired ingredient based on known measurements obtained in the field. Once the correlation between spectral characteristics and the levels of the component has been determined (i.e., once the regression model coefficients have been calculated), the model can be employed to estimate the component's levels for any other spectrum. The second approach includes semi-empirical methods that rely on training a regression model using a diverse range of in situ observations. Semi-empirical approaches are general models that utilize a huge database of in situ bio-optical observations to train the regression model. One such example is the OC3 bio-optical model (O'Reilly et al., 2019). Unlike empirical approaches, which have localized or site-specific training, this approach is different. In empirical approaches, the training occurs at the specific location in question.

The water quality assessment using remote sensing has been in demand as it helps to get the data from different ecosystems in different environmental conditions or when there are difficulties in field data collection (Lettenmaier et al., 2006; Doell et al., 2016). Apart from lakes, lagoons, and wetlands, the river ecosystems are also globally mapped with RS technology (Gleason and Durand,

2020). These vital inferences in our understanding of the RS observed data are enormously helpful in visually representing the earth's systems. RS data can be used as input for different kinds of modeling such as a hydrologic model for advanced studies in water resources and their management. For instance, Lin et al. (2019) studied new simulations of global hydrology which were taken from more than 10,000 river gauges and RS products to utilize RS for global hydrologic modeling. The advanced methods of RS play a very crucial role in water quality analysis that will eventually help in the analysis of climate change on a long-term basis as well as its prediction. This is emerging as a vital tool to monitor the anthropogenic impact on river ecosystems as well as lagoon ecosystems. This technology also helps to figure out the impact of floods on the water body and the population in the affected areas (Grimaldi et al., 2016).

The application of RS with spaceborne sensors utilized for water quality assessment is immensely helpful for primary data collection and the analysis of the health status of the ecosystem of interest. However, studies on this application are scant, especially in the tropical coastal ecosystems of India, and Chilika lagoon is no exception. There are only limited studies on Chilika lagoon in this respect, and hence it is important to explore the possibility of analyzing several parameters of water quality. The calibration and validation of the RS data with field-observed data, and establishing robust numerical models to monitor water quality remotely with a higher level of accuracy, are essential. The present study provides insights into the methodologies that have been followed so far for water quality analysis, from the conventional methods to advanced remote sensing methods in Chilika lagoon.

6.2 DECIDING SAMPLING LOCATIONS

The sampling location is decided according to the objective of the monitoring. When there is an intention of monitoring all the environmental parameters, the best idea is to create a grid of the ecosystem map for fixing sampling points at equal distance. If researchers want to study the impact of seagrass/macrophytes on water quality, sampling must be specifically selected in the areas covered with seagrass/macrophytes and some control sites without any patches of seagrass or macrophytes. Similarly, if the objective of the study is to understand the influence of riverine discharge on water quality, sampling points must be selected in all the rivers, the river confluence point in Chilika, and corresponding sampling points inside the Chilika lagoon too. In order to understand the role of the water quality of Chilika on the coastal Bay of Bengal, the sampling points should be decided considering a transect from river-Chilika-nearshore and offshore. Apart from the scientific view, there are also logistic arrangements that must be considered before deciding on the sampling locations with respect to different objectives. Sometimes, theoretically or gridded sampling stations are not approachable due to several reasons such as low depth, the luxurious growth of macrophytes in water hindering the propeller of motorized boats, ruined road connectivity, etc. The gridded Chilika map in Figure 6.1 shows 33 locations where samples for more than 30 environmental parameters were collected by the Chilika Development Authority for research in different aspects. However, for specific scientific experiments and studies, other locations are considered according to the objectives of the studies.

6.3 METHODOLOGY

6.3.1 Sampling and Analysis

Depending on the objective of the study or research questions to be answered, the environmental parameter to be considered under the program is decided. It may include microbial, biological, sediment quality, bathymetry, fishery, etc., and parameters relevant to physical oceanography. The usual

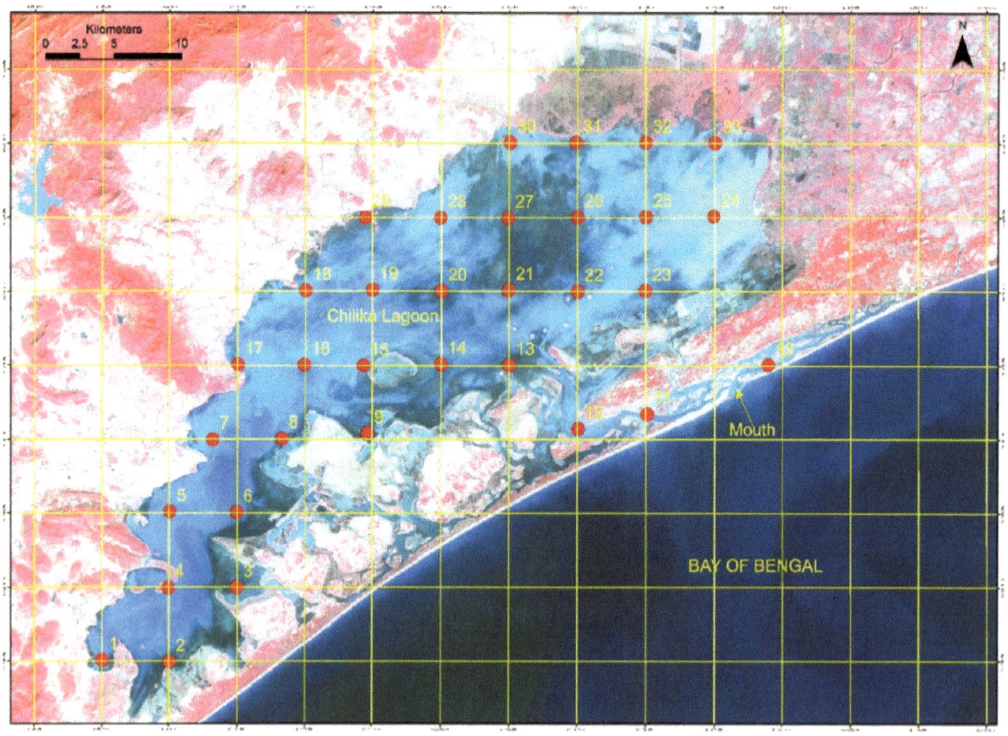

FIGURE 6.1 Gridding of Chilika lagoon for fixing sampling locations for the monitoring program by CDA to have samples at equal distance covering all sectors of the water body.

water quality parameters studied in order to monitor the health status of coastal or lagoon ecosystems, and helpful recommendations for management action plans, are listed in Table 6.1.

Scientists use a range of different techniques to put together results of water quality and related environmental parameters for interpretation. The results of such studies indicate the health of aquatic ecosystems. On-site measurements are the most precise for assessing water quality due to the equilibrium between water and its surroundings. It is usual practice to take measurements close to the water source while conducting these tests on-site. The sample starts to achieve chemical equilibrium with its new environment as soon as it is taken out of the water system.

Hence, stringent methodologies are followed for sample collection and preservation until transporting to the laboratory. There are several environmental parameters that need to be recorded at the sampling spot itself, such as air temperature, water temperature, wind speed, light intensity, depth, and transparency (Secchi disk depth). Most of the parameters are analyzed within 24 hours of sampling. Some samples, like biochemical oxygen demand (BOD), are analyzed after three days or five days of incubation according to the objective. All the water quality parameters can be analyzed following Grasshoff et al. (1999). The complete procedure for monitoring water quality is shown in Figure 6.2 as a flow diagram, starting from deciding sampling locations to reporting the results and findings.

After the collection of samples in the field (Figure 6.3) using field equipment setup (Figure 6.4), samples are brought to the laboratory with instrumentation facilities. Initially, samples are processed in the wet lab (Figure 6.5a) and finally brought to the instrumentation laboratory (Figure 6.5b) for analysis of hydrochemical parameters. Sophisticated instruments are used

TABLE 6.1
List of Basic Hydro-chemical Parameters Recommended for the Monitoring of Lagoon Ecosystems

Physico-chemical parameters	Nutrients	Biological/ Productivity	Carbon parameters	Pollution indicator parameters	Mineral constituents	Metals
Water temperature, Air temperature, Humidity, Wind speed, Light intensity (PAR-Quantum) depth, Water current transparency, Turbidity, Total suspended matter (TSM), Dissolved oxygen, DO%, ORP (Oxidation reduction potential) pH, Biochemical oxygen demand	Nitrite, Nitrate, Ammonia, Phosphate, Silicate, Total nitrogen (TN), Total phosphorus (TP), Total organic nitrogen, Total organic phosphorus	Pigments (Chlorophyll-a), Gross primary production (GPP), Net community production (NCP), Community respiration (CR), Bacterial respiration	Total organic carbon, Dissolved organic carbon, Particulate organic carbon	Pesticides, Petroleum, Hydrocarbons, Trace metals	Boron, Total hardness, Calcium, Magnesium, Sodium, Potassium, Chloride, Sulphate, Fluoride	Chromium (Cr) (total and hexavalent), Iron (Fe), Nickel (Ni), Copper (Cu), Zinc (Zn), Cadmium (Cd), Mercury (Hg), Lead (Pb)

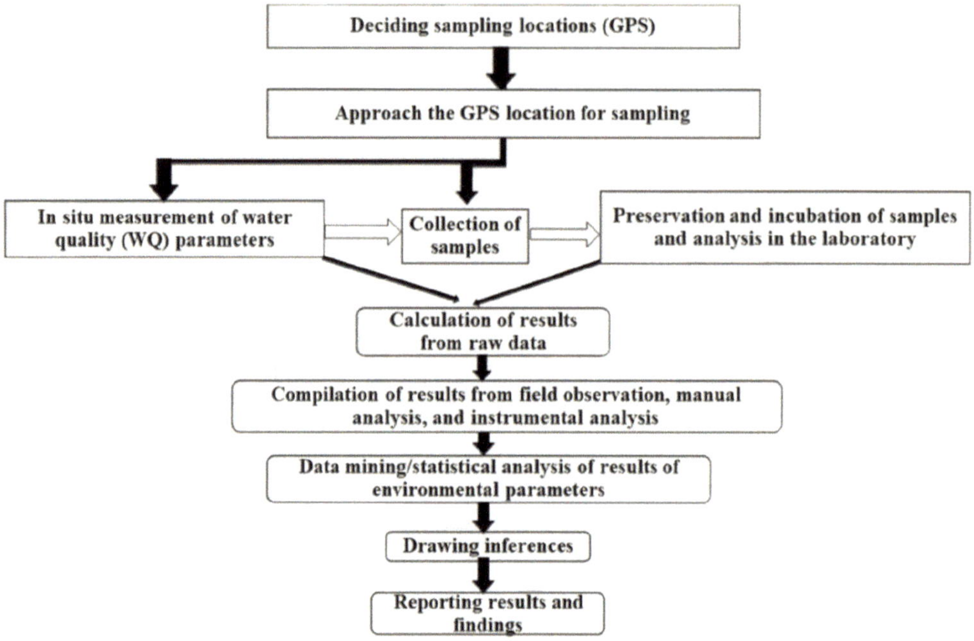

FIGURE 6.2 Flow diagram of the sampling methodology for Chilika water quality monitoring.

for the analysis of more complex parameters (Figure 6.6), for example, nutrient auto analyzer and spectrophotometer for nutrients/chlorophyll analysis, atomic absorption spectroscopy and cyclic voltammetry for metal and its speciation studies, gas chromatography (GC) for pesticide and greenhouse gas analysis, high-performance liquid chromatography (HPLC) for analysis of pigments of harmful algal blooms, polycyclic aromatic hydrocarbon, and total organic carbon analyzer for estimation of carbon load into the aquatic ecosystem. The data collected through the field survey are compared with the data derived through remote sensing following the pathways shown in Figure 6.7.

6.4 MONITORING OF WATER QUALITY PARAMETERS

6.4.1 Hydrochemical Parameters

Monitoring hydrochemical parameters is crucial for assessing the health and functioning of lagoon ecosystems. Lagoons, being dynamic coastal ecosystems, are influenced by various factors, including tidal movements, nutrient inputs, and anthropogenic activities.

Monitoring water temperature provides insights into seasonal variations and influences various biological and chemical processes. Temperature affects the solubility of gases, metabolic rates of organisms, and overall ecosystem dynamics. pH levels indicate the acidity or alkalinity of the water. Lagoon ecosystems are sensitive to changes in pH, which can affect the availability of nutrients and the behavior of aquatic organisms. Dissolved oxygen (DO) is essential for the survival of aquatic organisms. It is influenced by factors such as temperature, salinity, and the presence of photosynthesizing organisms. Monitoring DO levels helps assess the ecosystem's ability to support aquatic life. Salinity is a critical parameter in lagoon ecosystems, as it can vary due to tidal influence and freshwater inputs. It influences the distribution of species and

FIGURE 6.3 Photographs showing the sampling in the field (Chilika and surrounding rivers) and analysis in the laboratory (chemical and instrumentation Laboratory, WRTC, CDA).

ecosystem structure. Fluctuations in salinity can impact the osmoregulation of aquatic organisms. Monitoring nitrogen and phosphorus concentrations helps assess nutrient loading and eutrophication potential. Excessive nutrient inputs can lead to algal blooms, oxygen depletion, and other ecological imbalances.

TSS measures the concentration of particles suspended in the water. High TSS levels can impact water clarity, light penetration, and sediment transport. It is particularly important in lagoon ecosystems influenced by sedimentation and runoff (e.g., Chilika lagoon). Turbidity is a measure of water clarity and the presence of suspended particles. Elevated turbidity levels can affect light penetration, photosynthesis, and the distribution of aquatic organisms. Secchi disk depth provides a simple measure of water transparency. It involves lowering a disk into the water until it becomes invisible.

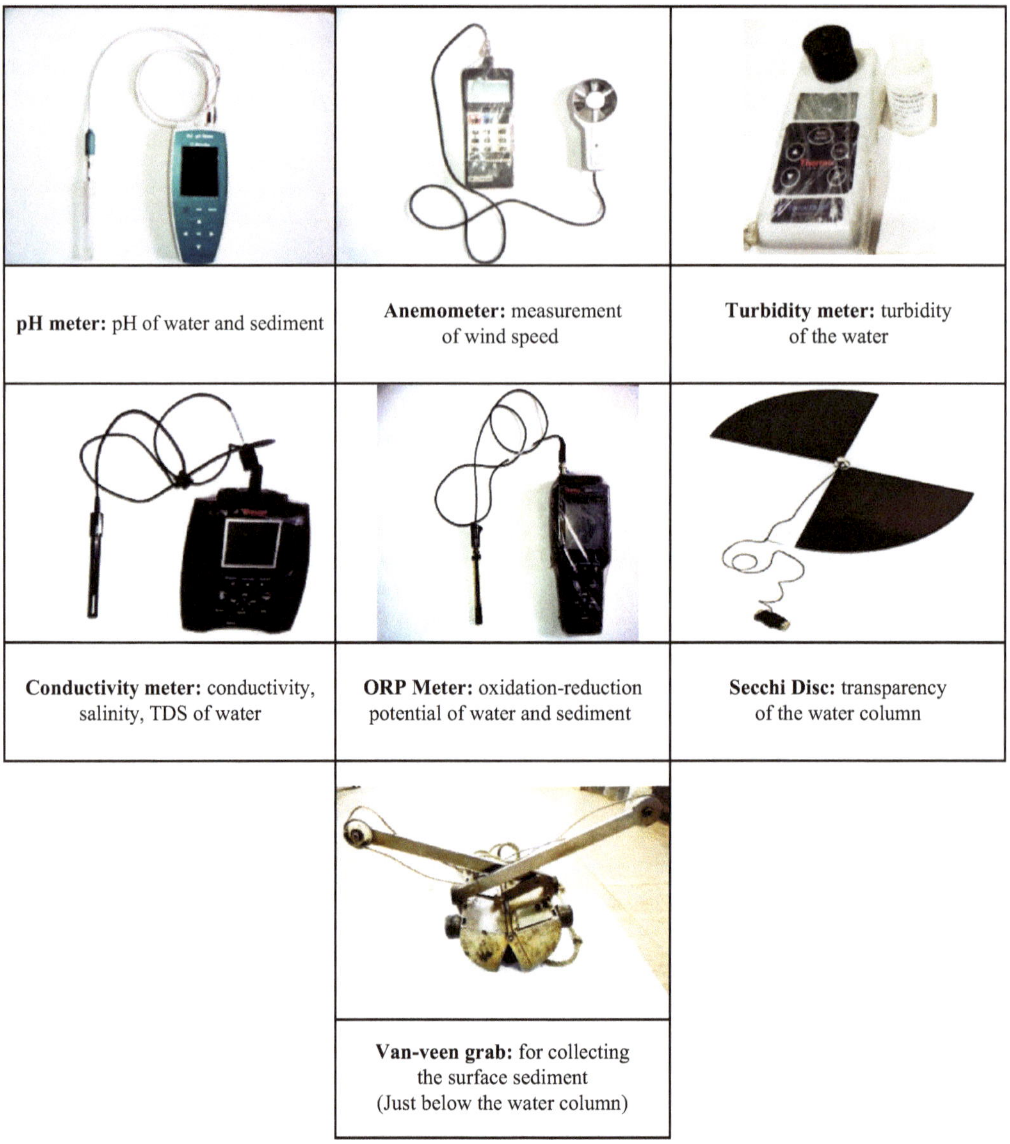

FIGURE 6.4 Scientific equipment used in the field and laboratory for sampling and analysis.

Changes in Secchi depth can indicate shifts in water clarity. Chlorophyll-a concentration is an indicator of phytoplankton biomass. Monitoring Chlorophyll-a levels helps assess the productivity of lagoon ecosystems and the potential for algal blooms. Redox potential measures the oxidation-reduction balance in the water. It is crucial for understanding biogeochemical processes, such as nutrient cycling and the availability of certain elements.

Monitoring these hydrochemical parameters provides valuable data for understanding the ecological status of lagoon ecosystems, identifying trends, and implementing effective management strategies. Monitoring programs should consider factors such as seasonal variations and the specific characteristics of the lagoon under investigation.

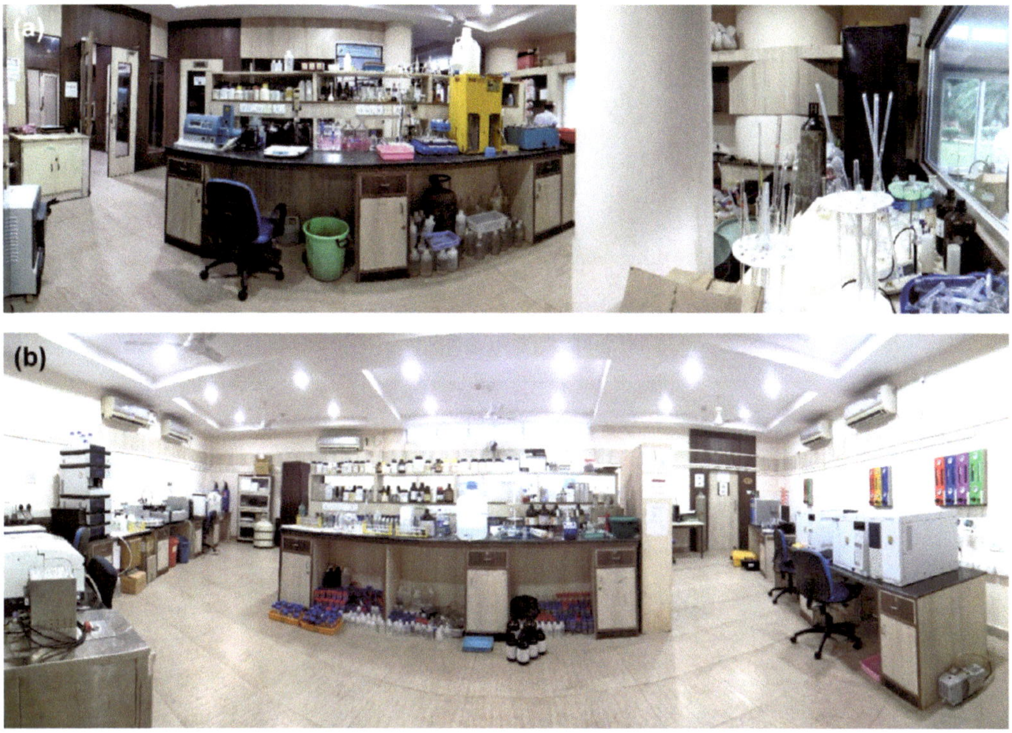

FIGURE 6.5 (a) Wet lab for analysis of physico-chemical parameters by manual method to be carried out immediately after sample collection from aquatic ecosystems. (b) Sophisticated instrumentation facility for the analysis of hydrochemical parameters after processing in the wet lab.

6.4.2 Pollution-Indicating Parameters

Monitoring concentrations of nitrogen and phosphorus is crucial to assessing nutrient loading. Excessive nutrients, often from agricultural runoff or wastewater discharge, can lead to eutrophication, algal blooms, and oxygen depletion. Low levels of dissolved oxygen can indicate oxygen depletion, a condition often associated with pollution. It may result from organic matter decomposition, especially in areas with high nutrient inputs. BOD measures the amount of oxygen consumed by microorganisms during the decomposition of organic matter. High BOD levels suggest the presence of pollutants and organic contaminants. Chemical oxygen demand (COD) measures the amount of oxygen required to chemically oxidize both organic and inorganic substances in water. Elevated COD levels can indicate the presence of pollutants and industrial discharges. High levels of total suspended solids can reduce water clarity, impact light penetration, and contribute to sedimentation. TSS is often associated with runoff and erosion. Changes in pH can affect the solubility and toxicity of various pollutants. Extreme pH levels may indicate the presence of acid mine drainage or other anthropogenic inputs. Monitoring heavy metal concentrations is important as they can be toxic to aquatic life and have long-term ecological impacts. Industrial discharges and runoff from urban areas are common sources of heavy metals. The presence of fecal coliforms or other pathogenic microorganisms can indicate contamination from sewage or agricultural runoff. Monitoring microbial indicators helps assess the safety of water for recreational and human use. Monitoring the presence of agricultural chemicals, such as pesticides and herbicides, helps assess the impact of runoff from agricultural areas on water quality. The presence of oil and grease in water can indicate pollution from urban runoff, industrial discharges, or oil spills. It can have detrimental effects on

FIGURE 6.6 Sophisticated instruments used in the laboratory to process and analyze hydrochemical parameters.

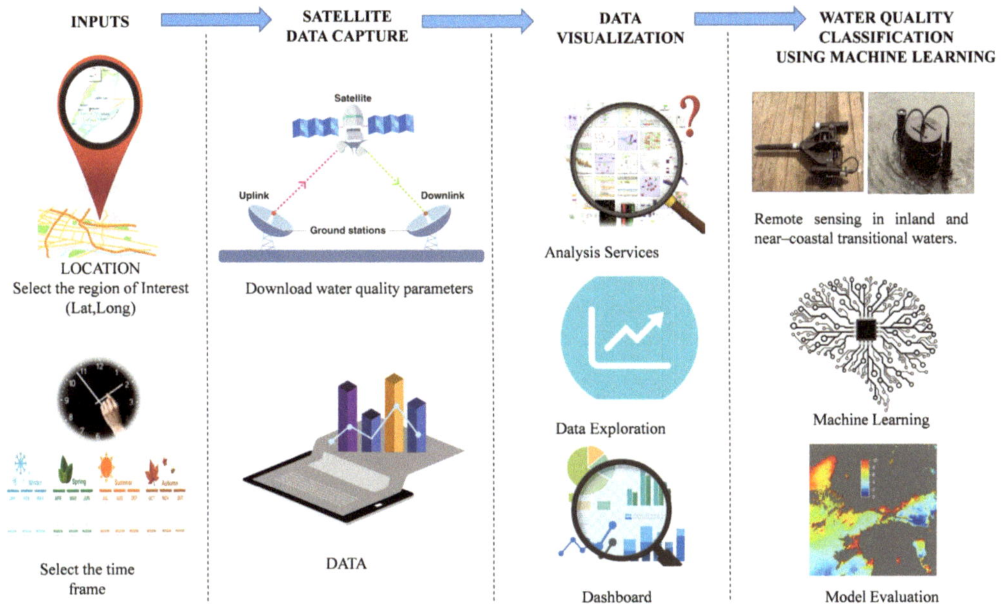

FIGURE 6.7 Pathways for water quality monitoring through remote sensing techniques.

aquatic ecosystems. TPH measurement provides an overall assessment of hydrocarbon contamination, including both crude oil and refined products (Acharya et al., 2022). Monitoring microplastics helps assess the impact of plastic pollution in lagoon ecosystems. Microplastics can have various ecological consequences and pose risks to aquatic organisms.

Regular monitoring of these pollution indicator parameters provides valuable information for understanding the sources and extent of pollution in lagoon ecosystems. Monitoring programs should be tailored to the specific characteristics and threats facing each lagoon and may involve seasonal or event-based sampling.

6.4.3 Water Quality Monitoring in Chilika Lagoon

As described in Section 6.4.2, the hydrological parameters as well as pollution indicator parameters are site specific; the parameters are to be chosen based on the values and threats of the particular lagoon or ecosystem. Apart from these parameters, several other parameters can be monitored, as listed in Table 6.1 as part of the monitoring program. Studies on Chilika revealed that the water quality alters significantly with respect to time and space. The major factors contributing to this variability are the mixing behavior of riverine freshwater and seawater in different seasons and the difference or extent of mixing in the different regions of the Chilika lagoon. As listed in Table 6.1, several parameters can be measured with traditional methods. However, only a few parameters can be measured or estimated through satellite data.

There are only limited studies on Chilika that focus on water quality assessment through a remote sensing approach. Earlier studies estimated Chlorophyll-a, colored dissolved organic matter (CDOM), and turbidity (Table 6.2). Apart from water quality, land use land cover parameters such as vegetation, water spread area, water body, and forest cover in the watershed area around the Chilika area are also studied (Table 6.3).

6.5 COLOURED DISSOLVED ORGANIC MATTER

Coloured dissolved organic matter (CDOM) is a composite of organic compounds that can originate from both land-based and water-based plants, as well as from phytoplankton and bacteria (Soomets

TABLE 6.2
Water Quality Analysis of Chilika Lagoon through Remote Sensing

Sl. No.	Satellite details/ Data source	Technique used	Parameter estimated through remote sensing	Reference
1	MODIS TERRA	SeaWiFS and Sea DAS software	Chlorophyll-a	Srichandan et al., 2015
2	FCC of EO-1 Hyperion and NDTI	Fast Line-of-sight Atmospheric Analysis of Hypercubes (FLAASH)	Turbidity	Garg et al., 2017
3	AVIRIS-NG	Quasi analytical algorithm	Colored dissolved and detrital matter	Sahay et al., 2019
4	AVIRIS-NG	Spectral angle mapper (SAM)	Turbidity	Chander et al., 2019
5	MODIS and LISS-I (IRS-IB)	SeaWiFS and Sea DAS software	TTS and Chl-a	Kumar et al., 2020
6	Copernicus Sentinel-5P satellite	NASA'MERRA-2	Chlorophyll and transparency, TSM	Mishra et al., 2020
7	Landsat-8 Operational Land Imager (OLI)	Open-source ACOLITE software	Cyanophyta and transparency	Tarafdar et al., 2021
8	AVIRIS-NG	Spectrum matching technique	Chlorophyll-a concentration (Chl-a) and colored dissolved organic matter (CDOM) + Detritus	Ali et al., 2021
9	FCC of EO-1 Hyperion	Fast Line-of-sight Atmospheric Analysis of Hypercubes (FLAASH)	Turbidity	Garg et al., 2024

AVIRIS-NG: Airborne Visible Infrared Imaging Spectrometer-Next Generation; SeaWiFS: Sea-viewing Wide Field-of-View Sensor; Sea DAS: Sea Data analysis system.

TABLE 6.3
Remote Sensing Studies on Land Use Land Cover (LULC) around Chilika Lagoon

Sl. No.	Satellite details	Technique used	Parameter estimated through remote sensing	Reference
1	Georeferenced Landsat	ArcGIS (10.1) and ERDAS	LULC	Sinha et al., 2020
2	Landsat-TM and Landsat-8	NDVI, EVI	Vegetation	Goswami et al., 2018
3	Landsat-8 (OLI) and Sentinel-2A	NDVI, EVI	Water body extraction	Naik and Anuradha, 2018
4	(IRS 1D and P6 LISS-III)	ERDAS	Vegetation and LULC	Panda and Mohanty, 2008
5	Landsat-TM		Chilika lagoon mouth dynamics	Goswami et al., 2019
6	Digital Shoreline Change Analysis System (DSAS)	ArcGIS (10.1)	LULC	Goswami and Choudhury 2018

et al., 2020). CDOM refers to the optically active fraction of the overall dissolved organic matter (DOM) in a body of water. It plays a significant role in influencing the ecological processes and biogeochemical cycles of freshwater ecosystems. CDOM serves as a convenient and efficient spectral tool for monitoring the transport and transformation of DOM in aquatic ecosystems (Zhang et al., 2011). This is possible because of its ability to absorb light and emit fluorescence (Zhou et al., 2017; Zhang et al., 2021). This attribute affects the quantity of light entering the water and serves as an indicator of the amount of carbon in lakes. Consequently, research on the carbon cycle and endeavors related to water treatment for various objectives can gain advantages from its input (Kutser et al., 2015).

By transformation, photo- or bio-degradation, CDOM influences the carbon cycle. Moreover, it influences how organic carbon ultimately mineralizes into CO2 or CH4 in the atmosphere, which has an impact on the global climate (Zhang et al., 2021). According to earlier research, dissolved organic carbon (DOC) in inland waters and CDOM are closely connected (Ward et al., 2017). Rapid spectral measurement of CDOM and studies of the link between CDOM and DOC are subjects at the core of research outlooks as DOC determinations are essential for calculating carbon budgets. Several earlier findings have demonstrated that CDOM can be utilized as a quick and reliable proxy for DOC assessment in a variety of aquatic environments (Kowalczuk et al., 2010). As CDOM can impact the biogeochemical cycle of nutrients, heavy metals, and organic contaminants, it is crucial to understand this concept (Zhang et al., 2021). As a bridge between carbon cycling and the environmental outcome of nutrients, heavy metals, and pollutants, CDOM is crucial. Additionally, understanding the dynamics of CDOM can assist in assessing water quality concerning climate change. This is because CDOM acts as a sink for carbon dioxide (Zhao and Song, 2018). So, it is essential to trace and explain CDOM's complex properties as well as its ecological and environmental significance in aquatic environments.

CDOM concentrations in inland waters can be determined using the absorption coefficients of CDOM at specific wavelengths (254, 280, 355, and 350 nm) by standard methods (Zhang et al., 2010; Yao et al., 2011; Xu et al., 2017; Zhou et al., 2017). The CDOM absorption coefficients at 250 nm and 365 nm can be compared to estimate the molecular weights of CDOM. Additionally, the aromaticity can be measured by measuring the specific UV absorbance at 254 nm. These methods were described by Spencer et al. (2009) and Zhou et al. (2016). CDOM, which stands for colored dissolved organic matter, is sometimes called fluorescent dissolved organic matter (FDOM) since it can fluoresce (Kalle, 1949). Recently, the use of parallel-factor analysis (PFA) in three-dimensional excitation and emission-matrix fluorescence (3DEEM) studies has become more common (Zhou et al., 2017). Fluorescence technology enables the monitoring of dynamic changes in the characteristics of CDOM. Although optical techniques are less accurate than high-resolution techniques such as nuclear magnetic resonance or Fourier transform ion cyclotron resonance mass spectrometry (Cortes-Francisco et al., 2014), they are easy to perform experimentally and require inexpensive analytical equipment. Consequently, the absorption and fluorescence spectra of CDOM have been extensively utilized in recent research to characterize different bodies of water (Liu et al., 2017; Song et al., 2017; Jiang et al., 2018; Ying et al., 2018; Song et al., 2019).

Over the past few decades, there have been significant advancements in the surveillance of water quality indicators for both inland and marine coastal areas. The utilization of ocean color sensors such as SeaWiFS (Sea-viewing Wide Field-of-View Sensor), MODIS (Moderate Resolution Imaging Spectroradiometer), and MERIS (Medium Resolution Imaging Spectrometer) has facilitated this achievement. Several empirical ocean color algorithms have been developed in recent years to retrieve CDOM within specific geographic areas (Matthews, 2011; Tehrani et al., 2013). Additionally, a limited number of universally applicable, worldwide empirical algorithms have been formulated. CDOM retrieval remotely relies on the assumption that the reflectance of the sea surface is roughly inversely proportional to the overall absorption coefficient. This coefficient can be divided into different contributions by particulate and dissolved elements, as explained by Morel and Prieur in 1977 and Gordon et al. in 1988.

There are very few studies focused on the application of remote sensing techniques to estimate the CDOM in the Chilika lagoon. For instance, Ali et al. (2021) derived CDOM from hyperspectral data using the spectrum matching technique (SMT). The accuracy was checked by comparing it with field observations done simultaneously when the data were recorded by a sensor mounted with AVIRIS-NG flight over the Chilika lagoon. Sahay et al. (2019) also mapped CDOM distribution in the lagoon using hyperspectral data of AVIRIS-NG. The CDOM absorption coefficient was retrieved using a quasi-analytical algorithm. The distribution in four different sectors of the Chilika lagoon (northern, central, southern, and outer channel) was analyzed. The study revealed the highest CDOM in the northern sector (having the highest CDOM absorption at 412 nm) as compared to the rest of the sectors. High CDOM in the northern sector is attributed to the inflow of terrestrial organic matter, as reported by earlier studies (Kanuri et al., 2013, 2018).

6.6 TOTAL SUSPENDED SOLIDS

TSS refers to both organic and mineral particles that are suspended in water and is directly related to the turbidity of the water as well as the Secchi depth (Soomets et al., 2020). The spatiotemporal monitoring of TSS is useful for evaluations of water quality and the management of lakes (Nechad et al., 2010). To date, several researchers have monitored TSS using conventional methods (Patra et al., 2016, Barik et al., 2017). There have been few attempts to examine the water quality of Chilika lagoon using satellite data (Sudhakar and Pal, 1993; Pal and Mohanty, 2002) and in situ remote sensing data (Rao et al., 1986). Unfortunately, TSS was left out of the studies listed above. For instance, using LISS-I (IRS-IB) data, Mohanty and Pal (2001) predicted TSS concentration using a coefficient (for the algorithm) available from other published literature. The chromaticity technique was used by Gupta (2013) with RESOURCESAT-1 AWiFS data to estimate TSS. As compared to this technique, MODIS data have several advantages such as no complexities for atmospheric correction, higher sensitivity, and being more cost-effective. Using these data, both short- and long-term monitoring of TSS in surface water can be carried out. The TSS monitoring also can be carried out by application of MODIS 250 m surface reflectance products and using the conventional relation between in situ measurements of TSS and atmospherically corrected reflectance in MODIS band 1 (Rrs at 645 nm).

In order to better quantify the TSS, the highest possible number of satellite images should be collected and analyzed for quality assessment. For instance, Kumar et al. (2020) collected 46 satellite datasets for each year and repeated it for 14 years. Efforts must be taken to select cloud-free images, which are expected to be obtained during the monsoon season. Solar angles can cause significant changes and adds uncertainty in remote sensing data; hence, enough replicates must be considered for field data collection to avoid errors in model output. Kumar et al. (2020) studied the TSS of Chilika lagoon using this setup and also validated the results by comparing them against published literature for similar lakes in China and the United States.

There are several factors that control the TSS in lagoon ecosystems. Lagoon ecosystems can experience heavy wind and rainfall during cyclones, which can increase the TSS level in water due to sediment churning in shallow waters. The re-suspension in the water column has the potential to alter the overall water quality of the lagoon. (Chen et al. 2009). It also can significantly influence the associated biological, chemical, and geomorphological processes (Miller et al., 2011). As compared to the northern and southern sectors and outer channel, the central sector TSS variation is intermediate as the number of rivers draining into this sector is higher than the southern sector and lower than the northern sector. Similar to Chilika lagoon, other ecosystems also experience the phenomenon of change in the TSS level as an impact of wind action. For instance, Chen et al. (2007) evidenced that the wind factor significantly controlled the TSS level in aquatic ecosystems. Several studies proved this through field surveys and manual analysis of TSS, and some studies utilized the RS technique to prove the same. The shallow Mobile Bay estuary, Alabama, was studied by Zhao et al. (2011) using MODIS images and described the TSS variability within a gap of two days with a change in wind action.

As part of the AVIRIS-NG program, Chander et al. (2019) employed the spectral angle mapper (SAM) approach along with AVIRIS-NG's semi-analytical simulation method to analyze hyperspectral reflectance data. Their objective was to simulate and monitor TSS, turbidity, and Chl-a in the lagoon. As per studies through RS as well as conventional methods, the TSS concentration in the northern sector has been consistently higher than in other sectors. Although the turbid water through riverine discharge is lowest in summer, the sediment churning in the shallow region of the northern sector increases the TSS level (Muduli and Pattnaik, 2020; Barik et al., 2017). In the Chilika lagoon, another factor that may influence the data quality is the interference of vegetation (emergent or submerged), which is recorded in different parts of the lagoon as shown in a map by Tripathi et al. (2022). A band ratio threshold of 1.5 ($B2/B1 > 1.5$) has been used by several studies (Kumar et al., 2020) in order to mask this vegetation. Similar care must be taken when there is the possibility of the influence of satellite data due to vegetation.

Garg et al. (2017) utilized a spectral similarity approach to examine the turbidity concentration and water quality in the lagoon. The analysis utilized the EO-1 Hyperion hyperspectral remote sensing image, and the data footprint encompassed the central and northern regions of the lagoon, where turbidity levels exhibit significant variability. The observed turbidity at each location fell within the estimated turbidity class range. Through remote sensing techniques, Mishra et al. (2021) identified excessive water turbidity (observed by the Copernicus Sentinel-5P satellite) as the primary factor that limits phytoplankton growth and hinders the establishment of blooms in the Chilika lagoon. The study conducted by Tarafdar et al. (2021) utilized satellite surveillance with Sentinel-3 OLCI to support field observations.

6.7 CHLOROPHYLL-A

The most common biological response to assessments of deteriorated water in lake ecosystems is an increase in primary production. This shows up as a rise in the quantities of Chlorophyll-a (Chl-a) (Xiao et al., 2007; Poikane et al., 2010). As a rapid and simple replacement for phytoplankton biomass, which has been employed for decades to determine the critical external nutrient loading levels of eutrophication, Chl-a has thus far concentrated on phytoplankton research (Janse et al., 2010; Kasprzak et al., 2008). Since Chl-a is regarded as a key parameter with significant consequences for lake eutrophication management, the accurate simulation and forecasting of Chl-a concentration have grown in importance in lake ecosystem research. It is common knowledge that a lake ecosystem's hydrology, geology, land cover, and climate factors heavily influence the ecosystem's structure and function (Martin et al., 2011). Climatic aspects, hydrologic variables, and meteorologic events all influence phytoplankton production, although nutrient concentrations remain one of the primary factors affecting Chl-a concentration (Arhonditsis et al., , 2003; Scheffer, 1989; Blindow et al., 1993).

The fluctuations in phytoplankton biomass and toxic blooms, in particular, are key stressors that affect the aquatic food web, aquatic ecosystems, nutrient cycling, and fish farming (Gower et al., 2005). Chl-a is the photosynthetic pigment present in all phytoplankton species. In order to monitor water quality, the Chl-a concentration is used as a proxy for the trophic condition of water bodies (Lillesand et al., 1983). Most studies followed the conventional methods to estimate the Chl-a in the waters of Chilika lagoon.

In a conventional method, the water samples are collected using a Niskin Sampler or bucket and filtered through 0.7 μm Whatman GF/F filter paper. They are preserved and transported carefully in liquid nitrogen or ice-cool condition to the laboratory for further analyses. In the laboratory, the Chl-a pigments are extracted into 90% acetone through overnight incubation at $0^{o}C$ and finally measured using an ultraviolet (UV)-visible spectrophotometer. There are also methods established for measurement using high-performance liquid chromatography (HPLC) techniques. Spectroscopic chlorophyll analysis follows destructive solvent extraction in conventional methods for quantifying chlorophyll (Porra et al., 1989). These procedures take a lot of time, call for removing the sample from the culture vessel, destroying it thereafter, and carefully disposing of the solvent waste that results. In time-course investigations, where the number of quantifiable parameters may be constrained by culture volume, the removal and destruction of samples might present challenges. Fluorescence is commonly used to detect and measure chlorophyll in plants and algae, including environmental samples (Wang et al., 2018). This technology is non-destructive and has been frequently applied for this purpose (Buschmann et al., 2000; Vincent, 1983). Digital imaging techniques have recently become a viable option for quick, non-destructive measurements of chlorophyll content. Particularly, the nearly widespread availability of smartphones has made it possible to use smartphone cameras in lab digital analysis (Rignon et al., 2016). A few studies have used digital imaging to determine the amount of chlorophyll in plant leaves, including those of maize (Friedman et al., 2016), potato plants (Gupta et al., 2012), and marine microalgal cultures (Su et al., 2008). The intensity of the red, green, and blue components of a pixel within a digital computer image is represented by the RGB color scale. Each color's intensity is measured on a scale from 0 to 255, with black having RGB values (0,0,0) and white having RGB values (255,255,255). Yet, numerous techniques reported so far call for either the use of sophisticated modeling software or intricate mathematical operations (Su et al., 2008; Dey et al., 2016; Friedman et al., 2016).

The microalga *Chlamydomonas reinhardtii* is also utilized to provide a quick and easy way to measure the chlorophyll concentration of algal cultures in situ without causing any damage, using the RGB scale to measure pixel color intensity from digital photos. Regarding the accuracy and repeatability of the procedure, the impact of bacterial contamination, sample amount, pH, and neutral lipid formation is also assessed. Comparing the method described here to other digital analysis methods for determining chlorophyll, it is noticeably more straightforward. The procedure is also non-destructive and doesn't require removing the sample from the culture vessel. The usage of only readily available software programs and minimal data modification are required for this strategy.

Recently, there has been limited use of remote sensing approaches to estimate Chl-a, which is a bio-optical parameter. Srichandan et al. (2015) processed the remote sensing data, utilizing SeaWiFS and Sea DAS software to find the variability and relationship between Chl-a and turbidity after a very severe cyclone "Phailin" event. Ali et al. (2021) obtained Chl-a values from hyperspectral data of fluids with complicated optical properties by employing a spectrum matching technique (SMT). The retrieved values of Chl-a were compared with the equivalent in situ measurements to identify the most suitable match. The retrieval was performed with semi-analytical inversion algorithms (SAA), namely the linear matrix inversion (LMI) and Garver-Siegal Maritorena (GSM) methods, but the results were deemed insufficient. The technique of spectral mixture analysis (SMA) was utilized on the AVIRIS-NG dataset, which is an airborne hyperspectral imaging system, to calculate the Chlorophyll-a concentration for each pixel in Chilika lagoon.

6.8 EVALUATION OF INFLUENCE OF RIVER DISCHARGE ON WATER QUALITY OF CHILIKA LAGOON

Similar to Chilika lagoon, the coastal water bodies are vulnerable to untreated water through riverine discharge. Hence, it is crucial to monitor the water quality of rivers draining into the ecosystem of interest. The nutrient and sediment input to Chilika lagoon can be greatly accelerated by human development in the watershed. The amount of nitrogen, phosphorus, sediment, and pollutants such as trace metals and pesticides deposited in lagoon ecosystems depends on the use of watershed land. Periodic storms followed by heavy rainfall result in a huge amount of discharge through 51 rivers and rivulets into the Chilika lagoon. Studies have revealed the significant impact of such precipitation on the nutrient flux into the Chilika lagoon and the consequent changes in water quality (Muduli et al., 2017). The sediment material originating from the watershed tends to nourish aquatic plants and algae due to the presence of phosphorus, nitrogen, and other vital nutrients that are bound to the entering particles. An elevated concentration of organic compound discharge might lead to a reduction in the dissolved oxygen content when decomposers respire and break down the organic waste. Sedimentation can also degrade the lagoon floor, negatively impacting aquatic insects, crabs, mussels, and other benthic organisms. The continuous input of sediments into the Chilika lagoon can make the basin shallower, resulting in a loss of water volume. This can also affect navigation and recreational use in the water. High levels of nutrient input can also lead to eutrophication, which in excess can lead to massive fish kills. Hence, it is very important to monitor the water quality of the river water too as part of the Chilika monitoring program.

River discharge, as one of the most essential climate variables, plays a vital role in altering the water quality of lagoon ecosystems (Muduli et al., 2017). Since many rivers are mostly ungauged, satellite technology has proved its potential to fill this gap with a combination of RS and corresponding in situ gauged discharge. Rivers draining into Chilika lagoon have been studied for more than two decades by manual in situ observations, which need to be studied through the RS approach. This could reduce manpower efforts as well as expenditure for frequent field trips during the monsoon season when the material flux discharge is at its highest level.

River discharge is calculated considering the product of river flow area and velocity, or the volume of water passing a specified point at any instant in time. The conventional method of *measuring* discharge involves a floater and a stopwatch. The distance moved in a given time period is the velocity of water (meter/second), and the area through which it is passed (m^2) which finally provides the discharge in the unit of m^3 per second (time). This method is obviously impractical for most riverine ecosystems. Hence, most discharge estimations are done using an acoustic Doppler current profiler (ADCP). However, these techniques estimate with specific errors, especially when the flows are very high or very low. The flux into the lagoon is estimated considering the concentration of material (e.g., nutrients, pollutants such as metals, pesticides, etc.). For instance, Muduli et al. (2017) estimated the nutrient flux after cyclone "Phailin" considering the average discharge data of each month multiplied by nutrient concentration. Prior to this study, Ganguly et al. (2015), Robin et al. (2016), and Patra et al. (2016) also focused on the impact of water quality in the lagoon due to the influence of water discharged through the 52 rivers and rivulets. However, no studies to date have used remote sensing to evaluate the water quality of the adjoining rivers and its impact on the lagoon.

River flow monitoring is crucial for managing water resources, assessing floods, and studying climate change. It is an essential factor in numerous scientific explanations and the management of water bodies. The estimation is primarily conducted indirectly by ground hydromonitoring networks, which assess flow velocity, water level, and cross-section area. In the past two decades, a significant quantity of satellites and sensors using visible and near infrared, microwaves, and other technologies have been utilized to measure and track water quality and its movement. In order to enhance the precision of river discharge retrieval, numerous investigations have concentrated on employing novel techniques and methodologies that aim to integrate data acquired from both ground-based and satellite observations. Nowadays, satellite sensors are considered a new source to monitor river discharge, which has become possible due to the large number of satellites launched during the last two decades. Among the satellite sensors, optical sensors and radar altimeters were found to be most appropriate to estimate river discharge. A combination of models with RS to estimate river discharge was first shown by Chen et al. (2010). In order to get more accurate results, the models need frequent calibration, and in this regard, RS was found to be an early source of independent calibration/validation data for such models (Sivapalan et al., 2003).

6.9 CHALLENGES AND OPPORTUNITIES IN WATER MONITORING WITH TRADITIONAL AND REMOTE SENSING TECHNIQUES

6.9.1 Traditional Water-Monitoring Techniques

Traditional monitoring methods often have limitations in spatial coverage and may not provide a comprehensive view of large water bodies or remote areas. Sample collection, transportation, and laboratory analysis can be labor-intensive and time-consuming, leading to delays in obtaining results. Frequent sampling may be cost-prohibitive, leading to gaps in data collection and potential issues in capturing short-term fluctuations in water quality. Traditional methods require substantial resources, including personnel, equipment, and laboratory facilities.

Traditional methods, when conducted properly, can provide highly accurate and precise measurements, especially for specific parameters. Laboratory-based analysis allows for a detailed assessment of various water quality parameters, including nutrients, metals, and contaminants. Ground-based monitoring provides a site-specific understanding of local conditions and facilitates targeted interventions.

6.9.2 Remote Sensing for Water-Monitoring Techniques

Satellite or aerial imagery may have limitations in capturing fine-scale spatial and spectral details needed for certain water quality parameters. Cloud cover and atmospheric conditions can limit the availability of clear satellite images, affecting the continuity of remote sensing data. Ensuring the accuracy of remote sensing data requires rigorous calibration and validation processes, often involving ground-based measurements. Interpreting remote sensing data requires specialized knowledge and tools, and it may be challenging to distinguish between different water quality indicators.

Remote sensing enables large-scale coverage, allowing for the monitoring of extensive water bodies and remote areas that are difficult to access. Satellite-based systems can provide real-time or near-real-time data, facilitating continuous monitoring and the detection of rapid changes. Remote sensing can be cost-effective for monitoring large geographical areas, offering a more economical approach compared to extensive ground-based sampling. Advances in technology provide multispectral and hyperspectral data, allowing for the assessment of various water quality parameters, including chlorophyll concentration and turbidity.

In practice, an integrated approach that combines the strengths of both traditional and remote sensing methods is often the most effective for comprehensive water monitoring. Leveraging the strengths of each method helps overcome their respective challenges and provides a more holistic understanding of water quality and dynamics.

6.10 SUMMARY

Chilika lagoon is the largest lagoon in Asia and a Ramsar site that has garnered international attention for studying and understanding ecosystem processes for their sustainable management. To safeguard the food value of Chilika fish, monitoring trace metal and petroleum hydrocarbon contamination should be added to basic water quality parameters, which are to be monitored periodically due to changing environmental conditions over time. Apart from pollution studies, the eutrophication status, nutrient biogeochemistry, and its flux in the Chilika lagoon also need to be monitored. Scientists are on the verge of monitoring the water quality of the Chilika lagoon utilizing advanced technologies in coordination with conventional monitoring methods. The conventional method helps monitor the lagoon in the long run but has certain limitations such as the requirement of manpower, laboratory setup, and funding for management. Although remote sensing cannot replace the conventional method, it can be used more frequently to collect data when field campaigns cannot be executed due to bad weather, windy conditions, or pandemic situations (like COVID-19). The application of remote sensing on Chilika lagoon is currently scant, and further studies on this aspect are essential. Even though only three parameters, Chl-a, TSS, and CDOM, are derived for Chilika lagoon, additional parameters such as nutrients and eutrophication status can also be derived through remote sensing using long-term data collected through conventional methods. These can be derived by coupling the monitoring data with modeling coefficients either obtained from literature or from field and lab experiments. This approach will be helpful to know the trend and the possible reasons underlying various ecosystem processes over different times and spaces. In a nutshell, the remote sensing approach, combined with the help of modeling and conventional data, would help scientists to monitor the health of the lagoon and formulate an action plan for sustainable management.

ACKNOWLEDGEMENT

The authors are grateful to the CDA for providing the required data, technical information, and GIS map of Chilika lagoon.

REFERENCES

Acharya, P., Muduli, P.R., Mishra, D.R., et al. 2022. Imprints of COVID-19 lockdowns on total petroleum hydrocarbon levels in Asia's largest brackish water lagoon. *Marine Pollution Bulletin,* 174, 113137. https://doi.org/10.1016/j.marpolbul.2021.113137.

Ali, S.Y., Gupta, A., Raman, M., Sahay, A., Motwani, G., Muduli, P.R., Krishna, A.V., Tirkey, A. 2021. Improved estimates of bio-optical parameters in optically complex water using hyperspectral remote sensing data. *International Journal of Remote Sensing*, 42(8), 3056–3073. https://doi.org/10.1080/01431161.2020.1865585.

Anderson, D.M., Glibert, P.M., Burkholder, J.M. 2002. Harmful algal blooms and eutrophication: Nutrient sources, composition, and consequences. *Estuaries*, 25, 704–726.

Arhonditsis, G., Karydis, M. and Tsirtsis, G., 2003. Analysis of phytoplankton community structure using similarity indices: a new methodology for discriminating among eutrophication levels in coastal marine ecosystems. *Environmental Management*, 31, 0619–0632.

Barik, S.K., et al. 2017. Spatio-temporal variability and the impact of Phailin on water quality of Chilika lagoon. *Continental Shelf Research,* 136, 39–56.

Blindow, I., Andersson, G., Hargeby, A. and Johansson, S., 1993. Long-term pattern of alternative stable states in two shallow eutrophic lakes. *Freshwater biology*, 30(1), 159–167. https://doi.org/10.1111/j.1365-2427.1993.tb00796.x

Buschmann, C., Langsdorf, G., Lichtenthaler, H. 2000. Imaging of the blue, green, and red fluorescence emission of plants: An overview. *Photosynthetica,* 38, 483–491.

Chander, S., Gujrati, A., Abdul Hakeem, K., et al. 2019. Water quality assessment of River Ganga and Chilika lagoon using AVIRIS-NG hyperspectral data. *Current Science,* 116, 1172–1181. https://doi.org/10.18520/cs/v116/i7/1172-1181.

Chen, B., Nam, S. N., Westerhoff, P. K., Krasner, S. W., & Amy, G. 2009. Fate of effluent organic matter and DBP precursors in an effluent-dominated river: A case study of wastewater impact on downstream water quality. *Water Research*, 43(6), 1755–1765. https://doi.org/10.1016/j.watres.2009.01.020

Chen, J.L., Wilson, C.R., Tapley, B.D. 2010. The 2009 exceptional Amazon flood and interannual terrestrial water storage change observed by GRACE. *Water Resources Research*, 46(12).https://doi.org/10.1029/2010WR009383

Chen, Z.H., Fang, H.W. and Liu, B., 2007. Numerical simulation of wind-induced motion in suspended sediment transport. *Journal of Hydrodynamics, 19(6),* 698–704.

Cortes-Francisco, N., et al. 2014. High-field FT-ICR mass spectrometry and NMR spectroscopy to characterize DOM removal through a nanofiltration pilot plant. *Water Research*, 67, 154–165.

Dey, A. K., Sharma, M., & Meshram, M. R. 2016. An analysis of leaf chlorophyll measurement method using chlorophyll meter and image processing technique. *Procedia computer science, 85,* 286–292.

Doell, D.L., Folmer, D.E., Lee, H.S., Butts, K.M., Carberry, S.E. 2016. Exposure estimate for FD&C colour additives for the US population. *Food Additives & Contaminants: Part A*, 33(5), 782–797. https://doi.org/10.1080/19440049.2016.1179536.

Friedman, J. M., Hunt Jr, E. R., & Mutters, R. G. 2016. Assessment of leaf color chart observations for estimating maize chlorophyll content by analysis of digital photographs. *Agronomy Journal, 108(2),* 822–829.

Ganguly, D., et al. 2015. Influence of nutrient input on the trophic state of a tropical brackish water lagoon. *Journal of Earth System Science,* 124(5), 1005–1017.

Garg, V., et al. 2017. Spectral similarity approach for mapping turbidity of an inland waterbody. *Journal of Hydrology,* 550, 527–537.

Garg, V., Dhote, P.R., Muduli, P.R., Thakur, P.K., Nikam, B.R., Aggarwal, S.P. 2024. Spatial spectral contextual image analysis approach to map water quality of inland waterbody using AVIRIS-NG datasets. *Advances in Space Research*, 73(2), 1549–1559. https://doi.org/10.1016/j.asr.2022.09.012.

Gleason, C.J., Durand, M.T. 2020. Remote sensing of river discharge: A review and a framing for the discipline. *Remote Sensing,* 12, 1107. https://doi.org/10.3390/rs12071107.

Gordon, H.R., Brown, O.B., Evans, R.H., Brown, J.W., Smith, R.C., Baker, K.S., Clark, D.K. 1988. A semi–analytic radiance model of ocean color. *Journal of Geophysical Research: Atmospheres*, 93, 10909–10924.

Goswami S, Samal R.N., Choudhury S.B. Coastal ecosystem monitoring using long-term satellite data records: A case study of Chilika Lake, Odisha. PeerJ Preprints; 2018. DOI: 10.7287/peerj.preprints.27448v1

Gower, J., King, S., Borstad, G., & Brown, L. 2005. Detection of intense plankton blooms using the 709 nm band of the MERIS imaging spectrometer. *International Journal of Remote Sensing, 26(9),* 2005–2012.

Grasshoff, K., et al. 1999. *Methods of Sea Water Analysis*, third ed. Verlag Chemie, Weinheim, 89–224.

Grimaldi, S., Li, Y., Pauwels, V.R.N., Walker, J.P. 2016. Remote sensing-derived water extent and level to constrain hydraulic flood forecasting models: Opportunities and challenges. *Surveys in Geophysics*, 37, 977–1034.

Gupta et al. 2008. Influence of net ecosystem metabolism in transferring riverine organic carbon to atmospheric CO_2 in a tropical coastal lagoon, Chilika Lake, India. *Biogeochemistry,* 87, 265–285.

Gurlin, D., Gitelson, A.A., Moses, W.J. 2011. Remote estimation of chl-a concentration in turbid productive waters—Return to a simple two-band NIR-red model? *Remote Sensing of Environment*, 115, 3479–3490.

Gupta D, Snehasish & Ibaraki, Yasuomi & Pattanayak, A. 2012. Development of a digital image analysis method for real-time estimation of chlorophyll content in micropropagated potato plants. Plant Biotechnology Reports. 7. https://doi.org/10.1007/s11816-012-0240-5.

Gupta, M. 2013. Chromaticity analysis of the Chilika lagoon for total suspended sediment estimation using RESOURCESAT-1 AWiFS data - A case study. *Journal of Great Lakes Research*, 39(4), 696–700. https://doi.org/10.1016/j.jglr.2013.09.021

Janse, J., Scheffer, M., Lijklema, L., Van Liere, L., Sloot, J., Mooij, W. 2010. Estimating the critical phosphorus loading of shallow lakes with the ecosystem model PCLake: Sensitivity, calibration and uncertainty. *Ecological Modelling*, 221(4), 654–665. https://doi.org/10.1016/j.ecolmodel.2009.07.023.

Jiang, T., et al. 2018. Influences of the alternation of wet-dry periods on the variability of chromophoric dissolved organic matter in the water level fluctuation zone of the Three Gorges Reservoir area, China. *Science of The Total Environment,* 636, 249–259.

Kalle, K. 1949. Fluoreszenz und Gelbstoff im Bottnischen und Finnischen Meerbusen. *Deutsche Hydrographische Zeitschrift,* 2, 117–124. https://doi.org/10.1007/BF02225972.

Kanuri, V.V., et al. 2013. Plankton metabolic processes and its significance on dissolved organic carbon pool in tropical brackish water lagoon. *Continental Shelf Research,* 61–62, 52–61.

Kanuri, V.V., et al. 2018. Bioavailable dissolved organic matter and its spatio-temporal variation in a river dominated tropical brackish water Lagoon, India. *Marine Pollution Bulletin,* 131, 460–467.

Kasprzak, P., Padisák, J., Koschel, R., Krienitz, L., Gervais, F. 2008. Chlorophyll a concentration across a trophic gradient of lakes: An estimator of phytoplankton biomass? *Limnologica*, 38(3–4), 327–338. https://doi.org/10.1016/j.limno.2008.07.002.

Kowalczuk, P., Zabocka, M., Sagan, S.A., Kuli´nski, K. 2010. Fluorescence measured in situ as a proxy of CDOM absorption and DOC concentration in the Baltic Sea. *Oceanologia*, 52(3), 331–344.

Kumar, A., Equeenuddin, S.M., Mishra, D.R. 2020. Long-term analysis of water quality in Chilika Lagoon and application of bio-optical models for cyclone impact Assessment. In Finlayson, C., Rastogi, G., Mishra, D., Pattnaik, A. (eds) *Ecology, Conservation, and Restoration of Chilika Lagoon, India. Wetlands: Ecology, Conservation and Management*, vol 6. Springer, Cham. https://doi.org/10.1007/978-3-030-33424-6_8.

Kutser, T., Verpoorter, C., Paavel, B., Tranvik, L.J. 2015. Estimating lake carbon fractions from remote sensing data. *Remote Sensing of Environment*, 157, 138–146.

Lettenmaier, D. P., & Famiglietti, J. S. 2006. Water from on high. *Nature, 444(7119),* 562–563. https://doi.org/10.1038/444562a

Lillesand, T.M., W.L. Johnson, R.L. Deuell, O.M. Lindstrom and D. E. Meisner. 1983. Use of Landsat data to predict the trophic state of Minnesota lakes. Photogram. Eng. Rem. S. 49, 219–229.

Lin, P., Pan, M., Beck, H.E., Yang, Y., Yamazaki, D., Frasson, R., David, C.H., Durand, M., Pavelsky, T.M., Allen, G.H., et al. 2019. Global reconstruction of naturalized river flows at 2.94 million reaches. *Water Resources Research*, 55, 6499–6516.

Liu, L.Z., Huang, Q., Zhang, Y.L., Qin, B.Q., Zhu, G.W. 2017. Excitation-emission matrix fluorescence and parallel factor analyses of the effects of N and P nutrients on the extracellular polymeric substances of Microcystis aeruginosa. *Limnologica,* 63, 18–26.

Maeda, E.E., Lisboa, F., Kaikkonen, L., Kallio, K., Koponen, S., Brotas, V., Kuikka, S. 2019. Temporal patterns of phytoplankton phenology across high latitude lakes unveiled by long-term time series of satellite data. *Remote Sensing of Environment*, 221, 609–620.

Martin, S., Soranno, P., Bremigan, M., Cheruvelil, K. 2011. Comparing hydrogeomorphic approaches to lake classification. *Environmental Management,* 48(5), 957–974. https://doi.org/10.1007/s00267-011-9740-2.

Matthews, M.W. 2011. A current review of empirical procedures of remote sensing in inland and near–coastal transitional waters. *International Journal of Remote Sensing,* 32, 6855–6899.

Miller, J.D., Schoonover, J.E., Williard, K.W. and Hwang, C.R., 2011. Whole catchment land cover effects on water quality in the lower Kaskaskia River watershed. *Water, Air, & Soil Pollution, 221,* 337–350.

Mishra, D.R., Kumar, A., Muduli, P.R., Acharyya, T., Acharya, P., Singh, S. and Rastogi, G., 2021. Landfall season is critical to the impact of a cyclone on a monsoon-regulated tropical coastal lagoon. Science of the Total Environment, 770, p.145235.

Mishra, D.R., Kumar, A., Muduli, P.R., et al. 2020. Decline in phytoplankton biomass along Indian coastal. *Remote Sensing,* 12, 2584. https://doi.org/10.3390/rs12162584.

Mohanty, B., et al. 2017. Sources and variability of petroleum hydrocarbon residues in sediments of Chilika Lagoon, East Coast of India. *Bulletin of Environmental Contamination and Toxicology,* 99(1), 100–107.

Mohanty, P.K., Pal, S.R. 2001. November. IRS data utilization for estimation of turbidity level in Chilika lagoon, east coast of India. In *Proceedings of 22nd Asian Conference on Remote Sensing* (pp. 5–9).

Morel, A., Prieur, L. 1977. Analysis of variations in ocean color. *Limnology and Oceanography,* 22, 709–722.

Muduli, P.R., et al. 2011. Spatial variation of hydrochemical characteristics in and around Port Blair Bay Andaman and Nicobar Islands, India. *World Applied Sciences Journal,* 13(3), 564–571.

Muduli, P.R., et al. 2012. Spatio-temporal variation of CO_2 emission from Chilika Lake, a tropical coastal lagoon, on the east coast of India. *Estuarine, Coastal and Shelf Science,* 113, 305–313.

Muduli, P.R., et al. 2013. Distribution of dissolved inorganic carbon and net ecosystem production in a tropical brackish water lagoon, India. *Continental Shelf Research,* 64, 75–87.

Muduli, P.R., et al. 2017. The impact of tropical cyclone 'Phailin' on the hydrology of Chilika Lagoon, India. *International Journal of Environmental Sciences & Natural Resources,* 4(2), 555632. https:doi.org/10.19080/IJESNR.2017.04.555632.

Muduli, P.R., Pattnaik, A. 2020. Spatio-temporal variation in physicochemical parameters of water in the Chilika Lagoon. https:doi.org/10.1007/978-3-030-33424-6_9.

Mukherjee, R., et al. 2018. Spatial variation of nitrogen uptake rates in the largest brackish water lagoon of Asia (Chilika, India). *Estuarine, Coastal and Shelf Science.* https://doi.org/10.1016/j.ecss.2018.01.012 (In press).

Naik, B.C., Anuradha, P.B. 2018. Water body extraction using images from two different satellites over Chilika Lake. *International Journal of Electronics Engineering, 10,* 140–145.

Nechad, B., Ruddick, K.G., Park, Y. 2010. Calibration and validation of a generic multisensor algorithm for mapping of total suspended matter in turbid waters. *Remote Sensing of Environment,* 114, 854–866.

Niroumand-Jadidi, M., Bovolo, F., Bruzzone, L. 2019. Novel spectra-derived features for empirical retrieval of water quality parameters: Demonstrations for OLI, MSI, and OLCI sensors. *IEEE Transactions on Geoscience and Remote Sensing*, 57, 10285–10300.

O'Reilly, J.E., Werdell, P.J. 2019. Chlorophyll algorithms for ocean color sensors—OC4, OC5 & OC6. *Remote Sensing of Environment*, 229, 32–47.

Pal, S. R., & Mohanty, P.K. 2002. Use of IRS-1B data for change detection in water quality and vegetation of Chilka Lagoon, East Coast of India. *International Journal of Remote Sensing, 23(6),* 1027–1042. https://doi.org/10.1080/01431160110076243

Parida, S., et al. 2017. Trace metal concentrations in euryhaline fish species from Chilika lagoon: human health risk assessment. *International Journal of Environmental Science and Technology,* 14(12), 2649–2660.

Patra, S., et al. 2016. Influence of suspended particulate matter on the nutrient biogeochemistry of a tropical shallow lagoon, Chilika, India. *Limnology*, 17(3), 223–238.

Patra, S., et al. 2017. Isotopic composition (C & N) of the suspended particles and N uptake by phytoplankton in a shallow tropical coastal lagoon. *Chemistry and Ecology,* 33(8), 708–724.

Poikāne, S., Alves, M., Argillier, C., van den Berg, M., Buzzi, F., Hoehn, E., de Hoyos, C., Karottki, I., Laplace-Treyture, C., Solheim, A., Ortiz-Casas, J., Ott, I., Phillips, G., Pilke, A., Pádua, J., Remec-Rekar, S., Riedmüller, U., Schaumburg, J., Serrano, M., Soszka, H., Tierney, D., Urbanič, G., Wolfram, G. 2010. Defining chlorophyll-a reference conditions in European Lakes. *Environmental Management,* 45(6), 1286–1298. https://doi.org/10.1007/s00267-010-9484-4.

Porra, R. J., Thompson, W. A., & Kriedemann, P. E. (1989). Determination of accurate extinction coefficients and simultaneous equations for assaying chlorophylls a and b extracted with four different solvents: verification of the concentration of chlorophyll standards by atomic absorption spectroscopy. *Biochimica et Biophysica Acta (BBA)-Bioenergetics*, 975(3), 384–394.

Rao, C.R.N., Stowe, I.L., Mcclain, E.P. 1989. Remote sensing of aerosols over the oceans using avhrr data theory, practice and applications. *International Journal of Remote Sensing*, 10(4–5), 743–749. https://doi.org/10.1080/01431168908903915.

Rao, D. P., S.K. Subramanian, and R. Sudarshana. 1986 "Geomorphic analysis of Chilka lake and adjoining area using remotely sensed data." *Journal of the Indian Society of Remote Sensing* 14, 53–61.

Rignon, J.P.G., Capuani, S., Fernandes, D.M., Guimarães, T.M. 2016. A novel method for the estimation of soybean chlorophyll content using a smartphone and image analysis. *Photosynthetica*, 54, 559–566. https://doi.org/10.1007/s11099-016-0214-x.

Robin, R.S., et al. 2016. CO_2 saturation and trophic shift induced by microbial metabolic processes in a river-dominated ocean margin (tropical shallow lagoon, Chilika, India). *Geomicrobiology Journal,* 33(6), 513–529.

Sahay, A., Gupta, A., Motwani, G., et al. 2019. Distribution of coloured dissolved and detrital organic matter in optically complex waters of Chilika lagoon, Odisha, India, using hyperspectral data of AVIRIS-NG. *Current Science,* 116, 1166–1171. https://doi.org/10.18520/cs/v116/i7/1166-1171.

Sarma, V.V.S.S., et al. 2010. Influence of river discharge on plankton metabolic rates in the tropical monsoon-driven Godavari estuary, India. *Estuarine, Coastal and Shelf Science*, 85, 515–524.

Scheffer. 1989. Alternative stable states in eutrophic, shallow freshwater systems: A minimal model, Hydrobiol. Bull., 23, pp. 73–83.

Sivapalan, M., Blöschl, G., Zhang, L., & Vertessy, R. 2003. Downward approach to hydrological prediction. *Hydrological processes, 17(11),* 2101–2111.

Song, K., et al. 2019. Characterization of CDOM in saline and freshwater lakes across China using spectroscopic analysis. *Water Research*, 150, 403–417.

Song, K.S., Zhao, Y., Wen, Z.D., Fang, C., Shang, Y.X. 2017. A systematic examination of the relationships between CDOM and DOC in inland waters in China. *Hydrology and Earth System Sciences,* 21(10), 5127–5141.

Soomets, T., Uudeberg, K., Jakovels, D., Brauns, A., Zagars, M., Kutser, T. 2020. Validation and comparison of water quality products in baltic lakes using sentinel-2 msi and sentinel-3 OLCI data. *Sensors,* 20, 742.

Spencer, R. G. M., Aiken, G.R., Butler, K. D., Dornblaser, M. M., Striegl, R. G., & Hernes, P. J. 2009. *Utilizing chromophoric dissolved organic matter measurements to derive export and reactivity of dissolved organic carbon exported to the Arctic Ocean : A case study of the Yukon River, Alaska. 36,* 1–6. https://doi.org/10.1029/2008GL036831

Srichandan, S., et al. 2015. Interannual and cyclone drove variability in phytoplankton communities of a tropical coastal lagoon. *Marine Pollution Bulletin,* 101(1), 39–52.

Su, C.-H., Fu, C.-C., Chang, Y.-C., Nair, G.R., Ye, J.-L., Chu, I.-M., et al. 2008. Simultaneous estimation of chorophyll a and lipid contents in microalgae by three-color analysis. *Biotechnology and Bioengineering,* 99, 1034–1039. https://doi.org/10.1002/bit.21623.

Sudhakar, S., Pal, D.K. 1993. Water quality assessment of the lake chilka. *International Journal of Remote Sensing*, 14(14), 2575–2579. https://doi.org/10.1080/01431169308904294.

Tarafdar, L., Kim, J.Y., Srichandan, S., et al. 2021. Responses of phytoplankton community structure and association to variability in environmental drivers in a tropical coastal lagoon. *Science of the Total Environment,* 783, 146873. https://doi.org/10.1016/j.scitotenv.2021.146873.

Tehrani, N., Sa, E., Osburn, C., Bianchi, T., Schaeffer, B. 2013. Chromophoric dissolved organic matter and dissolved organic carbon from sea–viewing wide field–of–view sensor (SEAWIFS), moderate resolution imaging spectroradiometer (MODIS) and meris sensors: Case study for the northern gulf of mexico. *Remote Sensing,* 5, 1439–1464.

Tripathi, G., Pandey, A.C., Parida, B.R. 2022. Flood hazard and risk zonation in North Bihar using satellite-derived historical flood events and socio-economic data. *Sustainability,* 14, 1472. https://doi.org/10.3390/su14031472.

Vincent, W. F. 1988. Microbial Ecosystems of Antarctica. Cambridge University Press, Cambridge, pp.304.

Wang, H., Zhu, R., Zhang, J., Ni, L., Shen, H., Xie, P. 2018. A novel and convenient method for early warning of algal cell density by chlorophyll fluorescence parameters and its application in a highland lake. *Frontiers in Plant Science,* 9, 869. https://doi.org/10.3389/fpls.2018.00869.

Ward, N.D., Bianchi, T.S., Medeiros, P.M., Seidel, M., Richey, J.E., Keil, R.G., & Sawakuchi, H.O. (2017). Where carbon goes when water flows: carbon cycling across the aquatic continuum. *Frontiers in Marine Science*, 4, 7.

Xiao, Y., Ferreira, J., Bricker, S., Nunes, J., Zhu, M., Zhang, X. 2007. Trophic assessment in Chinese coastal systems-review of methods and application to the Changjiang (Yangtze) Estuary and Jiaozhou Bay. *Estuaries and Coasts,* 30(6), 901–918. https://doi.org/10.1007/bf02841384.

Xu, J., Wang, Y., Gao, D., Yan, Z., Gao, C., & Wang, L. (2017). Optical properties and spatial distribution of chromophoric dissolved organic matter (CDOM) in Poyang Lake, China. *Journal of Great Lakes Research*, 43(4), 700–709.

Yao, X., et al. 2011. Resolving the variability of CDOM fluorescence to differentiate the sources and fate of DOM in Lake Taihu and its tributaries. *Chemosphere,* 82(2), 145–155.

Ying, Z., et al. 2018. Relationship changes between CDOM and DOC in the Songhua River affected by highly polluted tributary, Northeast China. *Environmental Science and Pollution Research International*, 5, 1–12.

Zhao, H., Chen, Q., Walker, N.D., Zheng, Q. and MacIntyre, H.L., 2011. A study of sediment transport in a shallow estuary using MODIS imagery and particle tracking simulation. *International Journal of Remote Sensing, 32(21),* 6653-–6671.

Zhang, Y., et al. 2010. Characteristics and sources of chromophoric dissolved organic matter in lakes of the Yungui Plateau, China, differing in trophic state and altitude. *Limnology and Oceanography,* 55(6), 2645–2659.

Zhang, Y., et al. 2021. Chromophoric dissolved organic matter in inland waters: Present knowledge and future challenges. *Science of the Total Environment,* 759, 143550.

Zhao, Y., Song, K.S. 2018. Relationships between DOC and CDOM based on the total carbon-specific fluorescence intensities for river waters across China. *Journal of Geophysical Research: Biogeosciences,* 123(8), 2353–2361.

Zhou, L., et al. 2017. Hydraulic connectivity and evaporation control the water quality and sources of chromophoric dissolved organic matter in Lake Bosten in arid northwest China. *Chemosphere,* 188, 608–617.

Zhou, Y., Jeppesen, E., Zhang, Y., Shi, K., Liu, X., & Zhu, G. 2016. Dissolved organic matter fluorescence at wavelength 275/342 nm as a key indicator for detection of point-source contamination in a large Chinese drinking water lake. *Chemosphere,144, 503–509.https://doi.org/https://doi.org/10.1016/j.chemosphere.2015.09.027*

7 Stream Channel Dynamics over Alluvial Fan Systems

An Overview

Manish Pandey
University Center for Research & Development (UCRD), Chandigarh University, Mohali, India, Department of Civil Engineering, Chandigarh University, Mohali, India

*Pinkey Bisht**
Wadia Institute of Himalayan Geology, Dehradun, Uttarakhand, India

7.1 INTRODUCTION

Alluvial fans are depositional environments found at the base/foothill of mountain ranges where sediment-laden streams emerge from steep slopes and spread out onto flatter terrain, characterizing subaerial plains/lacustrine plains/submarine plains (Bull 1977; Stanistreet and McCarthy 1993; Shanmugam 2016). The size of the terrestrial alluvial fan surface area ranges from a few square kilometers to more than thousands of square kilometers (León et al. 2016). Alluvial fan volume as a morphometric parameter is a widely accepted metric of fan size (Giles 2010). Hartley et al. (2010) classified fan-shaped sedimentary bodies, ranging from 1 to 700 km in length and predominantly shaped by fluvial and alluvial processes, as distributive fluvial systems (DFS) across three scales: (1) large-scale fluvial megafans, exceeding 100 km in radius and covering an area of 1000–100,000 km^2; (2) small-scale alluvial fans, with a radius of less than 30 km and an area smaller than 100 km^2; and (3) medium-scale fluvial fans, falling between the dimensions of the former two. The introduction of the DFS concept sparked significant research interest in fluvial fan dynamics, albeit accompanied by contentious debates. Subsequent discussions led some scholars to adopt "fluvial fan" as the preferred terminology for such formations post-2018. Geomorphologically, it is a triangular or fan-shaped geological feature formed from deposits of loose sediment (alluvium) carried by rivers or streams as they flow out of mountainous or hilly terrain onto a flatter plain (Ventra and Clarke 2018). The fan-shaped deposit is created as the flowing water loses energy and drops its sediment load due to reduced velocity and the sudden reduction slope gradient at the exit point of the valley or mountainous terrain (Anstey 1965; Kain et al. 2018). This sediment, which can consist of gravel, sand, and finer particles, spreads out in a fan-like pattern as it settles on the lower, more gently sloping land (Blair and McPherson 1994; Rees et al. 2014).

7.1.1 Definition of Alluvial Fan Systems

Alluvial fans (AF) (Figure 7.1a) differ from alluvial fan systems (AFS) (Figures 7.1b–d) (Pandey 2014). Harvey (2018) defines an alluvial fan as "sedimentary bodies of broadly conical form, deposited where transporting power suddenly decreases, as at mountain-front or tributary junction sites." We can say that an alluvial fan is an accumulation of sediments that spreads outwards from a

* Corresponding author.

 DOI: 10.1201/9781003354000-10

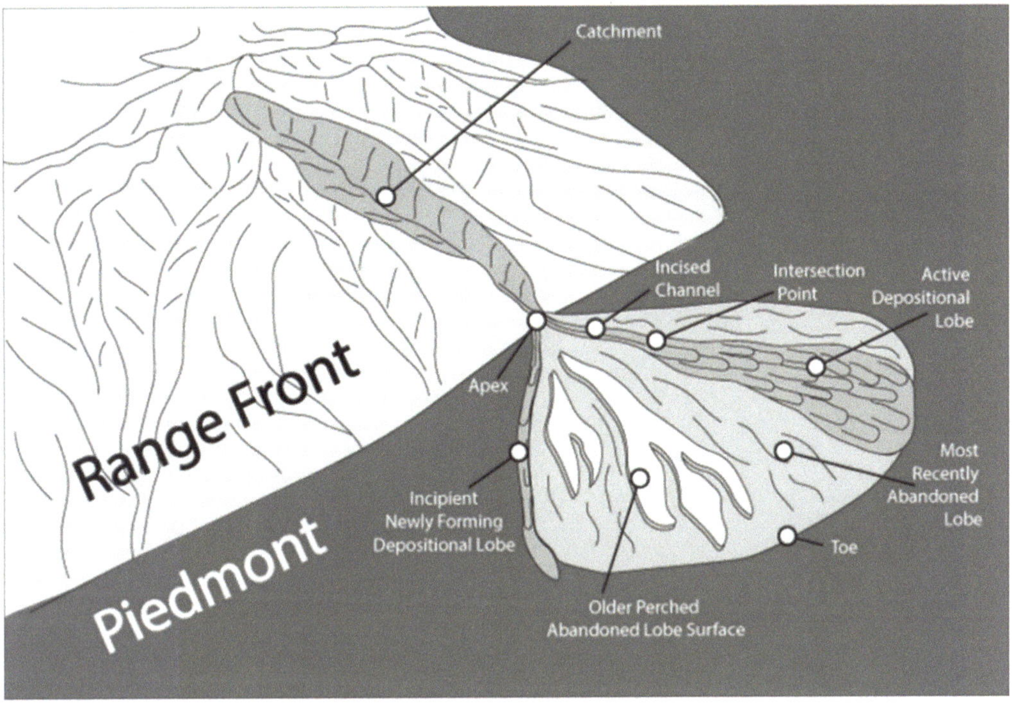

FIGURE 7.1A Sketch of a subaerial alluvial fan with its different components. Source: Iacobucci et al. (2024).

concentrated source, like a narrow canyon, and is characteristic of mountainous terrain in arid to semiarid climates (see Figures 7.1a–d). These fans range in size and form where flow emerges from a confined channel, reducing its carrying capacity and leading to sediment deposition (French et al. 1993). On the other hand, an alluvial fan system refers to the broader context of alluvial fans within a landscape. It encompasses the interconnected network of alluvial fans, their channels, and associated depositional features (Chant et al. 1999). Alluvial fan systems are influenced by factors like tectonic uplift, climatic changes, and variations in stream discharge, debris load, and surface gradient. They play a crucial role in sediment transfer from mountains to adjacent basins and can act as groundwater reservoirs in arid regions.

7.1.2 Importance of Understanding Stream Channel Dynamics over the AFS

Understanding stream channel dynamics or behavioral shifts in stream channel morphology, channel sediment load variability, avulsive behavior, etc., will help geomorphologists, hydrologists, sedimentologists, ecologists, and hazard managers and policymakers in the following manner:.

Understanding stream channel dynamics helps in comprehending the hydrological processes that govern sediment transport, erosion, and deposition within these systems (Scheinert et al. 2012; Ventra and Clarke 2018). This knowledge is essential for predicting flood risks, managing water resources, and designing infrastructure such as bridges and culverts. Stream channels within alluvial fans are the primary conduits for transporting sediment from upstream sources to downstream areas. Changes in stream channel dynamics, such as channel avulsion or aggradation, can affect sediment transport patterns and the distribution of sediment across the landscape (Scheinert et al. 2012). This has implications for landscape evolution, sediment budgets, and the formation of geological features. Stream channel behavior shifts over the alluvial fan surfaces over time help in

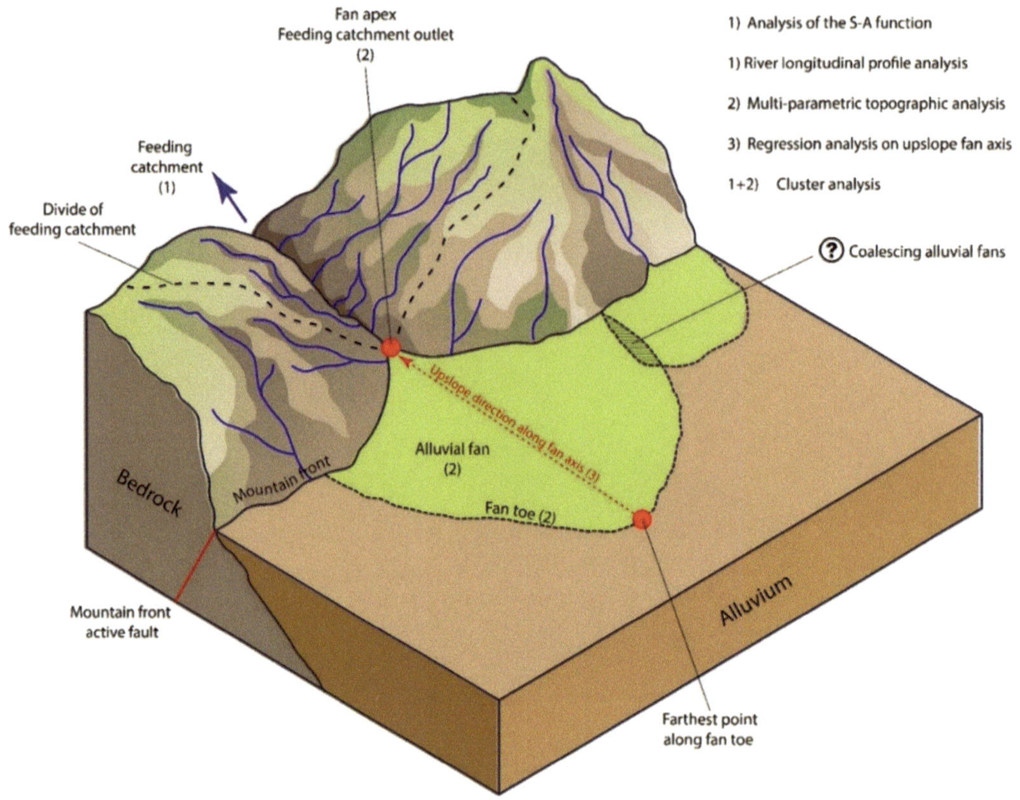

FIGURE 7.1B Sketch of a subaerial alluvial fan with its different components. Source: Iacobucci et al. (2024).

understanding the interplay of various geomorphological processes, including erosion, deposition, and channel migration. Understanding stream channel dynamics is essential for deciphering the geomorphological evolution of alluvial fan systems and the factors driving landscape change over different timescales (Han et al. 2023). Unique ecosystems flourishing over alluvial fans affect and get affected by stream channel dynamics (Lane et al. 2016). These ecosystems provide habitat for various species of plants and animals. Understanding stream channel dynamics helps in assessing the impact of environmental changes, such as land use alterations or climate change, on the ecological functioning of alluvial fan systems. Alluvial fan systems are often inhabited or utilized by human communities for various purposes such as agriculture, urban development, and recreation (Pandey 2014; Pandey et al. 2020). Understanding stream channel dynamics is critical for assessing and mitigating natural hazards such as flash floods and debris flows, which can pose risks to human life and infrastructure. Additionally, effective management of water resources and land use planning in alluvial fan areas (Liu 2004) requires a thorough understanding of stream channel dynamics.

7.1.3 Objectives of the Chapter

In this work, we present an overview of the alluvial and fluvial fan systems vis-à-vis stream channel dynamics over the fan systems leading to their evolution in subaerial and submarine environmental settings. The control factors influencing the alluvial fan system evolution are also outlined. Various processes working on the fan surfaces and alluvial fan system surfaces are also described. We also discuss different methods of studying stream channel dynamics over the alluvial fan

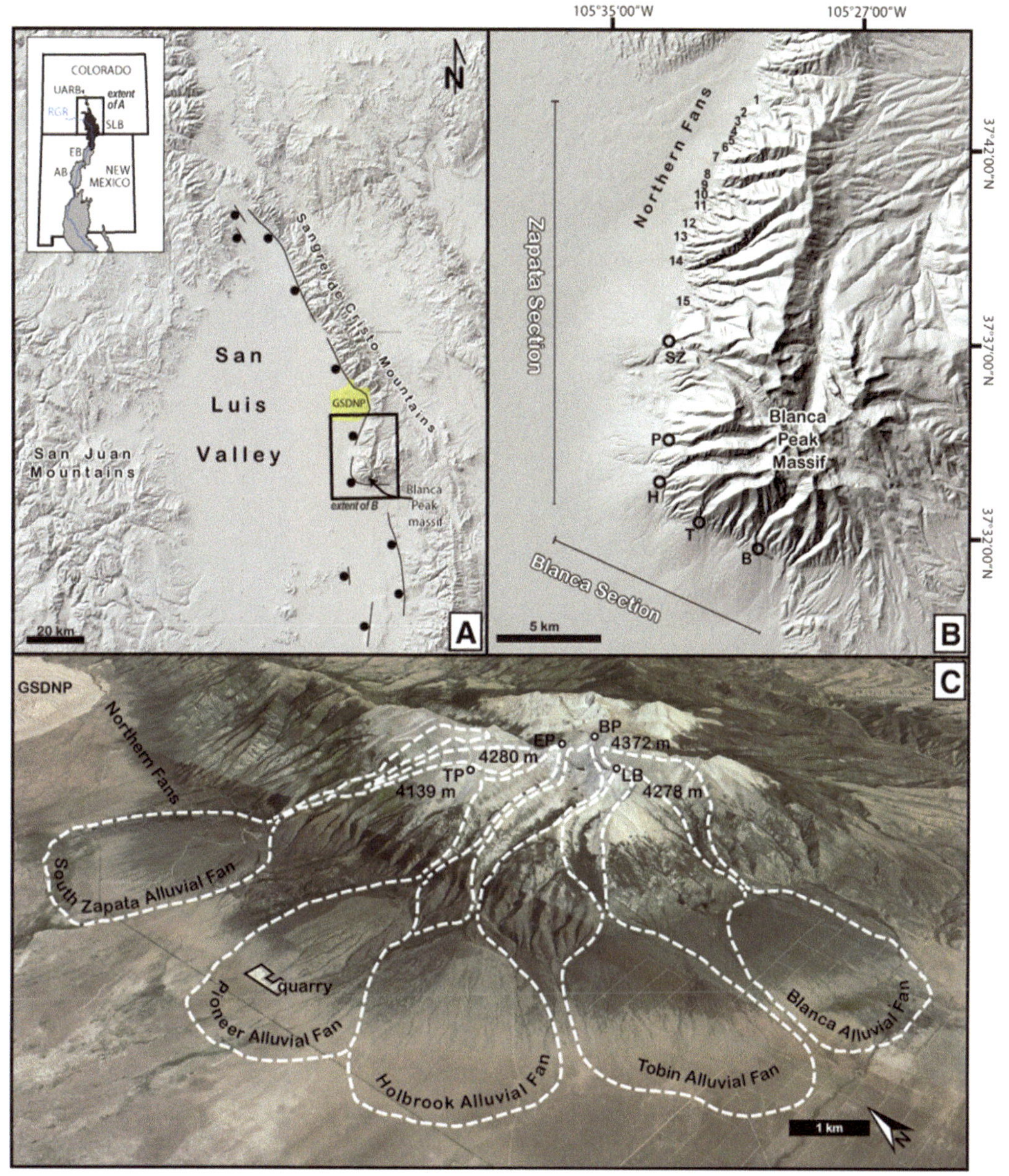

FIGURE 7.1C Sketch of a subaerial alluvial fan with its different components. Part A shows the location of the study area and how faults intersperse the mountain front; Part B zooms in to the boxed portion of the study area to portray a clear morphological perspective of the alluvial fans at a large scale; and Part C demonstrates alluvial fan boundaries marking the alluvial fan system. For details of the figure, consult the source work. Source: Nicovich et al. (2023).

surfaces, i.e., field-based methods, lab-based and mathematical modeling-based approaches, and also remote sensing-based approaches that works as complementary and supplementary, as well as independent methods to study landscape evolution that describe channel dynamics over the fans. Brief limitations of each of these approaches are also mentioned. And finally we identify the future directions in alluvial fan research and unanswered questions that will guide future research in this discipline.

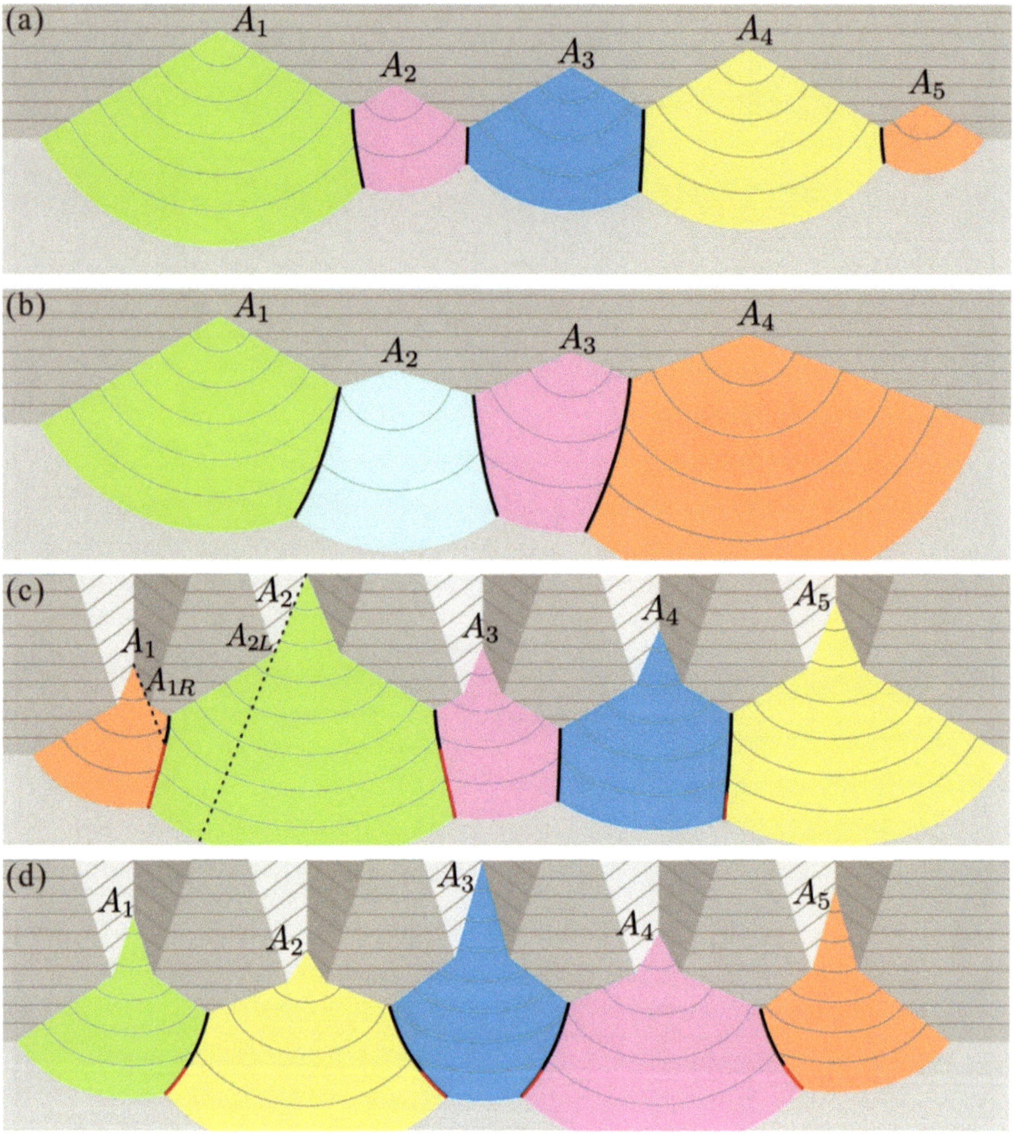

FIGURE 7.1D Sketch of a subaerial alluvial fan with its different components. The coalescing fans make up an alluvial fan system. Source: Chen and Capart (2022).

7.2 FORMATION AND CHARACTERISTICS OF ALLUVIAL FAN SYSTEM

7.2.1 Surface Processes Shaping Alluvial Fans

The evolution of alluvial fans is a complex interplay of deposition, erosion, and reorganization of sediment by various surface processes. Here's an extensive list:

7.2.1.1 Sediment Deposition

- **Streamflow Decrease:** As a stream exits a confined mountain channel and spreads out onto a flat surface, its velocity drops (Parker et al. 1998). This reduced capacity leads to the deposition of sediment sorted by size (Zhang et al. 2020).

1. **Proximal Fan:** Coarser materials like boulders and cobbles settle near the apex (Mack and Rasmussen 1984) due to their higher settling velocity.
2. **Medial Fan:** Sands and gravels are deposited further out as the stream loses energy (Blair and McPherson 1994; De Haas et al. 2014).
3. **Distal Fan:** Silts and clays, the finest materials, travel the farthest and settle on the outer edges of the fan (Blair 1987).

According to a sedimentation model (Koundi et al. 2022; Lee and Gihm 2023), alluvial fans and alluvial fan systems exhibit three distinct zones. Firstly, the proximal zone, located nearest to the sediment source, typically at the fan apex, features a steep slope and coarse sediment, often gravel, with feeder trunks branching into parallel channels. Secondly, the medial zone is characterized by bifurcating channels, parallel flow paths, and overbank sheet flow areas, with a less steep slope compared to the proximal zone and decreasing sediment size as channels branch frequently. Lastly, the distal zone, situated farthest from the sediment source, displays the least steep slope and finest sediment, where shallow channels result in wide-spreading sheetfloods on a floodplain-like surface unconfined by channel boundaries, with overall decreases in depth and gradient. It is important to note that the actual characteristics of these zones may vary due to factors such as local geology, climate, and human activity.

- **Flash Floods:** Intense, short-lived bursts of water transport large volumes of sediment. These can deposit coarse debris flows near the apex and finer materials further out, sometimes forming levees along channel margins (Jo et al. 1997; Gao et al. 2020).
- **Debris Flows:** Mudflows and landslides can deposit poorly sorted sediment on the fan surface, often altering existing channels (Larsen and Steel 1978; Blair and McPherson 1998; Blair 1999).
- **Biotic Activity:** Vegetation can trap and stabilize fine-grained sediments on the fan surface (Lane et al. 2016; Ventra et al. 2018), influencing future deposition patterns in different components of the fan system, viz. proximal fan, medial fan, and distal fan surfaces (from up-fan to down-fan).

7.2.1.2 Erosion and Reshaping

- **Stream Channel Avulsion:** Stream channel avulsion is a common process of channel shifting and sediment redistribution over alluvial fan systems (Field 2001; Leenman and Eaton 2021). Streams can abruptly change course due to sediment overloading, encountering obstacles, or seeking a steeper gradient. This initiates a new channel course and abandons older ones, reshaping the fan surface. Leenman and Eaton (2021) have studied various mechanisms through which stream avulsion occurs over alluvial fan surfaces.
- **Sheetfloods:** Widespread, shallow flows can erode finer materials from the fan surface, exposing coarser deposits underneath (Blair and McPherson 1994; Gao et al. 2020).
- **Rill and Gully Erosion:** Concentrated flows of water can cut into the fan surface, creating small channels (rills) that can grow into larger gullies (Brooks et al. 2009; Shellberg et al. 2013; Li et al. 2022). This can dissect existing channels and redistribute sediment.
- **Wind Erosion:** In arid environments, wind can remove finer particles from the fan surface, sorting the remaining sediment and creating features like sand dunes (Turner and Makhlouf 2002; Morgan et al. 2014).
- **Bioturbation:** Burrowing animals like rodents can disturb and redistribute surface sediments on the fan (Brierley et al. 1993; Ramos et al. 2002; Buatois et al. 2007).

7.2.1.3 Other Processes

- **Subsurface Water Movement:** Infiltration of water into the fan can lead to dissolution of soluble minerals and cementation of sediments, influencing the overall stability of the fan (Ghoneim and El-Baz 2007; Tchakerian and Pease 2015).

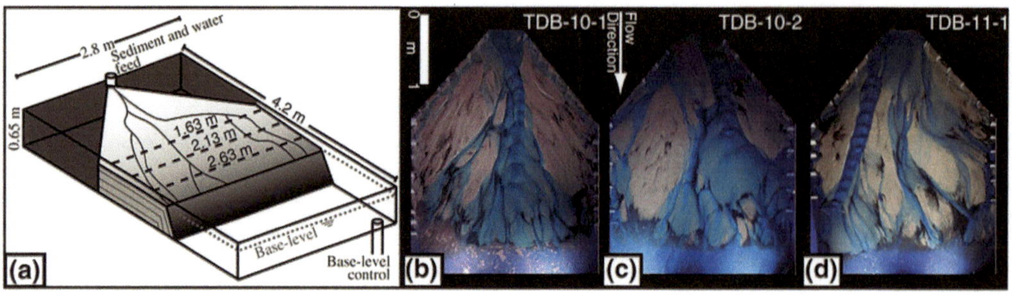

FIGURE 7.2A Illustrative diagram showcasing the Tulane Delta Basin facility. For details of the figure, consult the source work. Source: Straub and Esposito (2013).

- **Tectonic Activities:** Uplift of the source area due to tectonic activity can increase the gradient of the feeder stream, leading to renewed deposition and a steeper fan angle (Goswami et al. 2009; Bahrami 2013).
- **Climatic Change:** Changes in rainfall patterns can affect the frequency and intensity of floods, altering the rate of erosion and deposition on the fan surface (Colombera and Bersezio 2011; Vincent et al. 2022).

7.2.1.4 Interactions and Feedbacks

These processes often interact and create complex feedback loops. For instance, avulsion can expose a new area to erosion, while deposition in an avulsed channel can create a new, elevated surface. Understanding these interactions is important for predicting how alluvial fans will evolve over time. For instance, a recent paper by Vincent et al. (2022) investigates the role of secondary events in shaping alluvial fan morphology and determining channel behavior. The experiments conducted in the study show that longer secondary events result in larger fans with gentler gradients. These longer secondary events also lead to increased flow channelization and centralization between primary events. The morphologic changes caused by longer secondary events result in fewer avulsions during primary events, which occur later during the event. The study highlights the importance of understanding the duration of secondary events in determining fan morphology and flood hazards. Process interactions and resultant features in the case of the fan delta system are illustrated in Figures 7.2a and 7.2b.

7.2.2 Geomorphic Characteristics of Alluvial Fan Systems

An alluvial fan's key elements include the apex, the point where the feeder stream exits the mountains and begins depositing sediment (Drew 1873). Additionally, there are often incised channels carrying water across the fan, active depositional lobes where new sediment accumulates, and older surfaces representing past channel positions. Finally, some alluvial fans may also exhibit headward-eroding gullies.

7.2.3 Factors Influencing the Morphology of Alluvial Fans

Factors influencing the morphology of alluvial fans include:

a) Tectonic movements: Tectonic uplift and other tectonic influences play a significant role in determining the development, shape, structure, size, location, and thickness of alluvial fans. In Figure 7.3, types of alluvial fans in different tectonic settings are presented schematically.

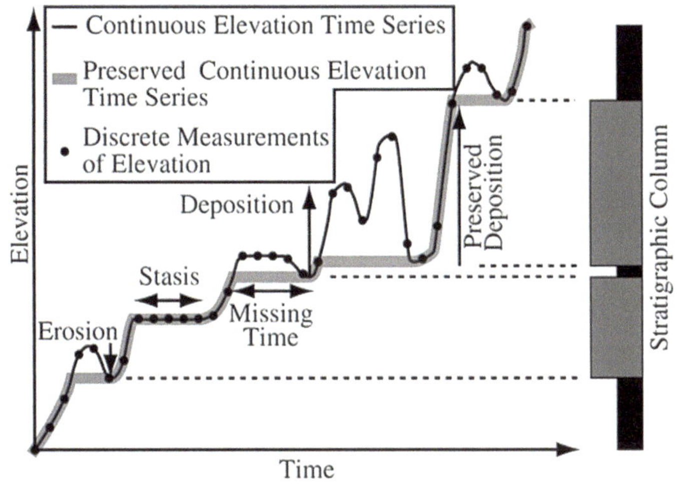

FIGURE 7.2B A diagrammatic representation depicting the assembly process of a stratigraphic column through elevation increments and factors influencing its completeness. Source: Straub and Esposito (2013).

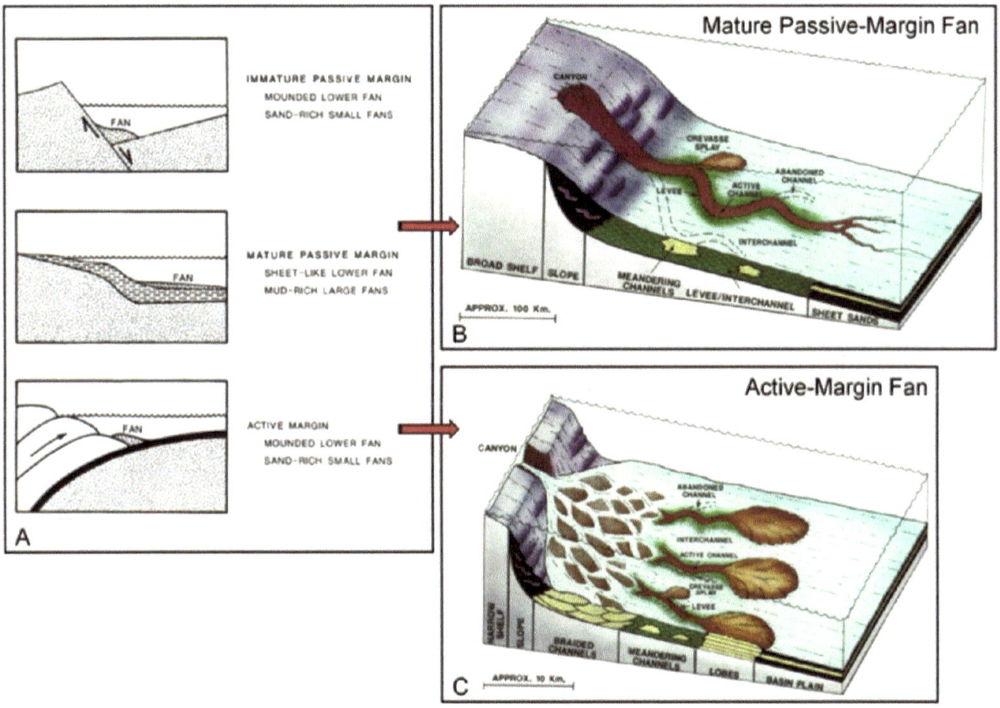

FIGURE 7.3 A. Three types of fan models demonstrate how tectonic settings affect fan morphology. B and C. Mature passive and mature active margin fans are schematically detailed. Source: Shanmugam (2016).

b) Sediment supply: The amount and type of sediment supplied to the fan influence its morphology and development. Figure 7.4 demonstrates effects of sediment supply on channel morphology.
c) Climatic: Climate conditions, such as precipitation patterns and temperature, can affect the formation and characteristics of alluvial fans.

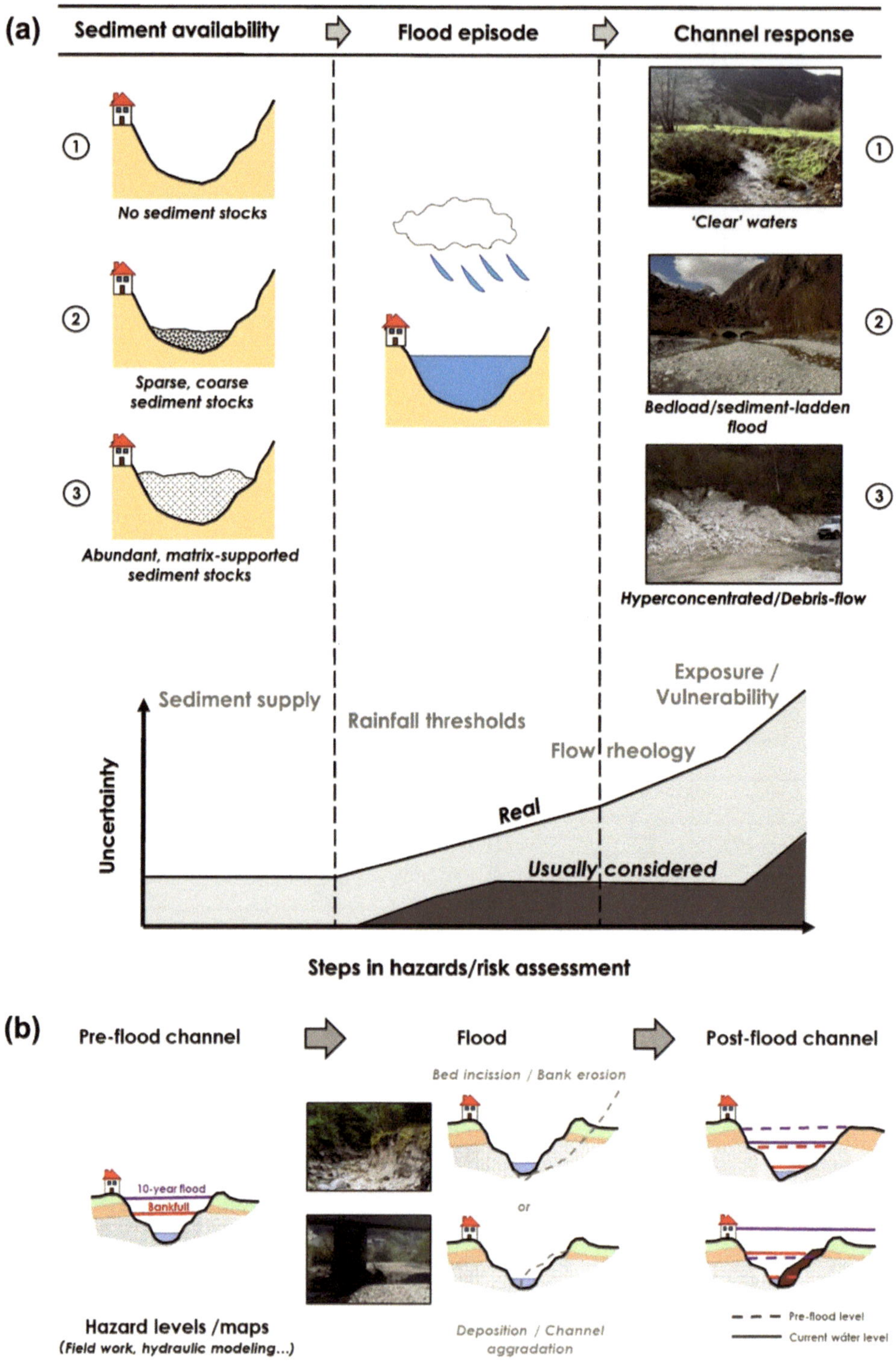

FIGURE 7.4 Effect of flood hazards on sediment mobilization in mountainous streams (3-a), and alluvial streams (3-b) and its effects on downslope alluvial fans. Source: Vázquez-Tarrío et al. (2024).

FIGURE 7.5A Field photos of sediment types and morphology of different aspects of alluvial fan in coastal region. Source: Walk et al. (2022).

d) Vegetation: The presence or absence of vegetation can impact channel stabilization, sediment transport, and overall ecosystem dynamics on alluvial fans.
e) Human activities: Human activities like mining, construction, urbanization, and agriculture can alter channel dynamics, stability, and the morphology of alluvial fans.

FIGURE 7.5B Morphological characterization of an alluvial fan in Chile . Source: Sun et al. (2023).

f) Runoff flow: Factors related to runoff flow, including river flow, flooding, and debris flows, can shape the morphology of alluvial fans.

It is notable that these factors interact in complex ways to shape the morphology and evolution of alluvial fans, highlighting the dynamic nature of these geomorphic features.

7.3 STREAM CHANNEL DYNAMICS IN ALLUVIAL FAN SYSTEMS

7.3.1 Overview of Stream Channel Processes in Alluvial Fan Systems

Alluvial fan systems are complex geomorphic environments characterized by a variety of processes that shape their formation and evolution (Bull 1977; Clarke 2015). The alluvium transported by rivers and streams flowing over the fan surface consists of a broad range of materials, including rock and shell fragments, gravel, sand, soil, and silt, with the size and speed of the water current determining the amount, type, and size of the alluvium (Le Hooke and Rohrer 1979; Nilsen 1982).

The formation of alluvial fans is influenced by several factors, including the presence of flowing water, a narrow channel between mountains, hills, or canyon walls, and a flat plain that opens up at the end of the narrow channel. Alluvial fans can form at any elevation as long as these three characteristics are present. They are typically triangular or cone-shaped, with two main parts: the apex and the apron (Nickel 1985). The apex refers to the tip of the alluvial fan, containing the largest pieces of alluvium, while the apron represents the fan-like or triangular portion of the alluvial fan, with smaller sediments like silt and sand deposited farther away from the water source (see Figures 7.5a and 7.5b).

Stream channel processes in alluvial fan systems involve both primary and secondary geomorphic events (El Asmi et al. 2023). Primary processes include major geomorphic events that build the fan, such as rockfalls, rock slides, rock avalanches, earth flows, colluvial slips, debris flows, incised-channel floods, and other mass wasting processes. Secondary processes are those that occur after the initial formation of the fan and include erosion, sediment transport, and deposition.

Alluvial fan systems are also influenced by climate and changes in base level, which can be as important as tectonic uplift in shaping these environments. For example, alluvial fans in the Himalayas show older fans entrenched and overlain by younger fans, with evidence of tectonic tilting about 45,000 years ago (Kumar et al. 2007) and an end to fan deposition about 20,000 years ago.

7.3.2 Will Streams Exhibit Similar or Different Behaviors across Alluvial Fan and Non-alluvial Fan Environments?

While alluvial fans encompass various categories, including alluvial fans, terminal fans, coastal fans, and fan deltas, stream behavior across these environments is not solely determined by their surface characteristics. Instead, it is primarily influenced by two to three key factors. The foremost factor is the slope gradient, followed by the sediment-water budget. High sediment-water ratios typically lead to braiding behavior in streams, while steep slope gradients also tend to induce braiding. In contrast, low slope gradients often result in meandering streams that gradually migrate across fan surfaces. In coastal fans, which ultimately evolve into fan deltas, marine influences become a third controlling factor in stream channel behavior. For instance, when a stream transitions from hilly terrain to a coastal plain, forming an alluvial fan, its behavior is influenced by the humid environment. However, in arid and semi-arid regions, alluvial fans are predominantly formed by debris flows and mud flows triggered by intense, short-duration rains. Yet, even in such environments, streams exhibit similar behaviors to those in other fan environments when subjected to comparable controlling factors.

7.3.3 Sediment Transport Dynamics

The sediment transport dynamics on alluvial fans are characterized by the transport of coarse sediments from the proximal fan to the distal fan, with finer grains being deposited last as they require less energy to remain in suspension. Debris flows are one of the most important formative processes for alluvial fans, capable of transporting large amounts of water and debris in a laminar flow, similar to a glacier. Sheet flows produce broad deposits that are clast-supported, with some imbrication of clasts.

Alluvial fans are also influenced by episodic sediment supply, which can affect fan morphology and channel dynamics. Experimental studies have shown that fans become steeper as the sediment supply (and, importantly, the sediment concentration) increases (Whipple et al. 1998; Allen 2008; Leenman and Eaton 2022). The duration of sediment supply oscillations also governs their impact on fan morphology and channel patterns.

7.3.4 Channel Migration and Avulsion

Channel migration and avulsion over alluvial fan systems are crucial processes that shape the formation and evolution of these dynamic landforms. Channel migration refers to the movement of a channel's position, dimensions, shape, and gradient in response to hydrologic, sedimentologic, and organic inputs (Pandey et al. 2021). Avulsion, on the other hand, is the process of a channel abandoning its previous course and establishing a new one.

Channel migration in alluvial fan systems is influenced by several factors, including the hydrologic regime, level of sediment supply, and other environmental factors (Heyvaert and Walstra 2016; Leenman and Eaton 2021; Edmonds et al. 2022). When a tributary abruptly enters a larger valley, it experiences a dramatic decrease in its sediment transport capacity, leading to rapid deposition of bed material and the formation of an alluvial fan. The channel migration zone (CMZ) typically encompasses the entire area of the alluvial fan (Rapp and Abbe 2003; Brummer et al. 2006; Alber and Piégay 2017), as the fan is built through a series of discrete avulsion events.

Avulsion is a key process in building alluvial fans, but it is also a formidable natural hazard. Avulsion events can occur at different scales, from small-scale sheet flows to large-scale debris flows. These events are influenced by the sediment supply, which can be episodic, and the duration of sediment supply oscillations. High-frequency topographic data can provide insights into the mechanisms of avulsion on alluvial fans.

7.3.5 Role of Vegetation in Channel Dynamics

Vegetation plays a crucial role in channel dynamics over alluvial fan systems. It assists in channel stabilization, influences sediment transport, and contributes to the overall ecosystem (Tal and Paola 2010; Gran et al. 2015; Ielpi et al. 2022).

The role of vegetation in channel stabilization is particularly important in alluvial fan systems (Michelini et al. 2017; Bahrami and Ghahraman 2019), which are characterized by a mobile boundary of relatively unconsolidated material. In the absence of vegetation, these systems may be more prone to erosion and instability, as there is less resistance to the movement of sediment and water (Ielpi and Lapôtre 2020; Hasson et al. 2023).

In addition to stabilizing channels, vegetation can also influence sediment transport dynamics. For example, plants can trap sediment particles as they flow through the channel, reducing the amount of sediment that is transported downstream (Ielpi and Lapôtre 2020). This can lead to the deposition of sediment in areas where the water flow is slower, such as in the vicinity of vegetation or in areas with lower channel gradients.

Vegetation can also affect the distribution of plant species in alluvial fan systems (Parker 1995; Bedford and Small 2008; Lane et al. 2016).

7.3.6 Impacts of Human Activities on Channel Dynamics

Human activities can have significant impacts on the channel dynamics of alluvial fan systems. Some of the key impacts include:

a) Sediment mining: Sediment mining in alluvial channels can lead to physical effects such as upstream incision (headcutting along the main river and tributaries) and downstream incision. This can result in channel bed degradation and changes in channel morphology (Robinson 2004; Smith et al. 2016).
b) Human activities, including in-stream mining, construction, and urbanization, can detrimentally affect riparian and aquatic habitats (Kondolf et al. 1987; Bennett 2004; Gumiero et al. 2015), infrastructure integrity (Schick et al. 1999; Santangelo et al. 2011), and channel stability. These activities contribute to channel instability, increased runoff, and sediment mobilization, ultimately impacting the ecosystem and biodiversity of alluvial fan systems.

c) Long-term impact on avulsion and fan development: Human activities can have long-term effects on avulsion and fan development (Brizga and Finlayson 1990), as demonstrated by different studies (Jones and Schumm 1999; Gregory 2006; Moodie and Nittrouer 2021).
d) Changes in land use: Human activities such as agriculture, urbanization, and mining can lead to changes in land use patterns, which can affect the distribution and characteristics of alluvial fans (Gómez-Villar and García-Ruiz 1997; Wang et al. 2022; Louzada et al. 2023).

7.4 CONTROLS ON STREAM CHANNEL DYNAMICS

The dynamics of stream channel behavior on alluvial fan surfaces are governed by various controls, which can be categorized into four groups: geological, climatic, tectonic, and vegetation controls.

1.1. *Geological Controls:* The type of bedrock in the catchment area heavily influences the primary processes of alluvial fans (Blair and McPherson 1994; Blair 1999). Variations in lithology lead to differences in sediment compositions and volumes due to weathering reactions. Bedrock prone to weathering into gravel and sand, but minimal mud, produces colluvium with low cohesion, leading to incised-channel and sheet floods on the alluvial fan system.

1.2. *Climatic Influences:* Climatic variables play a very important role in shaping stream behaviour in an alluvial fan setting. Precipitation, especially high-intensity storms in desert environments, triggers fan aggradation (Pryor et al. 2011). Temperature affects chemical weathering rates, with higher temperatures accelerating the process. Temperature gradients induced by orographic conditions may initially decrease weathering rates upslope due to cooler temperatures (Tucker and Slingerland 1997; Enzel et al. 2012), followed by increased weathering at higher altitudes (Lohse et al. 2020; Moidu et al. 2021).

1.3. *Tectonic Activity:* Alluvial fan formation and preservation are most favorable in tectonically active regions. Sediment yield increases exponentially with relief (Ahnert 1970), influenced by ongoing tectonic activity. Various aspects of tectonism, such as rates of uplift and lateral displacement, significantly influence fan shape and evolution. Extensional and translational tectonic settings are particularly conducive to fan development.

1.4. *Vegetation and Ecosystem Interactions*: Vegetation significantly influences sediment yield from a catchment. Plant cover enhances clay production through increased chemical weathering near roots (Lustig 1965) and strengthens sediment slope stability by increasing shear strength (Greenway 1987). However, vegetation can also contribute to long-term instability by steepening slopes, leading to slope failures following disturbances such as fires (Wells et al. 1987).

1.5. *Influence of Catchment Shape:* The morphology and development of a watershed can significantly influence sedimentary processes on an alluvial fan (Blair and McPherson 2009). Factors like side slopes, feeder channel profiles, relief, flash flood potential, and sediment storage capacity are affected by watershed characteristics. Variations in bedrock type and slope gradients can increase the likelihood of rockfalls, debris flows, and flash floods. The transport of silt to the fan is also influenced by sediment storage capacity. Additionally, watershed elevation affects precipitation patterns, while relief and surface area impact sediment volume during events. Orientation to sunlight and storm paths further affect transport, weathering, erosion, and fan accumulation.

7.5 METHODS OF STUDYING STREAM CHANNEL DYNAMICS IN ALLUVIAL FAN SYSTEMS

There are two main approaches to studying stream channel behavior in alluvial fan systems: a) field-based approaches; and b) laboratory- and modeling-based approaches. The remote sensing-based approach is hybrid in nature as it involves both field-based and lab-based approaches. Remote sensing also proves to be complementary and supplementary to the two approaches.

7.5.1 Field-Based Approaches

Studying stream channel dynamics over alluvial fans often requires a combination of field-based approaches to gather data on various aspects of the system. Alluvial fans are landforms created by the deposition of sediments carried by streams or rivers as they exit mountainous terrain on to flatter plains or valley floors. Here are some field-based approaches commonly used to investigate stream channel dynamics over alluvial fans:

a) Stream Gauging: Installing stream gauges along the channel can provide continuous measurements of streamflow, which are essential for understanding sediment transport dynamics and channel morphology changes over time.
b) Topographic Surveys: Conducting detailed topographic surveys using techniques such as total stations, GPS, or LiDAR (Light Detection and Ranging) can provide accurate elevation data, allowing researchers to track changes in channel elevation and morphology.
c) Sediment Sampling: Collecting sediment samples from the channel bed and banks can provide information on sediment composition, grain size distribution, and sediment transport rates, which are crucial for understanding sediment dynamics on the alluvial fan.
d) Cross-Section Surveys: Conducting cross-sectional surveys of the channel at regular intervals helps in quantifying channel geometry, including channel width, depth, and slope, which are essential parameters for understanding channel stability and sediment transport processes.
e) Sediment Tracing: Using tracers such as dyes, sediment markers, or isotopes can help track the movement of sediment within the channel and across the alluvial fan, providing insights into sediment transport pathways and rates.
f) Remote Sensing: Utilizing remote sensing techniques such as aerial photography, satellite imagery, or drones can provide high-resolution spatial data for mapping channel features, vegetation cover, and land use changes over the alluvial fan area.
g) Hydrological Monitoring: Installing rain gauges and soil moisture sensors in the study area can help quantify precipitation patterns, infiltration rates, and groundwater contributions to streamflow, which influence channel dynamics on alluvial fans.
h) Geomorphological Mapping: Conducting detailed geomorphological mapping of the study area helps identify landforms, sediment sources, and geomorphic processes shaping the alluvial fan landscape.
i) Flow Velocity Measurements: Using instruments such as current meters or Acoustic Doppler Velocimeters (ADVs) to measure flow velocities at different points along the channel helps assess flow dynamics and hydraulic conditions influencing channel morphology.
j) Paleoenvironmental Analysis: Collecting sediment cores or conducting stratigraphic analyses can provide insights into past environmental conditions, sediment deposition rates, and long-term changes in channel morphology on the alluvial fan.

These field-based approaches are often integrated with laboratory experiments, numerical modeling, and geospatial analysis to develop comprehensive understandings of stream channel dynamics over alluvial fans.

7.5.2 Remote Sensing Techniques

Remote sensing techniques offer valuable tools for investigating stream channel dynamics over alluvial fans by providing spatially extensive and temporally continuous data. Here are some remote sensing techniques commonly used for this purpose:

a) Aerial Photographs: Aerial photographs, whether obtained from airplanes or drones, provide high-resolution imagery that can be used to map channel features, vegetation cover, and land use changes over time. Time-series analysis of aerial photographs can reveal trends in channel migration, vegetation encroachment, and sediment deposition patterns.
b) Satellite Imagery: Satellite sensors, such as those aboard Landsat, Sentinel, and other Earth observation satellites, provide multispectral imagery with varying spatial and temporal resolutions. These data can be used to monitor changes in channel morphology, vegetation dynamics, and land cover across large areas, facilitating regional-scale assessments of alluvial fan dynamics.
c) Orthophotos: Orthorectified aerial or satellite images, which are geometrically corrected to remove distortions caused by terrain relief and sensor characteristics, are useful for accurate mapping of channel features, floodplain extents, and landform morphology on alluvial fans.
d) LiDAR: LiDAR data, whether acquired from airborne or terrestrial platforms, provides high-resolution elevation models that enable detailed mapping of channel topography, bank erosion features, and sediment deposition patterns. LiDAR-derived digital elevation models (DEMs) also support geomorphic analysis and hydrological modeling of stream channels on alluvial fans.
e) Structure from Motion (SfM): SfM photogrammetry, often implemented using drone imagery, allows for the creation of high-resolution 3D models of the landscape, including stream channels and adjacent terrain. These models facilitate the quantification of channel morphology, volumetric changes, and sediment transport dynamics over time.
f) Hyperspectral Imaging: Hyperspectral sensors capture images across hundreds of narrow spectral bands, allowing for detailed characterization of surface materials and vegetation types. Hyperspectral data can be used to identify sediment sources, map sedimentary deposits, and monitor vegetation dynamics along stream channels on alluvial fans.
g) RADAR Remote Sensing: Synthetic Aperture Radar (SAR) sensors onboard satellites provide all-weather imaging capabilities and penetrate through vegetation cover to reveal subsurface features. SAR data can be used to detect changes in channel morphology, monitor flood extents, and assess sediment transport dynamics on alluvial fans.
h) Thermal Infrared Imaging: Thermal infrared sensors measure emitted radiation from the Earth's surface, which can be used to detect variations in surface temperature related to channel flow, groundwater discharge, and sediment deposition processes on alluvial fans.

Remote sensing techniques prove to be complementary, supplementary, as well as independent. Integration of these remote sensing techniques with field observations, hydrological modeling, and GIS analysis enables comprehensive assessments of stream channel dynamics over alluvial fans at various spatial and temporal scales.

7.5.3 Modeling Approaches and Laboratory-Based Approaches

Investigating stream channel dynamics over alluvial fans often involves a combination of modeling approaches and laboratory-based experiments to understand the complex interactions between sediment transport, flow dynamics, and channel morphology. Here are some modeling and laboratory-based approaches commonly used for this purpose:

7.5.3.1 Modeling Approaches

a) Hydrodynamic Modeling: Numerical models, such as 1D, 2D, or 3D hydrodynamic models (e.g., HEC-RAS, MIKE 21, Delft3D), simulate flow patterns and velocities within stream channels over alluvial fans (Majerova et al. 2020). These models consider factors such as channel geometry, hydraulic roughness, and boundary conditions to predict water surface elevations, flow velocities, and sediment transport rates.
b) Sediment Transport Models: Models such as the Exner equation or the Einstein-Brownian motion equation simulate sediment transport processes, including bedload and suspended load transport, as well as bed erosion and deposition. These models help predict changes in channel morphology and sediment distribution over alluvial fans under varying flow conditions (Schlunegger et al. 2023; Zăinescu et al. 2023).
c) Geomorphological Models: Cellular automaton models, agent-based models, or landscape evolution models simulate long-term morphological changes in stream channels and alluvial fans due to sediment transport, erosion, deposition, and tectonic processes (Kenzler et al. 2023). These models can capture complex interactions between hydrology, sediment dynamics, and landscape evolution over geological timescales.
d) Hydrological Modeling: Distributed hydrological models (e.g., SWAT, HEC-HMS) simulate rainfall-runoff processes, infiltration, and surface runoff generation within watersheds feeding into alluvial fans. These models provide input data for hydrodynamic and

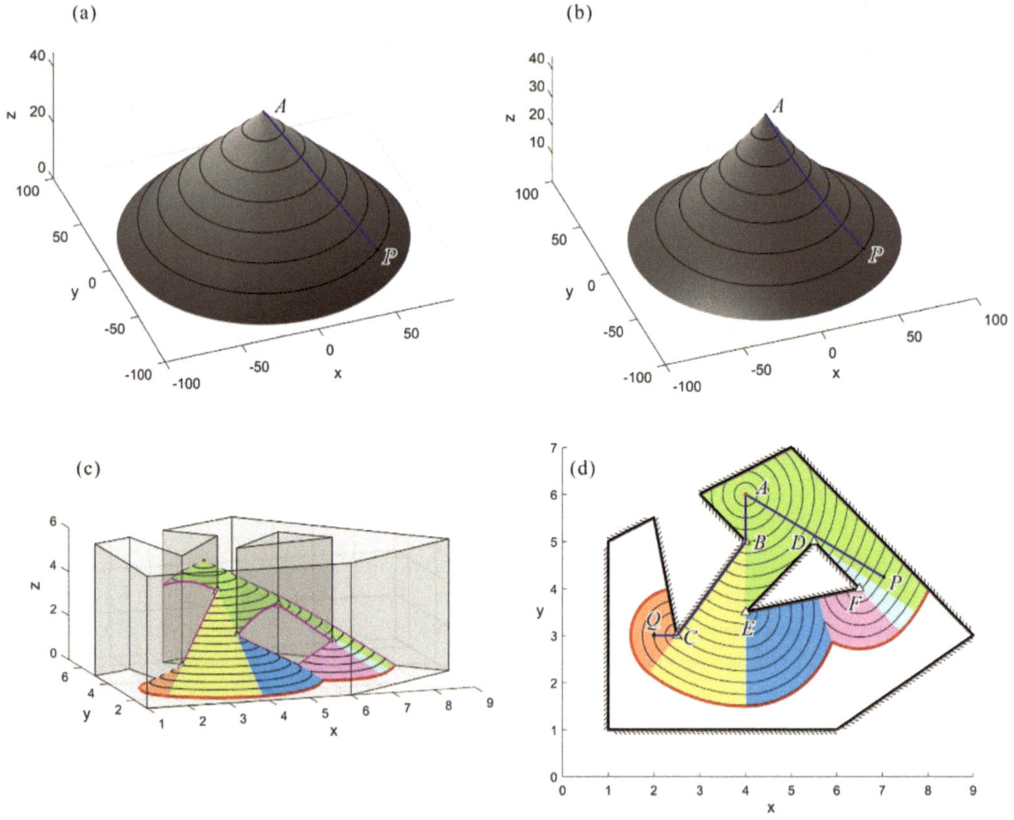

FIGURE 7.6 Fan morphology visualized through idealized models. For details of the parameters in the figure, consult the source work, figure 1. Source: Chen and Capart (2022).

sediment transport models, helping to understand the influence of upstream hydrology on stream channel dynamics (Lin et al. 2022; Guduru et al. 2023).

e) Numerical Experiments: Conducting numerical experiments using computational fluid dynamics (CFD) models (Bahoosh et al. 2022) allows for detailed investigations into flow structures, turbulence characteristics, and sediment transport mechanisms within stream channels over alluvial fans. These models provide insights into flow-sediment interactions and channel morphodynamics under different flow regimes. Figures 7.7, 7.8, and 7.9 demonstrate different scenarios of mathematical models, their results, and validation in the field (for more information, consult the source paper cited at the respective figures).

7.5.3.2 Lab-Based Approaches

a) Flume Experiments: Physical models constructed in laboratory flumes simulate stream channel processes under controlled conditions (Figures 7.10 and 7.11). Flume experiments allow researchers to study sediment transport, channel morphodynamics, and flow behavior at small scales, providing insights into the fundamental processes governing alluvial fan dynamics (Van Dijk et al. 2009).
b) Wave Tanks: Wave tanks equipped with sediment beds simulate wave action, sediment transport, and coastal processes relevant to alluvial fan environments. These experiments help understand the effects of wave-induced currents, wave shoaling, and sediment redistribution on channel morphology and sediment deposition patterns. A wave flume setup is shown in Figure 7.12.
c) Sediment Transport Studies: Laboratory experiments using sediment flumes or sediment transport facilities measure sediment transport rates, sediment sorting, and bedform development under controlled flow conditions. These experiments provide data for validating sediment transport models and understanding sediment dynamics in stream channels over alluvial fans.

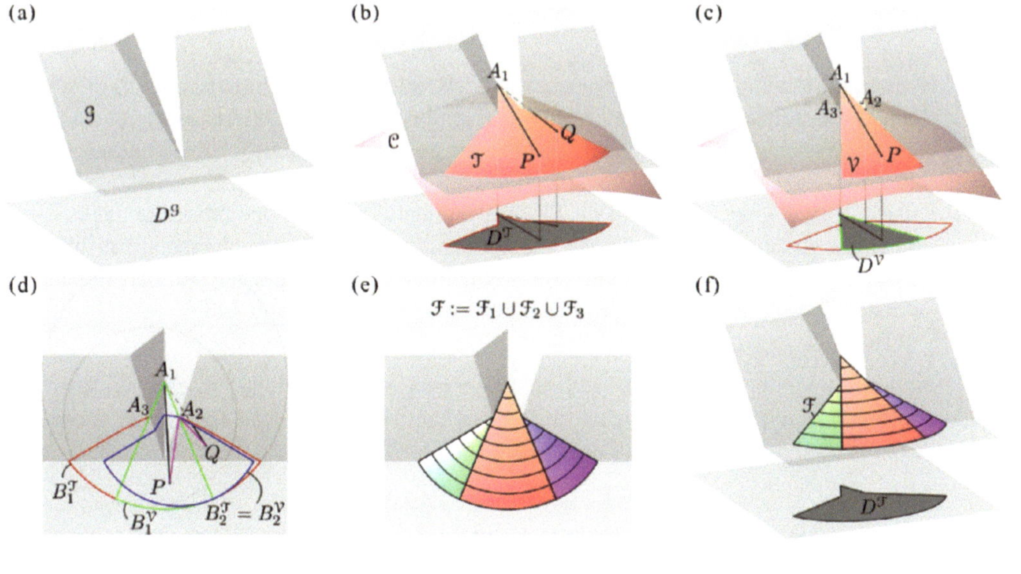

FIGURE 7.7 Coalescing fans (bajadas) along mountain fronts under various conditions: (a) Uniform slope fans on a flat mountain front, (b) Differently sloped fans on a flat mountain front, (c) Uniform slope fans at valley mouths, (d) Differently sloped fans at valley mouths. Plots display contour lines, fan origins, and analytical weld line paths. Dashed lines in (c) divide visible and invisible domains to fan. Source: Chen and Capart (2022).

FIGURE 7.8 Field sites in Taiwan's montane areas chosen for algorithm testing: (a, b) Zhuoshui River tributary fan, Central Taiwan; (c, d) Debris flow fan at Laonong River confluence with Pu-Tun-Pu-Nas tributary, southwestern Taiwan. Source: Chen and Capart (2022).

d) Hydraulic Modeling: Physical hydraulic models, such as scaled-down replicas of stream channels, simulate flow patterns, hydraulic jumps, and sediment transport processes in laboratory settings. These models help investigate channel stability, erosion control measures, and flood mitigation strategies for alluvial fan environments.
e) Particle Image Velocimetry (PIV): PIV experiments in laboratory flumes or tanks measure flow velocities and turbulence characteristics within stream channels over alluvial fans. PIV data provide insights into flow-sediment interactions and help validate numerical models of flow dynamics in natural channels (see Figure 7.13a and 7.13b).

By combining modeling approaches with laboratory-based experiments, researchers can gain a comprehensive understanding of stream channel dynamics over alluvial fans, from fundamental processes to long-term morphological evolution.

7.5.4 Challenges and Limitations of Each Method

Investigating stream channel dynamics over alluvial fans using field-based, remote sensing-based, modeling, and laboratory-based approaches comes with various challenges and limitations (Nordin Jr. 1975; Brooks et al. 2009; Güneralp et al. 2012; Pelletier et al. 2015).

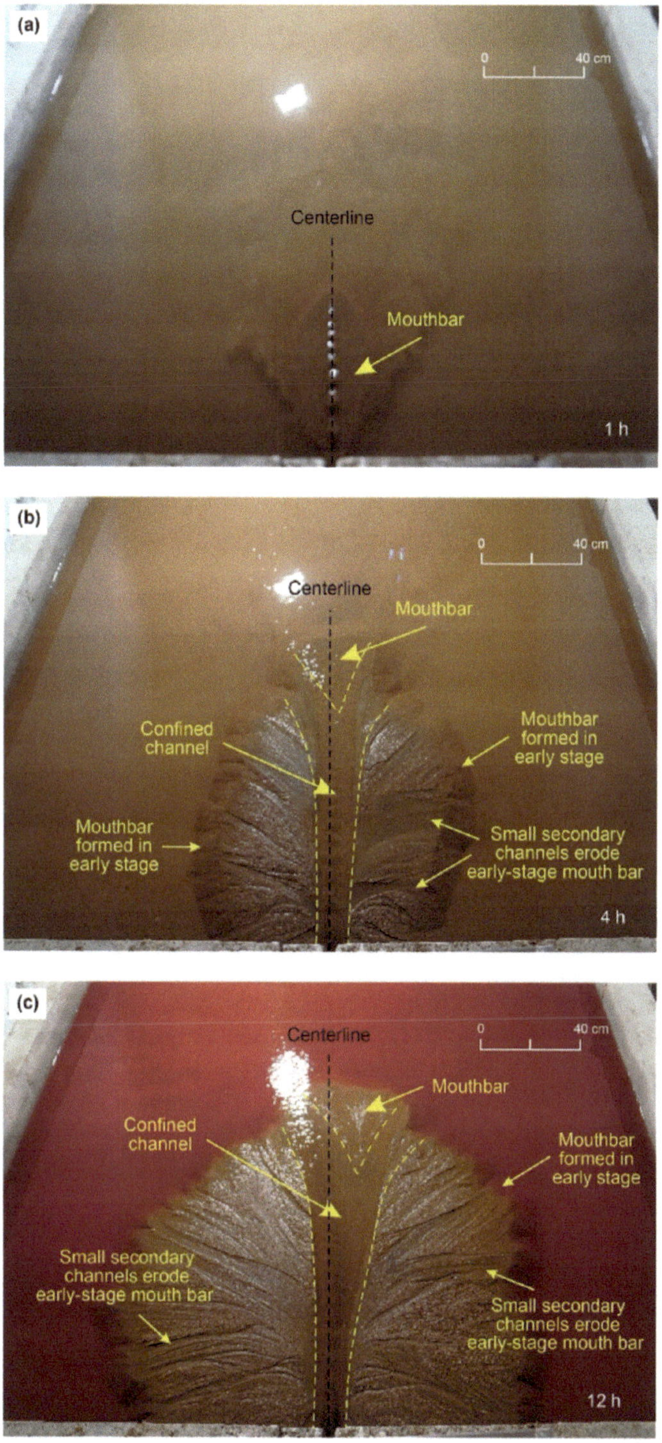

FIGURE 7.9 The sediment, transported efficiently within a confined channel flow with heightened transport capacity, settled rapidly, forming a fan delta lobe that extended quickly. Subsequently, a discernible mouth bar emerged along the centerline. Notably, mouth bars formed during periods of fully confined channelized flow are significantly larger than those formed during periods of multiple small secondary channelized flows. Source: Zhang et al. (2022).

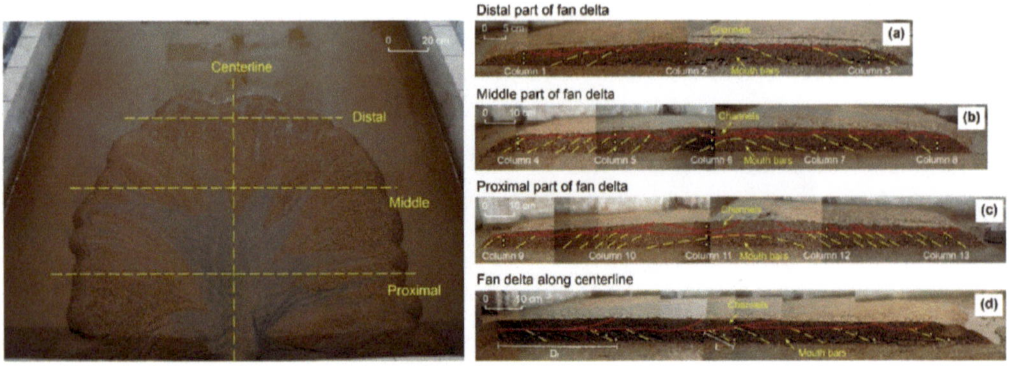

FIGURE 7.10 Samples were chosen for grain size analysis, and definitions were established for D_f and D_s. The yellow spots denote where samples were selected, while red and yellow lines represent the channels and foreset bedding of the mouth bar, respectively. Samples from the proximal, middle, and distal parts of the fan delta were evenly distributed both vertically and horizontally. Additionally, samples from the fan delta profile along the centerline were evenly distributed along the foreset bedding of mouth bars. D_f is defined as the horizontal distance between the alluvial river outlet and the end of foreset bedding, while L represents the length of foreset bedding. Source: Zhang et al. (2022).

FIGURE 7.11 Wave flume experiment setup. Source: Viriyakijja and Chinnarasri (2015).

a) Field-Based Approaches:

1. Accessibility: Alluvial fans can be located in remote or rugged terrain, making fieldwork logistically challenging and costly. Limited access to certain areas may restrict the collection of comprehensive data.
2. Temporal and Spatial Variability: Field measurements are often point-specific and may not capture the spatial and temporal variability of channel dynamics over the entire alluvial fan. This limitation can lead to uncertainties in data interpretation and modeling.
3. Data Quality and Resolution: Field measurements, such as topographic surveys or sediment sampling, may have inherent errors and uncertainties associated with equipment limitations, human error, or natural variability. Achieving high spatial and temporal resolution data across large study areas can be time-consuming and resource-intensive.
4. Environmental Constraints: Fieldwork can be subject to environmental constraints such as extreme weather conditions, seasonal variations in streamflow, and safety concerns related to working in remote or hazardous environments.

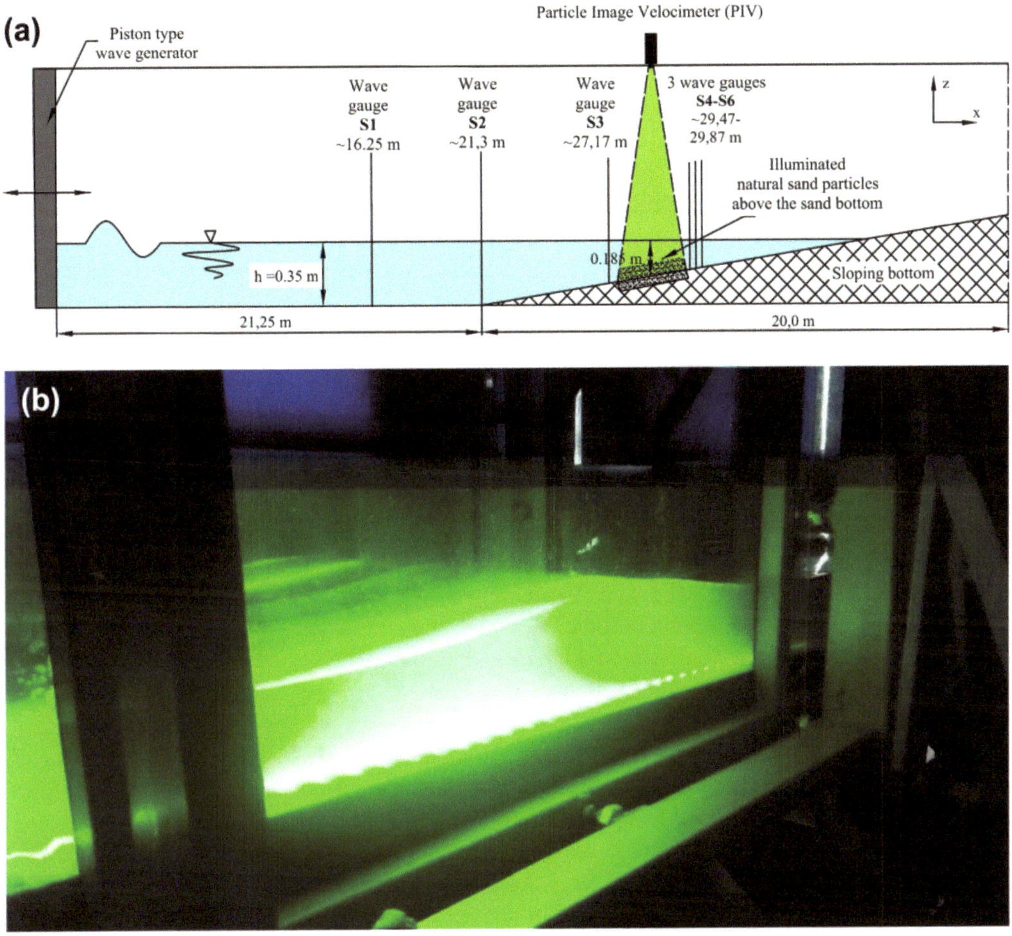

FIGURE 7.12 (a) PIV experimental setup. (b) PIV-based area measurement through laser light. Source: Stachurska et al. (2022).

5. Limited Observational Period: Field campaigns are typically conducted over relatively short time periods, which may not capture long-term morphological changes or rare events (e.g., extreme floods) that can significantly influence stream channel dynamics over alluvial fans.

b) Remote Sensing-Based Approaches:

1. Data Availability and Cost: Acquiring high-resolution remote sensing data (e.g., LiDAR, high-resolution satellite imagery) can be expensive, especially for large study areas or repeated acquisitions over time. The limited availability of free or open-access data can also pose challenges, particularly in developing regions.
2. Cloud Cover and Atmospheric Interference: Satellite imagery may be affected by cloud cover, atmospheric conditions, and sensor limitations, which can degrade data quality and hinder the interpretation of surface features, especially in areas with frequent cloud cover or dense vegetation.
3. Temporal Resolution: Satellite-based remote sensing platforms often have limited temporal resolution, with revisit times ranging from days to weeks. This limitation may constrain

the ability to monitor rapid changes in stream channel dynamics, particularly during dynamic events such as floods or channel avulsions.

4. Spatial Resolution: While advancements in remote sensing technology have improved spatial resolution, achieving sub-meter resolution over large areas remains challenging. Coarse spatial resolution can limit the ability to detect small-scale channel features and geomorphic processes on alluvial fans.
5. Data Processing and Interpretation: Processing and analyzing remote sensing data require specialized skills, software, and computational resources. Challenges may arise in data preprocessing, image classification, and validation, particularly for complex landscapes with heterogeneous land cover and topography.

c) Modeling and Lab-Based Approaches:

1. Simplifying Assumptions: Modeling approaches often involve simplifying assumptions and parameterizations to represent complex processes such as sediment transport, flow turbulence, and channel morphology. These simplifications may introduce uncertainties and biases in model predictions, particularly when applied to heterogeneous or poorly understood systems like alluvial fans.
2. Model Calibration and Validation: Validating models of stream channel dynamics requires field data for calibration and validation, which may be limited or unavailable for certain study areas. Uncertainties in model inputs and boundary conditions can affect the reliability and accuracy of model predictions.
3. Scale Dependency: Models may exhibit scale dependency, wherein processes operating at different spatial and temporal scales are not adequately represented. Upscaling or downscaling model outputs to match the scale of interest can introduce errors and uncertainties in predictions of stream channel dynamics over alluvial fans.
4. Computational Complexity and Resource Requirements: Numerical models of stream channel dynamics can be computationally intensive and require significant computational resources, expertise, and time to set up and run. Limited access to computational resources and technical expertise may hinder the application of complex models in some research contexts.
5. Laboratory Constraints: Laboratory experiments may not fully replicate natural conditions due to constraints such as scale limitations, simplified boundary conditions, and artificial substrates. These constraints can affect the transferability of laboratory findings to real-world alluvial fan environments and may require careful interpretation and validation against field observations.

7.6 FUTURE DIRECTIONS AND RESEARCH NEEDS

There are a number of unanswered questions related to stream channel behavior over alluvial fan environments that need to be answered and the main focus of research in the earth sciences dealing with alluvial fan research should be directed to some of the most important questions:

a) Gravel transport: How gravel transport or coarse clastic transport operates over fan surfaces and how far, on the fan surface, these coarse clastic materials travel over the fan surface needs to be focused on in future research.
b) Weathering: How, what type of weathering, and at what rate different weathering processes work on the alluvial fan surfaces.
c) Fluvial architecture: When streams create alluvial fans, what type of internal fluvial architecture do they form and how? And how do these internal fluvial architectures differ from braided and meandering river internal fluvial architectures?

d) Facies distribution: Since the sediment source is in the hinterland, how these facies get distributed along the alluvial fan long profile and what facies associations are created is a very important question that is not fully explored in alluvial fan research.
e) Economic value: Since alluvial fans are a very good source of groundwater in arid and semi-arid environments, and a number of mineral resources have been reported in these sedimentary environments, the economic value of alluvial fans should be explored (He et al. 2021; Saeid Pourmorad et al. 2021).
f) Archaeological significance and demography: since the alluvial fans provide habitable environments, there is a high probability of an archaeological treasure trove in these settings that needs to be explored.
g) Interaction of hill-slope processes and fluvial processes in alluvial fan settings is another not very well-understood topic of research that has potential and needs to be explored.
h) Decoupling the role of tectonics and that of climate in the alluvial fans resting in tectonically active settings is another very important topic of research.
i) Proportionate contributions of the axial river channel and other stream channels that flow over the alluvial fan surface in the evolution of the fan morphology are an unexplored topic that needs to be explored further.

7.7 CONCLUSION

This study highlights that the intricate dynamics shaping alluvial fan systems are governed by a multitude of factors, including sediment deposition, erosion, and reshaping processes influenced by geological, climatic, tectonic, and human-induced factors. These interactions give rise to diverse geomorphic features and stream channel dynamics within alluvial fans. Understanding these complexities is essential for predicting fan evolution and associated hazards accurately. Recent research underscores the significance of considering secondary processes in shaping fan morphology and flood behavior, highlighting the need for continued investigation. Various methodologies, including field-based, remote sensing-based, modeling, and laboratory-based approaches, contribute to our understanding of stream channel dynamics in alluvial fan environments. However, each method presents its own set of challenges, necessitating ongoing efforts to address unanswered questions and refine our understanding. By advancing our knowledge of these systems, we can better manage associated risks and contribute to landscape evolution studies.

ACKNOWLEDGEMENTS

The authors are grateful to their respective institutions for providing lab and library facilities, as well as peaceful sitting space, necessary for the completion of this work.

REFERENCES

Ahnert, F., 1970. Functional relationships between denudation, relief, and uplift in large, mid-latitude drainage basins. *American Journal of Science* 268, 243–263. https://doi.org/10.2475/ajs.268.3.243

Alber, A. and Piégay, H. 2017. Characterizing and modelling river channel migration rates at a regional scale: Case study of south-east France. *Journal of Environmental Management*, **202**, 479–493, https://doi.org/10.1016/j.jenvman.2016.10.055.

Allen, P.A. 2008. Time scales of tectonic landscapes and their sediment routing systems. *Geological Society, London, Special Publications*, **296**, 7–28, https://doi.org/10.1144/SP296.2.

Anstey, R. L. 1965. *Physical characteristics of alluvial fans.* US Department of Defense, Army Corps of Engineers. U.S. ARMY NATICK LABORATORIES, Natick, Massachusetts, USA, **109**.

Bahoosh, M., Kashi, E. and Shokrollahzadeh, S. 2022. The side stream and different spacers effects on the permeate water flux in forward osmosis process using computational fluid dynamics. *Chemical Engineering and Processing – Process Intensification*, **181**, 109113, https://doi.org/10.1016/j.cep.2022.109113.

Bahrami, S. 2013. Tectonic controls on the morphometry of alluvial fans around Danehkhoshk anticline, Zagros, Iran. *Geomorphology*, **180–181**, 217–230, https://doi.org/10.1016/j.geomorph.2012.10.012.

Bahrami, S. and Ghahraman, K. 2019. Geomorphological controls on soil fertility of semi-arid alluvial fans: A case study of the Joghatay Mountains, Northeast Iran. *CATENA*, **176**, 145–158, https://doi.org/10.1016/j.catena.2019.01.016.

Bedford, D.R. and Small, E.E. 2008. Spatial patterns of ecohydrologic properties on a hillslope-alluvial fan transect, central New Mexico. *CATENA*, **73**, 34–48, https://doi.org/10.1016/j.catena.2007.08.005.

Bennett, S.J. 2004. Effects of emergent Riparian vegetation on spatially averaged and turbulent flow within an experimental channel. *In*: *Riparian Vegetation and Fluvial Geomorphology,* 29–41, https://doi.org/10.1029/008WSA03.

Blair, T.C. 1987. Sedimentary processes, vertical stratification sequences, and geomorphology of the Roaring River alluvial fan, Rocky Mountain National Park, Colorado. *Journal of Sedimentary Research*, **57**, 1–18, https://doi.org/10.1306/212F8A8A-2B24-11D7-8648000102C1865D.

Blair, T.C. 1999. Sedimentology of the debris-flow-dominated Warm Spring Canyon alluvial fan, Death Valley, California. *Sedimentology*, **46**, 941–965, https://doi.org/10.1046/j.1365-3091.1999.00260.x.

Blair, T.C. and McPherson, J.G. 1994. Alluvial fans and their natural distinction from rivers based on morphology, hydraulic processes, sedimentary processes, and facies assemblages. *Journal of Sedimentary Research*, **64**, 450–489, https://doi.org/10.1306/D4267DDE-2B26-11D7-8648000102C1865D.

Blair, T.C. and McPherson, J.G. 1998. Recent debris-flow processes and resultant form and facies of the Dolomite alluvial fan, Owens Valley, California. *Journal of Sedimentary Research*, **68**, 800–818, https://doi.org/10.2110/jsr.68.800.

Blair, T.C. and McPherson, J.G. 2009. Processes and forms of alluvial fans. *In*: Parsons, A. J. and Abrahams, A. D. (eds) *Geomorphology of Desert Environments*, 413–467, https://doi.org/10.1007/978-1-4020-5719-9_14.

Brierley, G.J., Liu, K. and Crook, K.A.W. 1993. Sedimentology of coarse-grained alluvial fans in the Markham Valley, Papua New Guinea. *Sedimentary Geology*, **86**, 297–324, https://doi.org/10.1016/0037-0738(93)90027-3.

Brizga, S.O. and Finlayson, B.L. 1990. Channel avulsion and river metamorphosis: The case of the Thomson River, Victoria, Australia. *Earth Surface Processes and Landforms*, **15**, 391–404, https://doi.org/10.1002/esp.3290150503.

Brooks, A.P., Shellberg, J.G., Knight, J. and Spencer, J. 2009. Alluvial gully erosion: an example from the Mitchell fluvial megafan, Queensland, Australia. *Earth Surface Processes and Landforms*, **34**, 1951–1969, https://doi.org/10.1002/esp.1883.

Brummer, C.J., Abbe, T.B., Sampson, J.R. and Montgomery, D.R. 2006. Influence of vertical channel change associated with wood accumulations on delineating channel migration zones, Washington, USA. *Geomorphology*, **80**, 295–309.

Buatois, L.A., Uba, C.E., Mángano, M.G., Hulka, C. and Heubeck, C. 2007. Deep and intense bioturbation in continental environments: Evidence from miocene fluvial deposits of Bolivia. *In*: Bromley, R. G., Buatois, L. A., Mángano, G., Genise, J. F. and Melchor, R. N. (eds) *Sediment–Organism Interactions: A Multifaceted Ichnology,* https://doi.org/10.2110/pec.07.88.0123.

Bull, W.B. 1977. The alluvial-fan environment. *Progress in Physical Geography: Earth and Environment*, **1**, 222–270, https://doi.org/10.1177/030913337700100202.

Chant, L.J.D., Pease, P.P. and Tchakerian, V.P. 1999. Modelling alluvial fan morphology. *Earth Surface Processes and Landforms*, **24**, 641–652, https://doi.org/10.1002/(SICI)1096-9837(199907)24:7<641::AID-ESP979>3.0.CO;2-3.

Chen, T.-Y.K. and Capart, H. 2022. Computational morphology of debris and alluvial fans on irregular terrain using the visibility polygon. *Computers & Geosciences*, **169**, 105228, https://doi.org/10.1016/j.cageo.2022.105228.

Clarke, L.E. 2015. Experimental alluvial fans: Advances in understanding of fan dynamics and processes. *Geomorphology*, **244**, 135–145, https://doi.org/10.1016/j.geomorph.2015.04.013.

Colombera, L. and Bersezio, R. 2011. Impact of the magnitude and frequency of debris-flow events on the evolution of an alpine alluvial fan during the last two centuries: Responses to natural and anthropogenic controls. *Earth Surface Processes and Landforms*, **36**, 1632–1646, https://doi.org/10.1002/esp.2178.

De Haas, T., Ventra, D., Carbonneau, P.E. and Kleinhans, M.G. 2014. Debris-flow dominance of alluvial fans masked by runoff reworking and weathering. *Geomorphology*, **217**, 165–181, https://doi.org/10.1016/j.geomorph.2014.04.028.

Drew, F. 1873. Alluvial and lacustrine deposits and glacial records of the Upper-Indus Basin. *Quarterly Journal of the Geological Society*, **29**, 441–471.

Edmonds, D.A., Martin, H.K., et al. 2022. Rivers in reverse: Upstream-migrating dechannelization and flooding cause avulsions on fluvial fans. *Geology*, **50**, 37–41, https://doi.org/10.1130/G49318.1.
El Asmi, H., Gourari, L., et al. 2023. Primary and secondary sedimentary processes in debris-flow-dominated alluvial fan deposits within karstic setting: An example from the Middle Atlas-Sais foreland basin transition zone, Morocco. *Journal of African Earth Sciences*, **206**, 105028, https://doi.org/10.1016/j.jafrearsci.2023.105028.
Enzel, Y., Amit, R., et al. 2012. Late Quaternary weathering, erosion, and deposition in Nahal Yael, Israel: An "impact of climatic change on an arid watershed"1? *GSA Bulletin*, **124**, 705–722, https://doi.org/10.1130/B30538.1.
Field, J. 2001. Channel avulsion on alluvial fans in southern Arizona. *Geomorphology*, **37**, 93–104, https://doi.org/10.1016/S0169-555X(00)00064-7.
French, R.H., Fuller, J.E. and Waters, S. 1993. Alluvial fan: Proposed new process-oriented definitions for arid southwest. *Journal of Water Resources Planning and Management*, **119**, 588–598, https://doi.org/10.1061/(ASCE)0733-9496(1993)119:5(588).
Gao, C., Ji, Y., et al. 2020. Facies and depositional model of alluvial fan dominated by episodic flood events in arid conditions: An example from the Quaternary Poplar Fan, north-western China. *Sedimentology*, **67**, 1750–1796, https://doi.org/10.1111/sed.12684.
Ghoneim, E. and El-Baz, F. 2007. DEM-optical-radar data integration for palaeohydrological mapping in the northern Darfur, Sudan: Implication for groundwater exploration. *International Journal of Remote Sensing*, **28**, 5001–5018, https://doi.org/10.1080/01431160701266818.
Giles, P.T. 2010. Investigating the use of alluvial fan volume to represent fan size in morphometric studies. *Geomorphology*, **121**, 317–328, https://doi.org/10.1016/j.geomorph.2010.05.001.
Gómez-Villar, A. and García-Ruiz, J.M. 1997. The role of human activities in the development of alluvial fans. *Physics and Chemistry of the Earth*, **22**, 345–349, https://doi.org/10.1016/S0079-1946(97)00156-0.
Goswami, P.K., Pant, C.C. and Pandey, S. 2009. Tectonic controls on the geomorphic evolution of alluvial fans in the Piedmont Zone of Ganga Plain, Uttarakhand, India. *Journal of Earth System Science*, **118**, 245–259, https://doi.org/10.1007/s12040-009-0012-y.
Gran, K.B., Tal, M. and Wartman, E.D. 2015. Co-evolution of riparian vegetation and channel dynamics in an aggrading braided river system, Mount Pinatubo, Philippines. *Earth Surface Processes and Landforms*, **40**, 1101–1115, https://doi.org/10.1002/esp.3699.
Greenway DR. 1987. Vegetation and slope stability. In M. G. Anderson & K. S. Richards (Eds.), Slope stability, 187–230. New York: Wiley.
Gregory, K.J. 2006. The human role in changing river channels. *Geomorphology*, **79**, 172–191, https://doi.org/10.1016/j.geomorph.2006.06.018.
Guduru, J.U., Jilo, N.B., Rabba, Z.A. and Namara, W.G. 2023. Rainfall-runoff modeling using HEC-HMS model for Meki river watershed, rift valley basin, Ethiopia. *Journal of African Earth Sciences*, **197**, 104743, https://doi.org/10.1016/j.jafrearsci.2022.104743.
Gumiero, B., Rinaldi, M., Belletti, B., Lenzi, D. and Puppi, G. 2015. Riparian vegetation as indicator of channel adjustments and environmental conditions: The case of the Panaro River (Northern Italy). *Aquatic Sciences*, **77**, 563–582, https://doi.org/10.1007/s00027-015-0403-x.
Güneralp, İ., Abad, J.D., Zolezzi, G. and Hooke, J. 2012. Advances and challenges in meandering channels research. *Geomorphology*, **163–164**, 1–9, https://doi.org/10.1016/j.geomorph.2012.04.011.
Han, K., Wilson, J.L. and Emry, E. 2023. Autonomous generation of alluvial fans in landscape evolution models. *Earth Surface Processes and Landforms*, **48**, 2842–2863, https://doi.org/10.1002/esp.5663.
Hartley, A.J., Weissmann, G.S., Nichols, G.J. and Warwick, G.L. 2010. Large distributive fluvial systems: Characteristics, distribution, and controls on development. *Journal of Sedimentary Research*, **80**, 167–183, https://doi.org/10.2110/jsr.2010.016.
Harvey, A. 2018. Alluvial fans. *In*: *Reference Module in Earth Systems and Environmental Sciences*, https://doi.org/10.1016/B978-0-12-409548-9.11066-8.
Hasson, M., Marvin, M.C., Gunn, A., Lelpi, A. and Lapôtre, M.G.A. 2023. A depositional model for meandering rivers without land plants. *Sedimentology*, **70**, 2272–2301, https://doi.org/10.1111/sed.13121.
He, S., Wang, D., et al. 2021. Social-ecological system resilience of debris flow alluvial fans in the Awang basin, China. *Journal of Environmental Management*, **286**, 112230, https://doi.org/10.1016/j.jenvman.2021.112230.
Heyvaert, V.M.A. and Walstra, J. 2016. The role of long-term human impact on avulsion and fan development. *Earth Surface Processes and Landforms*, **41**, 2137–2152, https://doi.org/10.1002/esp.4011.

Lacobucci, G., Delchiaro, M., Troiani, F. and Nadali, D. 2024. Land-surface quantitative analysis for mapping and deciphering the construction processes of piedmont alluvial fans in the Anti-Lebanon Mountains. *Geomorphology*, **453**, 109148, https://doi.org/10.1016/j.geomorph.2024.109148.

Lelpi, A. and Lapôtre, M.G.A. 2020. A tenfold slowdown in river meander migration driven by plant life. *Nature Geoscience*, **13**, 82–86, https://doi.org/10.1038/s41561-019-0491-7.

Lelpi, A., Lapôtre, M.G.A., Gibling, M.R. and Boyce, C.K. 2022. The impact of vegetation on meandering rivers. *Nature Reviews Earth & Environment*, **3**, 165–178, https://doi.org/10.1038/s43017-021-00249-6.

Lustig, L. K. 1965. Clastic sedimentation in deep springs valley, California (Vol. 352). US Government Printing Office.

Jo, H.R., Rhee, C.W. and Chough, S.K. 1997. Distinctive characteristics of a streamflow-dominated alluvial fan deposit: Sanghori area, Kyongsang Basin (Early Cretaceous), southeastern Korea. *Sedimentary Geology*, **110**, 51–79, https://doi.org/10.1016/S0037-0738(96)00083-8.

Jones, L.S. and Schumm, S.A. 1999. Causes of Avulsion: An Overview. In Fluvial Sedimentology VI (eds N.D. Smith and J. Rogers). https://doi.org/10.1002/9781444304213.ch13

Kain, C.L., Rigby, E.H. and Mazengarb, C. 2018. A combined morphometric, sedimentary, GIS and modelling analysis of flooding and debris flow hazard on a composite alluvial fan, Caveside, Tasmania. *Sedimentary Geology*, **364**, 286–301, https://doi.org/10.1016/j.sedgeo.2017.10.005.

Kenzler, M., Gibb, M.A., Gehrmann, A., Deutschmann, A., Rother, H., Obst, K. and Hüneke, H. 2023. Identification of Quaternary alluvial-fan deposits (Rügen, SW Baltic Sea): Significance for recognition of syn-kinematic sedimentation in glacitectonic complexes. *Geomorphology*, **424**, 108558, https://doi.org/10.1016/j.geomorph.2022.108558.

Kondolf, G.M., Webb, J.W., Sale, M.J. and Felando, T. 1987. Basic hydrologic studies for assessing impacts of flow diversions on riparian vegetation: Examples from streams of the eastern Sierra Nevada, California, USA. *Environmental Management*, **11**, 757–769, https://doi.org/10.1007/BF01867243.

Koundi, N.E.E.-E., Gugliotta, M. and Mitten, A.J. 2022. Alluvial fans in the latest phase of the Messinian Salinity Crisis: An example from the Chaabet Et Tabbela Formation, northern Tunisia, Central Mediterranean. *Sedimentary Geology*, **442**, 106276, https://doi.org/10.1016/j.sedgeo.2022.106276.

Kumar, R., Suresh, N. and Sangode, S.J. 2007. Differential features of alluvial fans in the Pinjaur-Soan Dun, NW Himalaya, India: controlling factors. *Himalayan Geology*, **28**, 37–46.

Lane, S.N., Borgeaud, L. and Vittoz, P. 2016. Emergent geomorphic–vegetation interactions on a subalpine alluvial fan. *Earth Surface Processes and Landforms*, **41**, 72–86, https://doi.org/10.1002/esp.3833.

Larsen, V. and Steel, R.J. 1978. The sedimentary history of a debris-flow dominated, Devonian alluvial fan–a study of textural inversion. *Sedimentology*, **25**, 37–59, https://doi.org/10.1111/j.1365-3091.1978.tb00300.x.

Le Hooke, R.B. and Rohrer, W.L. 1979. Geometry of alluvial fans: Effect of discharge and sediment size. *Earth Surface Processes*, **4**, 147–166, https://doi.org/10.1002/esp.3290040205.

Lee, K. and Gihm, Y.S. 2023. Downstream changes in floodplain sedimentation and their effects on channel avulsion in stream-dominated alluvial fans: The Cretaceous Duwon Formation in the southern Korean Peninsula. *Sedimentary Geology*, **456**, 106473, https://doi.org/10.1016/j.sedgeo.2023.106473.

Leenman, A. and Eaton, B. 2021. Mechanisms for avulsion on alluvial fans: Insights from high-frequency topographic data. *Earth Surface Processes and Landforms*, **46**, 1111–1127, https://doi.org/10.1002/esp.5059.

Leenman, A.S. and Eaton, B.C. 2022. Episodic sediment supply to alluvial fans: implications for fan incision and morphometry. *Earth Surface Dynamics*, **10**, 1097–1114, https://doi.org/10.5194/esurf-10-1097-2022.

León, D.Z. de, Kershaw, S. and Mahan, S.A. 2016. Quaternary alluvial fans of Ciudad Juárez, Chihuahua, northern México: OSL ages and implications for climatic history of the region. *Boletín de la Sociedad Geológica Mexicana*, **68**, 111–128, https://doi.org/10.18268/BSGM2016v68n1a13.

Li, J., Zhao, C., et al. 2022. Gully erosion on alluvial fans can be mitigated by altering the hydrological connectivity between an alluvial fan and the contributing catchment: A study in the Lhasa River basin. *Land Degradation & Development*, **33**, 1170–1183, https://doi.org/10.1002/ldr.4176.

Lin, Q., Lin, B., Zhang, D. and Wu, J. 2022. Web-based prototype system for flood simulation and forecasting based on the HEC-HMS model. *Environmental Modelling & Software*, **158**, 105541, https://doi.org/10.1016/j.envsoft.2022.105541.

Liu, C.W. 2004. Decision support system for managing ground water resources in the Choushui River alluvial in Taiwan1. *JAWRA Journal of the American Water Resources Association*, **40**, 431–442, https://doi.org/10.1111/j.1752-1688.2004.tb01041.x.

Lohse, K.A., Gallo, E.L. and Meixner, T. 2020. Influence of climate and duration of stream water presence on rates of litter decomposition and nutrient dynamics in temporary streams and surrounding environments of Southwestern USA. *Frontiers in Water*, **2**, https://doi.org/10.3389/frwa.2020.571044.

Louzada, R.O., Roque, F. de O., Diniz, J.M.F. de S. and Bergier, I. 2023. River channel avulsion in the Taquari River megafan of the Brazilian Pantanal: Remote sensing and modeling reveal recent and future changes. *Applied Geography*, **155**, 102955, https://doi.org/10.1016/j.apgeog.2023.102955.

Mack, G.H. and Rasmussen, K.A. 1984. Alluvial-fan sedimentation of the Cutler Formation (Permo-Pennsylvanian) near Gateway, Colorado. *GSA Bulletin*, **95**, 109–116, https://doi.org/10.1130/0016-7606(1984)95<109:ASOTCF>2.0.CO;2.

Majerova, M., Neilson, B.T. and Roper, B.B. 2020. Beaver dam influences on streamflow hydraulic properties and thermal regimes. *Science of The Total Environment*, **718**, 134853, https://doi.org/10.1016/j.scitotenv.2019.134853.

Pandey, M., Ray, Y., Arora, A., Shukla, U. K., & Ranjan, S. (2021). Structural control on the landscape evolution of son alluvial fan system in Ganga foreland basin.(Pandey PC., & Sharma LK) , *Advances in remote sensing for natural resource monitoring*, 189–246. Wiley.

Michelini, T., Bettella, F. and D'Agostino, V. 2017. Field investigations of the interaction between debris flows and forest vegetation in two Alpine fans. *Geomorphology*, **279**, 150–164, https://doi.org/10.1016/j.geomorph.2016.09.029.

Moidu, H., Obedzinski, M., Carlson, S.M. and Grantham, T.E. 2021. Spatial patterns and sensitivity of intermittent stream drying to climate variability. *Water Resources Research*, **57**, e2021WR030314, https://doi.org/10.1029/2021WR030314.

Moodie, A.J. and Nittrouer, J.A. 2021. Optimized river diversion scenarios promote sustainability of urbanized deltas. *Proceedings of the National Academy of Sciences*, **118**, e2101649118, https://doi.org/10.1073/pnas.2101649118.

Morgan, A.M., Howard, A.D., et al. 2014. Sedimentology and climatic environment of alluvial fans in the martian Saheki crater and a comparison with terrestrial fans in the Atacama Desert. *Icarus*, **229**, 131–156, https://doi.org/10.1016/j.icarus.2013.11.007.

Nickel, E. 1985. Carbonates in alluvial fan systems. An approach to physiography, sedimentology and diagenesis. *Sedimentary Geology*, **42**, 83–104, https://doi.org/10.1016/0037-0738(85)90075-2.

Nicovich*, S.R., Schmitt, J.G., Gray, H.J., Klinger, R.E. and Mahan, S.A. 2023. Construction and modification of Debris-Flow Alluvial Fans as captured in the geomorphic and sedimentary record: Examples from the Western Sangre de Cristo Mountains, South-Central Colorado, https://doi.org/10.1130/2023.2561(01).

Nilsen, T. H. 1982. Alluvial fan deposits. Editors (Peter A. Scholle; Darwin Spearing). *Sandstone Depositional Environments . American Association of Petroleum Geologists American Association of Petroleum Geologists,* **31**. DOI: https://doi.org/10.1306/M31424C4

Nordin Jr., C.F. 1975. Erosion and sedimentation. *Reviews of Geophysics*, **13**, 458–460, https://doi.org/10.1029/RG013i003p00458.

Pandey, M., Ray, Y., Arora, A., Shukla, U.K. and Ranjan, S. 2021. Structural control on the landscape evolution of Son Alluvial Fan System in Ganga Foreland Basin. *In*: *Advances in Remote Sensing for Natural Resource Monitoring,* 189–246, https://doi.org/10.1002/9781119616016.ch12.

Pandey, M.K. 2014. Confluence dynamics of lower son river with river Ganga tectonic and climatic perspective. Unpublished Ph.D. Thesis, Banaras Hindu University, Varanasi-221005, India.

Parker, G., Paola, C., Whipple, K.X. and Mohrig, D. 1998. Alluvial fans formed by channelized fluvial and sheet flow. I: Theory. *Journal of Hydraulic Engineering*, **124**, 985–995, https://doi.org/10.1061/(ASCE)0733-9429(1998)124:10(985).

Parker, K.C. 1995. Effects of complex geomorphic history on soil and vegetation patterns on arid alluvial fans. *Journal of Arid Environments*, **30**, 19–39, https://doi.org/10.1016/S0140-1963(95)80036-0.

Pelletier, J.D., Brad Murray, A., et al. 2015. Forecasting the response of Earth's surface to future climatic and land use changes: A review of methods and research needs. *Earth's Future*, **3**, 220–251, https://doi.org/10.1002/2014EF000290.

Pourmorad, S., Harami, R.M., Solgi, A. and Aleali, M. 2021. Sedimentological, geochemical and hydrogeochemical studies of alluvial fans for mineral and environmental purposes (Case Study of Southwestern Iran). *Lithology and Mineral Resources*, **56**, 89–112, https://doi.org/10.1134/S0024490221010077.

Pryor, B.S., Lisle, T., Montoya, D.S. and Hilton, S. 2011. Transport and storage of bed material in a gravel-bed channel during episodes of aggradation and degradation: A field and flume study. *Earth Surface Processes and Landforms*, **36**, 2028–2041, https://doi.org/10.1002/esp.2224.

Ramos, E., Busquets, P. and Vergés, J. 2002. Interplay between longitudinal fluvial and transverse alluvial fan systems and growing thrusts in a piggyback basin (SE Pyrenees). *Sedimentary Geology*, **146**, 105–131, https://doi.org/10.1016/S0037-0738(01)00169-5.

Rapp, C. F., & Abbe, T. B. 2003. *A framework for delineating channel migration zones* (No. Ecology Publication# 30-06-027). http://www.ecy.wa.gov/pubs/0306027.pdf

Rees, A.J., Thomas, A.T., Lewis, M., Hughes, H.E. and Turner, P. 2014. Lithostratigraphy and palaeoenvironments of the Cambrian in SW Wales. *Geological Society, London, Memoirs*, **42**, 33–100, https://doi.org/10.1144/M42.2.

Robinson, E. 2004. Coastal changes along the coast of Vere, Jamaica over the past two hundred years: Data from maps and air photographs. *Quaternary International*, **120**, 153–161, https://doi.org/10.1016/j.quaint.2004.01.014.

Santangelo, N., Santo, A., Di Crescenzo, G., Foscari, G., Liuzza, V., Sciarrotta, S. and Scorpio, V. 2011. Flood susceptibility assessment in a highly urbanized alluvial fan: The case study of Sala Consilina (southern Italy). *Natural Hazards and Earth System Sciences*, **11**, 2765–2780, https://doi.org/10.5194/nhess-11-2765-2011.

Scheinert, C., Wasklewicz, T. and Staley, D. 2012. Alluvial fan dynamics – Revisiting the field. *Geography Compass*, **6**, 752–775, https://doi.org/10.1111/gec3.12004.

Schick, A.P., Grodek, T. and Wolman, M.G. 1999. Hydrologic processes and geomorphic constraints on urbanization of alluvial fan slopes. *Geomorphology*, **31**, 325–335, https://doi.org/10.1016/S0169-555X(99)00085-9.

Schlunegger, F., do Prado, A.H., Norton, K.P. and Delunel, R. 2023. On the mechanisms resulting in the formation of the Quaternary staircase terrace systems in the valleys of the western Andean margin of Peru. *Geomorphology*, **442**, 108923, https://doi.org/10.1016/j.geomorph.2023.108923.

Shanmugam, G. 2016. Submarine fans: A critical retrospective (1950–2015). *Journal of Palaeogeography*, **5**, 110–184, https://doi.org/10.1016/j.jop.2015.08.011.

Shellberg, J.G., Brooks, A.P., Spencer, J. and Ward, D. 2013. The hydrogeomorphic influences on alluvial gully erosion along the Mitchell River fluvial megafan. *Hydrological Processes*, **27**, 1086–1104, https://doi.org/10.1002/hyp.9240.

Smith, C.M.S., Bowie, M.H., et al. 2016. Punakaiki Coastal Restoration Project: A case study for a consultative and multidisciplinary approach in selecting indicators of restoration success for a sand mining closure site, West Coast, New Zealand. *CATENA*, **136**, 91–103, https://doi.org/10.1016/j.catena.2015.07.024.

Stachurska, B., Mahdavi-Meymand, A. and Sulisz, W. 2022. Machine learning methodology for determination of sediment particle velocities over sandy and rippled bed. *Measurement*, **197**, 111332, https://doi.org/10.1016/j.measurement.2022.111332.

Stanistreet, I.G. and McCarthy, T.S. 1993. The Okavango Fan and the classification of subaerial fan systems. *Sedimentary Geology*, **85**, 115–133, https://doi.org/10.1016/0037-0738(93)90078-J.

Straub, K.M. and Esposito, C.R. 2013. Influence of water and sediment supply on the stratigraphic record of alluvial fans and deltas: Process controls on stratigraphic completeness. *Journal of Geophysical Research: Earth Surface*, **118**, 625–637, https://doi.org/10.1002/jgrf.20061.

Sun, X., Matthias May, S., et al. 2023. Water-dispersible colloids distribution along an alluvial fan transect in hyper-arid Atacama Desert. *Geoderma*, **438**, 116650, https://doi.org/10.1016/j.geoderma.2023.116650.

Tal, M. and Paola, C. 2010. Effects of vegetation on channel morphodynamics: Results and insights from laboratory experiments. *Earth Surface Processes and Landforms*, **35**, 1014–1028, https://doi.org/10.1002/esp.1908.

Tchakerian, V. and Pease, P. 2015. Chapter 14 - The critical zone in desert environments. *In*: Giardino, J. R. and Houser, C. (eds) *Developments in Earth Surface Processes,* 449–472, https://doi.org/10.1016/B978-0-444-63369-9.00014-8.

Tucker, G.E. and Slingerland, R. 1997. Drainage basin responses to climate change. *Water Resources Research*, **33**, 2031–2047, https://doi.org/10.1029/97WR00409.

Turner, B.R. and Makhlouf, I. 2002. Recent colluvial sedimentation in Jordan: Fans evolving into sand ramps. *Sedimentology*, **49**, 1283–1298, https://doi.org/10.1046/j.1365-3091.2002.00497.x.

Van Dijk, M., Postma, G. and Kleinhans, M.G. 2009. Autocyclic behaviour of fan deltas: an analogue experimental study. *Sedimentology*, **56**, 1569–1589, https://doi.org/10.1111/j.1365-3091.2008.01047.x.

Vázquez-Tarrío, D., Ruiz-Villanueva, V., Garrote, J., Benito, G., Calle, M., Lucía, A. and Díez-Herrero, A. 2024. Effects of sediment transport on flood hazards: Lessons learned and remaining challenges. *Geomorphology*, **446**, 108976, https://doi.org/10.1016/j.geomorph.2023.108976.

Ventra, D. and Clarke, L.E. 2018. Geology and geomorphology of alluvial and fluvial fans: Current progress and research perspectives. *Geological Society, London, Special Publications*, **440**, 1–21, https://doi.org/10.1144/SP440.16.

Ventra, D., Abels, H.A., Hilgen, F.J. and de Boer, P.L. 2018. Orbital-climate control of mass-flow sedimentation in a Miocene alluvial-fan succession (Teruel Basin, Spain). *Geological Society, London, Special Publications*, **440**, 129–157, https://doi.org/10.1144/SP440.14.

Vincent, L.T., Eaton, B.C., Leenman, A.S. and Jakob, M. 2022. Secondary geomorphic processes and their influence on alluvial fan morphology, channel behavior and flood hazards. *Journal of Geophysical Research: Earth Surface*, **127**, e2021JF006371, https://doi.org/10.1029/2021JF006371.

Viriyakijja, K. and Chinnarasri, C. 2015. Wave flume measurement using image analysis. *Aquatic Procedia*, **4**, 522–531, https://doi.org/10.1016/j.aqpro.2015.02.068.

Walk, J., Bartz, M., Stauch, G., Binnie, A., Brückner, H. and Lehmkuhl, F. 2022. Weathering under coastal hyperaridity – Late Quaternary development of spectral, textural, and gravelometric alluvial fan surface characteristics. *Quaternary Science Reviews*, **277**, 107339, https://doi.org/10.1016/j.quascirev.2021.107339.

Wang, Z., Shu, L., Xu, P., Yin, X., Lu, C., Liu, B. and Li, Y. 2022. Influence of land use changes on the remaining available aquifer storage (RAAS): A case study of the Taoerhe alluvial-proluvial fan. *Science of The Total Environment*, **849**, 157848, https://doi.org/10.1016/j.scitotenv.2022.157848.

Wells, S. G., McFadden, L. D., & Dohrenwend, J. C. 1987. Influence of late Quaternary climatic changes on geomorphic and pedogenic processes on a desert piedmont, eastern Mojave Desert, California. Quaternary Research, 27(2), 130–146.

Whipple, K.X., Parker, G., Paola, C. and Mohrig, D. 1998. Channel dynamics, sediment transport, and the slope of alluvial fans: Experimental study. *The Journal of Geology*, **106**, 677–694, https://doi.org/10.1086/516053.

Zăinescu, F., van der Vegt, H., et al. 2023. The role of wind-wave related processes in redistributing river-derived terrigenous sediments in Lake Turkana: A modelling study. *Journal of Great Lakes Research*, **49**, 368–386, https://doi.org/10.1016/j.jglr.2022.12.013.

Zhang, K., Wu, S.-H., Wang, J.-J., Xu, Y.-J., Xu, Z.-H. and Zhang, J.-J. 2022. Vertical grain-size trend of mouth bar in lacustrine fan delta: Flume experiments. *Petroleum Science*, **19**, 1964–1977, https://doi.org/10.1016/j.petsci.2022.03.024.

Zhang, Y., Dai, X., Wang, M. and Li, X. 2020. The concept, characteristics and significance of fluvial fans. *Petroleum Exploration and Development*, **47**, 1014–1026, https://doi.org/10.1016/S1876-3804(20)60113-6.

8 Mapping Shoreline Changes in Two River Deltas of Greece Exploiting Geoinformation Technologies

George P. Petropoulos, Spyridon E. Detsikas, Ioanna Tzanavari and Efthimios Karymbalis
Department of Geography, Harokopio University of Athens, Kallithea, Athens, Greece

8.1 INTRODUCTION

Coastlines are globally recognized as vital environmental areas due to their significant role in supporting a wide range of activities, including tourism, agriculture, and fishery (Ghoneim et al., 2015; Salimi et al., 2021). They are dynamic ecosystems that undergo constant changes and interactions with various natural processes and human influences (Karymbalis et al., 2016; Aedla et al., 2015). Among the main physical factors influencing coastal erosion are the presence of easily erodible geological formations, the shortage of sediment delivery to the coast, and the oceanographic condition of the receiving basin, including wave activity, tides, and longshore currents (Maroukian and Karymbalis, 2004; Papoutsis et al., 2015). Anthropogenic factors include marine structures and other man-made shoreline protection structures (Kapsimalis et al., 2005) , as well as dam construction at the upper part of the catchment (Parcharidis et al. 2012; Karymbalis et al., 2022). These alterations can often bring catastrophic social, economic, and ecological impacts within the coastal zone, leading to loss of life, property damage, agricultural land degradation, and destruction of natural resources and habitats.

Monitoring shorelines is therefore vital not only for the protection of coastal structures from erosion and other processes that alter coastal stability, but also for sustainable development and coastal planning, safe navigation, management and safeguarding of coastal resources, and the overall environmental management of a coastal ecosystem. Rapid technological development in recent years has led to the creation of increasingly modern and innovative geoinformatics tools. Geographic Information Systems (GIS), Global Positioning Systems (GPS), and Remote Sensing (RS) have contributed significantly to the integrated management of the coastal zone. They are also cost-effective and efficient tools for mapping shoreline changes, providing reliable and at the same time qualitative results (Xia et al., 2021; Tsatsaris et al., 2021; Karymbalis et al., 2014).

In recent decades, RS methods have dominated the mapping of coastlines, always in conjunction with traditional techniques. Field observations have been and will continue to be needed and will not be completely replaced by RS. However, RS provides time- and cost-efficient data in inaccessible areas that are difficult to reach. In particular, RS has been used in a widespread way to monitor shoreline changes in the coastal zone (White and El-Asmar, 1999, Chen et al., 2005; Sarpa et al., 2016). RS is much more practical today compared to the surveying and aerial photography that were used in the past. Besides, a change in the shoreline is easily perceived by evaluating satellite images for different time intervals and time depths for a given area (Temiz and Durduran, 2016; Louati et al., 2014; Lan et al., 2016). Satellite/airborne images that collect data in the optical portion of the electromagnetic radiation (0.4–2.5 μm) are widely used today to study the coastal zone, as

DOI: 10.1201/9781003354000-11

they are quite simple to interpret and are cost-effective. In addition, satellite images such as those from Landsat TM (Thematic Mapper) and ETM+ (Enhanced Thematic Mapper), Sentinel, MODIS, etc., are ideal tools for mapping the coastline (Alesheikh et al., 2007; Pham et al., 2013; Mustafa et al., 2017). By combining data from archival sources on the position of a coastline in the past with contemporary sources that depict its present position, one can identify how the coastline has evolved over the years. This information can help obtain a better understanding of coastal stability dynamics, and the main drivers of coastal erosion and other processes (sediment deposition, sea level rise, tidal effects, etc.) can be mitigated.

In purview of the above, this study utilizes a range of geoinformation technologies to map the changes that occurred in the deltaic coastline of the Axios and Aliakmonas rivers, located in Northern Greece over a period of 25 years. As such, the individual objectives of the present study are: 1) to evaluate the application of a supervised image classification method based on machine learning to extract the shoreline of the study area from Landsat images, and 2) to calculate the shoreline displacement rate and the rates of erosion and deposition using GIS.

8.2 STUDY AREA

For the purposes of this study, the Axios and Aliakmonas river deltas were chosen, located in the northwestern part of the Aegean Sea, Greece (Figure 8.1). Axios is the second largest river in the Balkans with a total length of 380 km, of which only part (76 km) flows through Greek territory. On the other hand, Aliakmonas is the longest river in Greece with a total main channel length of 322 km.

The two rivers create a fertile deltaic complex with significant environmental, economic, and agricultural importance for Northern Greece. The wider study area is mainly characterized by intensive agricultural activity, with the most prominent cultivated crop being rice, while a smaller proportion is occupied by cereal crops, cotton, tree crops, vineyards, and other perennial plantations. The two river deltas create an important wetland ecosystem for Greece, protected under the

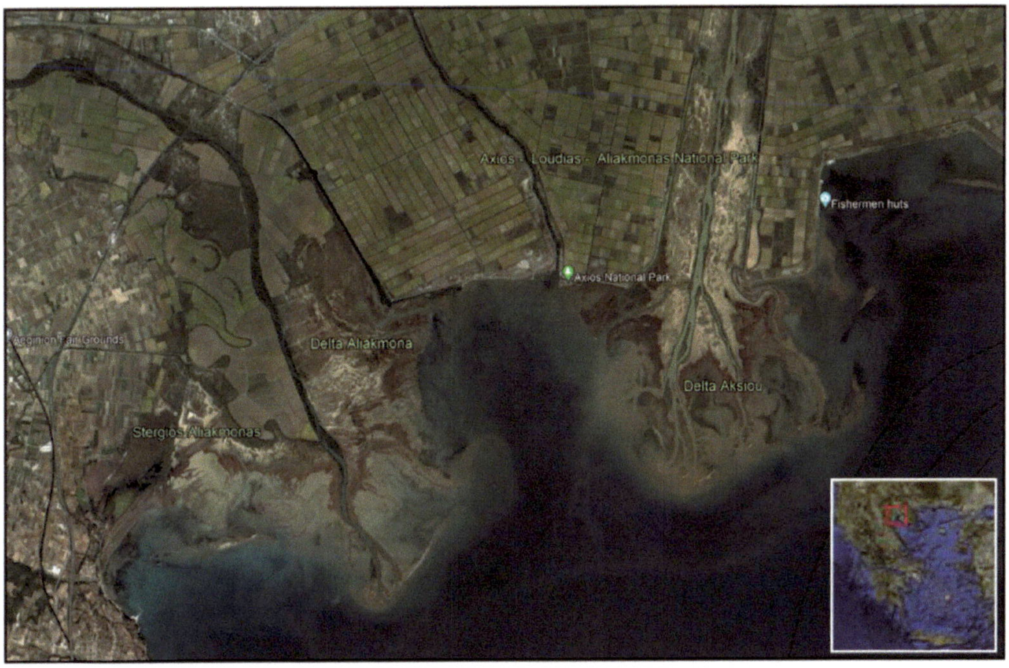

FIGURE 8.1 Axios and Aliakmonas deltas (Source: Google Earth).

NATURA2000 legislation, being the habitat for a number of protected fauna and flora species. This area is a dynamic cumulative coastal environment due to the activity of the two rivers, which are distinguished for discharging a large depositional load of water and sediment. This fact results in continuous changes of the shoreline that need to be monitored (Bedini E., 2007). Nevertheless, it has not been adequately studied by the scientific community, and the studies that have been conducted regarding shoreline variability are limited (Athanasiou, 2009; Demertzi et al., 2022). Therefore, it is useful to systematically monitor shoreline changes to ensure the sustainability of the sites.

8.2.1 Datasets

In this study, the extensive image archive of the Landsat satellite program was utilized to retrieve six multispectral satellite images from three different Landsat satellites, namely: Landsat 5 TM, Landsat 8 OLI, and Landsat 9. The Landsat satellite missions offer multispectral satellite data free of charge, providing a resolution ranging from 30 meters in the visible near-infrared and shortwave infrared (VISNIR and SWIR) spectrum to 100 meters in the thermal infrared (TIR) spectrum since 1972. These satellites have been extensively utilized for various environmental monitoring purposes, including the assessment of shoreline changes, as highlighted in previous studies (e.g. Bishop-Taylor et al., 2021; Cao et al., 2020) and are considered ideal for the spatiotemporal monitoring of wetlands.

The collected images span a period of 25 years, from 1995 to 2020. Specifically, in this study, "near-anniversary" cloud-free images were acquired between June and August for each year. All retrieved images were atmospherically and geometrically corrected Landsat Collection 2 – Level 2 products. All the data used herein was accessed at no cost from the United States Geological Survey platform Earth Explorer (https://earthexplorer.usgs.gov/). To facilitate the delineation of the study area, empirical line normalization was employed for all the images. This normalization technique utilized the 1995 image as a reference, aiming to correct variations in reflectance caused by solar conditions and detector degradation. Similar approaches have been applied in other published studies, such as the work of Petropoulos et al. (2015). Finally, the images were subsequently cropped to the extent of the two examined river deltas before any further analysis.

8.3 METHODOLOGY

The flowchart in Figure 8.2 presents step by step the process that was adopted for extracting the coastlines and performing the change detection of the coastal areas.

8.3.1 Coastline Extraction

The coastline extraction was implemented using two separate approaches: a semi-automated approach, utilizing the supervised classification method of Support Vector Machines (SVMs), and a manual approach, digitizing the shorelines using direct photo-interpretation, with the second approach serving as the "reference" dataset against which the output from the semi-automated approach was compared.

8.3.1.1 Semi-automated Approach

The semi-automated approach was implemented by applying a supervised pixel-based classification to the pre-processed satellite images, thereby automatically delineating water from land in two examined river deltas over the selected study period. In this scope, a Support Vector Machine was selected. SVM is a machine-learning supervised pixel-based classification technique originally formulated by Cortes and Vapnik (1995). It is a binary classification method that separates two classes of a training sample by searching for the optimal separating hyperplane that provides

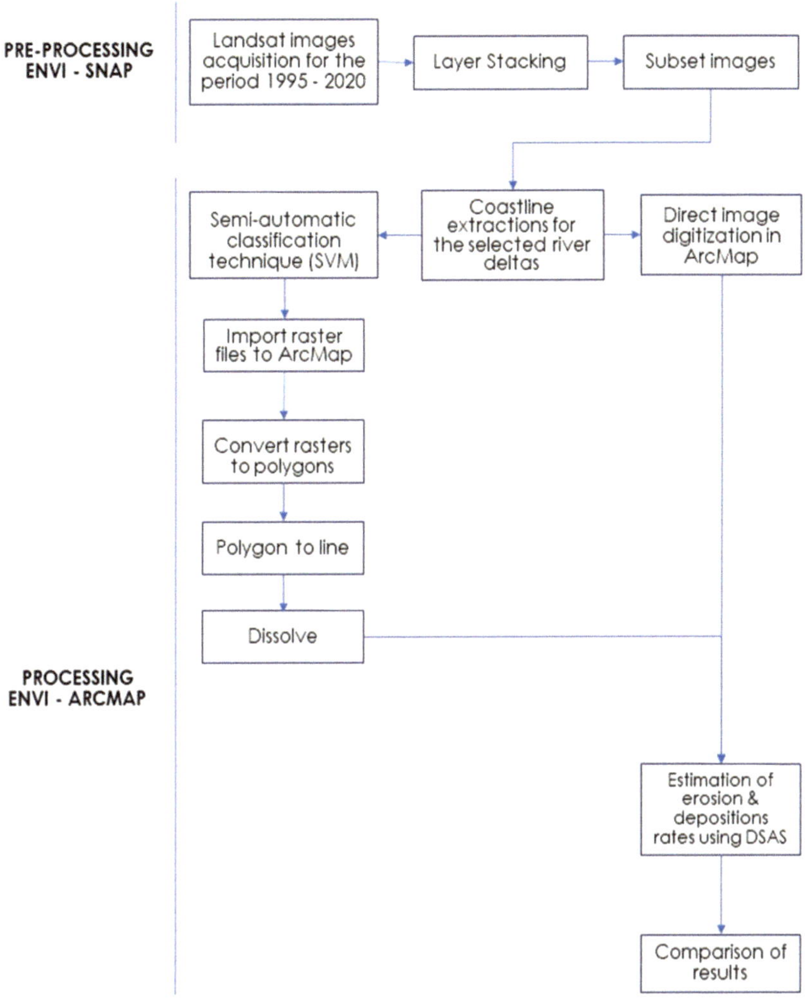

FIGURE 8.2 Flowchart of overall methodology.

the best generalization in a multidimensional space. The SVM classification method has generally shown to provide high accuracy results; it is also easy to implement and efficient in handling large data (Fragou et al., 2020).

The first step to implement SVM classification was to collect training and validation samples on which the algorithm is to be developed and evaluated. The classification scheme, which consisted of the classes "Water" and "Land", was formulated and for each category, a significant number of representative training and validation samples were obtained from each of the selected Landsat imagery adopting a random sampling approach. The spectral separability between the target map classes was assessed by estimating the Jeffries Matsushita and Transformed Divergence metrics in ENVI (version 5.3). The values of both metrics range from 0.0 to 2.0 where scores exceeding 1.9 represent well-separated spectral classes. If the separation value in a category is less than 1.0, then most likely we are dealing with regions of interest (ROIs) labeled in the same surface category. The spectral separability values obtained for each classified image achieved a satisfactory intraclasses spectral seperability as evidenced by the derived ROI Separability Reports that were ranging between 1.99 and 2.00.

After collecting training and validation samples, a supervised SVM classification technique was applied. During the SVM classification implementation, several hyperparameter had to be tuned, with one of the most important being the kernel type. In the present study, the radial basis function (RBF) was selected since it required the definition of minimum parameters for its execution and it has been shown to generally produce very good results in a wide range of classification studies (Chatziantoniou et al., 2017; Whyte et al., 2018; Kalogeropoulos et al., 2020; Srivastava et al., 2021; Popa et al., 2022). In terms of the Gamma parameter γ, the value of 0.167 was chosen, which is equal to the inverse of the number of spectral channels of each image (Petropoulos et al., 2014; Brown et al., 2018; Cass et al., 2019). Another crucial parameter in SVM implementation is Parameter C (penalty parameter). This expresses the error term and can be considered as the degree of correct classification or the degree of optimization that the algorithm must meet. The higher this parameter is, the lower the probability that the SVM optimizer will misclassify any point. In our case, a value of 100 (the maximum value that could be given) was set in order to minimize the probability of error. The last parameter to be set in SVM was the pyramid parameter, which was set to zero, allowing each image to be processed in its full analysis even though this would increase the computation time required.

The classification evaluation was performed by estimating the statistical metrics of the confusion matrix using the validation samples, including user's (UA) and producer's (PA) accuracies, Overall Accuracy (OA), and Kappa index (Kc) (Congalton and Green, 1999). Overall classification accuracy (overall accuracy) essentially shows the percentage of correctly classified pixels. The kappa coefficient (kappa coefficient) determines whether the results of the error matrix are significantly better than the random classification. In other words, it expresses the proportional reduction of the error generated by the classification process compared to the error of the random classification. In terms of User Accuracy, this indicates the probability that a pixel classified into a given category actually represents that category on the ground (Stehman, 2009). Finally, Producer Accuracy represents how well the reference data (training pixels) have been classified for a given type of cover (Pandey et al., 2019). After classifying the images into two categories (binary classification), the result obtained for each of them was saved in .tiff format and imported into ArcMap in order to extract in vector format the coastlines for each image corresponding to a reference year included in the analysis.

8.3.1.2 Manual Digitization Approach

In order to evaluate the derived shorelines from the SVM classification, reference shorines were produced using image photo-interpretation. Given that manual digitization is is a time-consuming procedure we limited ourselves to extracting the shorelines with this method only for the years 1995, 2005, and 2020. A false color composite of Landsat bands 7,4,2 was chosen to present the three satellite images, as it is already known to visibly separate the thin limits between the deltaic surface and water. The digitization was performed in ArcMap (version 10.5) software with the Editor tool.

8.3.2 Quantifying Shoreline Displacement and Erosion

Shoreline displacement results obtained from the two approaches were subsequently imported into a GIS environment to quantify the erosion and deposition. In this scope, the Digital Shoreline Analysis System (DSAS) tool was utilized. DSAS is an add-on to Esri ArcGIS (10.4–10.7+) that has been used in several studies found in the scientific literature to measure, quantify, compute, and monitor statistical rates of shoreline change from various historical sources and locations (Valderrama-Landeros and Flores-de-Santiago, 2019; Samra and Ali, 2020; Matin and Hasan, 2021; Rasha et al., 2021).

Specifically, this plugin allows the generation of cross sections positioned perpendicular to a baseline at a user-defined distance along the coast. It works by modeling the cross sections from a theoretical baseline at a distance from the last shoreline of the study and then calculating the convergence of each shoreline with each cross section (Abou Samra and Ali, 2020). This tool is user-friendly and has been widely used to study erosion and accretion phenomena along shorelines, and for this reason, it was chosen to be used for the purpose of this thesis. In particular, all six

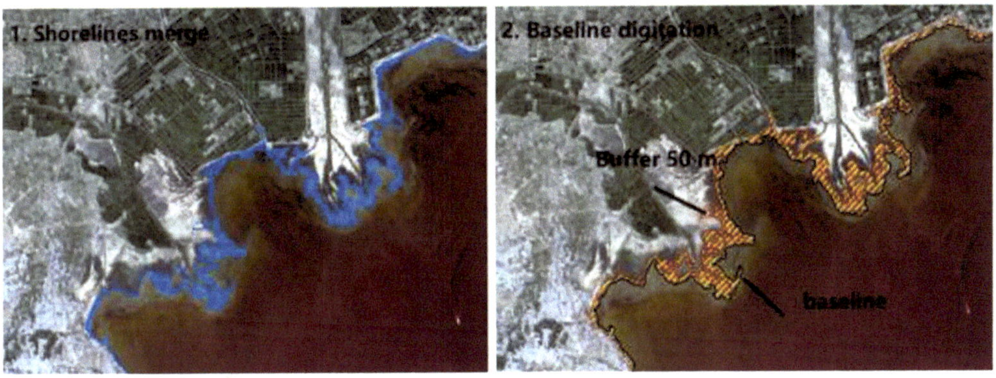

FIGURE 8.3 Shorelines merging and baseline digitization.

shorelines, which resulted from the implementation of the first technique (SVM-based) for each reference year, were imported into an ArcGIS environment in shapefile format. The six shorelines were merged into one shapefile using the merge command. The baseline was obtained by creating a 50 m buffer around the shorelines and digitizing the line defining the buffer (Figure 8.3). A similar procedure was followed for the three shorelines resulting from the implementation of the second technique (manual digitization).

Subsequently, the parameters for the shoreline and baseline were set, and the maximum search distance from the baseline and the maximum distance between the vertical cross sections were determined. A smoothing distance of 500 m was considered ideal for generating orthogonal cross sections to the shoreline across the study area and therefore for measuring appropriate shoreline change (U.S. Geological Survey, 2021). Once the vertical cross sections were created, the rate of shoreline displacement (erosion/deposition) was calculated over the entire study area using the Calculate rates tool, with the statistical method of linear regression (LRR – Linear Regression Rate) being selected.

8.4 RESULTS

8.4.1 Classification Accuracy Assessment

Overall, the classification of all six images studied was quite good, and the results obtained were satisfactory since, for all images, the percentage of overall classification accuracy was above 99%, and the Kappa index was greater than 0.997. The following tables detail the percentages of overall accuracy (OA) and Kappa index (Table 8.1) as well as User Accuracy (User Accuracy) and Producer Accuracy (Producer Accuracy) for each class (Table 8.2).

8.4.2 Erosion and Deposition Rates from DSAT Analysis

Using as data the shorelines extracted from the first approach, 618 cross sections were obtained, positioned perpendicular to the baseline, which capture the rate of change of the shoreline at each point. Negative values indicate erosion and positive values indicate deposition. The extreme values of the shoreline change rate observed were –3.00 m/yr (erosion) and +2.94 m/yr (deposition) for the whole reference period. In Table 8.3, the Deposition/Erosion Rates (m/yr) based on the two approaches are presented.

In Figure 8.4, the areas where high erosion was observed are shown in red and the areas with high deposition are shown in green. It is worth noting that in the areas in yellow, the rate of shoreline

TABLE 8.1
Percentages of Overall Accuracy and Kappa Index for Each Classified Image

Image	Overall Accuracy (%)	Kappa Index
23/06/1995	100	1
06/07/2000	99.87	0.997
20/07/2005	99.93	0.998
19/08/2010	100	1
01/08/2015	99.88	0.997
30/08/2020	99.93	0.998

TABLE 8.2
User Accuracy and Producer Accuracy Percentages for Each Classified Image

Image	Class	User Accuracy (%)	Producer Accuracy (%)
23/06/1995	land	100	100
	water	100	100
06/07/2000	land	100	99.77
	water	99.74	100
20/07/2005	land	100	99.88
	water	99.85	100
19/08/2010	land	100	100
	water	100	100
01/08/2015	land	99.93	99.90
	water	99.78	99.86
30/08/2020	land	99.95	99.89
	water	99.92	99.97

TABLE 8.3
Deposition/Erosion Rates(m/yr) Based on the Two Approaches

Deposition/Erosion Rates (m/yr) Based on Semi-automated Approach	Deposition/Erosion Rates (m/yr) Based on the Manual Approach
−3.00 / −1.50	−2.80 / −1.50
−1.49 / −0.50	−1.49 / −0.50
−0.49 / 0.50	−0.49 / 0.50
0.51/ 1.50	0.51 /1.50
1.51 / 2.94	1.51 / 2.98

displacement was very low, so it was assumed that neither deposition nor erosion occurred in these areas, i.e. practically no change was observed.

Similarly, using as data the coastlines extracted from the direct digitization method, 550 cross sections were obtained (Figure 8.5). The extreme values of the shoreline change rate observed in

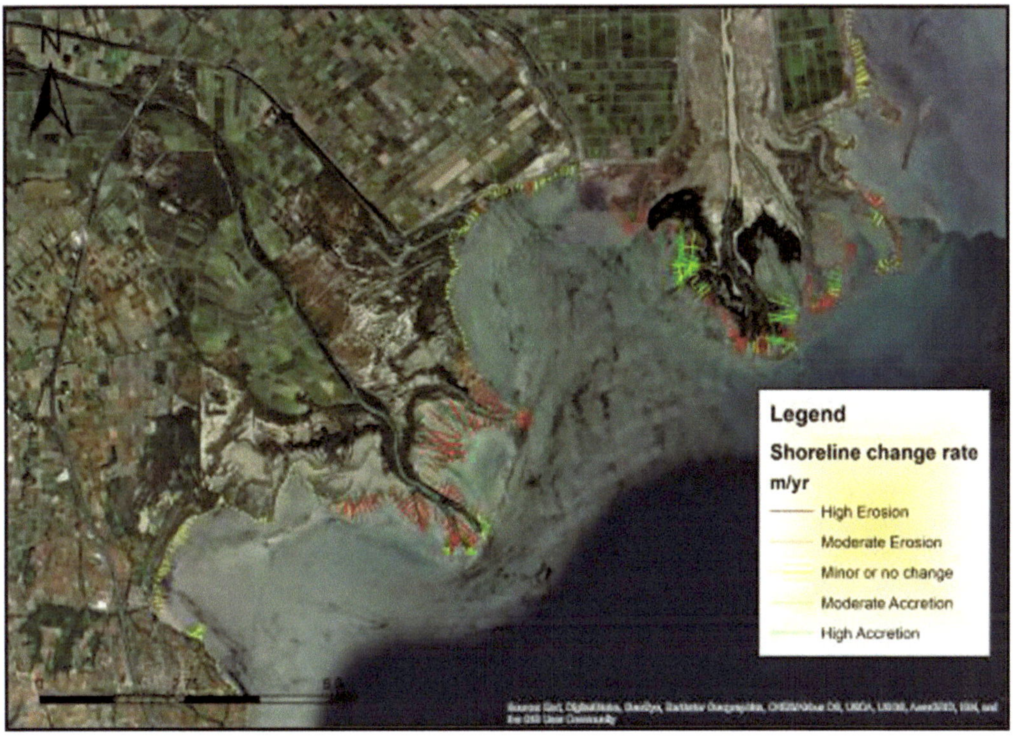

FIGURE 8.4 Shoreline displacement rate (1995–2020) on the basis of the results obtained from the SVM semi-automated approach.

FIGURE 8.5 Shoreline displacement rate (1995–2020) on the basis of the results obtained from the manual/direct digitization approach.

this case did not show a large deviation from the first technique; namely, the highest erosion rate was −2.8 m/yr, and +2.98 m/yr was the highest deposition rate for the whole reference interval.

8.5 DISCUSSION

The photo-interpretation approach for both study areas and the semi-automated approach only for the Aliakmonas River delta indicate the retreat of the deltaic coastlines, which confirms the general trend that has been reported in previous studies by Athanasiou (2009), Kapsimalis et al. (2005), and Petropoulos et al. (2015).

The calculation of the erosion and deposition rate by the DSAS tool, using as input the shorelines extracted by the SVM supervised classification method, had some discrepancies when compared with the results obtained by the manual digitization approach. By employing the semi-automated approach, the maximum erosion and deposition rates were higher than ±40 m/yr. On the contrary, the results of the second technique (manual digitization) gave clearly better results since no extreme values of erosion and deposition of ±40 m/yr were detected. The range of values in this case was between −2.8 and +2.98 m/yr. However, this range does not deviate much from the erosion and deposition values calculated from the implementation of the first technique (−3.0 – +2.94 m/yr, not including the extreme values), suggesting that both techniques gave the position of the shoreline relatively accurately. Despite the satisfactory results obtained from the semi-automatic image classification method (SVM) applied, the direct photo-interpretation, although a time-consuming process, gave safer and more accurate results overall.

The obtained results suggest a higher deposition rate for the Axios delta while in the Aliakmonas River delta erosion is the dominant process. According to the results, deposition in the Aliakmonas River occurs only in the southern mouth of the main distributary channel (up to 2.7 m/yr). The highest deposition values (2.74–2.94 m/yr) were observed at two points in the Axios River delta, while the highest erosion value (3 m/yr) was observed in the Aliakmonas delta. Deposition in the Aliakmonas River delta is found only in the southern part (estuary) of the main river (up to 2.05 m/yr). The highest deposition value (2.98 m/yr) was observed in the delta of the Axios River, while the highest erosion value (2.8 m/yr) was observed in the delta of the Aliakmonas River.

Insightful information was also provided by the statistics of the transect. The average deposition rate over the whole study area for the reference period (25 years) was calculated to be 1.01 m/yr, and the average erosion rate was 1.23 m/yr. The results obtained from the implementation of the two techniques (supervised SVM classification and on-screen digitization) showed that the deltas of the Aliakmonas and Axios rivers experienced geomorphological changes during the studied period (1995–2020) in some places while in others a relative stability was observed. Overall, the average erosion value calculated for the study area suggests significant short-term erosion (less than 30 years) due to human constructions related to coastal protection structures and sea level rise due to climate change. In fact, since the two deltas are located at low altitude, climate change is expected to increase erosion rates in a more adverse way in the future. Deposition, on the other hand, is mainly due to the high solid discharge (amount of sediment transported) combined with the wave processes (longitudinal currents) acting in the area. Both deltas have river processes as the dominant shaping factor, which explains the large amount of sediment transported and thus the high deposition rates. Comparatively, the two techniques implemented herein showed similar results in terms of erosion and deposition rates. Although manual digitization by photo-interpretation was used in a complementary way to verify the results obtained by the first technique (supervised SVM technique), it provided clearly better results compared to the semi-automated method.

Finally, although the study area has been examined by other researchers who have quantified the rate of erosion and deposition, comparisons with those studies should be made with care (Petropoulos et al., 2015; Demertzi et al., 2022). This is due to the different reference periods used and also due to the different types of results. In particular, this study presents the rate of shoreline change in the study area (in m/yr) for the period 1995–2020, while the research of Petropoulos et al.

presents the amount of erosion and deposition (area in km^2) for the periods 1984–1990, 1990–2003, 2003–2009, and overall, for the period 1984–2009. Similarly, the study by Demertzi et al. estimated the magnitude of erosion and deposition (area in km^2) for the periods 1986–2000, 2000–2014, and 2014–2022. In general, however, the results of these studies show a similar picture to the present study's findings in terms of where erosion and deposition are prevalent. Specifically, in the Axios river delta, deposition is greater than erosion mainly in the western and southern parts of the delta, while in the Aliakmon delta, erosion predominates with deposition being found only in the southern part (estuary) of the main river channel. In any case, discrepancies in results will clearly exist from study to study due to different reference data, differences in observation periods, different image analyses, and also due to the different techniques and tools implemented/used.

This study's findings provide valuable insights into the coastal dynamics and changes in the studied deltaic coastlines. Understanding erosion and deposition patterns is crucial for coastal management and planning, especially in areas susceptible to climate change impacts. The study can help coastal managers and policymakers make informed decisions regarding coastal protection structures and adaptation strategies. Although the present study focuses on specific study areas (Aliakmonas and Axios river deltas) in the Mediterranean basin, the findings and the developed methodological workflow can potentially be transferable to other landscapes with different climatic conditions. Deltas worldwide often exhibit similar geomorphological processes and respond to similar drivers, such as sediment transport and sea level rise. Therefore, the study results can provide a basis for comparison and analysis in other deltaic regions, using similar Earth observation (EO) datasets and methodologies. Overall, the study findings contribute to the scientific understanding of coastal dynamics, provide insights for coastal management, and highlight the importance of considering climate change impacts in coastal planning and protection efforts.

Current trends show an upward preference among researchers toward implementing classification also in cloud computing platforms such as the Google Earth Engine environment, with planetary-scale analysis capabilities. Earth Engine is regarded as a powerful and innovative tool for identifying and mapping shoreline changes since it has the ability to detect changes, map trends, and quantify variations in the Earth's surface. Hence, it is considered an ideal and innovative tool for detecting and mapping shoreline changes. Such investigations will further enhance the accuracy and capabilities of shoreline change assessments. Moreover, the presented methodological framework can be adapted to incorporate new sensors, including upcoming ones. As new satellite sensors and technologies emerge, researchers can explore their potential within this framework. By integrating advancements such as hyperspectral systems like EnMAP, a more comprehensive analysis of coastal processes can be achieved. Last but not least, the methodological framework employed in this study has the potential for broader applications beyond the specific study area. Coastal regions facing threats similar to erosion and deposition, such as permafrost regions, could benefit from adopting a similar approach.

8.6 CONCLUSIONS

In the present study, the rate of erosion and deposition at the Axios and Aliakmonas river deltas was estimated utilizing the Landsat archive for over 25 years. For this purpose, the deltaic fronts of the two rivers were extracted using a supervised classification and a direct digitization approach. The erosion and deposition rates were estimated using the DSAT software tool.

Results of the present study indicate that the rate of erosion estimated for the period under study, particularly in the Aliakmonas delta, was significantly high and is mainly due to anthropogenic activity but also to the rise in sea level due to climate change. The socioeconomic and environmental problems that can arise from the erosion of the adjacent area are diverse and need to be addressed and monitored. The two rivers discharge large amounts of sediment load, resulting in a large amount of deposition in the deltas. Especially in the delta of the Axios River, the deposition during the period under study is quite high, which may create significant problems that need to be addressed and monitored.

All in all, the methodological framework presented holds promise for application in other coastal areas facing different threats, including climate change, for example permafrost regions such as those found in the Arctic. Future research should explore the adaptability of the framework of new sensors and technologies to such regions (e.g. in the Arctic), including hyperspectral systems like EnMAP, to enhance the accuracy and capabilities of coastal monitoring. Continuously enriching a web map or platform that holds a range of EO data and offers the implementation of different techniques for mapping coastal areas and assessing their socioeconomic impacts would offer a valuable and up-to-date monitoring tool that could benefit both the areas under study as well as stakeholders, informing decision-making processes at the same time. In this context, the recently EU-funded project EO-PERSIST (https://eo-persist.eu/, accessed on 07/2023) aims to bridge this gap by developing an EO-based platform for assessing the socioeconomic impacts of climate change on permafrost areas located in the European Arctic. One of the key use cases in this context are coastal areas. The contribution of the methodological framework presented in the current study to EO-PERSIST's specific aims and objectives is under examination by our research team.

ACKNOWLEDGEMENT

In the present research study, the participation of all the co-authors, apart from Ioanna Tzanavari, has been financially supported by the project "EO-PERSIST" European Union's Horizon Europe Research and Innovation program HORIZON-MSCA-2021-SE-01-01 under grant agreement N. 101086386.

REFERENCES

Abou Samra R.M., Ali, R.R., 2020. Applying DSAS tool to detect coastal changes along Nile Delta, Egypt, *The Egyptian Journal of Remote Sensing and Space Sciences*, https://doi.org/10.1016/j.ejrs.2020.11.002.

Aedla R., Dwarakish G.S., Reddy D.V., 2015. Automatic shoreline detection and change detection analysis of Netravati-GurpurRivermouth using histogram equalization and adaptive thresholding techniques, *Aquatic Procedia*, 4, 563–570, ISSN 2214-241X, https://doi.org/10.1016/j.aqpro.2015.02.073.

Alesheikh A., Ghorbanali A., Nouri N., 2007. Coastline change detection using remote sensing, *International Journal of Environmental Science and Technology*, 4(1), 61–66, https://doi.org/10.1007/BF03325962.

Athanasiou H., 2009. Wetland habitat loss in Thessaloniki Plain (M.Sc. Dissertation 1990), Greece: University College, London.

Bedini E., 2007. Use of GIS and remote sensing to detect change along the coastline segment between Shkumbini and Semani rivers, Central Albania, *Geological Survey of Albania*, 40, https://doi.org/10.12681/bgsg.17218.

Bishop-Taylor R., Nanson R., Sagar S., Lymburner L., 2021. Mapping Australia's dynamic coastline at mean sea level using three decades of Landsat imagery, *Remote Sensing of Environment*, 267, 112734, ISSN 0034-4257, https://doi.org/10.1016/j.rse.2021.112734.

Brown R.A, Petropoulos G.P., Ferentinos K., 2018. Appraisal of the Sentinel-1 & 2 use in a large-scale wildfire assessment: A case study from Portugal's fires of 2017, *Applied Geography*, 100, 78–89, https://doi.org/10.1016/j.apgeog.2018.10.004.

Cao W., Zhou Y., Li R., Li X., 2020. Mapping changes in coastlines and tidal flats in developing islands using the full time series of Landsat images, *Remote Sensing of Environment*, 239, 111665, ISSN 0034-4257, https://doi.org/10.1016/j.rse.2020.111665.

Cass A., Petropoulos G.P., Ferentinos K.P., Pavlides A., Srivastava P.K., 2019. Exploring the synergy between Landsat and ASAR towards improving thematic mapping accuracy of optical EO data, *Applied Geomatics*, 1–12, https://doi.org/10.1007/s12518-019-00258-7.

Chatziantoniou A., Petropoulos G.P., Psomiadis E., 2017. Co-Orbital Sentinel 1 and 2 for LULC mapping with emphasis on wetlands in a Mediterranean setting based on machine learning, *Remote Sensing*, 9, 1–18, https://doi.org/10.1080/10106049.2017.1307460.

Chen S., Chen L., Liu Q., Li X., Tan Q., 2005. Remote sensing and GIS-based integrated analysis of coastal changes and their environmental impacts in Lingding Bay, Pearl River Estuary, South China, *Ocean & Coastal Management*, 48(1), 65–83, ISSN 0964-5691, https://doi.org/10.1016/j.ocecoaman.2004.11.004.

Congalton R., Green, K., 1999. *Assessing the accuracy of remotely sensed data: Principles and practices,* Boca Raton: CRC/Lewis Press.

Cortes C., Vapnik V., 1995. Support-vector networks, *Machine Learning,* 20, 273–297, https://doi.org/10.1007/BF00994018.

Demertzi I.I., Detsikas S.E., Tselka I., Petropoulos G.P., Karymbalis E., 2022. Deposition and erosion dynamics in Axios and Aliakmonas river deltas (Greece) with the use of Google Earth Engine and geospatial analysis tools. Chapter 5, pp: 91–108, In: *Geoinformatics for Geosciences. Advanced Geospatial Analysis using RS, GIS and Soft Computing.* Edited by Stathopoulos, N., A Tsatsaris & K. Kalogeropoulos, Publisher Elsevier, ISBN: 9780323989831.

Feyisa G.L., Meilby H., Fensholt R., Proud S., 2014. Automated water extraction index: A new technique for surface water mapping using Landsat imagery, *Remote Sensing of Environment*, 140, 23–35, ISSN 0034-4257, https://doi.org/10.1016/j.rse.2013.08.029.

Fragou S., Kalogeropoulos K., Stathopoulos N., Louka P., Srivastava P.K., Karpouzas S., Kalivas D.P., Petropoulos, G.P., 2020. Quantifying land cover changes in a Mediterranean environment using landsat TM and support vector machines, *Forests MDPI*, 11, 750769, https://doi.org/10.3390/f11070750.

Ghoneim E., Mashaly J., Gamble D., Halls J., AbuBakr M., 2015. Nile Delta exhibited a spatial reversal in the rates of shoreline retreat on the Rosetta promontory comparing pre- and post-beach protection, *Geomorphology*, 228, 1–14, ISSN 0169-555X, https://doi.org/10.1016/j.geomorph.2014.08.021.

Kalogeropoulos K., Stathopoulos N., Psaroginanis A., Pissias E., Louka P., Petropoulos G.P., Chalkias C., 2020. An integrated GIS-Hydro modeling methodology for surface runoff exploitation via small-scale reservoirs, *Water MDPI*, 12, 3182–3200, https://doi.org/10.3390/w12113182.

Kapsimalis V., Poulos S.E., Karageorgis A.P., Pavlakis P., Collins M., 2005. Recent evolution of a Mediterranean deltaic coastal zone: Human impacts on the Inner Thermaikos Gulf, NW Aegean Sea, *Journal of the Geological Society*, 162(6), 897908.

Karymbalis E., Chalkias C., Ferentinou M., Chalkias G., Magklara M., 2014. Assessment of the sensitivity of Salamina and Elafonissos islands to sea-level rise, *Journal of Coastal Research*, (70), 378–384, https://doi.org/10.2112/SI70-064.1.

Karymbalis E., Gaki-Papanastassiou K., Tsanakas K., Ferentinou M., 2016. Geomorphology of the Pinios River delta, Greece, *Journal of Maps*, 12, 12–21, https://doi.org/10.1080/17445647.2016.1153356.

Karymbalis E., Gallousi C., Cundy A., Tsanakas K., Gaki-Papanastassiou K., Tsodoulos I., Batzakis V.-D., Papanastassiou D., Liapis I., Maroukian H., 2022. Long-term spatial and temporal shoreline changes of the Evinos River delta, Gulf of Patras, Western Greece, *Zeitschrift für Geomorphologie*, 63/2–3, 141–155, https://doi.org/10.1127/zfg/2021/0684.

Lan P., Son T., Gunasekara K., Nhan N.T., Hien L.P., 2013. Application of remote sensing and GIS technology for monitoring coastal changes in estuary area of the Red river system, Vietnam, *Journal of the Korean Society of Surveying, Geodesy, Photogrammetry and Cartography*, 31(6–2), 529–538, http://dx.doi.org/10.7848/ksgpc.2013.31.6-2.529.

Louati M., Saïdi H., Zargouni F., 2014. Shoreline change assessment using remote sensing and GIS techniques: A case study of the Medjerda delta coast, Tunisia, *Arabian Journal of Geosciences*, http://dx.doi.org/10.1007/s12517-014-1472-1.

Maroukian H., Karymbalis E., 2004. Geomorphic evolution of the fan delta of the Evinos river in western Greece and human impacts during the last 150 years, *Zeitschrift für Geomorphologie,* 48/2, 201–217, http://dx.doi.org/0372-8854/04/0201.

Matin N., Hasan G.M.J., 2021. A quantitative analysis of shoreline changes along the coast of Bangladesh using remote sensing and GIS techniques, *Dhaka, Bangladesh: Catena - Elsevier B.V,* https://doi.org/10.1016/j.catena.2021.105185.

Mustafa M.T., Hassoon K.I., Hussain H.M., Abd M.H., 2017. Using water indices (NDWI, MNDWI, NDMI, WRI AND AWEI) to detect physical and chemical parameters by apply remote sensing and gis techniques, *International Journal Of Research -Granthaalayah*, 5(10), 117–128, https://doi.org/10.29121/Granthaalayah.V5.I10.2017.2289.

Pandey, P. C., Koutsias, N., Petropoulos, G. P., Srivastava, P. K., & Ben Dor, E. 2019. Land use/land cover in view of earth observation: data sources, input dimensions, and classifiers—a review of the state of the art. *Geocarto International, 36(9),* 957–988. https://doi.org/10.1080/10106049.2019.1629647.

Pandey P.C., Koutsias N., Petropoulos G.P., Srivastava P.K., Ben Dor E., 2021. Land use/land cover in view of earth observation: Data sources, input dimensions, and classifiers—a review of the state of the art, *Geocarto International*, 36(9), 957–988, https://doi.org/10.1080/10106049.2019.1629647.

Papoutsis I., Kontoes Ch., Loupasakis C., Kiratzi A., 2015. Geo-hazard monitoring in Northern Greece using Insar Techniques: The case study of Thessaloniki, Fringe 2015 Workshop, Italy, https://www.researchgate.net/publication/278026755.

Parcharidis I., Kourkouli P., Karymbalis E., Foumelis M., Karathanassi, V., 2012. Time series synthetic aperture radar interferometry for ground deformation monitoring over a small scale tectonically active deltaic environment (Mornos, Central Greece), *Journal of Coastal Research*, 29(2), 325–338, http://dx.doi.org/10.2112/JCOASTRES-D-11-00106.1.

Petropoulos G.P., Griffiths H.M., Kalivas D.P., 2014. Quantifying spatial and temporal vegetation recovery dynamics following a wildfire event in a Mediterranean landscape using EO data and GIS, *Applied Geography*, 50, 120–131, ISSN 0143-6228, https://doi.org/10.1016/j.apgeog.2014.02.006.

Petropoulos G., Kalivas D., Griffithsa H., Dimou P., 2015. Remote sensing and GIS analysis for mapping spatio-temporal changes of erosion and deposition of two Mediterranean river deltas: The case of the Axios and Aliakmonas rivers, Greece, *International Journal of Applied Earth Observation and Geoinformation,* 35, 217–228, http://dx.doi.org/10.1016/j.jag.2014.08.004.

Pham T.L., Son T.S., Gunasekara K., Thi Nhan N., Phu Hien L., 2013. Application of remote sensing and GIS technology for monitoring coastal changes in estuary area of the Red river system, Vietnam, *Journal of the Korean Society of Surveying, Geodesy, Photogrammetry and Cartography*, 31(6–2), 529–538, http://dx.doi.org/10.7848/ksgpc.2013.31.6-2.529.

Popa A.M., Onose D.A., Sandric I.C., Dosiadis E.A., Petropoulos G.P., Gavrilidis A.A., Faka A., 2022. Using GEOBIA and vegetation indices to assess small urban green areas in two climatic regions, *Remote Sensing*, 14(19), 4888, https://doi.org/10.3390/rs14194888.

Rasha M., Samra A., Ali R.R., 2021. Applying DSAS tool to detect coastal changes along Nile Delta, Egypt, *The Egyptian Journal of Remote Sensing and Space Science*, 24(3), Part 1, 463–470, ISSN 1110-9823, https://doi.org/10.1016/j.ejrs.2020.11.002.

Salimi S., Almuktar S.A., Scholz M., 2021. Impact of climate change on wetland ecosystems: A critical review of experimental wetlands, *Journal of Environmental Management*, 286, 112160, ISSN 0301-4797, https://doi.org/10.1016/j.jenvman.2021.112160.

Sarpa G., Ozcelik M., 2016. Water body extraction and change detection using time series: A case study of Lake Burdur, Turkey, *Journal of Taibah University for Science,* 11, 381–391, http://dx.doi.org/10.1016/j.jtusci.2016.04.005.

Srivastava P.K., Petropoulos G.P., Prasad R., Triantakonstantis D., 2021. Random Forests with Bagging and Genetic Algorithms coupled with least trimmed squares regression for soil moisture deficit using SMOS satellite soil moisture, *ISPRS International Journal of Geo-Information MDPI*, 10, 507–520, https://doi.org/10.3390/ijgi10080507.

Stehman S.V., 2009. Sampling designs for accuracy assessment of land cover, *International Journal of Remote Sensing*, 30(20), 5243–5272, https://doi.org/10.1080/01431160903131000.

Temiz F., Durduran S., 2016. Monitoring coastline change using remote sensing and GIS technology: A case study of Acıgöl Lake, Turkey, *Turkey, IOP Conf. Series: Earth and Environmental Science,* 44, 042033, https://doi.org/10.1088/1755-1315/44/4/042033.

Tsatsaris A., Kalogeropoulos K., Stathopoulos N., Louka P., Tsanakas K., Tsesmelis D.E., Krassanakis V., Petropoulos G.P., Pappas V., Chalkias C., 2021. Geoinformation technologies in support of environmental hazards monitoring under climate change: An extensive review, *ISPRS International Journal of Geo-Information*, 10, 94, https://doi.org/10.3390/ijgi10020094.

U.S. Geological Survey, 2021. Digital Shoreline Analysis System (DSAS) version 5.1 user guide, Reston, Virginia, https://doi.org/10.3133/ofr20211091.

Valderrama-Landeros L., Flores-de-Santiago F., 2019. Assessing coastal erosion and accretion trends along two contrasting subtropical rivers based on remote sensing data, *Ocean & Coastal Management*, 169, 58–67, ISSN 0964-5691, https://doi.org/10.1016/j.ocecoaman.2018.12.006.

White K., Asmar H.M., 1999. Monitoring changing position of coastlines using Thematic Mapper imagery, an example from the Nile Delta, *Geomorphology*, 29(1–2), 93–105, ISSN 0169-555X, https://doi.org/10.1016/S0169-555X(99)00008-2.

Whyte A., Feredinos K., Petropoulos, G.P. 2018. A new synergistic approach for monitoring wetlands using Sentinels -1 and 2 data with object-based machine learning algorithms, *Environmental Modelling & Software*, 104, 40–57, https://doi.org/10.1016/j.envsoft.2018.01.023.

Xia J., Luan G., Zhao F., Peng Z., Song L., Tan S., Zhao Z., 2021. Exploring the spatial–temporal analysis of coastline changes using place name information on Hainan Island, China, *ISPRS International Journal of Geo-Information,* 10(9), 609, https://doi.org/10.3390/ijgi10090609.

Part IV

Aquatic Vegetation Monitoring and Assessment

9 Application of Remote Sensing for Monitoring Aquatic Vegetation

A Case Study of Egyptian Coastal Wetlands Using MODIS/Terra Data

*Heba Bedair**
Department of Life Sciences, School of Natural Sciences, Shiv Nadar Institution of Eminence, Greater Noida, Uttar Pradesh, India, SILVIS Lab, Department of Forest and Wildlife Ecology, University of Wisconsin-Madison

Asmaa M. El-Makawy
Botany and Microbiology Department, Faculty of Science, Alexandria University, Alexandria Egypt

Haidy Nasser
Botany and Microbiology Department, Faculty of Science, Kafrelsheikh University, Kafrelsheikh Egypt

Marwa Waseem A. Halmy
Department of Environmental Sciences, Faculty of Science, Alexandria University, Alexandria, Egypt

9.1 INTRODUCTION

Aquatic vegetation (AV) is a vital biotic component of aquatic habitats and performs crucial ecological functions, including taking part in nutrient cycling, purification of water, and controlling harmful algal growth (Ou et al., 2008). Consequently, it is key in preserving the biological balance and biodiversity of aquatic environments like rivers and lakes (Fernandes et al., 2013). Aquatic vegetation can be grouped into emergent, floating, and submerged aquatic vegetation (EAV, FAV, and SAV respectively), where both FAV and EAV live above the water's surface while SAV lives underwater (Ou et al., 2008). Aquatic vegetation in many aquatic ecosystems including lakes, rivers, and wetlands is declining due to human-induced activities. Therefore, continuous monitoring of aquatic vegetation at different temporal and spatial scales is critical for understanding the processes that influence their growth, the changes in their distribution, and deciphering the relationship between aquatic vegetation and water quality (Wang et al., 2022).

* Corresponding author.

DOI: 10.1201/9781003354000-13

Remote sensing (RS) technologies allow efficient approaches for environmental monitoring and ecological evaluation (Chen et al., 2019; Fitton et al., 2021; Yang et al., 2022). Approaches based on remote sensing are increasingly being utilized as efficient tools in studies of aquatic ecosystems, often in integration with typical methods to overcome the constraints of traditional *in situ* approaches (Costa et al., 2021; Duffy et al., 2019; Free et al., 2020; Visser et al., 2013). For academics working at the local, regional, and global levels, RS is efficient in time and becoming more affordable (Ashraf et al., 2010; Free et al., 2020). Satellite images are increasingly used for identifying features because field surveys are costly and time-consuming (Pu et al., 2022). As data are collected from a distance, accessibility difficulties are reduced, allowing surveys of fragile or unsafe locations to be conducted while the operator stays at just one point of safety (Kalacska et al., 2017; Rhee et al., 2018). When gathered and processed properly, RS results can be employed in quantitative evaluation and to compare across time and location (Jia et al., 2017; Klemas, 2016). The accurate, quantifiable data collected by RS can be utilized to track the slow, progressive variations in many ecosystems. Furthermore, remote sensing can be the only feasible option to monitor remote and under-funded areas efficiently (Duffy et al., 2019).

Traditional techniques for monitoring SAV are frequently constrained by the aquatic environments and expensive costs, leading to a limited range of monitoring and negative consequences on SAV (Espel et al., 2020). Furthermore, inadequate data from a field survey could be incapable of representing the geographical and temporal dynamics of SAV in aquatic ecosystems. Thus, remote sensing technology has evolved into an efficient method for mapping SAV (Luo et al., 2017).

The remote sensing principle in observing vegetation entails utilizing sensors for estimating the reflected and emitted radiation by plants and their surroundings at different wavelengths (Wang et al., 2010). Healthy vegetation absorbs and reflects a variety of wavelengths in a distinct manner that can be measured by remote sensing sensors (Gausman, 1977). On the whole, a particular band or set of bands can be utilized to differentiate aquatic vegetation from water bodies; hence, multispectral sensor data has been broadly utilized for the extraction of aquatic vegetation from remotely sensed data (Villa et al., 2014). This has aided in the mapping of large-scale aquatic vegetation distributions as well as the measurement of intra-annual and inter-annual changes in aquatic vegetation in various aquatic habitats (Luo et al., 2016; Wang et al., 2019).

Several remote sensing techniques have been developed to map the spatial cover of aquatic vegetation. For example, Zhao et al. (2013) and Luo et al. (2014) employed Landsat TM and HJ-CCD images, respectively, to propose a decision tree model for recognizing EAV, FAV, and SAV in Taihu Lake. Brooks et al. (2015, 2019) and Visser et al. (2013) investigated the prospect of using optical remote sensing for the detection and mapping of SAV by looking at the possibilities for discriminating submerged macrophyte species based on spectral information. Moreover, Brooks et al. (2015) created SAV maps of the Laurentian Great Lakes in the United States using an approach based on Landsat data.

Dai et al. (2021) used a collection of programmed threshold approaches for accurate extraction of SAV under diverse water quality conditions from lakes in China's Yangtze Plain. Wang et al. (2022) developed an enhanced aquatic vegetation index for extraction of aquatic vegetation above the water surface in Taihu Lake in China. The vegetation index was calculated using the variation between two short-wave infrared (SWIR) bands and, when integrated with data on vegetation phenological characteristics, it contributed to the precise differentiation between EAV and FAV and the depiction of the seasonal and annual distributions of these vegetation types. Chen et al. (2023) constructed a 3D model and investigated the variation in the ecological settings on a regional scale using unmanned air vehicle (UAV) remote sensing technology.

Additionally, Sentinel-2 images were commonly used as an efficient source of data for aquatic vegetation monitoring. Liang et al. (2022) utilized a time series of Sentinel-2 images to study the phenological properties of SAV and FAV in Baiyangdian Lake, in the central north of the China Plain. The data were used to develop a phenology–pixel algorithm to efficiently extract SAV and reduce the effect of the fluctuations in water level and the occurrence of sparse FAV in the complex macrophyte-dominant lakes. The data were also used to detect when the growing season of SAV began and ended in the lake.

Sentinel-2 time series data were furthermore utilized in the assessment of the ability of machine-learning algorithms to accurately classify AV in temperate wetlands (e.g., Piaser & Villa, 2023). Rodríguez-Garlito et al. (2022) also devised a way to automatically detect the geolocation of invading aquatic plants in Guadiana, the second-longest river in Spain. To that goal, numerous categorization algorithms were used on publicly accessible multispectral satellite imagery acquired by the European Space Agency's Sentinel-2 satellite.

Furthermore, Ade et al. (2022) achieved genus- and species-level categorization between two invasive species of aquatic vegetation, *Eichhornia crassipes* (water hyacinth) and *Ludwigia* spp. (water primrose), in summer and fall utilizing machine-learning classifiers and Sentinel-2 multispectral satellite data. Gerardo and de Lima (2022) used Sentinel-2 satellite imagery to estimate vegetation and water indices for monitoring duckweeds (*Lemna minor*) in Lis River, one of the small rivers in Portugal. Pu et al. (2022) developed a method for efficiently extracting aquatic vegetation and algal blooms from the turbid water bodies of Chinese lakes using high spatial resolution images from Sentinel-2. The approach proved to be efficient in extracting aquatic vegetation and algal blooms, which are hard to distinguish using satellite imagery, particularly in turbid water bodies, due to the similarity of spectral characteristics between aquatic vegetation and algae.

9.2 MAIN FEATURES OF AV SPECTRAL SIGNATURES

The concepts underlying AV spectral features are quite similar to those underlying land vegetation. The absorption of radiation at specific wavelengths by the pigments of leaves, found in the palisade mesophyll (e.g., carotene and chlorophyll a and b), the scattering of near-infrared (NIR) in the spongy mesophyll, and the amount of water present in the plant are the primary factors affecting leaf reflectance. In the visible region, the principal bands of chlorophyll absorption are at 430 to 450 nm and 650 to 660 nm (Blanco, 2013). Chlorophyll a and b are the major pigments in the leaves of AV; thus, they absorb much of the red and blue wavelengths. The NIR reflectance of a normal healthy green leaf increases from 700 to 1200 nm. Healthy and green vegetation has high reflectance (40 to 60%) in the NIR, high transmittance (40 to 60%) through the leaves onto the underlying leaves, and low absorption (5 to 10%) (Blanco 2013). Vertically growing plants have a smaller leaf area and thus a smaller surface area to interact with the down-flowing radiation; yet broad-leaved plants and plants with a canopy have a larger effective reflecting area (Williams et al., 2003).

Indeed, it is beneficial to differentiate whether the AV is emergent or floating and submerged, as these various lifeforms behave differently in different conditions. The green part of the electromagnetic spectrum (570–578 nm) is thought to be the best to identify SAV, followed by the red and red edge regions. The main differences between AV species are due to pigment concentrations and cellular structure. In waters with higher quantities of dissolved and suspended particles, the green region allows more radiation to penetrate (Kirk, 1994). For water, the electromagnetic radiation absorption occurs in the optical spectrum region, causing dampening of the radiometric signals. Surface reflectance measurements for SAV are often less than 10% reflectance due to this phenomenon (Santos et al., 2016).

Thus, several factors should be accounted for when using remotely sensed data for the detection of AV. First, some AV can be found under a water column, even if that water column is thin (i.e., 0.5 m). The water column contains dissolved and suspended elements such as phytoplankton or salts in the water, and all these elements are affecting all light reaching submerged leaves (Rowan et al., 2021). Light wavelengths lower than 450 nm are significantly scattered. In contrast, NIR wavelengths are strongly absorbed, while wavelengths in between, in the visible zone (VIS), are variably influenced depending on the depth and composition of the water column (Kirk, 1994). The effects of surface reflectance, including glint, need to be addressed by considering the surface state, such as roughness, and the sensor's viewing angle to avoid complicating the analysis (Rowan & Kalacska,

2021). Optical remote sensing approaches applied for SAV are currently limited to shallow water ecosystems with clear to moderate waters, as both water and water column elements interfere with the spectral signature of aquatic objects (Rowan & Kalacska, 2021). As these waters are very translucent, the water column elements have an insignificant effect (Uudeberg et al., 2019).

Second, SAV is commonly covered up by a thin biofilm, in addition to a layer of detritus, bacteria, and epibionts. The thickness and makeup of the layer differ according to the prevailing flow regime, level of disturbances, and water quality (Rowan et al., 2021). Third, the spectral signature of an individual plant differs substantially according to the phenological phase at which it is detected, as the concentrations of the leaf pigments vary greatly during growth. Previous studies on estuarine SAV have revealed that the measurements obtained at a late stage in the growing season result in more accurate classification results (Klemas, 2013). While there are several studies conducted on the spectral separability of SAV, including both plants and macroalgae, which tend to be functionally and spectrally similar, most of these studies have concentrated on marine ecosystems and inadequately addressed freshwater ecosystems (Rowan et al., 2021). For example, Fyfe (2003) used spectroradiometer measurements to study three species of seagrass across different habitats and water conditions and revealed that they were spectrally separable at the species level as well as at the population level depending on the sampling geographic location. Another study by McIlwaine et al. (2019) was successful in detecting and separating eight macroalgae species using remotely sensed data. However, these investigations focused merely on SAV of marine ecosystems. One of the few studies in freshwater ecosystems by Brooks et al. (2019) used multiscale spectroradiometer data integrated with multispectral imagery to examine freshwater AV. The study distinguished only between individual samples, not classes, limiting the results' usefulness to broad-scale SAV mapping. Therefore, there is a need for maintaining spectral separability among freshwater SAV to achieve large-scale mapping and monitoring activities for freshwater ecosystems.

9.3 MAIN SATELLITES/SENSORS USED IN MONITORING AV

There are several advantages to using satellite observations, such as covering large regions, the availability of time series and data continuity, consistency and comparability, diversity of measurements, ability for processing using both traditional statistical methods and machine-learning approaches, and finally being mostly free and open access (Lafaye, 2017). Indeed, the choice of satellite and sensor depends on several factors, such as spatial resolution, temporal resolution of data acquisition, spectral range, revisit time, satellite mission longevity (e.g., the Landsat program provides the longest satellite data record since the 1970s), atmospheric and geographical conditions at the study area (for instance, coastal ecosystems in common tend to be narrow [beaches] or small [seagrass beds], and tropical zones usually have more cloud cover throughout the year), data cost, and specific research or monitoring objectives (Fu et al., 2020). There are several satellites and sensors used in monitoring aquatic vegetation, and the main ones are listed in Table 9.1.

Landsat: The Landsat series of satellites, such as Landsat 7 (4/15/1999–Present), Landsat 8 (2/1/2013–Present), and Landsat 9 (10/31/2021–Present), provide multispectral imagery that can be used to monitor aquatic vegetation. These satellites have sensors that capture data in various spectral bands, including red and near-infrared, which are particularly useful for vegetation monitoring (Claverie et al., 2018).

Sentinel-2: The Sentinel-2 satellites, such as Sentinel-2A (6/23/2015–Present), Sentinel-2B (3/7/2017–Present), and Sentinel-3A (2/16/2016–Present), operated by the European Space Agency (ESA), also provide multispectral data that can be used for monitoring AV. They have high-resolution optical sensors that capture data in multiple spectral bands, including bands sensitive to chlorophyll content and vegetation health (Claverie et al., 2018).

MODIS: The Moderate Resolution Imaging Spectroradiometer (MODIS) instrument, on board both Aqua and Terra satellites, captures data in multiple spectral bands, including red-edge and

TABLE 9.1
Characteristics of Frequently Used Sensors for AV

Satellites	Sensors	Resolution
Sentinel-2A and Sentinel-2B	Multi Spectral Imager (MSI)	290 km swath; band spatial resolution 10, 20, and 60 m; 5-day revisit
Sentinel-3A	Ocean and Land Color Instrument (OLCI)	1270 km swath; band spatial resolution 300 m; 27-day revisit
Landsat 7	Enhanced Thematic Mapper (ETM+)	185 km swath; band spatial resolution 15, 30, and 60 m; 16-day revisit
Landsat 8	Operational Land Imager (OLI)	185 km swath; band spatial resolution 15, 30, and 60 m; 16-day revisit
Terra and Aqua	Moderate Resolution Imaging Spectroradiometer (MODIS)	2330 km swath; band spatial resolution 250, 500, and 1000 m; 1–2-day revisit
Suomi NPP	Visible Infrared Imaging Radiometer Suite (VIIRS)	3040 km swath; band spatial resolution 375 and 750 m; 1–2-day revisit

near-infrared bands that are relevant for monitoring aquatic vegetation. MODIS data can be used to assess vegetation dynamics and productivity over large areas (Blanco, 2013).

HICO: The Hyperspectral Imager for the Coastal Ocean (HICO) is a sensor that captures hyperspectral data in the visible to near-infrared range. It is specifically designed for monitoring coastal and aquatic environments, including aquatic vegetation. HICO provides high-resolution and detailed spectral information for improved characterization of AV types (Davis et al., 2015).

WorldView-2 and WorldView-3: These high-resolution commercial satellites provide multispectral and hyperspectral data, respectively. They have sensors that capture data in multiple spectral bands, including bands suitable for AV monitoring. These satellites offer detailed imagery that allows for fine-scale monitoring of AV (Wilson et al., 2022).

CHRIS/PROBA: The Compact High-Resolution Imaging Spectrometer (CHRIS) aboard the Project for On-Board Autonomy (PROBA) satellite captures hyperspectral data. Its high spectral resolution makes it useful for monitoring AV and studying the biochemical composition of vegetation (Barnsley et al., 2004; Guanter et al., 2005).

PACE: The PACE (Plankton, Aerosol, Cloud, ocean Ecosystem) satellite is a future Earth observation mission developed by the National Aeronautics and Space Administration (NASA) in collaboration with other partners. PACE is designed to provide a comprehensive understanding of the Earth's ocean and atmosphere systems, with a particular focus on studying the interactions between marine ecosystems and the atmosphere. The primary goal of the PACE mission is to observe and monitor key aspects of the Earth's oceans, including phytoplankton, aerosols, and clouds (Werdell et al., 2019). PACE will carry multiple advanced instruments to collect data related to these components:

1. Ocean Color Instrument (OCI): OCI is a hyperspectral radiometer that will measure the spectral radiance of light reflected from the ocean surface. It will provide highly detailed information about the distribution and properties of phytoplankton, including their abundance, diversity, and physiological characteristics (Hasekamp et al., 2019).
2. Polarimeter: PACE will also include a polarimeter, which measures the polarization properties of light reflected by the Earth's surface and atmosphere. This instrument will help study aerosols and clouds, providing valuable data on their composition, size, and distribution (Ramer et al., 2019).

3. Multi-Angle Imager for Aerosols (MAIA): MAIA is an additional instrument on board PACE that focuses specifically on aerosol measurements. It will provide information on the types, sizes, and concentrations of aerosol particles in the atmosphere, helping to understand their impact on climate and air quality.
4. Clouds and the Earth's Radiant Energy System (CERES): The CERES instrument suite, which is already installed on several other satellites, will also be part of PACE. CERES measures the Earth's radiation budget and helps quantify the energy exchange between the Earth's surface, clouds, and the atmosphere (Ramer et al., 2019; Hasekamp et al., 2019).

SBG: The SBG (Surface Biology and Geology) satellite is in its early stages of development and will provide hyperspectral imagery in the visible and shortwave infrared, as well as multi- or hyperspectral imagery in the thermal infrared (Cawse-Nicholson et al., 2021).

9.4 MAIN SPECTRAL INDICES USED IN ASSESSING SUBMERGED AND EMERGENT AV

There are several vegetation indices (VIs) commonly used in general vegetation monitoring, including terrestrial or AV. VIs based on combinations of spectral bands are used as indicators of vegetation biomass, density, growth, and vigor (Malthus, 2017). In Tables 9.2 and 9.3 we present some of these general spectral indices utilized for monitoring vegetation.

9.4.1 Normalized Difference Vegetation Index

The Normalized Difference Vegetation Index (NDVI) is one of the oldest and most commonly used vegetation indices. It is a simple indicator for vegetation health detection and characterization of

TABLE 9.2
Spectral Indices Used in Monitoring Vegetation

Index	Formula	Reference
NDVI	$(R_{NIR} - R_{Red}) / (R_{NIR} + R_{Red})$	(Rouse et al., 1973)
EVI	$G (R_{NIR} - R_{RED}) / (R_{NIR} + C_1R_{RED} - C_2R_{BLUE} + C_3)$	(Huete et al., 1999)
NDWI	$(R_{NIR} - R_{SWIR}) / (R_{NIR} + R_{SWIR})$	(Gao, 1996)

TABLE 9.3
Spectral Indices Specifically Used in Monitoring Submerged and Emergent Aquatic Vegetation

Index	Formula	Reference
NDAVI	$(R_{NIR} - R_{Blue}) / (R_{NIR} + R_{Blue})$	(Villa et al., 2013a, b, 2014)
WAVI	$(1 + L) [(R_{NIR} - R_{BLUE}) / (R_{NIR} + R_{BLUE} + L)]$	(Villa et al., 2013a, b, 2014)
SVSI	$TC_1\text{-}TC_2$	(Luo et al., 2016)
AVI	$-\text{wetness} = \sum_{i=1}^{6} k(\text{»}_i)R(\text{»}_i)$	(Luo et al., 2023)

canopy growth or vigor. It is calculated from multispectral information through a normalization algorithm as the ratio between red and near-infrared bands:

$$NDVI = (R_{NIR} - R_{Red}) / (R_{NIR} + R_{Red})$$

NIR is a good indicator for vegetation health detection as healthy vegetation reflects NIR more than any other wavelength (Rouse et al., 1973; Tran et al., 2022). The values of NDVI range from −1 to 1. Mathematically, NDVI is considered the basis for other indices (Tran et al., 2022).

Even though NDVI was initially created for the analysis of terrestrial vegetation, it is a useful tool for studying some aspects of aquatic ecosystems, particularly in relation to:

1. Algal Blooms Monitoring: Algal blooms in water bodies can be evaluated and tracked using NDVI. Pigments found in algae can both absorb and reflect light, particularly red and near-infrared light. The spatial distribution of algal biomass can be inferred from the NDVI by examining the reflectance values in these bands. High NDVI values may indicate areas with dense algal growth, which can help in the early detection and control of harmful algal blooms (Zhou et al., 2022).
2. Submerged Vegetation Mapping: Submerged vegetation is important for biodiversity and ecosystem health in aquatic environments. Seagrass beds and other submerged aquatic plants can have their extent and vitality mapped and tracked using the NDVI. It is possible to identify areas with healthy vegetation by comparing the NIR and red reflectance, which can help with habitat assessments and conservation efforts (Chen et al., 2018).
3. Mapping and Monitoring of Wetlands: Wetlands are significant ecological regions that frequently have unique vegetation patterns. NDVI can be used to map and track the health of wetland ecosystems, in addition to highlighting areas with various vegetation types. Researchers can evaluate wetland dynamics, such as vegetation phenology, changes in land cover, and possible effects on ecosystem services, by examining changes in NDVI over time (Ashok et al., 2021).
4. Water Quality Assessment: NDVI can infer information about water conditions indirectly, even though it is not a direct indicator of water quality parameters. Water bodies' reflectance characteristics and light penetration can be affected by changes in water quality, such as turbidity or concentrations of suspended sediment. These alterations may have an impact on the NDVI values of nearby vegetation, revealing variations in water quality (Lin et al., 2021).

9.4.2 Enhanced Vegetation Index

The Enhanced Vegetation Index (EVI) is a greenness-modified index that enhances the capabilities of NDVI by including corrective factors for some atmospheric disturbances and canopy background noise (Huete et al., 1997; Villa et al., 2013a). It is used for assessing the vigor of vegetation and compensating for atmospheric disturbances; moreover, it has a high response to canopy structure (Huete et al., 1997; Malthus, 2017).

$$EVI = G\ (R_{NIR} - R_{RED}) / (R_{NIR} + C_1 R_{RED} - C_2 R_{BLUE} + C_3)$$

Some studies showed that EVI was more effective than NDVI for mapping emergent aquatic vegetation such as *Phragmites australis* beds (Villa et al., 2013a). It was also revealed that EVI provided a better estimate for biophysical properties of mangroves at high density than NDVI (Tran et al., 2021).

9.4.3 Normalized Difference Water Index

The Normalized Difference Water Index (NDWI) can be used to monitor AV with the function formulated by Gao (1996). This version of NDWI uses NIR and SWIR wavelengths to monitor the changes in leaf water content.

$$NDWI = (R_{NIR} - R_{SWIR}) / (R_{NIR} + R_{SWIR})$$

The reflectance in SWIR used in this index indicates variations in both the water content of the vegetation and the structure of the spongy mesophyll within the vegetation canopies. As a result, it's an effective tool for tracking the health and condition of AV. It is helpful in separating FAV, EAV, and SAV growth habits (Malthus, 2017; Chen et al., 2018).

In general, most spectral vegetation indices were traditionally intended for studies focusing on terrestrial vegetation rather than AV. Villa et al. (2014) introduced new spectral vegetation indices, the Normalized Difference Aquatic Vegetation Index (NDAVI) and the Water Adjusted Vegetation Index (WAVI), which targeted aquatic vegetation. The new spectral indices provide good performance in distinguishing aquatic vegetation from terrestrial vegetation.

9.4.4 Normalized Difference Aquatic Vegetation Index

The Normalized Difference Aquatic Vegetation Index (NDAVI) is a modified version of NDVI, as it replaces the visible red spectral band in the NDVI formula with the visible blue spectral band (Villa et al., 2013a, b).

$$NDAVI = (R_{NIR} - R_{Blue}) / (R_{NIR} + R_{Blue})$$

NDAVI is specifically used for aquatic vegetation mapping. Therefore, the background consists of water and wet terrain. It has proved its effectiveness and accuracy in monitoring the conservation status of *P. australis* beds (Villa et al., 2013b). It is effectively used in differentiating aquatic from terrestrial vegetation (Villa et al., 2015) and also mapping types of large AV communities (Villa et al., 2015; Wang et al., 2023).

9.4.5 Water-Adjusted Vegetation Index (WAVI)

The Water-Adjusted Vegetation Index (WAVI) is an efficient index for mapping aquatic vegetation and differentiating it from open water components and terrestrial vegetation (Malthus, 2017).

$$WAVI = (1+ L) [(R_{NIR} - R_{BLUE}) / (R_{NIR} + R_{BLUE} + L)]$$

where L is a correction factor to compensate for the background water effect ranging from 0 to 1 (Villa et al., 2014).

WAVI was developed to improve the detection of different macrophytes in the Mantua lakes, and it enabled extraction and classification of aquatic vegetation and revealed distinct multi-temporal behaviors of these macrophytes (Villa et al., 2015).

9.4.6 Submerged Vegetation Sensitive Index

The Submerged Vegetation Sensitive Index (SVSI) was formulated by integrating brightness and greenness obtained from the tasseled cap transformation (Luo et al., 2016). Its function is calculated as follows:

$$SVSI = TC_1\text{-}TC_2$$

where TC_1 and TC_2 represent the first and second components of the tasseled cap transformation, respectively, which are referred to as brightness and greenness. SVSI could be utilized to differentiate between SAV and water (Luo et al., 2017).

9.4.7 Aquatic Vegetation Index

The Aquatic Vegetation Index (AVI) is a newly established index that was derived from the tasseled cap transformation used to separate aquatic vegetation (AV) and non-aquatic vegetation (non-AV) as a wetness-based index (Luo et al., 2023). The equation is as follows:

$$\mathrm{AVI} = -\mathrm{wetness} = \sum_{i=1}^{6} k(\lambda_i)R(\lambda_i)$$

where $k(\lambda_i)$ is the wetness coefficient for band i and $R(\lambda_i)$ is the reflectance for band i of the image. $k(\lambda)$ is constant for a specific sensor, but it varies between different sensors. AVI is a very effective index to classify AV and non-AV because the wetness coefficients ($k(\lambda)$) depend on the wavelength of the bands (Luo et al., 2023). The values are positive and minor for the visible and near-infrared (VIS/NIR) bands. For the two short-wave infrared bands (SWIR1 and SWIR2), the values are negative.

The above-mentioned indices are just examples utilized in remote sensing studies for investigating AV. The selection of a particular index depends on the study and research goals, the availability of spectral bands, and the characteristics of the vegetation under investigation.

9.5 PROSPECT OF USING REMOTE SENSING IN MONITORING AV

The prospect and potential of using remote sensing in monitoring AV are significantly promising. Remote sensing has multiple advantages, making it more applicable in monitoring aquatic vegetation. Here are some potential prospects of applying remote sensing in aquatic vegetation monitoring:

9.5.1 Large-Scale Monitoring

Most commonly utilized methods of monitoring AV comprise field-based observations through diving, optical procedures that depend on underwater and aerial drones, commercial very high-resolution satellite imagery, or aerial photographs. While these techniques provide detailed data on the coverage and distribution of SAV, they have certain limitations. They only cover small areas and are not suitable for large-scale investigations, or the process of data collection is costly, in addition to lacking a systematic approach which hinders consistent, repeatable mapping of vegetation (Huber et al., 2022). All these limitations hinder the capability to fully understand the dynamics, distribution, monitoring, and change detection of underwater coastal communities (Rowan & Kalacska, 2021). Remote sensing enables collecting and gathering data over large water bodies and provides extensive coverage and spatial data at different scales, from local to regional, quickly and repeatedly. This capability facilitates effective and efficient monitoring of AV over a wide region (Huber et al., 2022).

9.5.2 Mapping and Classification

Remote sensing is utilized for the mapping and classifying of aquatic vegetation at various spatial scales. Various methods and techniques are established to separate vegetation at various levels, for instance, separating aquatic vegetation and non-aquatic vegetation, and separating floating aquatic vegetation, submerged aquatic vegetation, and emergent aquatic vegetation using different indices or a combination of them, and newly established algorithms and models (Malthus, 2017; Luo et al., 2023; Wang et al., 2023). Various spectral characteristics of the vegetation can be recognized and analyzed, enabling the classification of different species or detection of changes in vegetation cover over time.

9.5.3 Species Identification

Remote sensing approaches have the potential to identify the species of submerged aquatic vegetation and quantify their extent. Brooks et al. (2022) identified the invasive species *Myriophyllum spicatum* using an uncrewed aerial system (UAS)-enabled methodology, which showed better results relative to earlier studies.

9.5.4 Health Assessment

Remote sensing allows health assessment of aquatic vegetation by detecting changes in size, color, and density. It is more effective than manual or conventional procedures which are time-consuming, costly, and may be dangerous (Rowan & Kalacska, 2021). Studies that utilized high-resolution imagery integrated with a newly developed decision function have shown to be practical and efficient methods to dynamically and precisely monitor AV, specifically SAV at large regional scale. This approach could offer valuable assistance for the ongoing surveillance of ecosystem health (Chen et al., 2018).

9.5.5 Vegetation Cover and Biomass

Several simple vegetation indices, effective algorithms, and models were established and are available for quantitative and qualitative assessments of vegetation cover, growth dynamics, and vigor of aquatic vegetation (Villa et al., 2015; Xue & Su, 2017; Luo et al., 2023). The natural vegetation cover and biomass are considered ecosystem health indicators and can be mapped using high-resolution remotely sensed data (Klemas, 2013). Mapping aquatic vegetation cover, particularly submergent aquatic vegetation, is a significant indicator of environmental conditions for both marine and freshwater habitats (Lathrop et al., 2006). They are also utilized in the quantification of the total amount absorbed by SAV and the assessment of coastal sustainable development (Ni et al., 2023).

9.5.6 Habitat Mapping

Remote sensing can help in habitat mapping of AV, identifying suitable areas for a specific species, and recognizing critical habitats for conservation and management. By combining remotely sensed data with other environmental variables in habitat distribution modeling approaches, suitable conditions for the growth and distribution of vegetation can be predicted. LiDAR-derived topographic data were applied for mapping habitat suitability of AV in wetlands and river channels (Knight et al., 2009). Mapping habitat has a crucial role in defining habitat suitability for species at risk such as threatened and endangered species. This can help in guiding the most appropriate conservation actions, including the reintroduction and conservation of such species, or restoring suitable habitat so that at-risk populations can expand (Questad et al., 2014).

9.5.7 Change Detection

Remote sensing can detect changes in aquatic vegetation cover over time by comparing satellite images from different periods. It is difficult to detect long-term changes, such as shifts in species distribution or the expansion/contraction of vegetation patches due to natural or anthropogenic factors (Luo et al., 2016; Bhuyan, 2022). Remote sensing of AV has seen promising progress in the last decade. It has proved effective in overcoming the limitations of traditional methodologies (Malthus, 2017). It is very dynamic and updated, as newly established approaches, algorithms, and models are usually developed to monitor aquatic vegetation, including distributions, classification, identification, vigor, health, biomass, and cover of macrophytes.

9.6 CASE STUDY: ASSESSING CARBON SEQUESTRATION BY EMERGENT REED VEGETATION IN COASTAL WETLANDS USING MODIS/TERRA DATA

The value of wetlands, particularly coastal ones, as effective and efficient carbon sinks has been highlighted recently (Donato et al., 2011; McLeod et al., 2011). Wetlands remove carbon dioxide from the atmosphere through photosynthesis, where part of the captured carbon is stored within the above-ground and below-ground biomass of the vegetation system. The accumulation of carbon in these stores over long periods of time makes wetlands considerably rich carbon sinks (Murray et al., 2011). Recent studies have highlighted the crucial role coastal habitats play in sequestering carbon dioxide (Beaumont et al., 2014). It is now realized that the loss of natural carbon sinks such as wetlands means losing their critical role in regulating climate change (Erwin, 2009; Duarte et al., 2008; Rogers et al., 2019).

The northern coastal land of Egypt, extending for about 1000 km from Rafah in the east to Sallum in the west, is characterized by the existence of five lakes (Bardawil, Burullus, Edku, Manzala, and Mariut). The wetlands surrounding these lakes are of global importance for migratory birds and biodiversity conservation (Shaltout et al., 2019). Population and economic growth have resulted in extensive expansion in land conversion along the coastline. Consequently, wetlands suffer from degradation, habitat loss, and pollution. The following study focused on assessing the carbon sequestration potential of coastal wetlands surrounding the three main coastal lakes – Burullus, Manzala, and Bardawil – which are aligned from west to east along the Mediterranean coast of Egypt (Figure 9.1). The main goal of the study is to highlight how the carbon sequestration ecological role of the coastal wetlands of Egypt can contribute to the long-term mitigation effects for global warming and climate change.

9.6.1 EVI Time Series Data Acquisition and Preprocessing

Time-series data composed of 22 scenes of the enhanced vegetation index were used in the study. The EVI data used were extracted from MOD13Q1 V6.1 products that were acquired from MODIS/Terra in 2022. These scenes were extracted from the MODIS/Terra vegetation indices L3, acquired every 16 days at spatial resolution of 250 m (Didan, 2021). The area covering the investigated wetlands was extracted from the scenes. The EVI time series was used to capture the different phases of plant growth throughout the year.

9.6.2 Estimating Mean Annual Vegetation Carbon Sequestration

The estimation of the mean annual vegetation carbon stock using remotely sensed data was investigated in different studies (e.g., Sanga-Ngoie et al., 2012; Vicharnakorn et al., 2014; Batsaikhan et al., 2020; Li et al., 2020; Dai et al., 2020). The vegetation carbon sequestration of each pixel was estimated from the EVI as an indicator of biomass carbon stock throughout the year, following the approach used by Dai et al. (2020) as follows:

$$VCS=978.8*EVI2-54.07*EVI+67.11$$

where VCS is the carbon stock in g C/m^2 and EVI is the value of the enhanced vegetation index. The sum of positive differences in biomass carbon stocks over the investigated dates throughout a given year was calculated. The EVI time-series images were utilized for reconstructing the dynamics of biomass carbon stock across the year using the most appropriate model. The cumulative annual

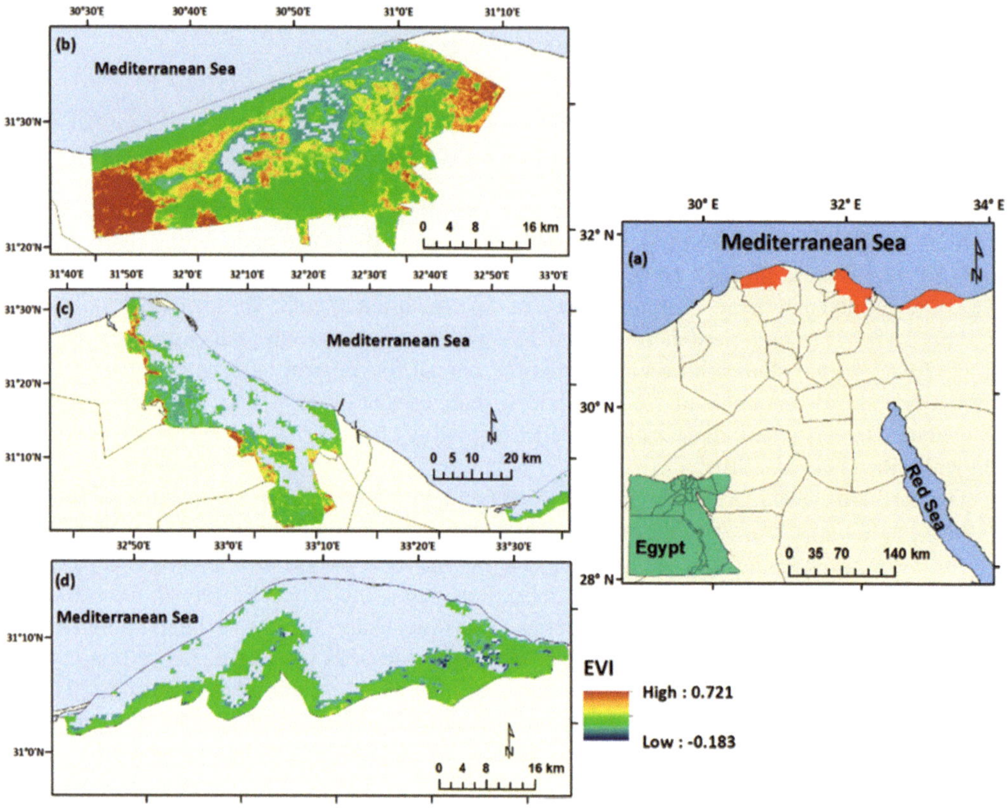

FIGURE 9.1 (a) Location of the studied coastal wetlands in the northeastern part of Egypt. The distribution of the values of the enhanced vegetation index (EVI) is presented for each wetland ordered from west to east: (b) Lake Burullus; (c) Lake Manzala; and (d) Lake Bardawil wetlands.

vegetation carbon sequestration per pixel was estimated as the sum of the positive differences in VCS over the available dates throughout the year.

9.6.3 Results and Discussion

Lake Burullus wetland was the highest in the potential vegetation carbon sequestration of three coastal wetlands, with a mean annual vegetation carbon sequestration of 98.04 g C/m^2 in 2022 and a standard deviation of 55.85 g C/m^2 (Figure 9.2, Table 9.4), followed by Manzala, then Bardawil wetlands. As a whole, Burullus wetland vegetation sequestered 1997.44 kgC/year, which is almost 50% of the collective potential vegetation carbon sequestration for the three wetlands. Overall, the vegetation in the three coastal wetlands sequestered 4019.76 kgC in 2022.

While Lake Burullus wetland attained the highest total cumulative annual vegetation carbon stock, Lake Manzala attained the highest mean cumulative annual vegetation carbon stock (Figure 9.3, Table 9.2). The proportion of the area contributing to the net cumulative vegetation carbon sequestration (VCS) in the studied wetlands varied considerably, being relatively higher in Burullus wetland compared to the other two wetlands (Figure 9.4), where over 75% of the wetland contributes to the cumulative VCS stock.

Significant spatial variation was noticed in carbon sequestration of the vegetation along the environmental gradient in the three studied wetlands, which can be attributed to fluctuations in the water level and the impact of human activities throughout the year. Overall, areas near the water proper in

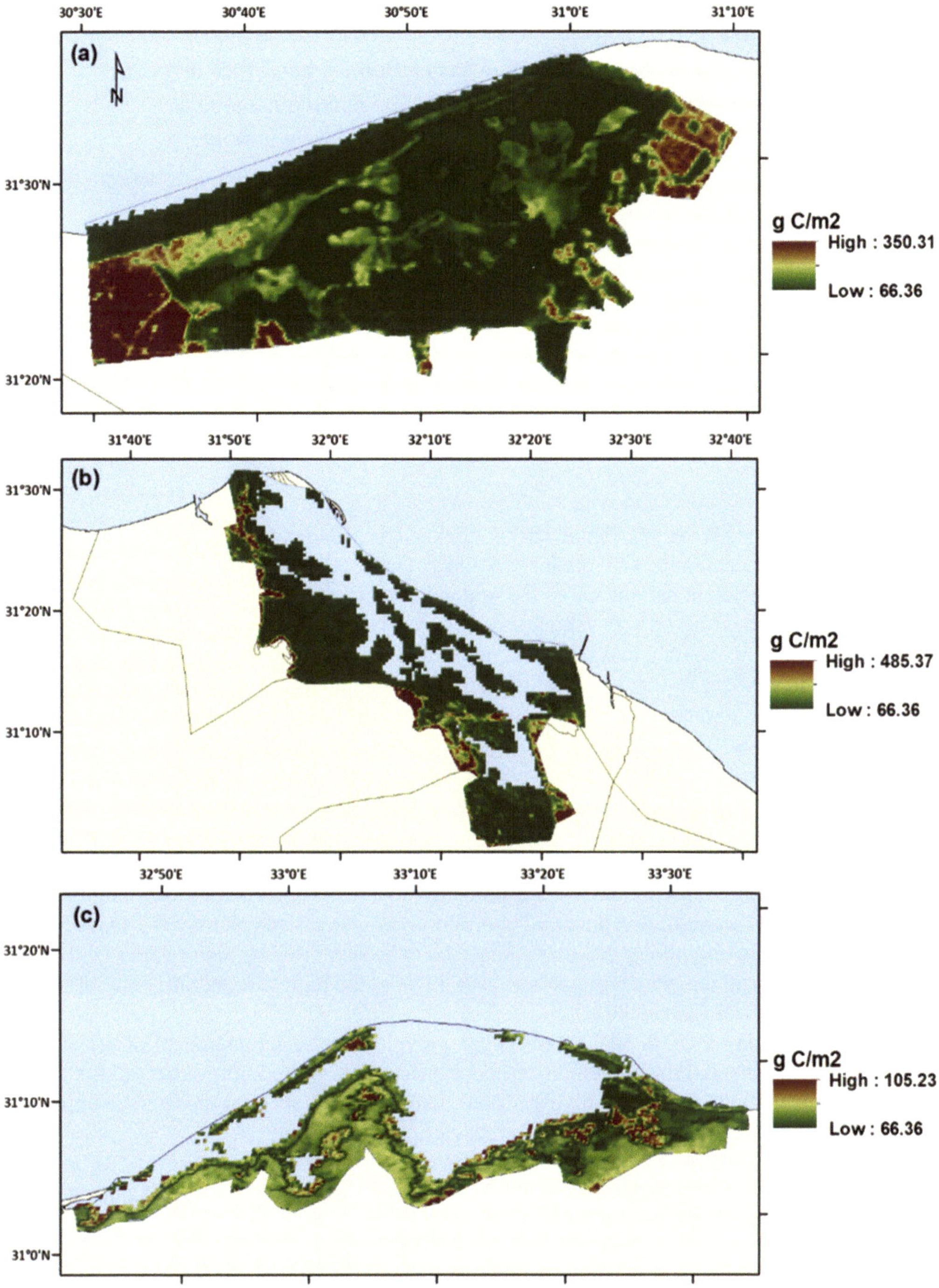

FIGURE 9.2 Spatial distribution of mean annual vegetation carbon stock in g C/m^{-2} of coastal wetlands of Egypt in 2022: (a) Lake Burullus; (b) Lake Manzala; and (c) Lake Bardawil wetlands.

TABLE 9.4

Mean ± SD and Annual Range of Vegetation Carbon Stock in g C/m² for Coastal Wetlands of Egypt in 2022. The cumulative annual vegetation carbon stock in g C/m² in 2022 is shown in bold.

Wetlands	Mean ± SD	Range	Total (kgC/year)
Burullus	98.04 ± 55.85	283.94	1997.44
	275.58 ± 343.75	**2810.53**	**4492.55**
Manzala	80.18 ± 35.14	419.01	1274.18
	295.37 ± 461.87	**6975.68**	**1946.59**
Bardawil	69.84 ± 3.38	38.87	748.14
	37.64 ± 75.53	**1048.55**	**221.66**

each wetland attained lower vegetation carbon stock compared to those near the borders of the wetlands. This was in accordance with other studies (e.g., Dai et al., 2020). This can be attributed to the variation in the vegetation pattern across the wetlands, where areas near the water proper tend to be less vegetated and intermixed with saltmarshes and mudflats that are usually sporadically vegetated as a result of water inundation and salt stress due to closeness to the influx from the Mediterranean Sea. Areas on the edges of the wetlands, particularly the southern edges, tend to have higher vegetation cover due to lower salinity and flooding stress, in addition to higher nutrient input from the discharge of domestic effluents.

The outcomes are consistent with the results from other wetlands with comparable vegetation communities. However, the amount of vegetation carbon stock was lower in the investigated wetlands because, overall, these wetlands are part of an arid region that is generally characterized by lower vegetation cover.

Although remotely sensed data provide immense sources of information that can allow us to evaluate the carbon sequestration potential of different ecosystems and monitor the changes in their carbon stocks, it is worth noting that these data need to be supported by high-quality field observations and ground truthing, which might be constrained by the high cost and difficulty of fieldwork in some circumstances (Didan, 2021).

The use of 16-day MODIS EVI products can provide an efficient approach for the evaluation of the wetlands' potential for vegetation carbon sequestration. More improvement to the approach through the availability of data of higher spatial, temporal, and spectral resolution, supported by field observations on the eco-physiological processes, is still needed.

Previous studies on these wetlands focused on assessing the sediment organic carbon (SOC) stock (e.g., Eid & Shaltout, 2013; Eid et al., 2017) and the impact of land use change on the soil carbon storage potential (e.g., Abd El-Hamid & Hafiz, 2022). It was revealed that the increase in agricultural lands led to a decrease in the area covered by vegetation by 50%, thus reducing carbon storage by 4508 mg C/km² in Manzala (Abd El-Hamid & Hafiz, 2022).

9.7 ADVANTAGES, CHALLENGES, AND OPPORTUNITIES IN CARBON SEQUESTRATION MAPPING

One way to assess the efficacy of different carbon storage techniques is through remote sensing. Through the analysis of remotely sensed data, scientists can investigate the effects of various

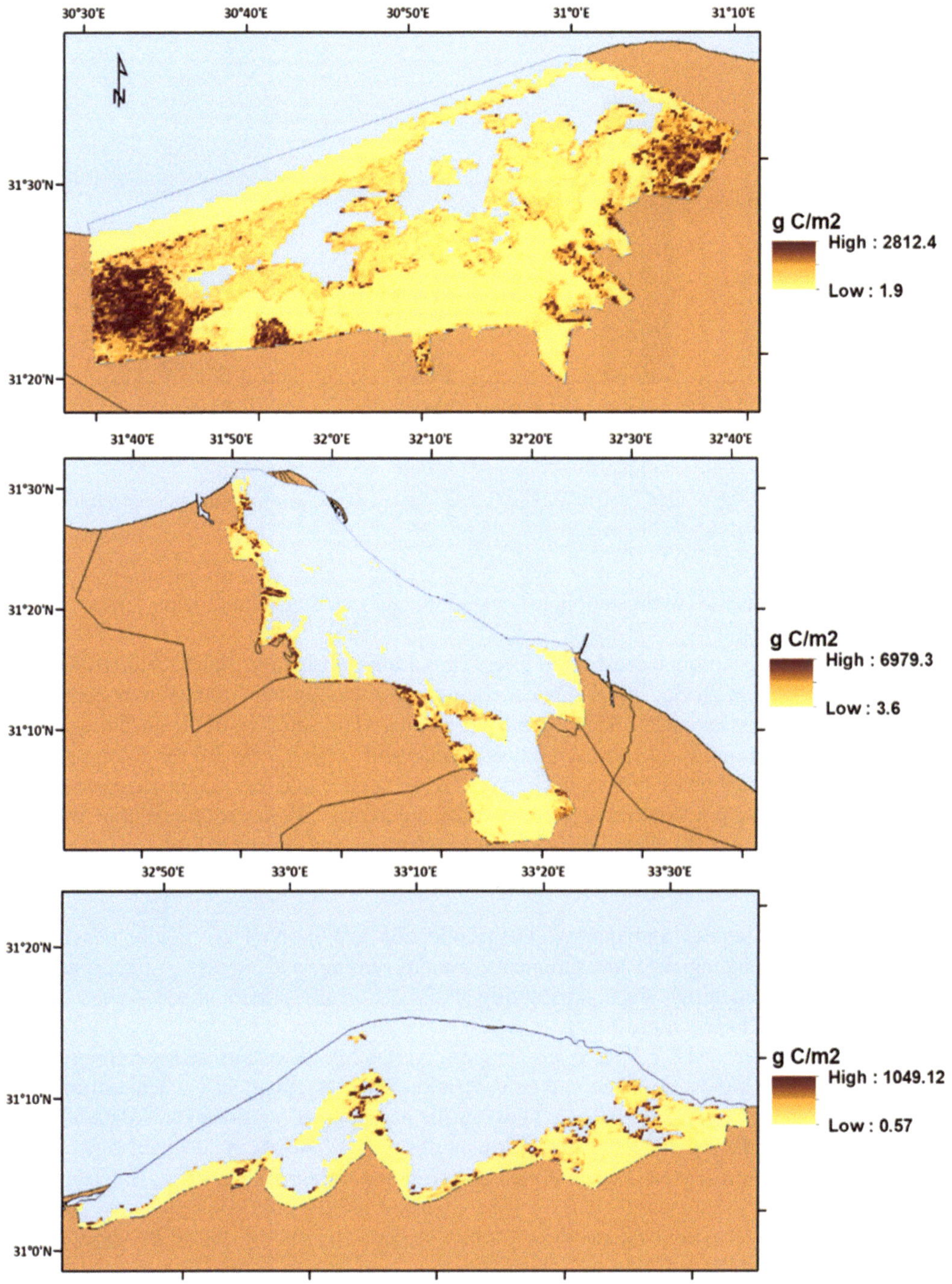

FIGURE 9.3 The spatial distribution of the cumulative annual vegetation carbon stock in g C/m^2 of coastal wetlands of Egypt in 2022: (a) Lake Burullus; (b) Lake Manzala; and (c) Lake Bardawil wetlands.

management strategies on the amount of carbon sequestered in an ecosystem. Developing and refining cost-effective and efficient carbon storage strategies requires this invaluable information (Jat et al., 2022). Mapping carbon sequestration helps in quantifying carbon stocks and monitoring changes over time. These data are crucial for assessing the effectiveness of climate change

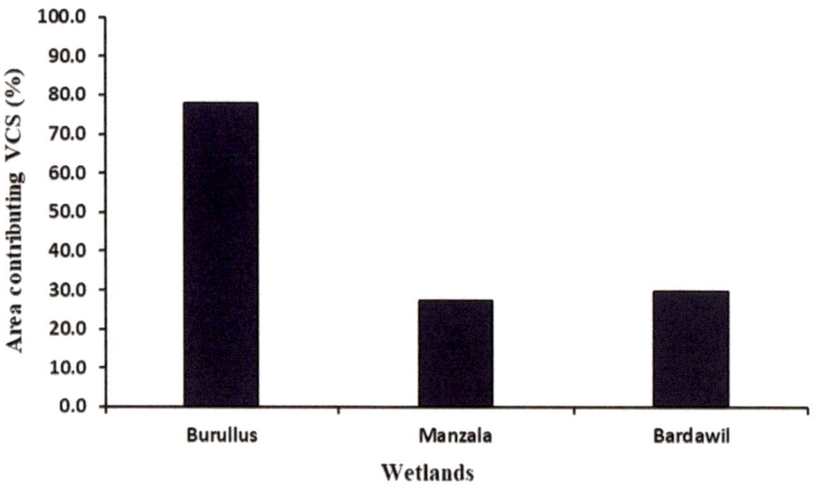

FIGURE 9.4 Percentage of area contributing to the net cumulative vegetation carbon sequestration (VCS) in the studied wetlands

mitigation efforts, such as reforestation, afforestation, and sustainable land management practices (Zhuang et al., 2022).

Indeed, mapping carbon sequestration offers important information about the distribution and fluctuations of carbon stocks in various ecosystems. It advances management and conservation strategies by assisting scientists and policymakers in their understanding of the dynamics and processes related to carbon. Furthermore, finding regions with a high potential for carbon storage is made possible by mapping carbon sequestration. The protection and restoration of ecosystems like wetlands, grasslands, and forests that greatly contribute to carbon sequestration can be prioritized using this information to direct conservation efforts (Araya-Lopez et al., 2023).

However, there are challenges in carbon sequestration mapping such as:

1. (1) Data availability and quality: Data collection and monitoring require substantial resources, including field measurements, remote sensing, and modeling. Ensuring data consistency, reliability, and comparability across different regions and ecosystems is also a challenge (Zhuang et al., 2022).
2. (2) Scale and resolution: Spatial and temporal variability at various scales, ranging from local to global, must be taken into consideration when mapping carbon sequestration. It can be difficult to strike a balance between the necessity of high-resolution data and the viability of large-scale assessments. Additionally, more thorough and detailed data collection techniques might be needed to map carbon sequestration at finer scales (Lanceman et al., 2022).
3. (3) Complex ecosystem interactions: Carbon sequestration is influenced by various factors, including climate, land use, vegetation type, and soil characteristics. Understanding the complex interactions among these factors and accurately modeling their impact on carbon sequestration is a challenge. Integrating multiple data sources and interdisciplinary approaches is necessary to account for these complexities (Zhuang et al., 2022).

9.8 CONCLUSION

Remote sensing technology provides effective approaches for environmental monitoring and ecological evaluation. Approaches based on remote sensing are increasingly being utilized as efficient tools in studies of aquatic ecosystems, often in integration with conventional techniques to overcome the constraints of traditional *in situ* approaches. These methods provide extensive coverage and spatial data for large water bodies at different scales quickly and repeatedly. Several remote sensing techniques have been developed to map the spatial cover of aquatic vegetation types. While several studies have been conducted on the spectral separability of SAV, including both plants and macroalgae, which tend to be functionally and spectrally similar, most of these studies were concentrated on marine ecosystems and inadequately addressed freshwater ecosystems. More techniques and approaches need to be developed to aid in the accurate extraction, detection, classification, and monitoring of AV in freshwater ecosystems from remotely sensed data. Although there are several vegetation indices that can be used in general vegetation monitoring, including terrestrial or AV, most of these indices were traditionally intended for studies focusing on terrestrial vegetation rather than AV. The selection of a particular index depends on the study and research goals, the availability of spectral bands, and the characteristics of the vegetation under investigation. More studies are needed for the development of new spectral vegetation indices that target aquatic vegetation, in addition to enhancing the existing ones to improve their performance in distinguishing AV from the terrestrial vegetation. Remote sensing has multiple advantages, making it more applicable in monitoring aquatic vegetation. The prospect and potential of using remote sensing in monitoring AV are significantly promising and advancing. For example, the illustrated case has shown that a high-quality, high-frequency VI time series could be utilized for estimating the carbon sequestration of vegetation in herbaceous wetlands, highlighting the importance of wetlands as key ecosystems that could mitigate the impact of climate change and global warming.

REFERENCES

Abd El-Hamid, H. T., & Hafiz, M. A. (2022). Modeling of carbon sequestration with land use and land cover in the northeastern part of the Nile Delta, Egypt. *Arabian Journal of Geosciences*, 15(14), 1267.

Ade, C., Khanna, S., Lay, M., Ustin, S. L., & Hestir, E. L. (2022). Genus-level mapping of invasive floating aquatic vegetation using Sentinel-2 satellite remote sensing. *Remote Sensing*, 14(13), 3013.

Araya-Lopez, R., de Paula Costa, M. D., Wartman, M., & Macreadie, P. I. (2023). Trends in the application of remote sensing in blue carbon science. *Ecology and Evolution*, 13(9), e10559.

Ashok, A., Rani, H. P., & Jayakumar, K. V. (2021). Monitoring of dynamic wetland changes using NDVI and NDWI based landsat imagery. *Remote Sensing Applications: Society and Environment*, 23, 100547.

Ashraf, S., Brabyn, L., Hicks, B. J., & Collier, K. (2010). Satellite remote sensing for mapping vegetation in New Zealand freshwater environments: A review. *New Zealand Geographer*, 66(1), 33–43.

Barnsley, M. J., Settle, J. J., Cutter, M. A., Lobb, D. R., & Teston, F. (2004). The PROBA/CHRIS mission: A low-cost smallsat for hyperspectral multiangle observations of the earth surface and atmosphere. *IEEE Transactions on Geoscience and Remote Sensing*, 42(7), 1512–1520.

Batsaikhan, B., Lkhamjav, O., Batsaikhan, G., Batsaikhan, N., & Norovsuren, B. (2020). Carbon stock estimation using remote sensing data and field measurement in haloxylon ammodendron dominant winter cold desert region of mongolia. *ISPRS Annals of the Photogrammetry, Remote Sensing and Spatial Information Sciences*, 3, 9–17.

Beaumont, N. J., Jones, L., Garbutt, A., Hansom, J. D., & Toberman, M. (2014). The value of carbon sequestration and storage in coastal habitats. *Estuarine, Coastal and Shelf Science*, 137, 32–40.

Bhuyan, J. M. & Sharma P. (2022). A study on the water quality and the change detection of aquatic vegetation of Satajaan beel, North Lakhimpur, Assam. *Research Square*, preprint, 1–20. DOI: https://doi.org/10.21203/rs.3.rs-2302522/v2

Blanco, A. (2013). Remote sensing techniques for monitoring aquatic vegetation. Doctoral Thesis. *George Mason University.* Theses, 2013. 3562260.

Brooks, C. N., Grimm, A. G., Marcarelli, A. M., & Dobson, R. J. (2019). Multiscale collection and analysis of submerged aquatic vegetation spectral profiles for Eurasian watermilfoil detection. *Journal of Applied Remote Sensing*, 13(3), 037501–037501.

Brooks, C., Grimm, A., Marcarelli, A. M., Marion, N. P., Shuchman, R., & Sayers, M. (2022). Classification of Eurasian Watermilfoil (Myriophyllum spicatum) using drone-enabled multispectral imagery analysis. *Remote Sensing*, 14(10). https://doi.org/10.3390/rs14102336

Brooks, C., Grimm, A., Shuchman, R., Sayers, M., & Jessee, N. (2015). A satellite-based multi-temporal assessment of the extent of nuisance Cladophora and related submerged aquatic vegetation for the Laurentian Great Lakes. *Remote sensing of Environment*, 157, 58–71.

Cawse-Nicholson, K., Townsend, P. A., Schimel, D., Assiri, A. M., Blake, P. L., Buongiorno, M. F., ... SBG Algorithms Working Group. (2021). NASA's surface biology and geology designated observable: A perspective on surface imaging algorithms. *Remote Sensing of Environment*, 257, 112349.

Chen, C., Chen, Y., Jin, H., Chen, L., Liu, Z., Sun, H., Hong, J., Wang, H., Fang, S., & Zhang, X. (2023). 3D model construction and ecological environment investigation on a regional scale using UAV remote sensing. *Intelligent Automation & Soft Computing*, 37, 1655–1672.

Chen, C., Fu, J., Zhang, S., & Zhao, X. (2019). Coastline information extraction based on the tasseled cap transformation of Landsat-8 OLI images. *Estuarine, Coastal and Shelf Science*, 217, 281–291.

Chen, Q., Yu, R., Hao, Y., Wu, L., Zhang, W., Zhang, Q., & Bu, X. (2018). A new method for mapping aquatic vegetation especially underwater vegetation in Lake Ulansuhai using GF-1 satellite data. *Remote Sensing*, 10(8), 1279.

Claverie, M., Ju, J., Masek, J. G., Dungan, J. L., Vermote, E. F., Roger, J. C., ... Justice, C. (2018). The Harmonized Landsat and Sentinel-2 surface reflectance data set. *Remote Sensing of Environment*, 219, 145–161.

Costa, V., Serôdio, J., Lillebø, A. I., & Sousa, A. I. (2021). Use of hyperspectral reflectance to non-destructively estimate seagrass Zostera noltei biomass. *Ecological Indicators*, 121, 107018.

Dai, X., Yang, G., Liu, D., & Wan, R. (2020). Vegetation carbon sequestration mapping in herbaceous wetlands by using a MODIS EVI time-series data set: A case in Poyang lake wetland, China. *Remote Sensing*, 12(18), 3000.

Dai, Y., Feng, L., Hou, X., & Tang, J. (2021). An automatic classification algorithm for submerged aquatic vegetation in shallow lakes using Landsat imagery. *Remote sensing of Environment*, 260, 112459.

Davis, C. O., Corson, M., Lucke, R., Arnone, R., & Gould, R. (2015). The Hyperspectral Imager for the Coastal Ocean (HICO): Sensor and data processing overview. International Ocean-Colour Coordinating Group. http://www.ioccg.org/sensors/Davis_HICO_IOCCG-15.pdf.

Didan, K. (2021). MODIS/Terra vegetation indices 16-Day L3 global 250m SIN Grid V061 [Data set]. NASA EOSDIS Land Processes Distributed Active Archive Center. Accessed 2023-10-31 from https://doi.org/10.5067/MODIS/MOD13Q1.061

Donato, D. C., Kauffman, J. B., Murdiyarso, D., Kurnianto, S., Stidham, M., & Kanninen, M. (2011). Mangroves among the most carbon-rich forests in the tropics. *Nature Geoscience,* 4(5), 293.

Duarte, C. M., Dennison, W. C., Orth, R. J. W., & Carruthers, T. J. B. (2008). The charisma of coastal ecosystems: Addressing the imbalance. *Estuaries Coasts*, 31, 233–238.

Duffy, J. E., Benedetti-Cecchi, L., Trinanes, J., Muller-Karger, F. E., Ambo-Rappe, R., Boström, C., Buschmann, A. H., Byrnes, J., Coles, R. G., & Creed, J. (2019). Toward a coordinated global observing system for seagrasses and marine macroalgae. *Frontiers in Marine Science*, 6, 317.

Eid, E. M., & Shaltout, K. H. (2013). Evaluation of carbon sequestration potentiality of Lake Burullus, Egypt to mitigate climate change. *The Egyptian Journal of Aquatic Research*, 39(1), 31–38.

Eid, E. M., Keshta, A. E., Shaltout, K. H., Baldwin, A. H., El-Din, S., & Ahmed, A. (2017). Carbon sequestration potential of the five Mediterranean lakes of Egypt. *Fundamental and Applied Limnology*, 190, 87–96.

Erwin, K. L. (2009). Wetlands and global climate change: The role of wetland restoration in a changing world. *Wetlands Ecology and Management*, 17(1), 71.

Espel, D., Courty, S., Auda, Y., Sheeren, D., & Elger, A. (2020). Submerged macrophyte assessment in rivers: An automatic mapping method using Pléiades imagery. *Water Research*, 186, 116353.

Fernandes, M. R., Aguiar, F. C., Silva, J. M., Ferreira, M. T., & Pereira, J. M. (2013). Spectral discrimination of giant reed (Arundo donax L.): A seasonal study in riparian areas. *ISPRS Journal of Photogrammetry and Remote Sensing*, 80, 80–90.

Fitton, J. M., Rennie, A. F., Hansom, J. D., & Muir, F. M. (2021). Remotely sensed mapping of the intertidal zone: A Sentinel-2 and Google Earth Engine methodology. *Remote Sensing Applications: Society and Environment*, 22, 100499.

Free, G., Bresciani, M., Trodd, W., Tierney, D., O'Boyle, S., Plant, C., & Deakin, J. (2020). Estimation of lake ecological quality from Sentinel-2 remote sensing imagery. *Hydrobiologia*, 847, 1423–1438.

Fu, W., Ma, J., Chen, P., & Chen, F. (2020). Remote sensing satellites for digital earth. In: Guo, H., Goodchild, M.F., Annoni, A. (eds) *Manual of Digital Earth*. Springer, Singapore. https://doi.org/10.1007/978-981-32-9915-3_3

Fyfe, S. K. (2003). Spatial and temporal variation in spectral reflectance: Are seagrass species spectrally distinct?. *Limnology and Oceanography*, 48(1, part2), 464–479.

Gao, B. C. (1996). NDWI—A normalized difference water index for remote sensing of vegetation liquid water from space. *Remote Sensing of Environment*, 58(3), 257–266. https://doi.org/10.1016/S0034-4257(96)00067-3

Gausman, H. W. (1977). Reflectance of leaf components. *Remote Sensing of Environment*, 6(1), 1–9.

Gerardo, R., & de Lima, I. P. (2022). Monitoring duckweeds (Lemna minor) in small rivers using sentinel-2 satellite imagery: Application of vegetation and water indices to the Lis River (Portugal). *Water*, 14(15), 2284.

Guanter, L., Alonso, L., & Moreno, J. (2005). First results from the PROBA/CHRIS hyperspectral/multiangular satellite system over land and water targets. *IEEE Geoscience and Remote Sensing Letters*, 2(3), 250–254.

Hasekamp, O. P., Fu, G., Rusli, S. P., Wu, L., Di Noia, A., aan de Brugh, J., ... van Amerongen, A. (2019). Aerosol measurements by SPEXone on the NASA PACE mission: Expected retrieval capabilities. *Journal of Quantitative Spectroscopy and Radiative Transfer*, 227, 170–184.

Huber, S., Hansen, L. B., Nielsen, L. T., Rasmussen, M. L., Sølvsteen, J., Berglund, J., Paz von Friesen, C., Danbolt, M., Envall, M., Infantes, E., & Moksnes, P. (2022). Novel approach to large-scale monitoring of submerged aquatic vegetation: A nationwide example from Sweden. *Integrated Environmental Assessment and Management*, 18(4), 909–920. https://doi.org/10.1002/ieam.4493

Huete, A. R., Justice, C., & van Leeuwen, W. (1999). Modis vegetation index algorithm theoretical basis v3. algorithm—Theoretical. *Basis Document*, 3, 295–309.

Huete, A. R., Liu, H. Q., Batchily, K., & Van Leeuwen, W. (1997). A comparison of vegetation indices over a global set of TM images for EOS-MODIS. *Remote Sensing of Environment*, 59(3), 440–451. https://doi.org/10.1016/S0034-4257(96)00112-5

Jat, M. L., Chakraborty, D., Ladha, J. K., Parihar, C. M., Datta, A., Mandal, B., ... Gerard, B. (2022). Carbon sequestration potential, challenges, and strategies towards climate action in smallholder agricultural systems of South Asia. *Crop and Environment*, 1(1), 86–101.

Jia, Q., Cao, L., Yésou, H., Huber, C., & Fox, A. D. (2017). Combating aggressive macrophyte encroachment on a typical Yangtze River lake: Lessons from a long-term remote sensing study of vegetation. *Aquatic Ecology*, 51, 177–189.

Kalacska, M., Chmura, G., Lucanus, O., Bérubé, D., & Arroyo-Mora, J. (2017). Structure from motion will revolutionize analyses of tidal wetland landscapes. *Remote Sensing of Environment*, 199, 14–24.

Kirk, J. T. (1994). *Light and Photosynthesis in Aquatic Ecosystems*. Cambridge University Press.

Klemas, V. (2013). Remote sensing of emergent and submerged wetlands: An overview. *International Journal of Remote Sensing*, 34(18), 6286–6320. https://doi.org/10.1080/01431161.2013.800656

Klemas, V. V. (2016). Remote sensing of submerged aquatic vegetation. In: Finkel, C., Makowski, C. (eds) *Seafloor Mapping along Continental Shelves: Research and Techniques for Visualizing Benthic Environments*, Springer, 125–140.

Knight, J. M., Dale, P. E. R., Spencer, J., & Griffin, L. (2009). Exploring LiDAR data for mapping the micro-topography and tidal hydro-dynamics of mangrove systems: An example from southeast Queensland, Australia. *Estuarine, Coastal and Shelf Science*, 85(4), 593–600. https://doi.org/10.1016/J.ECSS.2009.10.002

Lafaye, M. (2017). Benefit assessment of the application of satellite Earth observation for society and policy: Assessing the socioeconomic impacts of the development of downstream space-based Earth observation applications. In: Onoda, M., Younf, O. (eds)*Satellite Earth Observations and Their Impact on Society and Policy*, Springer, 93–103.

Lanceman, D., Sadat-Noori, M., Gaston, T., Drummond, C., & Glamore, W. (2022). Blue carbon ecosystem monitoring using remote sensing reveals wetland restoration pathways. *Frontiers in Environmental Science*, 10, 924221.

Lathrop, R. G., Montesano, P., & Haag, S. (2006). A multi-scale segmentation approach to mapping seagrass habitats using airborne digital camera imagery. *Photogrammetric Engineering and Remote Sensing*, 72(6), 665–675. https://doi.org/10.14358/PERS.72.6.665

Li, C., Li, M., & Li, Y. (2020). Improving estimation of forest aboveground biomass using Landsat 8 imagery by incorporating forest crown density as a dummy variable. *Canadian Journal of Forest Research*, 50(4), 390–398.

Liang, S., Gong, Z., Wang, Y., Zhao, J., & Zhao, W. (2022). Accurate monitoring of submerged aquatic vegetation in a macrophytic lake using time-series Sentinel-2 images. *Remote Sensing*, 14(3), 640.

Lin, H., Shu, G., Xiping, Y., Yan, L., Guokun, C., & Sha, G. (2021). Spatial differentiation analysis of water quality in Dianchi Lake based on GF-5 NDVI characteristic optimization. *Journal of Spectroscopy*,1-11.

Luo, J., Duan, H., Ma, R., Jin, X., Li, F., Hu, W., Shi, K., & Huang, W. (2017). Mapping species of submerged aquatic vegetation with multi-seasonal satellite images and considering life history information. *International Journal of Applied Earth Observation and Geoinformation*, 57, 154–165. https://doi.org/10.1016/j.jag.2016.11.007

Luo, J., Li, X., Ma, R., Li, F., Duan, H., Hu, W., Qin, B., & Huang, W. (2016). Applying remote sensing techniques to monitoring seasonal and interannual changes of aquatic vegetation in Taihu Lake, China. *Ecological Indicators*, 60, 503–513. https://doi.org/10.1016/j.ecolind.2015.07.029

Luo, J., Ma, R., Duan, H., Hu, W., Zhu, J., Huang, W., & Lin, C. (2014). A new method for modifying thresholds in the classification of tree models for mapping aquatic vegetation in Taihu Lake with satellite images. *Remote Sensing*, 6(8), 7442–7462.

Luo, J., Ni, G., Zhang, Y., Wang, K., Shen, M., Cao, Z., Qi, T., Xiao, Q., Qiu, Y., Cai, Y., & Duan, H. (2023). A new technique for quantifying algal bloom, floating/emergent and submerged vegetation in eutrophic shallow lakes using Landsat imagery. *Remote Sensing of Environment*, 287, 113480. https://doi.org/10.1016/j.rse.2023.113480

Malthus, T. J. (2017). Bio-optical modeling and remote sensing of aquatic Macrophytes. In *Bio-optical Modeling and Remote Sensing of Inland Waters*. Elsevier Inc. https://doi.org/10.1016/B978-0-12-804644-9.00009-4

Mcilwaine, B., Rivas Casado, M., & Leinster, P. (2019). Using 1st derivative reflectance signatures within a remote sensing framework to identify macroalgae in marine environments. *Remote Sensing*, 11(6), 704.

McLeod, E., Chmura, G. L., Bouillon, S., Salm, R., Björk, M., Duarte, C. M., et al. (2011) A blueprint for blue carbon: Toward an improved understanding of the role of vegetated coastal habitats in sequestering CO_2. *Frontiers in Ecology and the Environment*, 9(10), 552–560.

Murray, B. C., Pendleton, L., Jenkins, W. A., & Sifleet, S. (2011). Green payments for blue carbon. Economic Incentives for Protecting Threatened Coastal Habitats. Nicholas Institute, Duke University.

Ni, T. N. K., Thang, L. Van, Chi, L. T. T., Tuan, L. C., Hang, H. T. M., Uyen, N. T., Binh, N. H., Jamet, C., & Tin, H. C. (2023). Quantifying biomass and canopy cover of submerged aquatic vegetation on an offshore island. *Regional Studies in Marine Science*, 66(September), 103117. https://doi.org/10.1016/j.rsma.2023.103117

Ou, K., Lin, H., Cui, G., & Xia, W. (2008). Emergent plants and their garden applications. *Journal of Anhui Agricultural Sciences*, 36(20), 8556–8558.

Piaser, E., & Villa, P. (2023). Evaluating capabilities of machine learning algorithms for aquatic vegetation classification in temperate wetlands using multi-temporal Sentinel-2 data. *International Journal of Applied Earth Observation and Geoinformation*, 117, 103202.

Pu, J., Song, K., Lv, Y., Liu, G., Fang, C., Hou, J., & Wen, Z. (2022). Distinguishing algal blooms from aquatic vegetation in Chinese lakes using Sentinel 2 image. *Remote Sensing*, 14(9), 1988.

Questad, E. J., Kellner, J. R., Kinney, K., Cordell, S., Asner, G. P., Thaxton, J., Diep, J., Uowolo, A., Brooks, S., Inman-Narahari, N., Evans, S. A., & Tucker, B. (2014). Mapping habitat suitability for at-risk plant species and its implications for restoration and reintroduction. *Ecological Applications*, 24(2), 385–395. https://doi.org/10.1890/13-0775.1

Remer, L. A., Knobelspiesse, K., Zhai, P. W., Xu, F., Kalashnikova, O. V., Chowdhary, J., ... Torres, O. (2019). Retrieving aerosol characteristics from the PACE mission, part 2: Multi-angle and polarimetry. *Frontiers in Environmental Science*, 7, 94.

Rhee, D. S., Kim, Y. D., Kang, B., & Kim, D. (2018). Applications of unmanned aerial vehicles in fluvial remote sensing: An overview of recent achievements. *KSCE Journal of Civil Engineering*, 22, 588–602.

Rodríguez-Garlito, E. C., Paz-Gallardo, A., & Plaza, A. (2022). Automatic detection of aquatic weeds: A case study in the Guadiana River, Spain. *IEEE Journal of Selected Topics in Applied Earth Observations and Remote Sensing*, 15, 8567–8585.

Rogers, K., Kelleway, J. J., Saintilan, N., Megonigal, J. P., Adams, J. B., Holmquist, J. R., et al. (2019). Wetland carbon storage controlled by millennial-scale variation in relative sea-level rise. *Nature*, 567(7746), 91.

Rouse, J. W., Hass, R. H., Schell, J. A., & Deering, D. W. (1973). Monitoring vegetation systems in the great plains with ERTS. The Third ERTS Symposium. NASA SP-351, 309–317. https://ntrs.nasa.gov/citations/19740022614

Rowan, G. S., & Kalacska, M. (2021). A review of remote sensing of submerged aquatic vegetation for non-specialists. *Remote Sensing*, 13(4), 623. https://doi.org/10.3390/rs13040623

Rowan, G. S., Kalacska, M., Inamdar, D., Arroyo-Mora, J. P., & Soffer, R. (2021). Multi-scale spectral separability of submerged aquatic vegetation species in a freshwater ecosystem. *Frontiers in Environmental Science*, 9, 760372.

Sanga-Ngoie, K., Iizuka, K., & Kobayashi, S. (2012). Estimating CO2 sequestration by forests in Oita Prefecture, Japan, by combining Landsat ETM+ and ALOS satellite remote sensing data. *Remote Sensing*, 4(11), 3544–3570.

Santos, M. J., Khanna, S., Hestir, E. L., Greenberg, J. A., & Ustin, S. L. (2016). Measuring landscape-scale spread and persistence of an invaded submerged plant community from airborne remote sensing. *Ecological Applications*, 26(6), 1733–1744.

Shaltout, K., El-Bana, M., & Galal, T. (2019). Coastal lakes as hot spots for plant diversity in Egypt. In: Negm, A., Bek, M., Abdel-Fattah, S (eds) *Egyptian Coastal Lakes and Wetlands: Part II.* (pp. 129–146). Springer, Cham.

Tran, T. V., Reef, R., & Zhu, X. (2022). A review of spectral indices for mangrove remote sensing. *Remote Sensing*, 14(19). https://doi.org/10.3390/rs14194868

Tran, T. V., Tran, D. X., Nguyen, H., Latorre-Carmona, P., & Myint, S. W. (2021). Characterising spatiotemporal vegetation variations using LANDSAT time-series and Hurst exponent index in the Mekong River Delta. *Land Degradation and Development*, 32(13), 3507–3523. https://doi.org/10.1002/ldr.3934

Uudeberg, K., Ansko, I., Põru, G., Ansper, A., & Reinart, A. (2019). Using optical water types to monitor changes in optically complex inland and coastal waters. *Remote Sensing*, 11(19), 2297.

Vicharnakorn, P., Shrestha, R. P., Nagai, M., Salam, A. P., & Kiratiprayoon, S. (2014). Carbon stock assessment using remote sensing and forest inventory data in Savannakhet, Lao PDR. *Remote Sensing*, 6(6), 5452–5479.

Villa, P., Bresciani, M., Bolpagni, R., Pinardi, M., & Giardino, C. (2015). A rule-based approach for mapping macrophyte communities using multi-temporal aquatic vegetation indices. *Remote Sensing of Environment*, 171, 218–233. https://doi.org/10.1016/j.rse.2015.10.020

Villa, P., Bresciani, M., Braga, F., & Bolpagni, R. (2013a). Mapping aquatic vegetation through remote sensing data : A comparison of vegetation indices performances. 6th EARSeL Workshop on Remote Sensing of the Coastal Zone, 7–8 June, 10–15.

Villa, P., Mousivand, A., & Bresciani, M. (2014). Aquatic vegetation indices assessment through radiative transfer modeling and linear mixture simulation. *International Journal of Applied Earth Observation and Geoinformation*, 30, 113–127. https://doi.org/10.1016/j.jag.2014.01.017

Visser, F., Wallis, C., & Sinnott, A. M. (2013). Optical remote sensing of submerged aquatic vegetation: Opportunities for shallow clearwater streams. *Limnologica*, 43(5), 388–398.

Wang, H., Li, Y., Zeng, S., Cai, X., Bi, S., Liu, H., Mu, M., Dong, X., Li, J., & Xu, J. (2022). Recognition of aquatic vegetation above water using shortwave infrared baseline and phenological features. *Ecological Indicators*, 136, 108607.

Wang, J., Sammis, T. W., Gutschick, V. P., Gebremichael, M., Dennis, S. O., & Harrison, R. E. (2010). Review of satellite remote sensing use in forest health studies. *The Open Geography Journal*, 3(1), 28–42.

Wang, S., Gao, Y., Li, Q., Gao, J., Zhai, S., Zhou, Y., & Cheng, Y. (2019). Long-term and inter-monthly dynamics of aquatic vegetation and its relation with environmental factors in Taihu Lake, China. *Science of the Total Environment*, 651, 367–380.

Wang, Y., Gong, Z., & Zhou, H. (2023). Long-term monitoring and phenological analysis of submerged aquatic vegetation in a shallow lake using time-series imagery. *Ecological Indicators*, 154(February), 110646. https://doi.org/10.1016/j.ecolind.2023.110646

Werdell, P. J., Behrenfeld, M. J., Bontempi, P. S., Boss, E., Cairns, B., Davis, G. T., ... Remer, L. A. (2019). The Plankton, Aerosol, Cloud, ocean Ecosystem mission: Status, science, advances. *Bulletin of the American Meteorological Society*, 100(9), 1775–1794.

Williams, D. J., Rybicki, N. B., Lombana, A. V., O'Brien, T. M., & Gomez, R. B. (2003). Preliminary investigation of submerged aquatic vegetation mapping using hyperspectral remote sensing. In *Coastal Monitoring through Partnerships: Proceedings of the Fifth Symposium on the Environmental Monitoring and Assessment Program (EMAP)* Pensacola Beach, FL, USA, April 24–27, 2001 (pp. 383–392). Springer Netherlands.

Wilson, K. L., Wong, M. C., & Devred, E. (2022). Comparing Sentinel-2 and WorldView-3 imagery for coastal Bottom habitat mapping in Atlantic Canada. *Remote Sensing*, 14(5), 1254.

Xue, J., & Su, B. (2017). Significant remote sensing vegetation indices: A review of developments and applications. *Journal of Sensors*, 2017. https://doi.org/10.1155/2017/1353691

Yang, G., Huang, K., Sun, W., Meng, X., Mao, D., & Ge, Y. (2022). Enhanced mangrove vegetation index based on hyperspectral images for mapping mangrove. *ISPRS Journal of Photogrammetry and Remote Sensing*, 189, 236–254.

Zhao, D., Lv, M., Jiang, H., Cai, Y., Xu, D., & An, S. (2013). Spatio-temporal variability of aquatic vegetation in Taihu Lake over the past 30 years. *PLoS One*, 8(6), e66365.

Zhou, T., Ni, C., Zhang, M., & Xia, P. (2022). Assessing spatial and temporal distribution of algal blooms using Gini Coefficient and Lorenz Asymmetry Coefficient. *Frontiers in Environmental Science*, 10, 810902.

Zhuang, Q., Shao, Z., Gong, J., Li, D., Huang, X., Zhang, Y., ... Wu, S. (2022). Modeling carbon storage in urban vegetation: Progress, challenges, and opportunities. *International Journal of Applied Earth Observation and Geoinformation*, 114, 103058.

10 Monitoring Coastal Chlorophyll-a Concentration in the Coast of North Africa Using Remote Sensing and Machine Learning

*Mubaraka S. Algharian*i
Libyan Authority for Scientific Research, Tripoli, Libya

Nezha Mejjad
National Center for Energy, Sciences and Nuclear Techniques, Rabat, Morocco

Ehab M. Sagar
Department of Soil and Water, Faculty of agriculture, University of Tripoli, Tripoli, Libya

Heba Bedair
Botany Department, Faculty of Science, Tanta University, Tanta, Egypt

10.1 INTRODUCTION

Chlorophyll-a is used to assess water quality as it is a valuable indicator of primary production levels in coastal and marine waters in response to nutrient and light availability (Lathuilière et al., 2008; Alvarez-Fernandez & Riegman, 2014; Colella et al., 2016; Kwon et al., 2018; Kotta & Kitsiou, 2019; del Carmen Jiménez-Quiroz et al., 2021; Maslukah et al., 2021). Chlorophyll-a concentration is also a recognized sign of the ecological health of the aquatic setting. The study of chlorophyll-a abundance and distribution reveals the natural variations of biological productivity (low/mid/high) in marine areas (EEA, 2005). Indeed, monitoring chlorophyll provides information related to phytoplankton abundance mainly because it plays an essential role as a primary producer of the marine environment (Field et al., 1998; Vargas et al., 2006).

In addition to being the base of marine ecosystem food webs (Vargas et al., 2006), phytoplankton play a crucial role in the global carbon cycle, presenting an essential driver of the planet's climate (Bedair et al., 2023a, b). Indeed, substantial attention has been given to the phytoplankton subject in terms of theoretical, experimental, and observational studies and analysis; mainly, it is a fundamental contributor to biogeochemical cycles (Litchman et al., 2015). Thus, any changes in the phytoplankton community due to the deficiency or excess nutrient input into coastal waters from land-based origins and/or anthropogenic sources lead to unwanted effects such as eutrophication (Malone et Newton, 2020; Maslukah et al., 2021). In this sense, the monitoring and analysis of chlorophyll allow the assessment of marine and coastal water quality to implement adequate strategies

 DOI: 10.1201/9781003354000-14

and management projects for the sustainable growth of socio-economic and environmental sectors. Numerous monitoring studies were carried out in coastal and marine waters to address the effects of eutrophication, where the change of chlorophyll concentrations over time was the suitable candidate to monitor water quality and ecosystem functioning (Boyer et al., 2009; Delorenzo et al., 2012; Harvey et al., 2015; Nazeer et Nichole, 2016; Silveira Kupssinskü et al., 2020; Lehmann et al., 2021). Different methods and techniques were used for Chl-a monitoring, such as in situ biogeochemical analysis of samples and remote sensing using data satellites.

Since chlorophyll can be used as a proxy for phytoplankton, its presence in coastal water leads to selective absorption of light by chlorophyll-a pigment, causing changes in coastal water color that can be identified by ocean color remote sensing. However, traditional in situ sampling provides measured data for several chemical and biological water variables and is the only approach allowing confirmation of the presence of toxic compounds in coastal water with relatively high certainty. Remote sensing (RS) is a cost-effective approach to alternative traditional field-based survey methods (Cherif et al., 2020; Pepenfus et al., 2020; Azzirgue et al., 2021, 2022), as these methods can be costly in terms of labor, equipment, and travel. In addition, the traditional techniques based on field surveys may not always cover spatial and temporal water quality conditions (Schaeffer et al., 2013). Furthermore, these traditional methods do not permit researchers to map chlorophyll-a concentration on a large spatial scale (Chen et al., 2013). Instead, remote sensing can provide more representative spatial and temporal data on marine and coastal water quality. Besides, RS can potentially improve the efficiency of in situ sampling by targeting resource allocation (Pepenfus et al., 2020). Therefore, the main objective of this study is to monitor Chl-a concentration variation on the coast of North African countries and its potential effect on sea surface temperature (SST) by using a machine-learning model.

10.2 STUDY AREA DESCRIPTION

The African continent is divided into four distinct regions: North Africa, East Africa, West Africa, and South Africa. It is surrounded by various water bodies, including oceans and seas: in the north, the Mediterranean; the Red Sea along the Sinai Peninsula and the Suez Canal to the northeast; the Indian Ocean to the east and southeast; and in the west, the Atlantic Ocean (Figure 10.1).

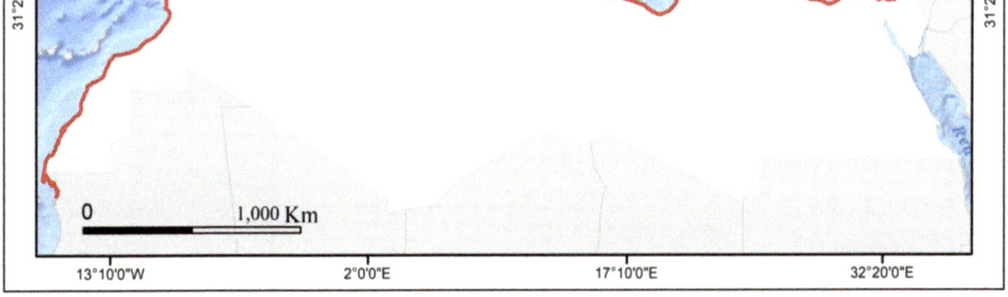

FIGURE 10.1 Map of the study area, North African coastal region.

The study area is located in the North African region; it is about 6000 km long and hosts five countries: Morocco, Algeria, Tunisia, Libya, and Egypt. This region presents the Mediterranean façade of Africa (African Mediterranean Sea). The Mediterranean Sea is connected to the Atlantic Ocean by the Strait of Gibraltar, and it is one of the most important and valued seas in the world mainly because its coasts are rich in many natural resources of ecological and socio-economic interest. Noting that one-third of the Mediterranean population lives in coastal cities, this suggests that the Mediterranean Sea is under pressure from human activities, including industrial and domestic activities, causing overfishing, overtourism, urbanization, and natural habitat loss (Bedair, 2020; Bedair et al., 2020, 2022, 2023a, 2023b, 2023c; Mejjad et al., 2022; Shaltout & Bedair, 2022).

10.3 MATERIALS AND METHODS

The satellite images used in this study were obtained from the Google Earth Engine platform for a time period (Table 10.1). The GCOM-C/SGLI L3 Chl-a algorithm was developed based on phytoplankton's photosynthetic pigment concentration (chlorophyll-a) in the sea surface layer. The GCOM-C satellite carries the Global Imager (GLI) sensor, which succeeded the ADEOS-II (MIDORI II) mission. The GLI is an Imaging Radiometer that measures radiation across a wide spectral range, from near-ultraviolet to thermal infrared (380 nm–12 μm), using 19 channels. It enables long-term and continuous global observation, collecting data to understand the mechanisms underlying fluctuations in the Earth's radiation budget. With a ground observation width of over 1000 km, GCOM-C can provide global observations approximately every two days at mid-latitudes near Japan. The Second-Generation Global Imager (SGLI) on GCOM-C offers higher resolution compared to similar global sensors and incorporates polarized observation and multi-angle observation capabilities (Murakami, 2020).

The NOAA Optimum Interpolation Sea Surface Temperature (SST) dataset is an analysis that combines observations from various platforms, including satellites, ships, buoys, and Argo floats, on a regular global grid. It generates a spatially complete map of sea surface temperatures by interpolating and filling in data gaps. The methodology includes bias adjustment of satellite and ship observations, referenced to buoys, to account for differences between platforms and sensor biases. The Version 5.2 SST dataset (PFV52) provides global, twice-daily, 4 km resolution sea surface temperature data (NOAA, 2008).

10.3.1 Machine-Learning Process Using GEE

A machine-learning process using GEE can monitor and map the Chl-a concentrations over large areas without requiring environmental chlorophyll-a measurement technology. The steps followed for the collection of GCOM-C/SGLI L3 and OISST/V2 data ingestion into GEE are summarized in Figure 10.2. It shows our workflow flowchart divided into two processes: one for estimating Chl-a

TABLE 10.1
List of Files Used in the Data Display Case

Sensor/Satellite	Data file	Resolution
GCOM-C/SGLI L3	Chlorophyll-a Concentration (V2)	1000 km (1/24 deg)
OISST Info (NOAA/CDR/OISST/V2-1)	Sea Surface Temperature (SST)	24-km grid (1/4 deg)

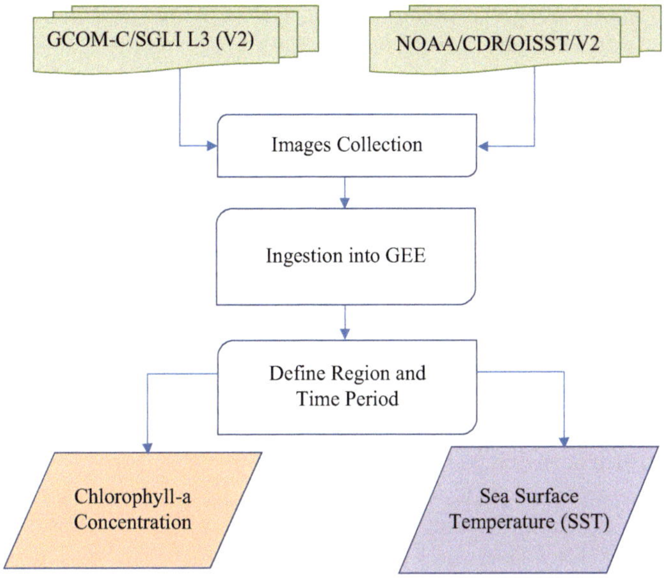

FIGURE 10.2 Flowchart of the process to monitor chlorophyll-a concentration and SST using GEE.

concentration based on the GCOM-C/SGLI L3 algorithm, and the other for applying the NOAA/CDR/OISST/V2-1 algorithm to determine SST. The first step consists of data retrieval from the providers, followed by determining the study area and the period of the satellite images. Then, data showing Chl-a concentrations and SST were downloaded for May, August, and November during the period extending from 2018 to 2021.

10.4 RESULTS AND DISCUSSION

10.4.1 Statistical Characteristics of Chlorophyll-a Concentrations

The statistical analysis of the collected data related to chlorophyll-a concentrations on the coast of North Africa is shown in Table 10.2. The maximum (Max), minimum (Min), average (Mean),

TABLE 10.2

Statistics of Chl-a Concentration (mg/m^3) on the Coast of North Africa (May, August, and November) 2018–2021

Year	Month	Min	Max	Mean	SD
2018	May	350.20	584.30	216.92	69.78
	Aug	251.64	395.52	178.60	69.17
	Nov	478.01	952.92	262.69	127.60
2019	May	300.13	556.35	210.25	59.76
	Aug	202.17	268.67	157.14	45.52
	Nov	302.41	494.63	215.18	58.00
2020	May	222.56	364.12	167.80	44.88
	Aug	208.91	317.43	152.63	64.99
	Nov	323.52	714.51	177.12	75.85
2021	May	309.31	437.34	193.78	64.14
	Aug	300.78	460.49	202.67	78.05
	Nov	391.33	723.32	214.00	55.29

and standard deviation (SD) values exhibited seasonal variations of Chl-a concentrations, where the highest values were recorded during autumn (November) and the lowest values in the summer period (August). This result is in good accordance with D'Ortenzio and Ribera d'Alcalà's (2009) findings, which highlighted very low Chl-a concentrations in the Mediterranean waters during the summer period compared to the winter season. A clear decrease in Chl-a concentrations was shown during 2019 and 2020. The average concentration indicates a noticeable decline in Chl-a concentration during 2020. Noting that the Mediterranean Sea is known as an oligotrophic environment, the shutdown of human activities (e.g. wastewater flux and reduced fishing) during COVID-19 is the possible reason for the decrease of chlorophyll-a concentrations, which, in turn, led to an intensification in oligotrophy character of the studied area (Sala et al., 2022).

10.4.2 Chlorophyll-a Concentrations Analysis

Based on chlorophyll concentrations, the abundance of phytoplankton is influenced by both seasonal variations, which occur over a short period, and climatic changes, which happen over a longer duration. The productivity of phytoplankton, in terms of primary production, can be influenced indirectly by climatic factors. For instance, changes in water temperatures and surface winds can impact the mixing of water layers and the availability of nutrients, thereby affecting the overall primary production by phytoplankton (Kavak & Karadogan, 2012). The Chl-a concentrations measured between 2018 and 2021 during May, August, and November show high values mainly during autumn (November), while the Chl-a concentrations remain low during May and August from 2018 to 2021. The results are in good accordance with studies done in the Mediterranean region (Colella et al., 2016), where the high levels of Chl-a were recorded during October–November while the lowest values were measured during the summer period (August). A box plot comparing chlorophyll concentrations on the coast of North Africa during the period 2018–2021 is illustrated in Figure 10.3; the median value is represented by the middle line. The top and bottom of the box correspond to the third quartile (75th percentile) and the first quartile (25th percentile) respectively. The vertical lines extending from the box indicate the 10th and 90th percentiles of the data distribution. Any asterisks present in the plot indicate extreme outliers (Solheim et al., 2008). The lowest values of the Chl-a were observed during August and the greatest ones in November, with increasing outliers during August.

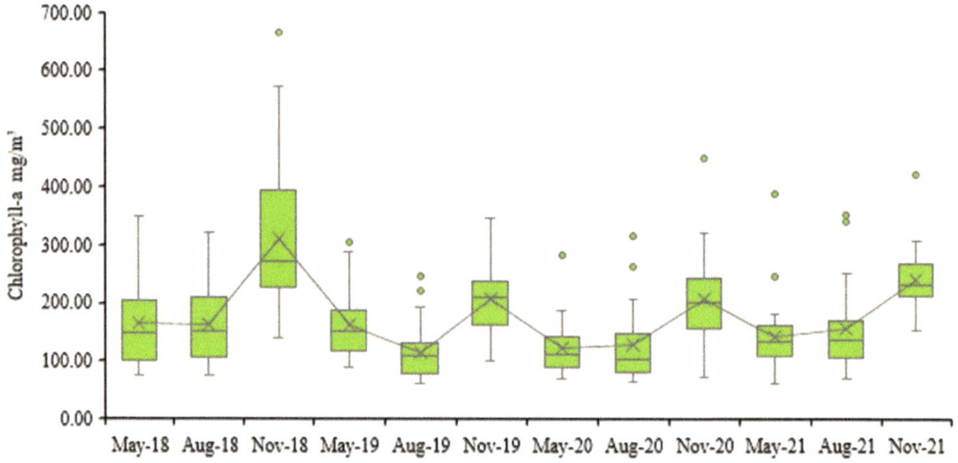

FIGURE 10.3 Box plot for chlorophyll-a concentrations observed on the coast of North Africa during a period extending from 2018 to 2021.

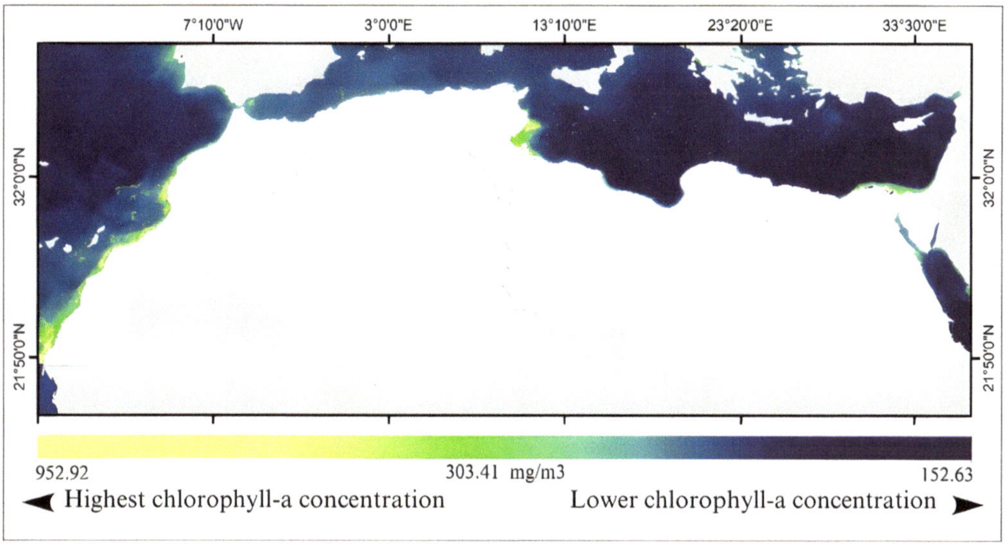

FIGURE 10.4 Chlorophyll-a concentration determined on the coast of North Africa (mg/m^3) during the time period 2018–2021.

The total chlorophyll concentration (mg/m^3) on the coast of North Africa for the period 2018–2021 is shown in Figure 10.4. The concentrations ranged from 152.63 to 952.92 mg/m^3, where chlorophyll-a distribution among Mediterranean countries is inferior to 303.41 mg/m^3. The Mediterranean is known as the most oligotrophic sea in the world because of the poor concentration of phosphate (Thingstad et al., 2005), so it is defined as a Low Nutrient–Low-Chlorophyll system (Durrieu de Madron et al., 2011), which is indeed confirmed by the obtained results shown in the map (Figure 10.4). The chlorophyll map shows milligrams of chlorophyll per cubic meter of seawater each month. Zones presenting low chlorophyll amounts, designating a very low amount of phytoplankton, are shown in blue. Zones with high chlorophyll concentrations, meaning a significant amount of phytoplankton is growing, are presented in yellow. The highest chlorophyll concentrations, where tiny surface-dwelling ocean plants are thriving, are in cold waters or in places where ocean currents bring cold water to the surface, such as the shores of continents, as is the case on Morocco's beaches overlooking the Atlantic Ocean (Hu et al., 2012). The Chl-a concentrations decrease from east to west, which was previously reported by Siokou-Frangou et al. (2010), meaning that an increasing gradient of oligotrophy from east to west and south to north characterizes the Mediterranean Sea. Additionally, throughout a large portion of its territory, the Eastern Mediterranean Sea exhibits non-blooming characteristics and fluctuations in chlorophyll-a concentrations from its lower concentrations in the summer to the highest values in the winter; only a few areas have high Chl-a concentrations, such as the southeastern part of Tunisia precisely in the Gulf of Gabes.

The Gulf of Gabes is highly productive (Ben Brahim et al., 2010; Béjaoui et al., 2019) and considered an anomaly in the Mediterranean basin, where it contributes around 40% of the Tunisian fish production (DGPA, 2015). It is also worth noting that the source of enrichment by nutrients in the Gulf of Gabes is the Saharan dust, which is rich in phosphorus (Taieb & Ben Brahim, 2014). High concentrations of Chl-a were also observed on the coastal waters of Egypt, particularly the Nile estuary, indicating significant nutrient transport. It should be noted that the increase in nitrogen in the Nile estuary is probably associated with the use of fertilizers in agriculture and livestock. At the same time, industrial and urban wastewater discharges and agricultural activities present the main possible sources of phosphorus (Colella et al., 2016).

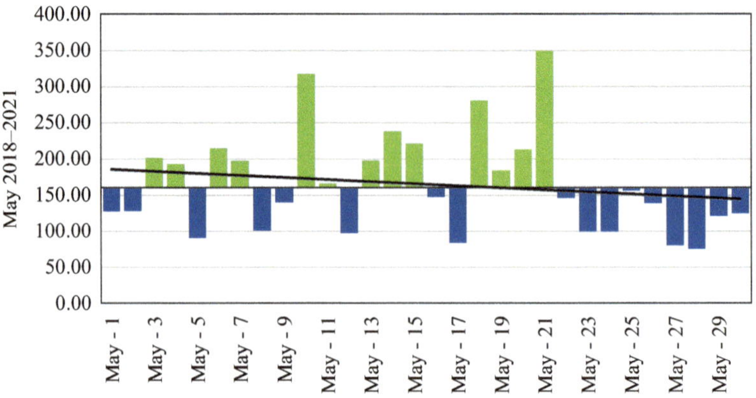

FIGURE 10.5 Trends of daily chlorophyll-a concentration observed on the coast of North Africa during May 2018–2021.

10.4.3 Seasonal Variability of SST and Chlorophyll-a

Figure 10.5 shows the daily mean time series of Chl-a concentration on the coast of North Africa during May 2018–2021, where the green color indicates an increase above average, while the blue indicates a concentration lower than average. Indeed, most of the high concentrations above average were recorded during mid-May, and it is worth noting that the concentration ratios below the average are the most dominant. This explains the decreasing tendency of the general line, as shown in 10.5.

The daily mean time series of Chl-a concentration on the coast of North Africa from August 2018 to 2021 is depicted in Figure 10.6, where the green color denotes an increase over the average, and the blue color denotes a concentration below normal. The concentrations were at their peak around the start of August, and it is essential to note that most concentrations are below the average, which caused the overall trend line to be downward.

Moreover, Figure 10.7 shows the daily mean time series of Chl-a concentration on the coast of North Africa during November 2018–2021, where green columns indicate an increase above average, while those shown in blue indicate concentrations lower than average. The highest concentrations were recorded from mid to late November, and it is worth noting that they are dominant, which made the general trend line ascend. Chlorophyll concentrations are often higher after

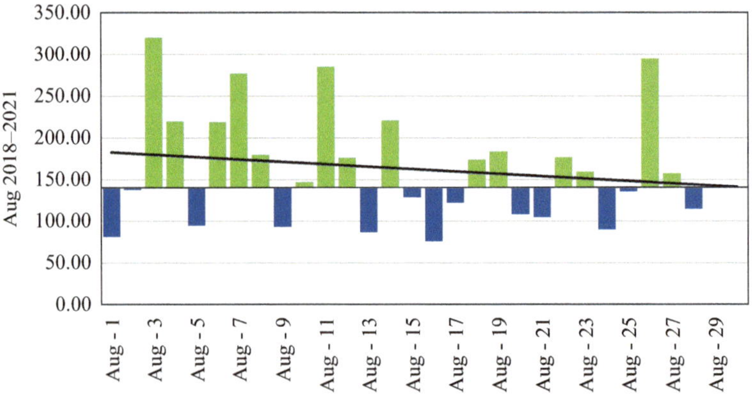

FIGURE 10.6 Trends of daily chlorophyll-a concentration observed on the coast of North Africa during August 2018–2021.

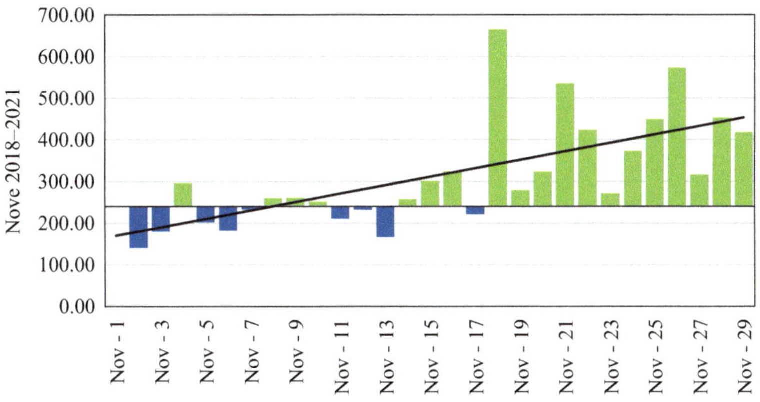

FIGURE 10.7 Trends of daily chlorophyll-a concentration on the coast of North Africa during November 2018–2021.

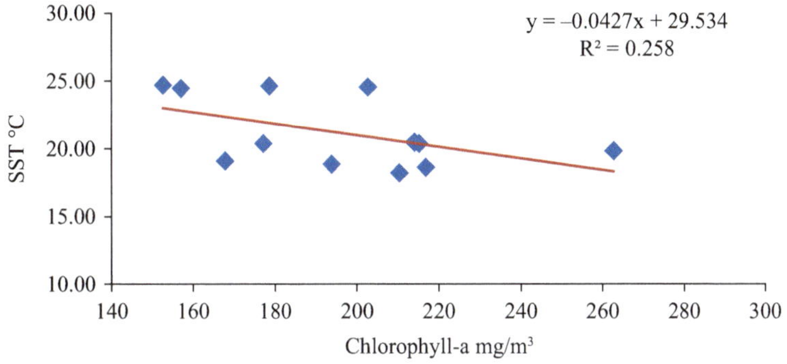

FIGURE 10.8 Linear regression of SST to chlorophyll-a concentration for the 2018–2021 period.

rainfall, particularly if the rain has flushed nutrients into the water. Thus, it is natural for chlorophyll levels to fluctuate over time and according to seasons.

In order to ascertain the association between Chl-a (chlorophyll-a) and SST (sea surface temperature), we computed a correlation coefficient by analyzing the respective datasets of Chl-a and SST. Figure 10.8 shows the correlation coefficient distribution for Chl-a and SST on the coast of North Africa during 2018–2021. We found that between SST and Chl-a, there was a negatively inverse correlation, with the coefficient of correlation at R = 0.51. This means that there is an inverse relationship between Chl-a concentrations and SST, implying that an increase in SST causes a decrease in Chl-a concentrations and vice versa (Navarro & Ruiz, 2006). This phenomenon can be explained by the occurrence of upwelling, which occurs when colder water from the lower layers rises to the surface and promotes the growth of phytoplankton blooms. According to the coefficient of determination (R2 = 0.258), SST accounts for 25.8% of the observed variation in Chl-a concentrations. Other factors such as ocean currents, wind patterns, salinity, runoff, and rainfall are likely to influence the remaining 74.2% (Nurdin et al., 2013).

The SST varies seasonally (see Figure 10.9): it is cold in spring (May) but warm in summer (August) and autumn (November). As reported by previous studies, the SST is impacted by many factors, such as monsoons, tides, and precipitation (Kako et al., 2016; Johnson & Toy, 2017; Nurdin et al., 2013). Chl-a concentration reaches its maximum in November, with concentrations decreasing after winter. Figure 10.9 gives the monthly average SST and Chl-a concentrations. As the green color indicates in the figure, November has the highest Chl-a concentration and August the lowest. As the yellow curve in the figure shows, August has the highest SST and May the lowest. In

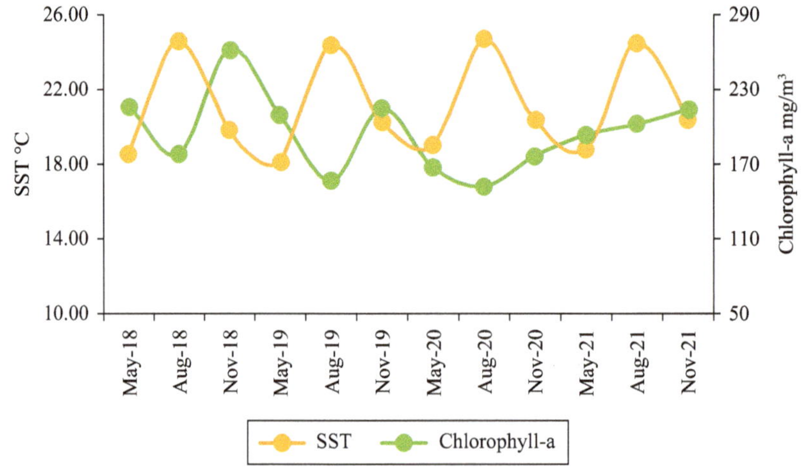

FIGURE 10.9 Comparison between sea surface temperature (SST) and chlorophyll-a concentration for the 2018–2021 period.

conclusion, the comparison between sea surface temperature and coastal chlorophyll concentration over the period 2018–2021 indicates that the higher the sea surface temperature, the lower the chlorophyll concentrations, in the sense of an inverse relationship between them (Chenxu Ji et al., 2018).

10.5 CHALLENGES AND OPPORTUNITIES IN RETRIEVING CHL-A OR SST FOR MONITORING COASTAL CHLOROPHYLL-A CONCENTRATION

While there are challenges in retrieving chlorophyll-a concentration and SST in coastal areas, there are also opportunities for improvement through the development of tailored algorithms, synergy between multiple sensors, integration of in situ data, and advancements in remote sensing technology. These efforts can contribute to more accurate and comprehensive monitoring of coastal processes and ecosystem dynamics. For instance, when retrieving chlorophyll-a concentration or SST from satellite imagery, atmospheric interference can affect the accuracy of the measurements. Atmospheric constituents, such as aerosols, can absorb or scatter the incoming radiation, leading to errors in the retrieved values. Appropriate atmospheric correction methods need to be applied to account for these effects (Lehmann et al., 2021). Further, coastal regions are characterized by complex coastal processes, including nearshore circulation, freshwater input from rivers, and tidal interactions. These factors can lead to variations in water properties and affect the retrieval of chlorophyll-a concentration or SST. Coastal areas often exhibit high spatial heterogeneity, with variations in chlorophyll-a concentration and SST occurring at small scales. The relatively coarse resolution of some satellite sensors may limit the ability to capture these fine-scale variations accurately (Kupssinskü et al., 2020).

However, combining data from multiple satellite sensors can provide complementary information and improve the accuracy of chlorophyll-a and SST retrievals. For example, using a combination of visible, near-infrared, and thermal bands from different sensors can enhance the retrieval algorithms and reduce uncertainties associated with individual sensors. Developing and refining retrieval algorithms specific to coastal regions can help address the challenges associated with coastal complexity. These algorithms can incorporate additional information, such as coastal bathymetry, water quality parameters, or ancillary data, to improve the accuracy of chlorophyll-a or SST retrievals in coastal areas. Integration of in situ measurements, such as water samples or buoys, can help validate and calibrate satellite-derived chlorophyll-a and SST values (Nazeer et Nichole, 2016; Silveira Kupssinskü et al., 2020). In situ data provide ground truth information and can be used to assess the accuracy of the satellite retrievals, leading to improved monitoring and understanding of coastal processes. Advances in satellite sensor technology, such as higher spatial resolution and improved radiometric and spectral capabilities, offer new opportunities for monitoring coastal chlorophyll-a and SST. New sensors, such as those on the Sentinel-2 or upcoming NASA's Plankton, Aerosol, Cloud, ocean Ecosystem (PACE) mission, provide enhanced capabilities for coastal monitoring applications (Lehmann et al., 2021).

10.6 CONCLUSIONS

This study obtained Chl-a concentration and SST data through GCOM-C/SGLI L3 and OISST/V2, respectively. The monitoring of Chl-a concentration in coastal water was performed using machine-learning and remote sensing techniques. Analysis of the two variables shows that Chl-a concentrations are highly correlated with SST in seawater locations along the coast of North Africa. There were seasonal variations of Chl-a concentrations where the highest values were recorded during November (autumn) while the lowest values were registered in August (summer). In fact, phytoplankton employs chlorophyll and other light-capturing pigments to facilitate the process of photosynthesis. The growth of phytoplankton is contingent upon the availability of sunlight, temperature, and nutrient concentrations. As cold waters generally exhibit higher nutrient levels compared to warm waters, phytoplankton tends to thrive and exhibit greater abundance in colder aquatic environments. An inverse correlation between SST and Chl-a was obtained, with the correlation coefficient at $R = 0.51$. Furthermore, spatial distribution variations were noticed from the east to the west part, where an obvious impoverishment in Chl-a was detected in the eastern part, mainly in Libyan waters and both the deltaic and western parts of the Egyptian coastal region. In addition, this study pointed out that ML and the GEE platform can be relied on to analyze data and monitor Chl-a concentration and SST in marine coasts over large areas.

REFERENCES

Alvarez-Fernandez, S., & Riegman, R. (2014). Chlorophyll in North Sea coastal and offshore waters does not reflect long term trends of phytoplankton biomass. *Journal of Sea Research*, 91, 35–44.

Azzirgue, E. M., Salmoun, F., Cherif, E. K., Ait Tchakoucht, T., & Mejjad, N. (2021). Determination of the Jouamaa groundwater quality using Physico-chemical water analysis and Geographic Information System. *E3S Web Conf.*, 314 (2021), 07006. https://doi.org/10.1051/e3sconf/202131407006.

Azzirgue, E. M., Salmoun, F., Cherif, E. K., Ait Tchakoucht, T., & Mejjad, N. (2022). Using machine learning approaches to predict water quality of Ibn Battuta Dam (Tangier, Morocco). In Proceedings of the 2nd International Conference on Big Data, Modelling and Machine Learning (BML 2021), pages 350–354.

Bedair, H. (2020). Composition and pattern of wild trees and shrubs in the Egyptian flora. M. Sc. Thesis. Botany Department, Faculty of Science, Tanta University, Tanta, Egypt.

Bedair, H., Shaltout, K., Ahmed, D., Sharaf El-Din, A., & El-Fahhar, R. (2020). Characterization of the wild trees and shrubs in the Egyptian Flora. *Egyptian Journal of Botany*, 60(1), 147–168.

Bedair, H., Shaltout, K. H., El-Din, A. S., El-Fahhar, R., & Halmy, M. W. (2022). Characterization of Mediterranean endemics in the Egyptian flora. In *Anales del Jardín Botánico de Madrid* (Vol. 79, No. 2, p. 4). Real Jardín Botánico.

Bedair, H., Alghariani, M. S., Omar, E., Anibaba, Q. A., Remon, M., Bornman, C., ... Alzain, H. M. (2023a). Global warming status in the African Continent: Sources, challenges, policies, and future direction. *International Journal of Environmental Research*, 17(3), 45.

Bedair, H., Shaltout, K. & Halmy, M.W.A. (2023b). A critical inventory of the mediterranean endemics in the egyptian flora. *Biodivers Conserv* **32**, 1327–1351. https://doi.org/10.1007/s10531-023-02555-5

Bedair, H., Shaltout, K. & Halmy, M.W.A. (2023c). Stacked machine learning models for predicting species richness and endemism for Mediterranean endemic plants in the Mareotis subsector in Egypt. *Plant Ecol* **224**, 1113–1126. https://doi.org/10.1007/s11258-023-01366-6

Béjaoui, B., Ismail, S. B., Othmani, A., Hamida, O. B. A. B. H., Chevalier, C., Feki-Sahnoun, W., & Hassen, M. B. (2019). Synthesis review of the Gulf of Gabes (eastern Mediterranean Sea, Tunisia): Morphological, climatic, physical oceanographic, biogeochemical and fisheries features. *Estuarine, Coastal and Shelf Science*, 219, 395–408.

Ben Brahim, M., Hamza, A., Hannachi, I., Rebai, A., Jarboui, O., Bouain, A., & Aleya, L. (2010). Variability in the structure of epiphytic assemblages of Posidonia oceanica in relation to human interferences in the Gulf of Gabes, Tunisia. *Marine Environmental Research*, 70, 411–421.

Boyer, J. N., Kelble, C. R., Ortner, P. B., & Rudnick, D. T. (2009). Phytoplankton bloom status: Chlorophyll a biomass as an indicator of water quality condition in the southern estuaries of Florida, USA. *Ecological Indicators*, 9(6), S56–S67.

Chen, J., Zhang, M., Cui, T., & Wen, Z. (2013). A review of some important technical problems in respect of satellite remote sensing of chlorophyll-a concentration in coastal waters. *IEEE Journal of Selected Topics in Applied Earth Observations and Remote Sensing,* 6(5), 2275–2289.

Cherif, E. K., Vodopivec, M., Mejjad, N., Esteves da Silva, J. C., Simonovič, S., & Boulaassal, H. (2020). COVID-19 pandemic consequences on coastal water quality using WST Sentinel-3 data: Case of Tangier, Morocco. *Water,* 12(9), 2638.

Colella, S., Falcini, F., Rinaldi, E., Sammartino, M., & Santoleri, R. (2016). Mediterranean Ocean colour chlorophyll trends. *PLoS ONE,* 11(6), e0155756. https://doi.org/10.1371/journal.pone.0155756.

De Madron, X. D., Guieu, C., Sempere, R., Conan, P., Cossa, D., D'Ortenzio, F., Estournel, C., Gazeau, F., Rabouille, C., Stemmann, L., Bonnet, S., et al. (2011). Marine ecosystems' responses to climatic and anthropogenic forcings in the Mediterranean. *Progress in Oceanography*, 91, 97–166. https://doi.org/10.1016/j.pocean.2011.02.003.

Del Carmen Jiménez-Quiroz, M., Martell-Dubois, R., Cervantes-Duarte, R., & Cerdeira-Estrada, S. (2021). Seasonal pattern of the chlorophyll-a in a coastal lagoon from the southern Baja California (Mexico), described with in situ observations and MODIS-Aqua imagery. *Oceanologia*, 63(3), 329–342.

DeLorenzo, M. E., Thompson, B., Cooper, E., Moore, J., & Fulton, M. H. (2012). A long-term monitoring study of chlorophyll, microbial contaminants, and pesticides in a coastal residential stormwater pond and its adjacent tidal creek. *Environmental Monitoring and Assessment*, 184(1), 343–359.

DGPA. (2015). Annuaire des Statistiques des Pêches en Tunisie (Année 2015). Direction Générale de la Pêche et de l'Aquaculture, Tunis.

European Environment Agency (EEA). (2005). Source apportionment of nitrogen and phosphorus inputs into the aquatic environment. EEA Report no. 7. 48pp

Field, C. B., Behrenfeld, M. J., Randerson, J. T., & Falkowski, P. (1998). Primary production of the biosphere: Integrating terrestrial and oceanic components. *Science,* 281, 237–240.

Harvey, E. T., Kratzer, S., & Philipson, P. (2015). Satellite-based water quality monitoring for improved spatial and temporal retrieval of chlorophyll-a in coastal waters. *Remote Sensing of Environment*, 158, 417–430.

Hu, C., Lee, Z., & Franz, B. (2012). Chlorophyll aalgorithms for oligotrophic oceans: A novel approach based on three-band reflectance difference. *Journal of Geophysical Research*, 117, C01011. https://doi.org/10.1029/2011JC007395.

Ji, C., Zhang, Y., Cheng, Q., Tsou, J.-Y., Jiang, T., & San Liang, X. (2018). Evaluating the impact of sea surface temperature (SST) on spatial distribution of chlorophyll-a concentration in the East China Sea. *International Journal of Applied Earth Observation and Geoinformation*, 68, 252–261. ISSN 1569-8432. https://doi.org/10.1016/j.jag.2018.01.020.

Johnson, R. H., & Toy, M. D. (2017). SST gradients and the East Asian early-summer monsoon. In The Global Monsoon System: Research and Forecast (pp.3–12).

Kako, S. I., Nakagawa, T., Takayama, K., Hirose, N., & Isobe, A. (2016). Impact of Changjiang River discharge on sea surface temperature in the East China Sea. *Journal of Physical Oceanography*, 46(6), 1735–1750.

Kavak, M. T., & Karadogan, S. (2012). The relationship between sea surface temperature and chlorophyll concentration of phytoplanktons in the Black Sea using remote sensing techniques. *Journal of Environmental Biology,* 33(2 Suppl), 493–498. PMID: 23424855.

Kotta, D., & Kitsiou, D. (2019). Chlorophyll in the Eastern Mediterranean Sea: Correlations with environmental factors and trends. *Environments,* 6(8), 98. https://doi.org/10.3390/environments6080098.

Kwon, Y. S., Baek, S. H., Lim, Y. K., Pyo, J., Ligaray, M., Park, Y., & Cho, K. H. (2018). Monitoring coastal chlorophyll-a concentrations in coastal areas using machine learning models. *Water,* 10, 1020. https://doi.org/10.3390/w10081020.

C5).Lathuilière, C., V. Echevin, and M. Lévy (2008), Seasonal and intraseasonal surface chlorophyll-a variability along the northwest African coast, *J. Geophys. Res.*, 113, C05007, doi:10.1029/2007JC004433.

Lehmann, M. K., Schütt, E. M., Hieronymi, M., Dare, J., & Krasemann, H. (2021). Analysis of recurring patchiness in satellite-derived chlorophyll a to aid the selection of representative sites for lake water quality monitoring. *International Journal of Applied Earth Observation and Geoinformation*, 104, 102547.

Litchman, E., de Tezanos Pinto, P., Edwards, K. F., Klausmeier, C. A., Kremer, C. T., & Thomas, M. K. (2015). Global biogeochemical impacts of phytoplankton: a trait-based perspective. *Journal of Ecology*, 103(6), 1384–1396.

Lyche Solheim, A., Rekolainen, S., Moe, S. J., et al. (2008). Ecological threshold responses in European lakes and their applicability for the Water Framework Directive (WFD) implementation: Synthesis of lakes results from the REBECCA project. *Aquatic Ecology,* 42, 317–334. https://doi.org/10.1007/s10452-008 -9188-5.

Malone, T. C., & Newton, A. (2020). The globalization of cultural eutrophication in the coastal ocean: Causes and consequences. *Frontiers in Marine Science.* https://doi.org/10.3389/fmars.2020.00670.

Maslukah, L., Setiawan, R. Y., Nurdin, N., Zainuri, M., Wirastriya, A., Helmi, M. (2021). Estimation of Chlorophyll a Phytoplankton in the Coastal Waters of Semarang and Jepara for Monitoring the Eutrophication Process using MODIS-Aqua Imagery and Conventional Methods. Journal of Ecological Engineering, 22(1), 51–59. https://doi.org/10.12911/22998993/128874

Mejjad, N., Rossi, A., & Pavel, A. B. (2022). The coastal tourism industry in the Mediterranean: A critical review of the socio-economic and environmental pressures & impacts. *Tourism Management Perspectives*, 44, 101007.

Murakami, H. (January 2020). ATBD of GCOM-C chlorophyll-a concentration algorithm (Version 2). https://suzaku.eorc.jaxa.jp/GCOM_C/data/ATBD/ver2/V2ATBD_O3AB_Chla_Murakami.pdf.

Navarro, G., & Ruiz, J. (2006). Spatial and temporal variability of phytoplankton in the Gulf of Cádiz through remote sensing images. *Deep Sea Research Part II: Topical Studies in Oceanography*, 53(11–13), 1241–1260. ISSN 0967-0645. https://doi.org/10.1016/j.dsr2.2006.04.014.

Nazeer, M., & Nichol, J. E. (2016). Development and application of a remote sensing-based Chlorophyll-a concentration prediction model for complex coastal waters of Hong Kong. *Journal of Hydrology*, 532, 80–89.

Nurdin, S., Mustapha, M. A., & Lihan, T. (2013). The relationship between sea surface temperature and chlorophyll-a concentration in fisheries aggregation area in the archipelagic waters of spermonde using satellite images. *AIP Conf. Proc.*, 1571, 466–472. https://doi.org/10.1063/1.4858699.

Papenfus, M., Schaeffer, B., Pollard, A. I., et al. (2020). Exploring the potential value of satellite remote sensing to monitor chlorophyll-a for US lakes and reservoirs. *Environmental Monitoring and Assessment,* 192, 808. https://doi.org/10.1007/s10661-020-08631-5.

Reynolds, R. W., Banzon, V. F., & NOAA CDR Program. (2008). NOAA optimum interpolation 1/4 degree daily sea surface temperature (OISST) analysis, version 2. [indicate subset used]. NOAA National Centers for Environmental Information. https://doi.org/10.7289/V5SQ8XB5.

Sala, M. M., Peters, F., Sebastián, M., Cardelús, C., Calvo, E., Marrasé, C., Massana, R., Pelejero, C., Sala-Coromina, J., Vaqué, D., & Gasol, J. M. (2022). COVID-19 lockdown moderately increased oligotrophy at a marine coastal site. *Science of The Total Environment*, 812, 151443. https://doi.org/10.1016/j.scitotenv.2021.151443.

Schaeffer, B. A., Schaeffer, K. G., Keith, D., Lunetta, R. S., Conmy, R., & Gould, R. W. (2013). Barriers to adopting satellite remote sensing for water quality management. *International Journal of Remote Sensing,* 34(November), 7534–7544. https://doi.org/10.1080/01431161.2013.823524.

Shaltout, K., & Bedair, H. (2022). Diversity, distribution and regional conservation status of the Egyptian tree flora. *African Journal of Ecology*, 60(4), 1155–1183.

Silveira Kupssinskü, L., Thomassim Guimarães, T., Menezes de Souza, E., Zanotta, D. C., Roberto Veronez, M., Gonzaga Jr, L., & Mauad, F. F. (2020). A method for chlorophyll-a and suspended solids prediction through remote sensing and machine learning. *Sensors*, 20(7), 2125. https://doi.org/10.3390/s20072125

Siokou-Frangou, I., Christaki, U., Mazzocchi, M. G., Montresor, M., Ribera d'Alcalà, M., Vaqué, D., & Zingone, A. (2010). Plankton in the open Mediterranean Sea: A review. *Biogeosciences*, 7, 1543–1586. https://doi.org/10.5194/bg-7-1543-2010.

Taieb, D., & Ben Brahim, A. (2014). Chemistry and mineralogy studies of PM10 atmospheric aerosols in the Gulf of Gabès, South Tunisia. *International Journal of Energy Technology and Policy*, 10, 125–143.

Thingstad, T. F., Krom, M. D., Mantoura, R. F. C., Flaten, G. F., Groom, S., Herut, B., Kress, N., Law, C. S., Pasternak, A., Pitta, P., Psarra, S., et al. (2005). Nature of phosphorus limitation in the ultraoligotrophic eastern Mediterranean. *Science*, 309(5737), 1068–1071. https://doi.org/10.1126/science.1112632.

Vargas, C. A., Escribano, R., & Poulet, S. (2006). Phytoplankton food quality determines time windows for successful zooplankton reproductive pulses. *Ecology*, 8, 2992–2999.

11 Aquatic Vegetation Species Identification and Mapping Using Multisource Data-Review

*Smrutisikha Mohanty, Prem C. Pandey**
Department of Life Sciences, School of Natural Sciences, Shiv Nadar Institution of Eminence, India, Department of Life Sciences, School of Natural Sciences, Shiv Nadar Institution of Eminence, India

Prashant K. Srivastava
Institute of Environment and Sustainable Development, Banaras Hindu University, Varanasi, India

11.1 INTRODUCTION

Wetlands are important natural resources that offer a wide range of ecological functions to both plants and animals (Amani et al., 2018). Their advantages come from natural hydrological and biogeochemical processes occurring in these environments (Ivory et al., 2019). To identify wetlands from other types of land cover, wetland vegetation is especially essential, suited to grow in hydric soil (Fu et al., 2017). Wetland ecosystem stability is highly dependent on the distribution pattern and growth stage of the plants (Short et al., 2016). The most obvious sign of wetland deterioration and succession is aquatic vegetation (Medina, 1996; Wang et al., 2022). A critical step for wetland management is to safeguard and recover the threatened ecosystems by accurately documenting the distribution, quality, and quantity of wetland vegetation (Henderson and Lewis, 2008). Wetlands are now frequently brought up in conversations about climate change since they hold 12% of the world's carbon reserves (Erwin, 2009).

Plants that are commonly found in water are known as aquatic vegetation. These plants are found either in freshwater or in marine water ecosystems such as ponds, lakes, rivers, and oceans (Den Hartog, 1981; Hellquist, 1993). Aquatic plants are characterized depending on their types. They are commonly divided into four types: submerged, floating, emergent, and shoreline vegetation (Madsen and Wersal, 2017; Larson et al., 2023). All this vegetation together forms a diverse aquatic habitat for a wide variety of aquatic species (Nagelkerken et al., 2008). Submerged plants are completely covered by water. They typically reach a depth of 2 to 10 feet or more (Gettys et al., 2014). Plants that float on the surface of water with floating roots or unrooted are known as floating vegetation (Harrell and Hochheimer, 1989). One of the most prevalent varieties of vegetation is emergent vegetation. Here, the roots grow underwater, and the shoots protrude above the surface of the water (Short et al., 2016). Shoreline vegetation indicates plants that prefer to grow near the shore. Algae, a type of small plant also known as phytoplankton, is found in water bodies and is the main plant present in the open water region of lakes (Bhateria and Jain, 2016).

* Corresponding author.

 DOI: 10.1201/9781003354000-15

Monitoring aquatic vegetation is essential to provide information on the ecological state of wetlands and connect them to ecosystem functionality (Silva et al., 2008; Herrera-Silveira and Morales-Ojeda, 2009). Water quality degradation is frequently related to the removal of aquatic vegetation and the rise in algal blooms (Coffey et al., 2019). As a result, it has been widely employed as a water quality and nutrient-loading indicator of aquatic ecosystems (Srivastava et al., 2008, Roberts et al., 1995). For the management of lakes and water resources, it is crucial to estimate the spread of aquatic flora accurately and efficiently (Ritchie et al., 2003). The ability of remote sensing techniques to support wetland monitoring over the past years has significantly increased (Gallant, 2015; Sun et al., 2016). This technology overcomes the drawbacks of conventional practices and has emerged as the most efficient way to collect spatio-temporal data on wetland vegetation (Kuenzer et al., 2011). It allows us to determine the extent of aquatic vegetation, biochemical and biophysical parameters, and the presence of invasive species (Radosevich et al., 2007). Currently, the information about aquatic plants is extracted using machine-learning methods such as random forest, support vector machine, etc. (Tian et al., 2016; Liu et al., 2018). The present study provides a comprehensive review of aquatic vegetation mapping by incorporating multisource data. It gives an overview of bibliometric analysis using specified keywords with a co-occurrence map, a thematic map, and scientific production of countries. This review helps to conduct more precise research and take appropriate measures for aquatic vegetation management.

11.2 NEED FOR AQUATIC VEGETATION MAPPING

Aquatic vegetation is a key component of the wetland ecosystem (Silva et al., 2008). It serves as a source of food and habitat for fish, waterfowl, moose, other mammals, and occasionally even for humans. Small fish can find a home among submerged foliage where they can hide from predators (Gettys et al., 2014; Mer et al., 2016). Native plants offer several benefits, including spawning grounds for aquatic species and animal viewing for the locals and visitors (Alikhani et al., 2021). Aquatic vegetation provides important ecological and socioeconomic services, such as water quality, increasing water clarity, stabilizing sediments, and further reducing the rate of nutrient cycling (Zhang et al., 2016). Not all aquatic plants are invasive and need to be removed (Hussner et al., 2017). Invasive floating aquatic vegetation is a significant global problem because of its detrimental effects on delicate wetland ecosystem processes such as hydrology and nutrient cycling (Rejmankova, 2011; Desta et al., 2012). It often has an economic impact on fisheries, hydropower production, and transportation services, endangering ecosystem biodiversity and function (Geist and Hawkins, 2016; Sousa, 2011). Degradation of habitat, which harms a variety of aquatic species and reduces ecosystem service functions, is another consequence of a decline in the biomass and diversification of aquatic vegetation (Schultz and Dibble, 2012). Climate change, increased land usage, and the introduction of exotic species all put stress on aquatic ecosystems (Husson et al., 2016). Hence, there is a need to keep track of changes to aquatic ecology as they are also linked to sustainable Development Goals (SDGs), directly or indirectly (Mohanty et al., 2024a).

11.3 MULTISOURCE DATA FOR AQUATIC VEGETATION MONITORING

Field-based monitoring approaches are either unavailable or logistically problematic in aquatic ecosystems (Dronova et al., 2021). Remote sensing offers a holistic strategy to monitor both aquatic vegetation and weeds across a wide geographic area (Hestir et al., 2008; Santos et al., 2009). The remote sensing approach must be precise, continuous over time and space, and account for the intrinsic environmental heterogeneity of a system (Hestir et al., 2008). Currently, unmanned aerial vehicles, radar, multispectral and hyperspectral sensors, with spatio-temporal variations, are available for identifying and mapping aquatic vegetation (Klemas, 2013; Doughty and Cavanaugh, 2019). Aquatic vegetation cover and temporal dynamics have both benefited from the applications of multispatial and temporal satellite images (Otera, 2010).

11.4 USE OF MULTISPECTRAL DATA FOR VEGETATION MONITORING

Due to sensor limitations, current missions were unable to provide data at the resolutions needed to adequately assess the features and processes of freshwater ecosystems (Bresciani et al., 2020). This resulted in limitations in satellite remote sensing of inland water systems. With the advancement of remote sensing technologies, the availability of low-resolution and high-frequency MODIS satellite images, the Landsat Thematic Mapper (TM), Enhanced Thematic Mapper Plus (ETM+), and QuickBird, IKONOS, and other high-resolution images has increased (Xie et al., 2008; Wang et al., 2010; Huang and Asner, 2009). Resolution of remote sensing images in terms of spatial, spectral, radiometric are essential along with machine learning techniques for accurate classificaiton to achieve high accuracy of Land Use Land Cover (Pandey et al., 2021) which may be further utilised in aquatic vegetation classification and discrimination. SPOT-5, with a resolution of 10 m, has achieved substantially higher precision in mapping submerged and emergent vegetation as well as tracking their ecological conditions (Kumar and Sinha, 2014). The near-infrared and red-edge bands are sensitive to vegetation spectral features (Fu et al., 2021). To map the invasive floating vegetation, a study was carried out in the Delta areas, California, using Sentinel-2 data (10 m resolution) and a random forest model. The results obtained an overall accuracy of 90% with species-level discrimination (Ade et al., 2022). Another study used Gao Fen 1 (GF-1) satellite data for mapping aquatic vegetation in Ulansuhai Lake, China, with a concave-convex decision function and concluded better accuracy for submerged aquatic vegetation classification and extraction (Chen et al., 2018). For a wide range of applications, marsh vegetation has been extensively mapped and characterized using optical and Synthetic Aperture Radar (SAR) data (Dabboor and Brisco, 2018).

11.5 VEGETATION MAPPING USING HYPERSPECTRAL DATA

Hyperspectral sensors have overcome numerous spectral constraints of multispectral satellite imaging for monitoring and mapping changes in vegetation cover (Almalki et al., 2022). The visible, near-infrared, mid-infrared, and short-wave infrared spectrums are all represented by bands on the hyperspectral sensors. As a result, spatial analysis and spectral signatures at various wavelengths offer a thorough understanding of the objects (Jiao et al., 2019). The advancement of hyperspectral remote sensing technologies and data collection enables more precise vegetation classification and variation in studies (Almalki et al., 2022). Hyperspectral remote sensing images can distinguish more minute distinctions among objects by providing more spectral information, greater spectral resolution, and spectral integration properties (Varshney and Arora, 2004; Khorram et al., 2012). A collection of vegetation signature libraries is typically needed to extract vegetation species from hyperspectral data (Zhang and Xie, 2012). However, in the majority of cases, the spectral signature library is built utilizing the field data combined with hyperspectral data (Williams et al., 2003). Thus, vegetation mapping with hyperspectral imaging needs to be carefully planned to get synchronized field data for accurate imaging signature creation (Xie et al., 2008). For example, a study conducted on the Sacramento-San Joaquin Delta for mapping two invasive aquatic species using HyMap hyperspectral imagery data concluded that for finer spatial scale, high map accuracy was recorded for both of the species, while for coarser spatial scale the identification of submerged aquatic species is difficult (Underwood et al., 2006). A similar study was carried out to identify the aquatic plants in the Tahiku Lake basin by developing a decision tree algorithm, and the result showed an accurate identification of different plant species and their distribution (Mu et al., 2023). A spectroradiometer is a handheld portable instrument used to record ground-based hyperspectral measurements. The spectral library is created by using this instrument (Figure 11.1a). With a resolution of less than 3 nm at 700 nm, the instrument measures radiation intensity in 750 consecutive channels over the spectral range of 325 to 1075 nm. Continuum removal is a method for analyzing and isolating certain absorption properties in a spectrum (Figure 11.1b) (Huang et al., 2004; Clark and Roush, 1984). This technique is frequently used to map various vegetation types and evaluate

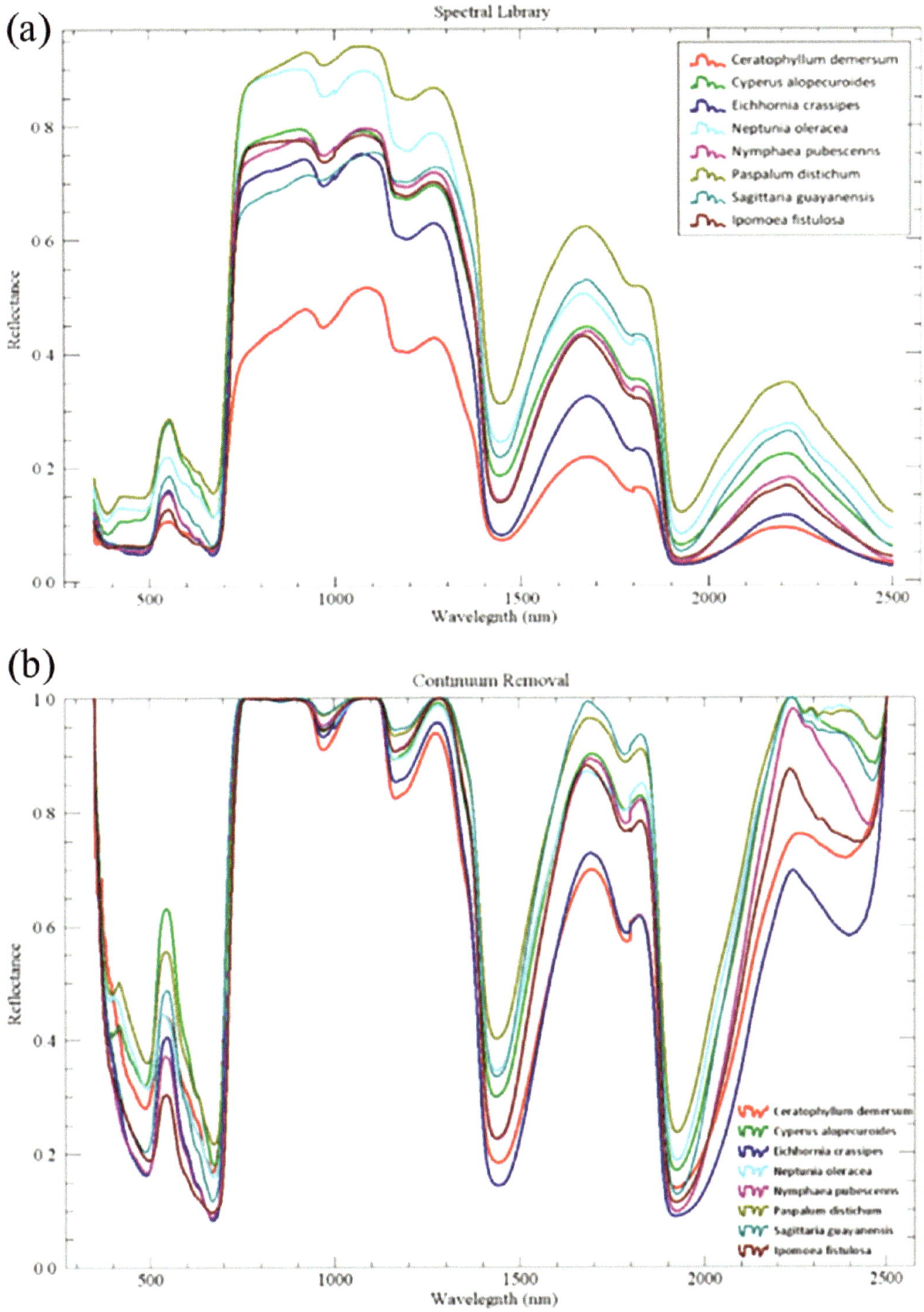

FIGURE 11.1 (a) Spectral library of invasive aquatic species. (b) Continuum removal.

their health, analyze data from water bodies, and examine mineral absorption properties (Khanna et al., 2013; Watanabe et al., 2013). To calculate the continuum of each absorption feature, straight-line segments are joined to the absorption band limits (Pandey et al., 2018). The equation is given as:

$$R_{(\lambda)cr} = \frac{R_{(\lambda)}}{C_{(\lambda)}}$$

where $R_{(\lambda)cr}$ is the continuum removed spectra, $R_{(\lambda)}$ is the original spectra, and $C_{(\lambda)}$ is the continuum curve.

11.6 ROLE OF UNMANNED AERIAL VEHICLES FOR AQUATIC VEGETATION MAPPING

Unmanned Aerial Vehicle (UAV) use for invasive and aquatic species identification has gained popularity nowadays (Ahmad et al., 2020; Tahsin, 2019). It is a cost-effective method of obtaining remote sensing images at any time (Yao et al., 2019). UAVs with cameras capable of collecting images with less than 1 cm resolution are useful for mapping vegetation and identifying different plant species (Bertacchi et al., 2019). Instead of being used for flight schedules, UAVs can be utilized to obtain fine spatial resolution data with a temporal resolution more suited to application needs (Zweig et al., 2015). To differentiate between different species, drones can be deployed selectively with optimal weather such as low wind, optimal sun angle, and longer sun exposure (Jeziorska, 2019; Zeng et al., 2017). When compared to other methods, UAVs typically have lower costs, can capture images from inaccessible areas, and provide data from the area of interest (Flynn and Chapra, 2014; Hardin and Jensen, 2011). In an aquatic ecosystem, drones are utilized for mapping submerged vegetation (Watts et al., 2012; Chabot et al., 2018). Drone data are utilised to generate spectral indices and implement sensitivity models to predict sensitivity of chlorophyll, where author incorporated machine learning to classify the vegetation cover to discriminate aquatic vegetation (Mohanty et al., 2024b). A study conducted on Seymour Lake to monitor aquatic vegetation by combining drone imagery with supervised classification found that the classification accuracy for emergent vegetation is 92% and for submerged vegetation is 84%, which is considered good (Chabot et al., 2018). A similar type of study was carried out for vegetation mapping of the Karst wetland, using UAV RGB images with multiple SegNet deep learning network fusion, and concluded that the result shows better classification accuracy for vegetation mapping (Deng et al., 2022).

11.7 VEGETATION MAPPING USING MICROWAVE REMOTE SENSING

Synthetic Aperture Radar (SAR) is capable of detecting smooth open water surfaces and distinguishing between distinct wetland plants and canopy structures (Tsyganskaya et al., 2016). The key characteristics of radar, particularly in tropical environments, include its ability to penetrate clouds and, to a lesser extent, rain, as well as day and night operability (Dubeau et al., 2017). Unique data on surface hydrology and vegetation cover of herbaceous wetlands can be obtained by the SAR system. It can penetrate vegetation canopies to capture understory conditions (White et al., 2015). RADARSAT-2 satellite with a full polarimetric C-band SAR sensor is used to monitor the biophysical properties of wetland vegetation (Solanki et al., 2022; Upakankaew et al., 2022). The SAR-captured images are utilized to identify submerged vegetation, as reported by various literature (Niculescu et al., 2020; Horritt et al., 2003). Polarimetric SAR images can offer more comprehensive scattering data and improve the capacity to distinguish between wetland vegetation (Van Beijma et al., 2014). With the advancement of SAR technology, quad polarization data have steadily been employed to extract aquatic vegetation (Yao et al., 2022). A study was carried out in Vembanad Lake to detect water hyacinths by employing Sentinel-1 imagery, showing a distinct identification of water hyacinths with a change detection rate of ~95% (Simpson et al., 2022). A similar study was performed in the Amazon Lowland to monitor flooded vegetation using RADARSAT-2 data, concluding a good vegetation classification accuracy of 86% (Canisius et al., 2019).

11.8 SCIENTOMETRIC ANALYSIS

Scientometric analysis has recently become popular in scientific research (Frehse et al., 2016). It enables the researcher to recognize and access scientific facts with greater objectivity and accuracy (Luo et al., 2022). The bibliometric data are taken into account to create a network model. The bibliometric analysis is carried out using multiple scientific search engines such as PubMed, Web of Science, Scopus, and Dimension. The different combinations of keywords were searched on 25 July 2023 using the Scopus database presented in Table 11.1. A total of 4936 documents were found with the keywords "aquatic vegetation" OR "aquatic vegetation monitoring" OR "aquatic vegetation mapping" OR "aquatic vegetation identification" OR "aquatic vegetation discrimination" OR

TABLE 11.1
Author Searched Keywords in Scopus Database for Bibliometric Analysis

Keywords	Total Documents	Period	Publications
"aquatic vegetation"	18,833 (Total Literature), 17,859 (English Literature)	1934–2023	Only English language including (i) Article, (ii) Conference paper, (iii) Review, (iv) Book chapter, (v) Book
"aquatic vegetation" AND "Remote Sensing"	42,16 3,987 (English Literature)	1973–2023	Only English language including (i) Article, (ii) Conference paper, (iii) Review, (iv) Book chapter, (v) Book
"aquatic vegetation mapping" AND "Remote Sensing" OR "GIS" OR "earth observation"	160, 157 (English Literature)	1990–2023	Only English language including (i) Article, (ii) Conference paper, (iii) Review, (iv) Book chapter, (v) Book
"aquatic vegetation" OR "aquatic vegetation monitoring" OR "aquatic vegetation mapping" OR "aquatic vegetation identification" OR "aquatic vegetation discrimination" OR "aquatic vegetation management" OR "aquatic vegetation health" OR "aquatic vegetation species composition" OR "aquatic vegetation assessment" OR "aquatic vegetation distribution" AND "multisource data" OR "geospatial technology" OR "Remote sensing" OR "GIS" OR "multispectral data" OR "Hyperspectral" OR "UAV" OR "earth observation" OR "drone" OR "Radar"	49,36 4,671 (English Literature)	1990–2023	Only English language including (i) Article. (ii) Conference paper, (iii) Review, (iv) Book chapter, (v) Book

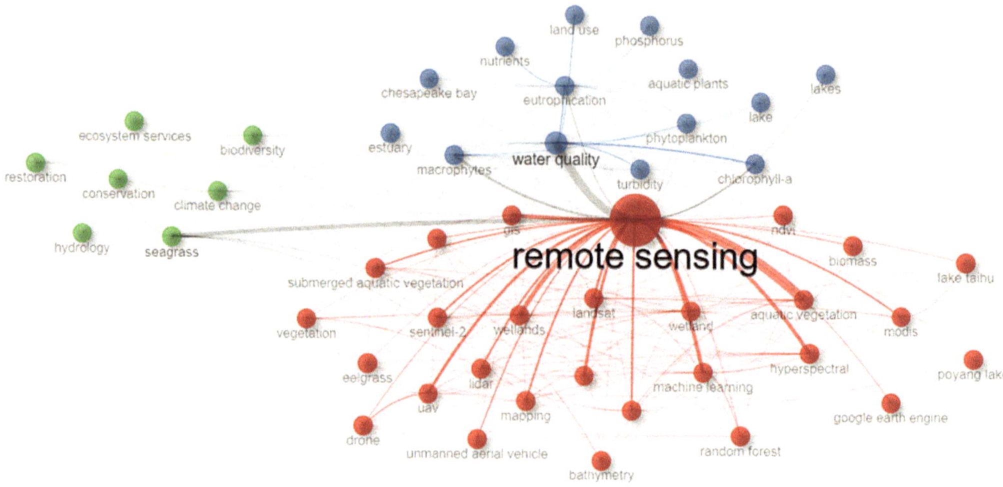

FIGURE 11.2 Co-occurrence map.

"aquatic vegetation management" OR "aquatic vegetation health" OR "aquatic vegetation species composition" OR "aquatic vegetation assessment' OR "aquatic vegetation distribution" AND "multisource data" OR "geospatial technology" OR "Remote sensing" OR "GIS" OR "multispectral data' OR "Hyperspectral" OR "UAV" OR "earth observation" OR "drone" OR "Radar", of which the total number in literature restricted to the English language was 4671. The publications only included research articles, conference papers, reviews, book chapters, and books in the period from 1990 to 2023. The searched documents were downloaded in BibTex format and further analyzed in R studio using the bibliometrix package.

11.9 CO-OCCURRENCE MAP

A map of keywords with links, co-occurrences, and density based on the frequency of co-occurrence is illustrated in Figure 11.2. The size of the keyword nodes presents the frequency, while its location is shown by co-occurrence (Zhang et al., 2022). The clusters are depicted on the map in a way that demonstrates how frequently they co-occur in different sources. In the present study, the clusters are represented in different color codes. The interconnected nodes indicate the correlation between the keywords (Pan et al., 2023). The higher the correlation, the closer nodes are present. "Remote sensing" is a frequently used word, establishing a close relationship between water quality and seagrass. Hence, the thicker the line, the better correlation is obtained (Colares et al., 2020). Cluster 1 is comprised of red color, while clusters 2 and 3 consist of blue and green colors respectively. The betweenness and closeness of "remote sensing", "wetlands", "aquatic vegetation" and "gis" in cluster 1 are 3.939 to 528.736 and 0.0121 to 0.0204 respectively. The keywords "water quality", "eutrophication", "aquatic plants", and "macrophytes" are presented in cluster 2, having a betweenness of 0.531 to 69.789 and closeness ranging from 0.0113 to 0.0161. In cluster 3, the keywords "seagrass", "climate change", "biodiversity", and "restoration" occurred with betweenness ranging from 2.420 to 26.874 and closeness of 0.0120 to 0.0136.

11.10 THEMATIC MAP

The typological theme is displayed in a two-dimensional diagram known as a thematic map. Using co-word analysis, themes in the research domain are produced by locating the keyword clusters

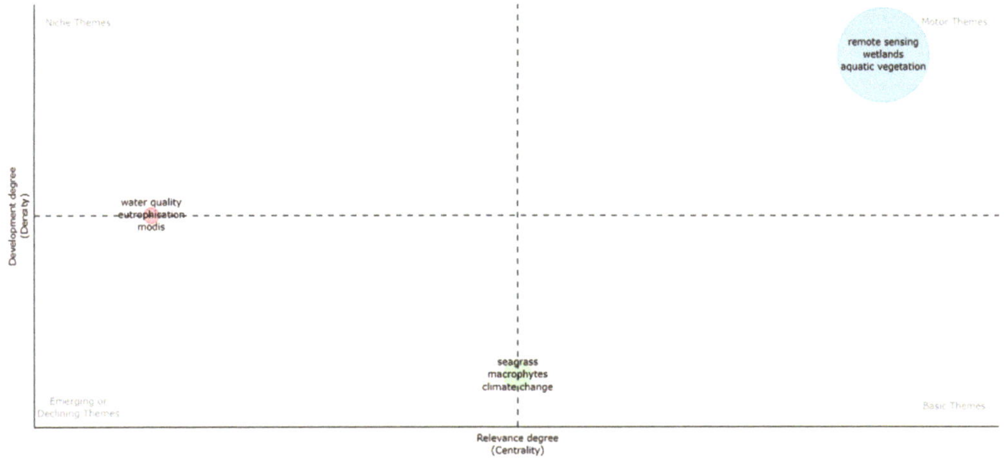

FIGURE 11.3 Thematic map using author-generated keywords.

(Cao et al., 2021). The two dimensions are development degree (density) and relevance degree (centrality) (Faghihinia et al., 2021). Four quadrants each reflect one of the topics (Pandey and Pandey, 2023). A bubble on the map designates each subject. Niche themes, motor themes, fundamental themes, and emerging or declining themes together make up the four quadrants (Figure 11.3). The motor themes are remote sensing, wetlands, and aquatic vegetation, indicating that these issues are addressed more frequently and serve as significant thematic drivers for future research. In basic and declining themes, the keywords "seagrass", "macrophytes", and "climate change" lie on the central dividing line, close to the relevance degree. Water quality, eutrophication, and MODIS are the themes located close to the density and on the line separating niche themes and emerging themes.

11.11 TREND TOPICS

The trend topics in the field of aquatic vegetation monitoring are presented in Figure 11.4 from 2001 to 2023. The most frequently used topic is "remote sensing", from 2013 to 2021, having a frequency of 587. With increasing years, different trending topics have evolved. "Water quality" is the second most used term (2015–2021) with a frequency of 192. During 2020–2023, the topics that are frequently used in publications are "sentinel-2", "machine learning", "hyperspectral", "vegetation indices", and "nature-based solutions".

11.12 SCIENTIFIC PRODUCTION OF COUNTRY

Figure 11.5 depicts the geographical distribution of research output. It is a representation of the nation's scientific production. The biggest contribution in terms of work is reported in the United States (USA) with 4077 publications and highlighted in dark blue. The area in gray illustrates that no publication occurred in that region. After the United States, China has the majority of scientific output (2911), followed by Australia (692), Canada (683), and Italy (539). India contributes a total of 393 scientific productions. The top 25 countries of the world in terms of their number of scientific productions are given in Table 11.2.

11.13 COUNTRY PRODUCTION OVER TIME

The countries' overall scientific production from 1997 to 2021 is shown in Figure 11.6. The highest number of article publications is reported in the United States (USA) followed by China, Canada,

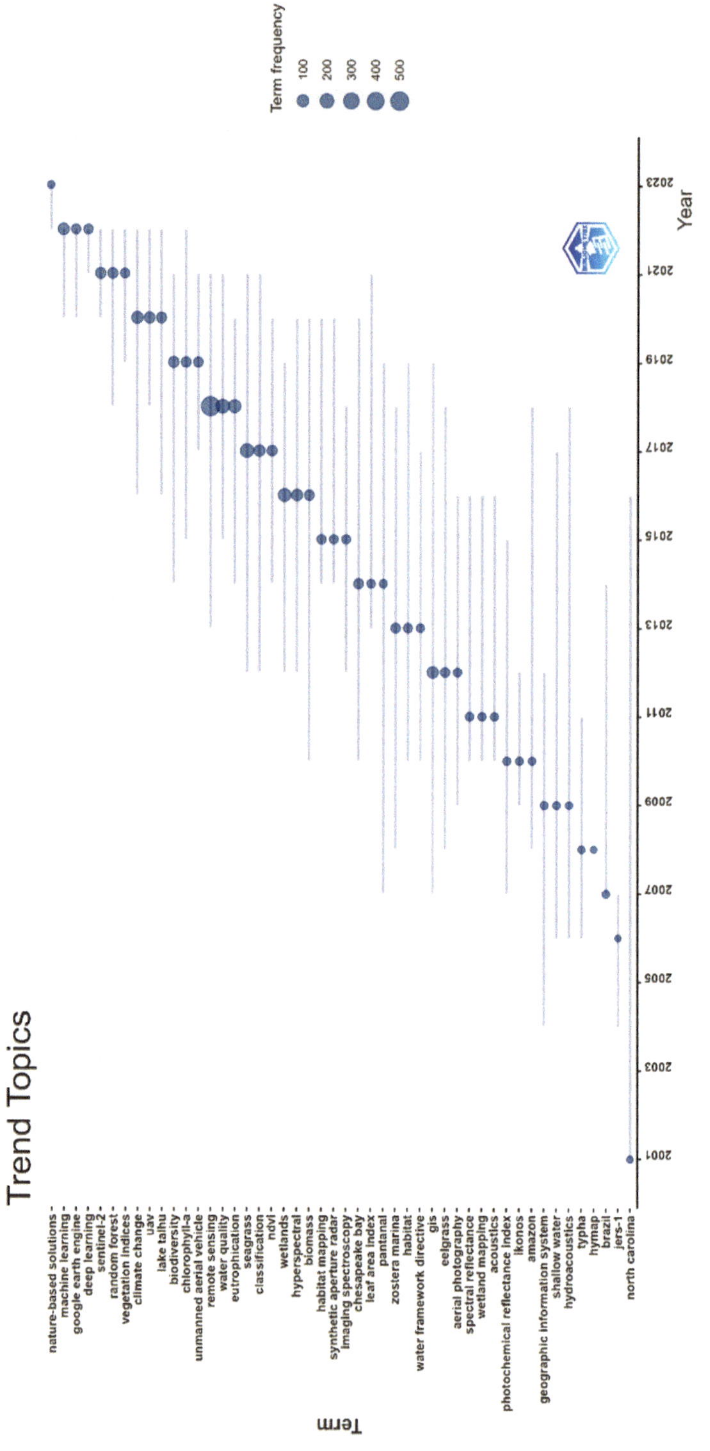

FIGURE 11.4 Trend topic analysis using author-generated keywords.

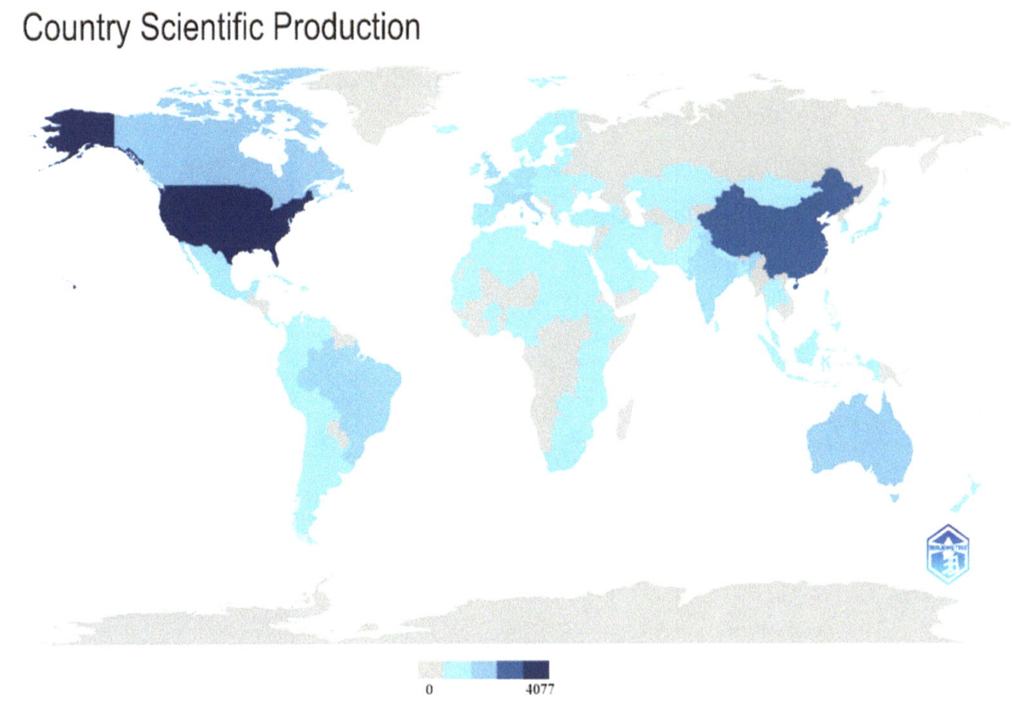

FIGURE 11.5 Scientific production of countries in the Scopus database

Australia, and Italy. The X-axis represents the year and the number of articles is shown on the Y-axis. The graph demonstrates the evolution of articles over time of a particular subject matter from different country levels.

11.14 CONCLUSION

Changes in vegetation cover are a global environmental phenomenon that can occur naturally or as a result of human activity. This may result in the transformation of naturally vegetated areas into bare areas, affecting the environmental system. Hence, the early detection of plant cover changes and the evaluation of their severity on a regional level become crucial. The present study provides an overview of aquatic vegetation, its types, and the need for aquatic vegetation mapping. This study also includes the role of various multisource datasets such as multispectral data, hyperspectral data, drone data, and SAR data for aquatic vegetation identification. Individual satellites and remote sensing data have their specific roles in providing information on different vegetation types. This study gives a comprehensive review of scientometric analysis using the Scopus database. The keywords are searched from 1934 to 2023 and include a co-occurrence map, a thematic map, trend topics, and scientific production across countries. More elaborate work must be carried out country-wise on aquatic vegetation monitoring. The fusion of multisource data provides better accuracy for aquatic vegetation mapping. Thus, in future research, combinations of multiple datasets can be employed for a better and more precise output.

Acknowledgements

The authors are thankful to Shiv Nadar Institution of Eminence (Deemed to be University) for continuous help and support. Authors would like to extend our deepest gratitude to the National

TABLE 11.2
Top 25 Countries Showing the Frequency of Scientific Production from 1970 to 2023

Country	Frequency
USA	4077
China	2911
Australia	692
Canada	683
Italy	539
Brazil	505
UK	481
Germany	393
India	393
France	383
Netherlands	281
Spain	270
Japan	227
Mexico	150
South Africa	142
Poland	141
Sweden	139
Portugal	131
Denmark	119
Finland	111
Argentina	99
Greece	96
South Korea	92
Indonesia	90
New Zealand	90

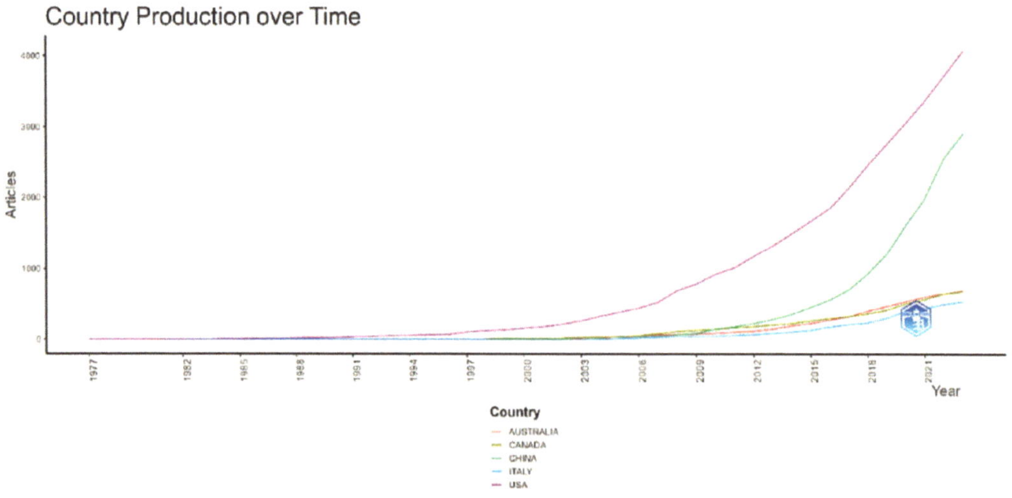

FIGURE 11.6 Countries' production over time (1997–2021).

Geospatial Program (erstwhile, NRDMS), DST Government of India (NGP/TPN/30705–2019(G)) for their generous financial assistance to Ms Smrutisikha Mohanty.

REFERENCES

ADE, C., KHANNA, S., LAY, M., USTIN, S. L. & HESTIR, E. L. 2022. Genus-level mapping of invasive floating aquatic vegetation using Sentinel-2 satellite remote sensing. *Remote Sensing,* 14, 3013.

AHMAD, A., ORDOÑEZ, J., CARTUJO, P. & MARTOS, V. 2020. Remotely piloted aircraft (RPA) in agriculture: A pursuit of sustainability. *Agronomy,* 11, 7.

ALIKHANI, S., NUMMI, P. & OJALA, A. 2021. Urban wetlands: A review on ecological and cultural values. *Water,* 13, 3301.

ALMALKI, R., KHAKI, M., SACO, P. M. & RODRIGUEZ, J. F. 2022. Monitoring and mapping vegetation cover changes in arid and semi-arid areas using remote sensing technology: A review. *Remote Sensing,* 14, 5143.

AMANI, M., SALEHI, B., MAHDAVI, S. & BRISCO, B. 2018. Spectral analysis of wetlands using multisource optical satellite imagery. *ISPRS Journal of Photogrammetry and Remote Sensing,* 144, 119–136.

BERTACCHI, A., GIANNINI, V., DI FRANCO, C. & SILVESTRI, N. 2019. Using unmanned aerial vehicles for vegetation mapping and identification of botanical species in wetlands. *Landscape and Ecological Engineering,* 15, 231–240.

BHATERIA, R. & JAIN, D. 2016. Water quality assessment of lake water: A review. *Sustainable Water Resources Management,* 2, 161–173.

BRESCIANI, M., PINARDI, M., FREE, G., LUCIANI, G., GHEBREHIWOT, S., LAANEN, M., PETERS, S., DELLA BELLA, V., PADULA, R. & GIARDINO, C. 2020. The use of multisource optical sensors to study phytoplankton spatio-temporal variation in a Shallow Turbid Lake. *Water,* 12, 284.

CANISIUS, F., BRISCO, B., MURNAGHAN, K., VAN DER KOOIJ, M. & KEIZER, E. 2019. SAR backscatter and InSAR coherence for monitoring wetland extent, flood pulse and vegetation: A study of the Amazon lowland. *Remote Sensing,* 11, 720.

CAO, T., HAN, D. & SONG, X. 2021. Past, present, and future of global seawater intrusion research: A bibliometric analysis. *Journal of Hydrology,* 603, 126844.

CHABOT, D., DILLON, C., SHEMROCK, A., WEISSFLOG, N. & SAGER, E. P. 2018. An object-based image analysis workflow for monitoring shallow-water aquatic vegetation in multispectral drone imagery. *ISPRS International Journal of Geo-Information,* 7, 294.

CHEN, Q., YU, R., HAO, Y., WU, L., ZHANG, W., ZHANG, Q. & BU, X. 2018. A new method for mapping aquatic vegetation especially underwater vegetation in Lake Ulansuhai using GF-1 satellite data. *Remote Sensing,* 10, 1279.

CLARK, R. N. & ROUSH, T. L. 1984. Reflectance spectroscopy: Quantitative analysis techniques for remote sensing applications. *Journal of Geophysical Research: Solid Earth,* 89, 6329–6340.

COFFEY, R., PAUL, M. J., STAMP, J., HAMILTON, A. & JOHNSON, T. 2019. A review of water quality responses to air temperature and precipitation changes 2: Nutrients, algal blooms, sediment, pathogens. *JAWRA Journal of the American Water Resources Association,* 55, 844–868.

COLARES, G. S., DELL'OSBEL, N., WIESEL, P. G., OLIVEIRA, G. A., LEMOS, P. H. Z., DA SILVA, F. P., LUTTERBECK, C. A., KIST, L. T. & MACHADO, Ê. L. 2020. Floating treatment wetlands: A review and bibliometric analysis. *Science of the Total Environment,* 714, 136776.

DABBOOR, M. & BRISCO, B. 2018. Wetland monitoring and mapping using synthetic aperture radar. *Wetlands Management - Assessing Risk and Sustainable Solutions,* 1, 13.

DEN HARTOG, C. Aquatic plant communities of poikilosaline waters. Salt Lakes: Proceedings of the International Symposium on Athalassic (Inland) Salt Lakes, held at Adelaide, Australia, October 1979, 1981. Springer, 15–22.

DENG, T., FU, B., LIU, M., HE, H., FAN, D., LI, L., HUANG, L. & GAO, E. 2022. Comparison of multiclass and fusion of multiple single-class SegNet model for mapping karst wetland vegetation using UAV images. *Scientific Reports,* 12, 13270.

DESTA, H., LEMMA, B. & FETENE, A. 2012. Aspects of climate change and its associated impacts on wetland ecosystem functions: A review. *Journal of American Science,* 8, 582–596.

DOUGHTY, C. L. & CAVANAUGH, K. C. 2019. Mapping coastal wetland biomass from high resolution unmanned aerial vehicle (UAV) imagery. *Remote Sensing,* 11, 540.

DRONOVA, I., KISLIK, C., DINH, Z. & KELLY, M. 2021. A review of unoccupied aerial vehicle use in wetland applications: Emerging opportunities in approach, technology, and data. *Drones,* 5, 45.

DUBEAU, P., KING, D. J., UNBUSHE, D. G. & REBELO, L.-M. 2017. Mapping the Dabus wetlands, Ethiopia, using random forest classification of Landsat, PALSAR and topographic data. *Remote Sensing,* 9, 1056.

ERWIN, K. L. 2009. Wetlands and global climate change: The role of wetland restoration in a changing world. *Wetlands Ecology and management,* 17, 71–84.

FAGHIHINIA, M., XU, Y., LIU, D. & WU, N. 2021. Freshwater biodiversity at different habitats: Research hotspots with persistent and emerging themes. *Ecological Indicators,* 129, 107926.

FLYNN, K. F. & CHAPRA, S. C. 2014. Remote sensing of submerged aquatic vegetation in a shallow non-turbid river using an unmanned aerial vehicle. *Remote Sensing,* 6, 12815–12836.

FREHSE, F. D. A., BRAGA, R. R., NOCERA, G. A. & VITULE, J. R. S. 2016. Non-native species and invasion biology in a megadiverse country: Scientometric analysis and ecological interactions in Brazil. *Biological Invasions,* 18, 3713–3725.

FU, B., WANG, Y., CAMPBELL, A., LI, Y., ZHANG, B., YIN, S., XING, Z. & JIN, X. 2017. Comparison of object-based and pixel-based Random Forest algorithm for wetland vegetation mapping using high spatial resolution GF-1 and SAR data. *Ecological Indicators,* 73, 105–117.

FU, B., XIE, S., HE, H., ZUO, P., SUN, J., LIU, L., HUANG, L., FAN, D. & GAO, E. 2021. Synergy of multi-temporal polarimetric SAR and optical image satellite for mapping of marsh vegetation using object-based random forest algorithm. *Ecological Indicators,* 131, 108173.

GALLANT, A. L. 2015. The challenges of remote monitoring of wetlands. *Remote Sensing, 7(8), 10938–10950.*

GEIST, J. & HAWKINS, S. J. 2016. Habitat recovery and restoration in aquatic ecosystems: Current progress and future challenges. *Aquatic Conservation: Marine and Freshwater Ecosystems,* 26, 942–962.

GETTYS, L.A., HALLER W. T. and BELLAUD, M. (2014). Biology and control of aquatic plants. *A best management practices handbook, 3rd edn. Aquatic Ecosystem Restoration Foundation, Marietta.*

HARDIN, P. J. & JENSEN, R. R. 2011. Introduction—Small-scale unmanned aerial systems for environmental remote sensing. *GIScience & Remote Sensing,* 48, 1–3.

HARRELL, R. M. & HOCHHEIMER, J. N. 1989. White water lily: aquatic plant identification and management. workbook, series 1, pp 1-4, Univeristy of Maryland System, USA.

HELLQUIST, C. B. 1993. Taxonomic considerations in aquatic vegetation assessments. *Lake and Reservoir Management,* 7, 175–183.

HENDERSON, F. M. & LEWIS, A. J. 2008. Radar detection of wetland ecosystems: A review. *International Journal of Remote Sensing,* 29, 5809–5835, USDA Forest Service.

HERRERA-SILVEIRA, J. A. & MORALES-OJEDA, S. M. 2009. Evaluation of the health status of a coastal ecosystem in southeast Mexico: Assessment of water quality, phytoplankton and submerged aquatic vegetation. *Marine Pollution Bulletin,* 59, 72–86.

HESTIR, E. L., KHANNA, S., ANDREW, M. E., SANTOS, M. J., VIERS, J. H., GREENBERG, J. A., RAJAPAKSE, S. S. & USTIN, S. L. 2008. Identification of invasive vegetation using hyperspectral remote sensing in the California Delta ecosystem. *Remote Sensing of Environment,* 112, 4034–4047.

HORRITT, M., MASON, D., COBBY, D., DAVENPORT, I. & BATES, P. 2003. Waterline mapping in flooded vegetation from airborne SAR imagery. *Remote Sensing of Environment,* 85, 271–281.

HUANG, C.-Y. & ASNER, G. P. 2009. Applications of remote sensing to alien invasive plant studies. *Sensors,* 9, 4869–4889.

HUANG, Z., TURNER, B. J., DURY, S. J., WALLIS, I. R. & FOLEY, W. J. 2004. Estimating foliage nitrogen concentration from HYMAP data using continuum removal analysis. *Remote Sensing of Environment,* 93, 18–29.

HUSSNER, A., STIERS, I., VERHOFSTAD, M., BAKKER, E., GRUTTERS, B., HAURY, J., VAN VALKENBURG, J., BRUNDU, G., NEWMAN, J. & CLAYTON, J. 2017. Management and control methods of invasive alien freshwater aquatic plants: A review. *Aquatic Botany,* 136, 112–137.

HUSSON, E., ECKE, F. & REESE, H. 2016. Comparison of manual mapping and automated object-based image analysis of non-submerged aquatic vegetation from very-high-resolution UAS images. *Remote Sensing,* 8, 724.

IVORY, S. J., MCGLUE, M. M., SPERA, S., SILVA, A. & BERGIER, I. 2019. Vegetation, rainfall, and pulsing hydrology in the Pantanal, the world's largest tropical wetland. *Environmental Research Letters,* 14, 124017.

JEZIORSKA, J. 2019. UAS for wetland mapping and hydrological modeling. *Remote Sensing,* 11, 1997.

JIAO, L., SUN, W., YANG, G., REN, G. & LIU, Y. 2019. A hierarchical classification framework of satellite multispectral/hyperspectral images for mapping coastal wetlands. *Remote Sensing,* 11, 2238.

KHANNA, S., SANTOS, M. J., USTIN, S. L., KOLTUNOV, A., KOKALY, R. F. & ROBERTS, D. A. 2013. Detection of salt marsh vegetation stress and recovery after the Deepwater Horizon oil spill in Barataria Bay, Gulf of Mexico using AVIRIS data. *PloS One,* 8, e78989.

KHORRAM, S., KOCH, F. H., VAN DER WIELE, C. F. & NELSON, S. A. 2012. *Remote sensing.* Springer Science & Business Media.

KLEMAS, V. 2013. Remote sensing of emergent and submerged wetlands: An overview. *International Journal of Remote Sensing,* 34, 6286–6320.

KUENZER, C., BLUEMEL, A., GEBHARDT, S., QUOC, T. V. & DECH, S. 2011. Remote sensing of mangrove ecosystems: A review. *Remote Sensing,* 3, 878–928.

KUMAR, L. & SINHA, P. 2014. Mapping salt-marsh land-cover vegetation using high-spatial and hyperspectral satellite data to assist wetland inventory. *GIScience & Remote Sensing,* 51, 483–497.

LARSON, D. M., CARHART, A. M. & LUND, E. M. 2023. Aquatic vegetation types identified during early and late phases of vegetation recovery in the Upper Mississippi River. *Ecosphere,* 14, e4468.

LIU, T., ABD-ELRAHMAN, A., MORTON, J. & WILHELM, V. L. 2018. Comparing fully convolutional networks, random forest, support vector machine, and patch-based deep convolutional neural networks for object-based wetland mapping using images from small unmanned aircraft system. *GIScience & Remote Sensing,* 55, 243–264.

LUO, Q., HU, K., LIU, W. & WU, H. 2022. Scientometric analysis for spatial autocorrelation-related research from 1991 to 2021. *ISPRS International Journal of Geo-Information,* 11, 309.

MADSEN, J. D. & WERSAL, R. M. 2017. A review of aquatic plant monitoring and assessment methods. *Journal of Aquatic Plant Management,* 55, 1–12.

MEDINA, A. L. 1996. Native aquatic plants and ecological condition of southwestern wetlands and riparian areas. In Desired future conditions for southwestern riparian systems: Bringing interests and concerns together. In USDA Forest Service, Rocky Mountain Forest and Range Experiment Station, General Technical Report, pp. 329–335. Fort Collins, CO.

MER, R., VADHER, K. & MER, D. 2016. Importance of freshwater aquatic weeds: A review. *Advancements in Life Sciences,* 5, 2027–2034.

MOHANTY, S., PANDEY, P. C., PANDEY, M., SRIVASTAVA, P. K., & DWIVEDI, C. S. (2024a). Wetlands contribution and linkage to support SDGs, its indicators and targets-A critical review. *Sustainable Development.* Online https://onlinelibrary.wiley.com/doi/full/10.1002/sd.2948

MOHANTY, S., PANDEY, P. C., SINGH, P., DUGESAR, V., & SRIVASTAVA, P. K. (2024b). Vegetation discrimination based on chlorophyll prediction in Marshy wetland using Unmanned Aerial Vehicles. *Aquatic Conservation: Marine and Freshwater Ecosystems, 34*(5), e4170

MU, S., YOU, K., SONG, T., LI, Y., WANG, L. & SHI, J. 2023. Identification for the species of aquatic higher plants in the Taihu Lake basin based on hyperspectral remote sensing. *Environmental Monitoring and Assessment,* 195, 989.

NAGELKERKEN, I., BLABER, S., BOUILLON, S., GREEN, P., HAYWOOD, M., KIRTON, L., MEYNECKE, J.-O., PAWLIK, J., PENROSE, H. & SASEKUMAR, A. 2008. The habitat function of mangroves for terrestrial and marine fauna: A review. *Aquatic Botany,* 89, 155–185.

NICULESCU, S., BOISSONNAT, J.-B., LARDEUX, C., ROBERTS, D., HANGANU, J., BILLEY, A., CONSTANTINESCU, A. & DOROFTEI, M. 2020. Synergy of high-resolution radar and optical images satellite for identification and mapping of wetland macrophytes on the Danube Delta. *Remote Sensing,* 12, 2188.

OTERA, K. 2010. *Monitoring seagrass: An investigation with multi-temporal satellite imagery in Boullanger Bay, Tasmania.* University of Tasmania.

PAN, S., ZHANG, W., LI, Y., ZHANG, C., ZHANG, H., NIU, L. & WANG, L. 2023. Trends and prospects in the Yangtze River Basin research: A bibliometric analysis. *River Research and Applications,* 39, 134–148.

PANDEY, P. C., MANEVSKI, K., SRIVASTAVA, P. K. & PETROPOULOS, G. P. 2018. The use of hyperspectral earth observation data for land use/cover classification: Present status, challenges and future outlook. *Hyperspectral Remote Sensing of Vegetation,* 1st ed.; Thenkabail, P., Ed. CRC Press, 147–173.

PANDEY, P. C. & PANDEY, M. 2023. Highlighting the role of agriculture and geospatial technology in food security and sustainable development goals. *Sustainable Development, 31*(5), 3175–3195.

PANDEY, P. C., KOUTSIAS, N., PETROPOULOS, G. P., SRIVASTAVA, P. K., & BEN DOR, E. (2021). Land use/land cover in view of earth observation: Data sources, input dimensions, and classifiers—A review of the state of the art. *Geocarto International,* 36(9), 957–988.

RADOSEVICH, S. R., HOLT, J. S. & GHERSA, C. M. 2007. *Ecology of weeds and invasive plants: relationship to agriculture and natural resource management.* John Wiley & Sons.

REJMANKOVA, E. 2011. The role of macrophytes in wetland ecosystems. *Journal of Ecology and Environment,* 34, 333–345.

RITCHIE, J. C., ZIMBA, P. V. & EVERITT, J. H. 2003. Remote sensing techniques to assess water quality. *Photogrammetric Engineering & Remote Sensing,* 69, 695–704.

ROBERTS, J., CHICK, A., OSWALD, L. & THOMPSON, P. 1995. Effect of carp, Cyprinus carpio L., an exotic benthivorous fish, on aquatic plants and water quality in experimental ponds. *Marine and Freshwater Research,* 46, 1171–1180.

SANTOS, M. J., KHANNA, S., HESTIR, E. L., ANDREW, M. E., RAJAPAKSE, S. S., GREENBERG, J. A., ANDERSON, L. W. & USTIN, S. L. 2009. Use of hyperspectral remote sensing to evaluate efficacy of aquatic plant management. *Invasive Plant Science and Management,* 2, 216–229.

SCHULTZ, R. & DIBBLE, E. 2012. Effects of invasive macrophytes on freshwater fish and macroinvertebrate communities: The role of invasive plant traits. *Hydrobiologia,* 684, 1–14.

SHORT, F. T., KOSTEN, S., MORGAN, P. A., MALONE, S. & MOORE, G. E. 2016. Impacts of climate change on submerged and emergent wetland plants. *Aquatic Botany,* 135, 3–17.

SILVA, T. S., COSTA, M. P., MELACK, J. M. & NOVO, E. M. 2008. Remote sensing of aquatic vegetation: theory and applications. *Environmental Monitoring and Assessment,* 140, 131–145.

SIMPSON, M. D., AKBARI, V., MARINO, A., PRABHU, G. N., BHOWMIK, D., RUPAVATHARAM, S., DATTA, A., KLECZKOWSKI, A., SUJEETHA, J. A. R. & ANANTRAO, G. G. 2022. Detecting water hyacinth infestation in kuttanad, india, using dual-pol Sentinel-1 SAR imagery. *Remote Sensing,* 14, 2845.

SOLANKI, J. B., KUMARI, R., DAS, A. K., MAURYA, P. & LELE, N. 2022. Retrieval of mangrove forest properties using synthetic aperture radar. In Pandey, Prem C., and Paul Arellano, (eds.), *Advances in Remote Sensing for Forest Monitoring,* 107–125. Wiley.

SOUSA, W. 2011. Hydrilla verticillata (Hydrocharitaceae), a recent invader threatening Brazil's freshwater environments: A review of the extent of the problem. *Hydrobiologia,* 669, 1–20.

SRIVASTAVA, J., GUPTA, A. & CHANDRA, H. 2008. Managing water quality with aquatic macrophytes. *Reviews in Environmental Science and Bio/Technology,* 7, 255–266.

SUN, T., LIN, W., CHEN, G., GUO, P. & ZENG, Y. 2016. Wetland ecosystem health assessment through integrating remote sensing and inventory data with an assessment model for the Hangzhou Bay, China. *Science of the Total Environment,* 566, 627–640.

TAHSIN, S. (2019). Investigation of Coastal Vegetation Dynamics and Persistence in Response to Hydrologic and Climatic Events Using Remote Sensing. Electronic Theses and Dissertations. 6822.https://stars.library.ucf.edu/etd/6822.

TIAN, S., ZHANG, X., TIAN, J. & SUN, Q. 2016. Random forest classification of wetland landcovers from multi-sensor data in the arid region of Xinjiang, China. *Remote Sensing,* 8, 954.

TSYGANSKAYA, V., MARTINIS, S., TWELE, A., CAO, W., SCHMITT, A., MARZAHN, P. & LUDWIG, R. 2016. A fuzzy logic-based approach for the detection of flooded vegetation by means of synthetic aperture radar data. *The International Archives of the Photogrammetry, Remote Sensing and Spatial Information Sciences,* 41, 371–378.

UNDERWOOD, E., MULITSCH, M., GREENBERG, J., WHITING, M., USTIN, S. & KEFAUVER, S. 2006. Mapping invasive aquatic vegetation in the Sacramento-San Joaquin Delta using hyperspectral imagery. *Environmental Monitoring and Assessment,* 121, 47–64.

UPAKANKAEW, K., NINSAWAT, S., VIRDIS, S. G. & SASAKI, N. 2022. Discrimination of mangrove stages using multitemporal Sentinel-1 C-Band Backscatter and Sentinel-2 data—A case study in Samut Songkhram Province, Thailand. *Forests,* 13, 1433.

VAN BEIJMA, S., COMBER, A. & LAMB, A. 2014. Random forest classification of salt marsh vegetation habitats using quad-polarimetric airborne SAR, elevation and optical RS data. *Remote Sensing of Environment,* 149, 118–129.

VARSHNEY, P. K. & ARORA, M. K. 2004. *Advanced image processing techniques for remotely sensed hyperspectral data.* Springer Science & Business Media.

WANG, H., LI, Y., ZENG, S., CAI, X., BI, S., LIU, H., MU, M., DONG, X., LI, J. & XU, J. 2022. Recognition of aquatic vegetation above water using shortwave infrared baseline and phenological features. *Ecological Indicators,* 136, 108607.

WANG, K., FRANKLIN, S. E., GUO, X. & CATTET, M. 2010. Remote sensing of ecology, biodiversity and conservation: A review from the perspective of remote sensing specialists. *Sensors,* 10, 9647–9667.

WATANABE, F. S. Y., IMAI, N. N., ALCÂNTARA, E. H., DA SILVA ROTTA, L. H. & UTSUMI, A. G. 2013. Signal classification of submerged aquatic vegetation based on the hemispherical–conical reflectance factor spectrum shape in the yellow and red regions. *Remote Sensing,* 5, 1856–1874.

WATTS, A. C., AMBROSIA, V. G. & HINKLEY, E. A. 2012. Unmanned aircraft systems in remote sensing and scientific research: Classification and considerations of use. *Remote Sensing,* 4, 1671–1692.

WHITE, L., BRISCO, B., DABBOOR, M., SCHMITT, A. & PRATT, A. 2015. A collection of SAR methodologies for monitoring wetlands. *Remote Sensing,* 7, 7615–7645.
WILLIAMS, D. J., RYBICKI, N. B., LOMBANA, A. V., O'BRIEN, T. M. & GOMEZ, R. B. Preliminary investigation of submerged aquatic vegetation mapping using hyperspectral remote sensing. Coastal Monitoring through Partnerships: Proceedings of the Fifth Symposium on the Environmental Monitoring and Assessment Program (EMAP) Pensacola Beach, FL, April 24–27, 2001, 2003. Springer, 383–392.
XIE, Y., SHA, Z. & YU, M. 2008. Remote sensing imagery in vegetation mapping: a review. *Journal of Plant Ecology,* 1, 9–23.
YAO, H., FU, B., ZHANG, Y., LI, S., XIE, S., QIN, J., FAN, D. & GAO, E. 2022. Combination of hyperspectral and quad-polarization SAR images to classify marsh vegetation using stacking ensemble learning algorithm. *Remote Sensing,* 14, 5478.
YAO, H., QIN, R. & CHEN, X. 2019. Unmanned aerial vehicle for remote sensing applications—A review. *Remote Sensing,* 11, 1443.
ZENG, C., RICHARDSON, M. & KING, D. J. 2017. The impacts of environmental variables on water reflectance measured using a lightweight unmanned aerial vehicle (UAV)-based spectrometer system. *ISPRS Journal of Photogrammetry and Remote Sensing,* 130, 217–230.
ZHANG, C. & XIE, Z. 2012. Combining object-based texture measures with a neural network for vegetation mapping in the Everglades from hyperspectral imagery. *Remote Sensing of Environment,* 124, 310–320.
ZHANG, Y., LIU, X., QIN, B., SHI, K., DENG, J. & ZHOU, Y. 2016. Aquatic vegetation in response to increased eutrophication and degraded light climate in Eastern Lake Taihu: Implications for lake ecological restoration. *Scientific Reports,* 6, 23867.
ZHANG, Y., YOU, X., HUANG, S., WANG, M. & DONG, J. 2022. Knowledge atlas on the relationship between water management and constructed wetlands—A bibliometric analysis based on citeSpace. *Sustainability,* 14, 8288.
ZWEIG, C. L., BURGESS, M. A., PERCIVAL, H. F. & KITCHENS, W. M. 2015. Use of unmanned aircraft systems to delineate fine-scale wetland vegetation communities. *Wetlands,* 35, 303–309.

12 Investigation of Seasonal Variability of Chlorophyll-a in the Arabian Sea

Suchita Pandey, Suneet Dwivedi and Sudhir Kumar Singh*
Centre of Atmospheric and Ocean Studies, University of Allahabad, Prayagraj, Uttar Pradesh, India

12.1 INTRODUCTION

The geographical setting of the Arabian Sea (AS) inhibits the thermocline ventilation in the north (Wiggert et al., 2006). The seasonal reversal of winds and associated changes in the oceanic currents are responsible for pronounced biogeochemical variability over the basin. Summer and winter blooms in the AS are caused by the summer and winter monsoonal winds. The seasons of the South-West Monsoon (SWM) and North-East Monsoon (NEM) have a significant impact on the spatiotemporal variability of biogeochemical indices in the AS. Levy et al.'s (2007) study of the summer and winter blooms showed how the monsoon-driven horizontal and vertical circulation is closely connected to the regional structure of the blooms. The climatological and inter-annual variability of biogeochemical parameters over the basin has been explored by Sharada et al. (2015) through simulations with Tracers of Phytoplankton and Allometric Zooplankton (TOPAZ) during 1949–2009. The model captures significant spatial and seasonal features of chlorophyll and primary productivity. In a different study, the one-way nested Indian Ocean regional model developed by Rahaman et al. (2014) was scrutinized and shown to capture well the seasonal variation of sea surface height (SSH), temperature, salinity, and mixed layer depth (MLD). The reversing wind fields of the northern hemisphere along the western boundary of the AS is the most striking physical response, where Somali currents flow equatorward during NEM and poleward during SWM. Winter bloom occurs across the northern and central AS as a result of convective mixing driven by cool, dry, northeasterly winds during the NEM (Banse and McClain, 1986; Madhupratap et al., 1996; Wiggert et al., 2000). However, during the SWM, a noteworthy phytoplankton bloom occurs off the coast of Oman (Brock and McClain, 1992; Latasa and Bidigare, 1998) and Somalia (Veldhuis et al., 1997; Hitchcock et al., 2000). This bloom may spread more than 500 km away from the coast (Wiggert et al., 2006). This is due to a combination of coastal upwelling, horizontal advection, and Ekman pumping. Ravichandran et al. (2012) examined the time series analysis of oxygen and Chl-a from Argo float data in the Southeast Asian region. They discovered that while there is nutrient entrainment from nutrient-rich subsurface water in the near-surface layer, the subsurface chlorophyll maximum (SCM) vertical fluxes also play a major role in the mixed layer blooms. Research has also demonstrated that low Fe/N ratios (Wiggert et al., 2006) and excess zooplankton feeding (Hambright et al., 2001) can restrict the phytoplankton bloom during SWM, while it is unclear how much iron restriction affects phytoplankton output (Piketh et al., 2000; Wiggert et al., 2006). There is evidence of a high rate of nitrogen (N_2) fixation in surface water of the Indian Ocean (IO) (Capone et al., 1998) and it was shown that 30–40% of euphotic zone nitrate (NO_3) is derived from N_2 fixation in AS (Brandes et al., 1999).

* Corresponding author.

DOI: 10.1201/9781003354000-16

Using a reduced gravity physical model in conjunction with a four-component ecosystem model, McCreary et al. (1996) investigated the processes that govern the yearly cycle of biological component activity in the AS. In their study, Swathi et al. (2000) employed a modular ocean model in conjunction with a seven-component nitrogen-based model and a chemical model that included dissolved inorganic carbon and alkalinity. They demonstrated that, as a result of nutrient availability, there is an increase in biological output during SWM in the southwest AS.

In the current study, the variability of Chl-a in the AS was examined about the variability of other biogeochemical parameters, including nitrate, phytoplankton, zooplankton, and detritus, as well as physical parameters, including salinity, ocean temperature, winds, and surface currents.

12.2 DATA, MODEL SETUP, AND VALIDATION

12.2.1 Model Forcing for Climatological Simulation

For the climatological simulation, the model is initiated from rest, and boundary conditions for temperature, salinity, dissolved oxygen, phosphate, silicate, and nitrate are derived from the World Ocean Atlas (WOA) monthly climatology with 1° × 1° spatial resolution (Levitus et al., 2010). Tidal constituents (both phase and amplitude) are obtained from the Oregon State University global model of ocean tides: TPX07 (Egbert and Erofeeva, 2002). Comprehensive Ocean-Atmosphere Data Set (COADS) at 0.5° spatial resolution (Da Silva et al., 1994) is used as monthly climatology for atmospheric forcing (e.g., wind stress, air temperature, air density, relative humidity, specific humidity, as well as air-sea fluxes viz. evaporation, precipitation, and surface heat fluxes). Global Digital Elevation Model (ETOPO2) at 2-min spatial resolution (Smith and Sandwell, 1997) is used as bathymetry data. Chl-a data is derived from the Sea-viewing Wide Field-of-view Sensor (SeaWiFS) monthly climatology with 1/4° spatial resolution (http://oceancolor.gsfc.nasa.gov).

12.2.2 ROMS Overview

For a deeper comprehension of the AS's physical and biogeochemical characteristics, the ROMS model was employed. ROMS for Eddy Permitting System allows for high-resolution customization across the zone of interest. It is a free-surface, three-dimensional representation of terrain. It solves the incompressible, primitive equations based on Boussinesq and hydrostatic approximations by using sigma coordinates in the vertical and curvilinear coordinates in the horizontal (Haidvogel and Beckmann, 1999; Malanotte-Rizzoli et al., 2000; Penven et al., 2001; Di Lorenzo et al., 2004; Shchepetkin and McWilliams, 2005). The ROMS model obtains mathematical feedback between the barotropic momentum and continuity equations and the baroclinic momentum and tracer equations via a split-explicit time step technique (Shchepetkin and McWilliams, 2005). This model makes use of an Arakawa C grid.

12.2.3 Model Setup and Validation

The model domain in the AS is extended from 40°E to 80°E longitudes and 5°S to 30°N latitudes. While the northern boundary is closed, the lateral boundaries on the eastern, western, and southern sides are open. The model has been set for high-resolution numerical analysis with a grid resolution of 1/7 degree (~15 km). A ten-year climatological run with a 300-second time step was conducted. The number of vertical levels is 32 (Figure 12.1). A total of 280 and 255 number of grids were used in the zonal and meridional directions, respectively. The vertical stretching parameters for surface (Θ_s) and bottom (Θ_b) are assigned as 7 and 0.1, respectively, to improve vertical layers (Song and Haidvogel, 1994; Sil, 2011). The model is forced with tides from all major ten components (M2, S2, N2, K2, K1, O1, P1, Q1, Mf, and Mm), river runoff from major rivers of the AS (Narmada,

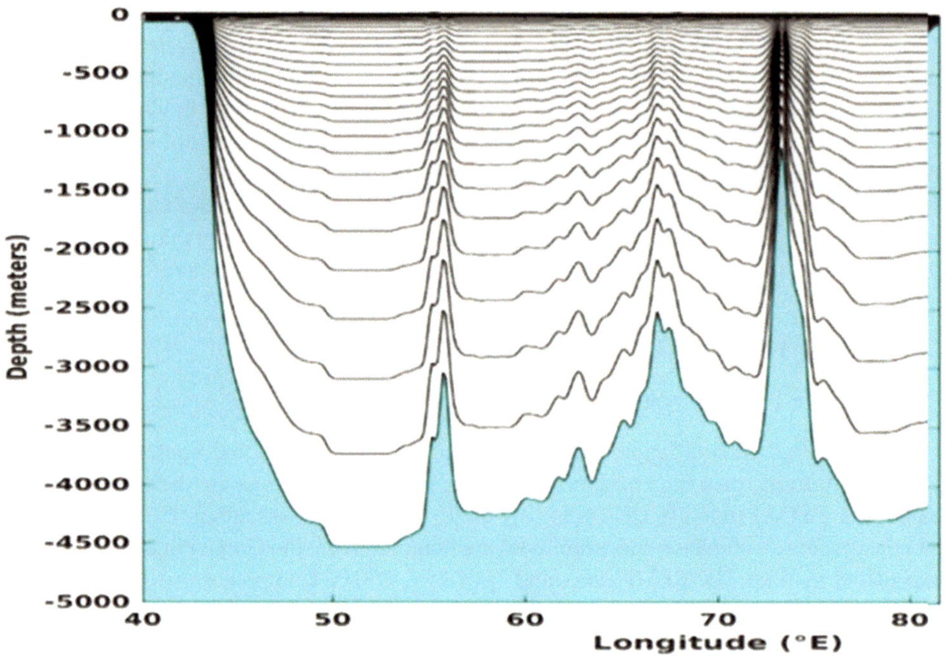

FIGURE 12.1 Model cross-section along the 0.0066°S latitude showing the depth of vertical sigma levels. Note: Model resolution increases as the thickness decreases to resolve mixed layer stability.

TABLE 12.1
List of Model Parameters

Domain	40°E–80°E, 5°S–30°N (AS)
Horizontal resolution	1/7 (~15 km)
Number of vertical levels in the grid (N)	32
Number of grid points in the zonal direction	280
Number of grid points in the meridional direction	255
Stretching parameters of vertical S-coordinate	$\Theta_s = 7$, $\Theta_b = 0.1$, hc = 250
Model baroclinic time step (dt)	300 seconds
Number of barotropic time steps (NDTFAST) within dt	60 seconds
Mean density used in Boussinesq equation	ρ_0 =1025 kg/m^3
h_{min}	20 m

Tapi, Sindhu, Purna, Periyar, and Sabarmati rivers), atmospheric forcing, and air-sea fluxes. Model parameters are listed in Table 12.1.

12.2.4 Ecosystem Model

The four-component NPZD ecosystem model is combined with ROMS. Nitrogen is the only limiting nutrient, and there is only one operational type of phytoplankton in the model, along with zooplankton, detritus, and nitrate. Nitrate, ammonium, detritus, phytoplankton, zooplankton, and an active phytoplankton carbon-to-chlorophyll ratio are among the seven coupled partial differential

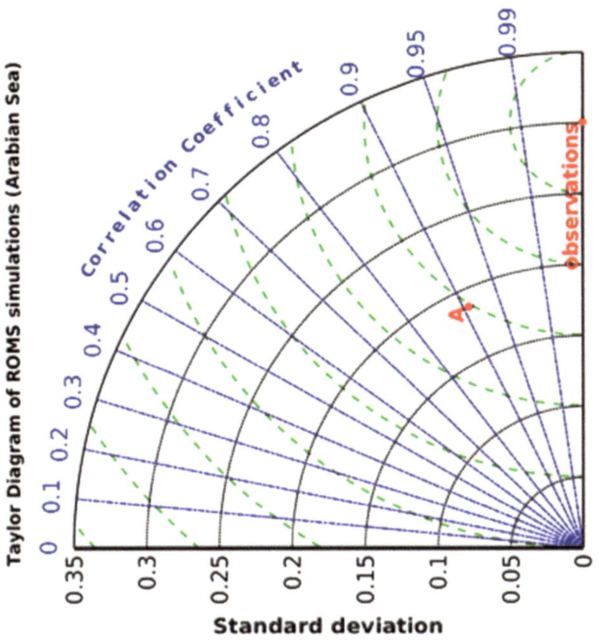

FIGURE 12.2 Taylor diagram represents the correlation coefficient for basin-averaged surface Chl-a values for the Arabian Sea derived from model and observation showing a correlation of 0.9.

equations that make up the model (Fasham et al., 1990). These scalar values are all diffused and advected at ROMS-determined rates. Phytoplankton may sink, and the majority of particulate mass movement is linked to the sinking of bigger particles.

12.3 RESULTS AND DISCUSSION

ROMS simulation with climatological forcing for winds, air-sea fluxes, and river runoff, as well as the incorporation of tidal parameters, deciphers the seasonal variations of Chl-a over the AS. The basin-averaged surface Chl-a for AS derived from satellite observation showed good agreement with climatological model simulation (Figure 12.2). The correlation of model-simulated surface Chl-a with observations was 0.9 for AS. The correlation value suggests that the surface Chl-a simulations generated by the ROMS model are realistic.

12.3.1 Seasonal Chl-a Variability and Associated Physical and Biogeochemical Parameters

It is important to know the seasonal Chl-a variability, nitrate, and other biogeochemical parameters as well as physical parameters at the surface and in the inner layer of the ocean. Following this, particular months representing the summer monsoon (July) and winter monsoon (January) are taken to describe the seasonal trend of different parameters. The variation of Chl-a in spatial and temporal domains and its vertical distribution in the upper 200 m of the water column along different sections in the central, eastern, and western AS is studied during the winter and summer monsoons. The Chl-a variability is investigated vis-à-vis variations in ocean temperature (Temp), salinity (Salt), winds, mixed layer depth (MLD), isothermal depth (ILD), currents, nitrate (NO_3), phytoplankton (PHYTO), zooplankton (Zoo), detritus (DET), and oxygen (O_2).

12.3.2 Central Arabian Sea

'Section A' in the central AS is chosen along 64°E and 11° to 21°N (Figure 12.3).

12.3.2.1 Summer Monsoon

Figure 12.3 showed the Chl-a variability in the central AS along 'Section A'. The Chl-a of the upper 20 m is high (~0.5 mg/m^3) between 13°N and 17°N and it increases to approximately 1 mg/m^3 between 17°N and 19°N. Below 20 m depth up to 80 m, the Chl-a value becomes 1 mg/m^3 between 14°N and 21°N, while the SCM is found at 45 m depth with a concentration of 1.3 mg/m^3 at 19°N (Figure 12.3e). The temperature and salinity higher than 25°C and 36 PSU, respectively, are observed in the upper 80 m of the water column (except along 18°N–19°N). The shoaling of the upper thermocline occurs along 17°N (Figure 12.3b). North of 15°N and 19°N, the Arabian Sea high salinity water (ASHSW) exists up to 100 m, and the shoaling and deepening occurs under the influence of southwest monsoon winds. The bloom exists along 17°N and north of it with a subsurface depth of 45 m. The shoaling of SCM is observed from 80 m to 60 m at 17°N to 19°N. The AS has an oxygen minimum zone.

The upper 40 m of the water column is devoid of oxygen, while the oxygen concentration increases (200 mMol O_2 m^{-3}) below 40 m north of 17°N. The oxygen concentration decreases with depth. Figure 12.4 describes the variability of nitrate, phytoplankton, zooplankton, and detritus during the summer monsoon. From 11°N to 17°N, the NO_3 concentration is 0.5 mMol N m^{-3} and

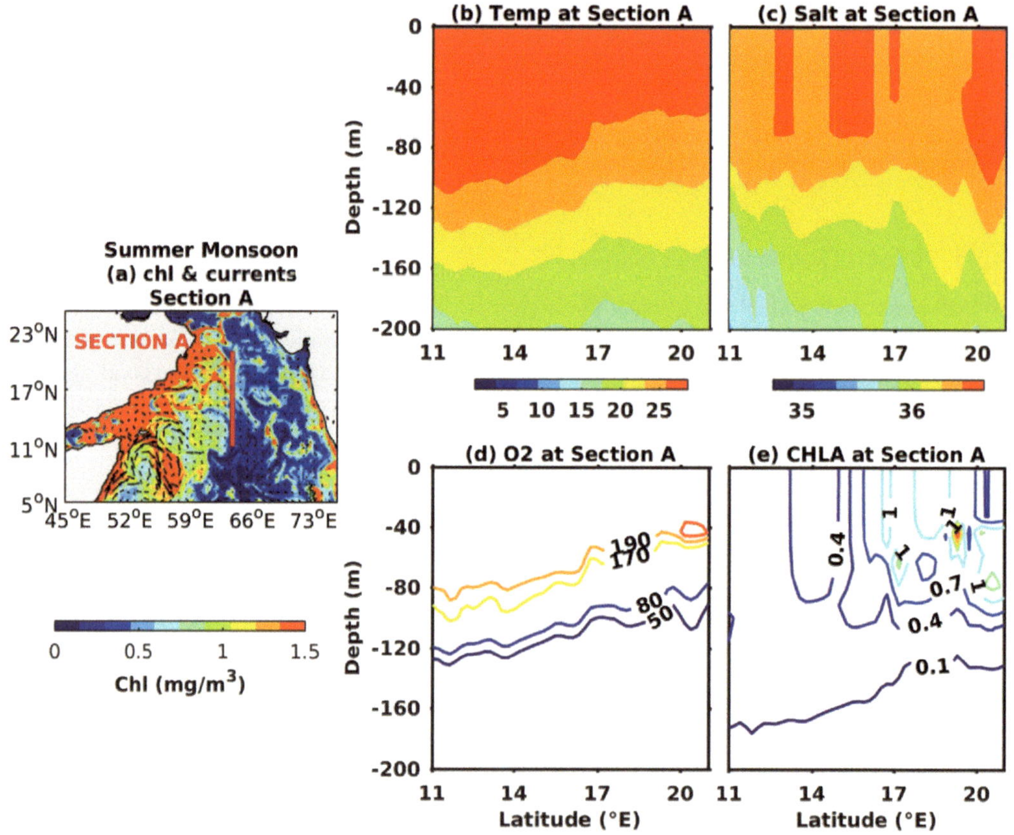

FIGURE 12.3 Variability of (a) surface Chl-a with surface currents and the vertical distribution of (b) oceanic temperature, (c) salinity, (d) oxygen, and (e) Chl-a during the summer monsoon along central (64°E) AS at Section A.

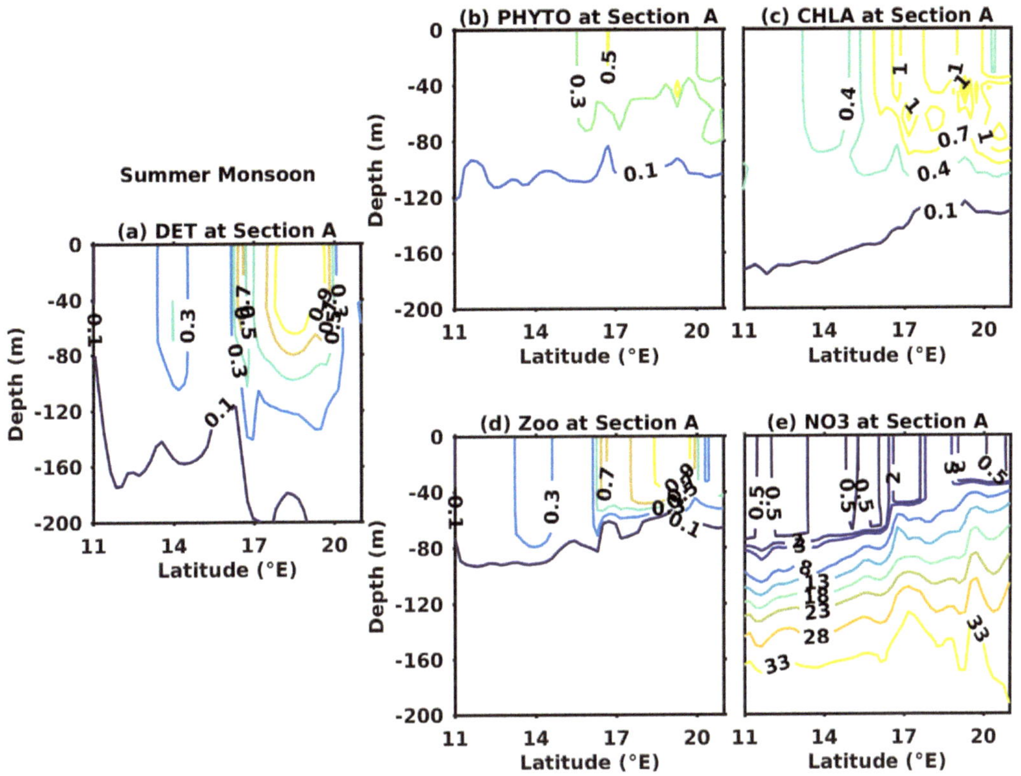

FIGURE 12.4 Vertical distribution of (a) detritus, (b) phytoplankton, (c) Chl-a, (d) zooplankton, and (e) nitrate during the summer monsoon along the central AS (64°E) at Section A.

nitracline is deep up to 80 m depth while northward shoaling of the nitracline occurs from 17°N, from 60 m to 40 m. The NO_3 values are high, from 2 to 3 mMol N m^{-3}, and nitracline from north of 17°N (Figure 12.4e). NO_3 is increasing with depth. The basin average climatology shows that very strong southwesterly winds occur in the AS. It is the region of the Findlater jet, which is a strong jet of southwesterly winds, and its axis is found between 16°N and 17°N and 64°E (Figure 12.5). The cyclonic wind stress curl associated with the north of the Findlater jet drives upward Ekman pumping (EP). This EP initiates upwelling and shoaling of the nitracline and thermocline north of 17°N (Prassana Kumar et al., 2009). Another important feature in this region is the mesoscale eddies that play a crucial role in increasing productivity in the upper layers of the ocean by supplying nutrients. Lateral advection of Chl-a and NO_3 from the western AS (through extended filament) to the central AS increases productivity.

Phytoplankton concentration is high north of 16°N and reaches a maximum (0.5 mMol N m^{-3}) along 17°N up to a depth of 40 m. The availability of nutrients is responsible for high Chl-a concentration. The zooplankton, which mainly feeds on phytoplankton of the euphotic layer and detritus, shows a correspondence with the detrital concentration in the interior of the ocean. North of 17°N, the concentration of zooplankton and detritus increases. The detrital concentration ranges between 0.5 and 0.7 mMol N m^{-3} in the upper 80 m north of 17°N. North of 17°N, the concentration of zooplankton varies from 0.3 to 0.7 mMol N m^{-3} in the upper 80 m of the water column. The shoaling of zooplankton occurs toward the north. The MLD varies from 50 to 100 m from north to south of the section. The MLD is deep (~60 m) along 16°N–17°N. The ILD is deep (~150 m) along the section (Figure 12.5).

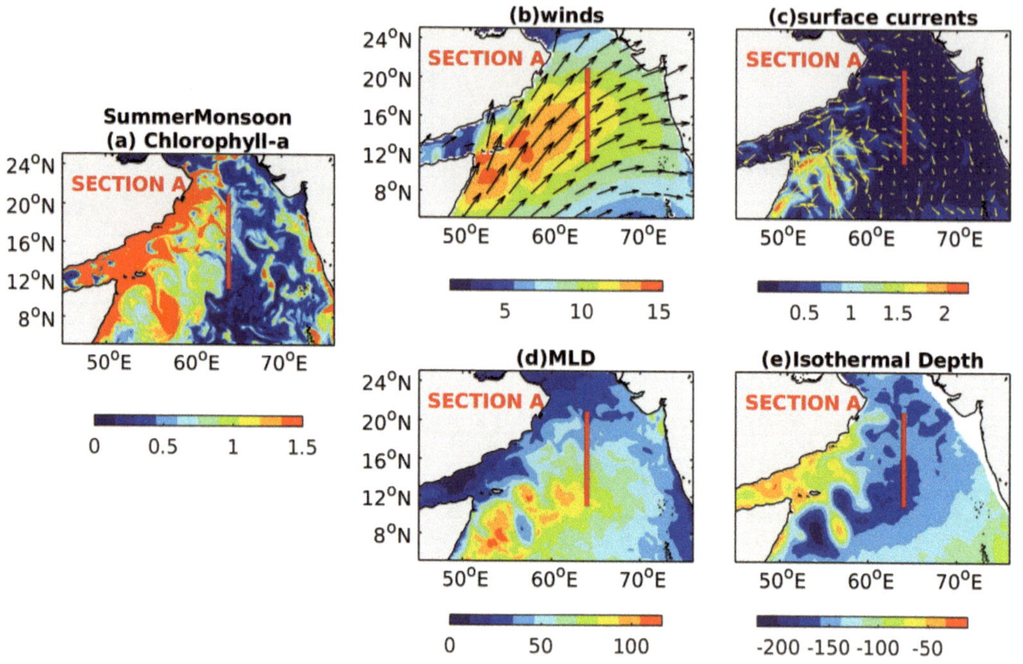

FIGURE 12.5 Spatial distribution of (a) Chl-a, (b) winds, (c) surface currents, (d) MLD, and (e) ILD during the summer monsoon along the central AS (64°E) at Section A.

12.3.2.2 Winter Monsoon

In the winter monsoon season, the SST becomes minimal, while the evaporation becomes maximal. The heat flux is negative and a net evaporative heat loss is observed in the AS. As a result, convective mixing is induced, which drives deep mixed layer depth and erodes the nitracline (which takes nitrate to the upper surface and increases biological activity). The upper 60 m of the water column shows a high Chl-a concentration ranging between 0.4 and 1 mg/m^3. The bloom region (Chl-a > 1 mg/m^3) occurs north of 16°N from the surface to 60 m depth (Figure 12.6). The Chl-a concentration declines with depth. The vertical thermal structure suggests that the upper 80 m of the water column has a temperature close to 25°C. During the northeast monsoon season, the evaporative cooling driven by dry air from the northeast results in the formation of ASHSW with salinity greater than 36 PSU. Such a situation exists up to 100 m in depth. This is seen from the vertical structure of salinity where there is a sharp decrease of 1 PSU from 37 PSU to 36 PSU along 17°N up to a depth of 80 m. The nitracline is deep up to 80–90 m with a high concentration (2–3 mMol N m^{-3}) of NO_3, which increases with depth. The O_2 is high with a concentration of 200 mMol O_2 m^{-3} from 60 to 80 m depth. The phytoplankton concentration (0.3 mMol N m^{-3}) is less compared to zooplankton (0.5 mMol N m^{-3}) due to excessive grazing over phytoplankton growth in the upper 70 m of the water column during NEM. Detritus concentration is also high (0.5 mMol N m^{-3}) at all latitudes up to 100 m depth (Figure 12.7). Winds are less strong in this season compared to the SWM season, but due to convective mixing, a deeper MLD (~80–90 m) is formed from south toward northern latitudes compared to the MLD of the SWM season. Similarly, in the central AS, the ILD (<150 m) during the NEM remains shallower than in the SWM (Figure 12.8).

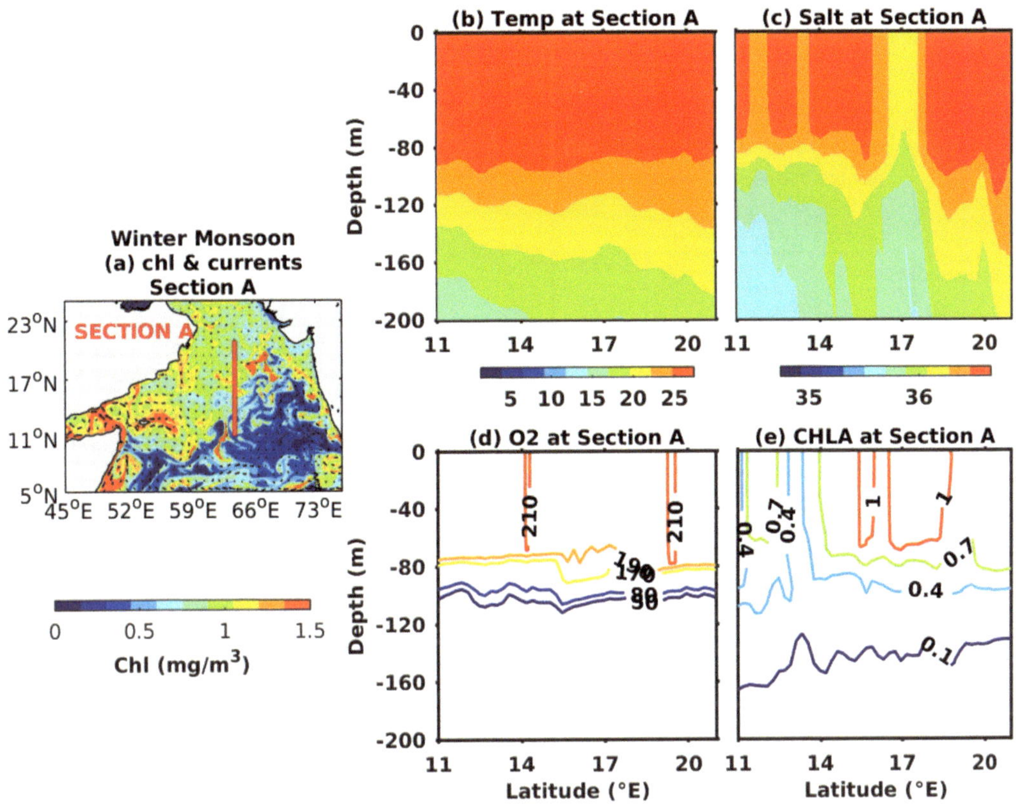

FIGURE 12.6 Variability of (a) surface Chl-a with surface currents and the vertical distribution of (b) oceanic temperature, (c) salinity, (d) oxygen, and (e) Chl-a during the winter monsoon along the central AS (64°E) at Section A.

12.3.3 Eastern Arabian Sea

The eastern side of the AS is less productive than the western AS. The eastern AS cross-section is taken between 74°E and 67°E and 12°N and 22°N along the eastern side of the AS. This section is named 'Section B'.

12.3.3.1 Summer Monsoon

The temperature >25°C was observed above 40 m. Salinity increases between 12°N (34 psu) and north of 16°N (>36 psu). The upper 20 m of the water column lacks oxygen, and oxygen concentration increases below this depth (Figure 12.9). Vertical distribution of detritus, phytoplankton, Chl-a, zooplankton, and nitrate during the SWM along the eastern AS (74°E to 67°E and 12°N to 22°N) is observed at Section B (Figure 12.10). Strong winds along the coast drive upwelling along the southwest coast of India (Figure 12.11). High values of NO_3 reach the surface layers through vertical advection and drive high Chl-a via phytoplankton production. Kelvin waves radiate as westward propagating Rossby waves into the AS, causing relatively high Chl-a values in the Laccadive Sea (Figure 12.11). The blooms around India and the Laccadive Sea follow the propagation of Rossby waves. The maximum concentration of Chl-a and NO_3 is found along 12°N–15°N and 17°N–19°N in the Laccadive Sea via vertical advection. The SCM at 70 m depth occurs with a high Chl-a concentration of 1.1 mg/m³ at 17°N. There is a doming of thermocline and nitracline occurring at 17°N. The subsurface bloom is found at this latitude with 80 m depth. The NO_3 is high between 12°N and 15°N and north of 18°N in the upper 40 m. Phytoplankton and zooplankton show correspondence

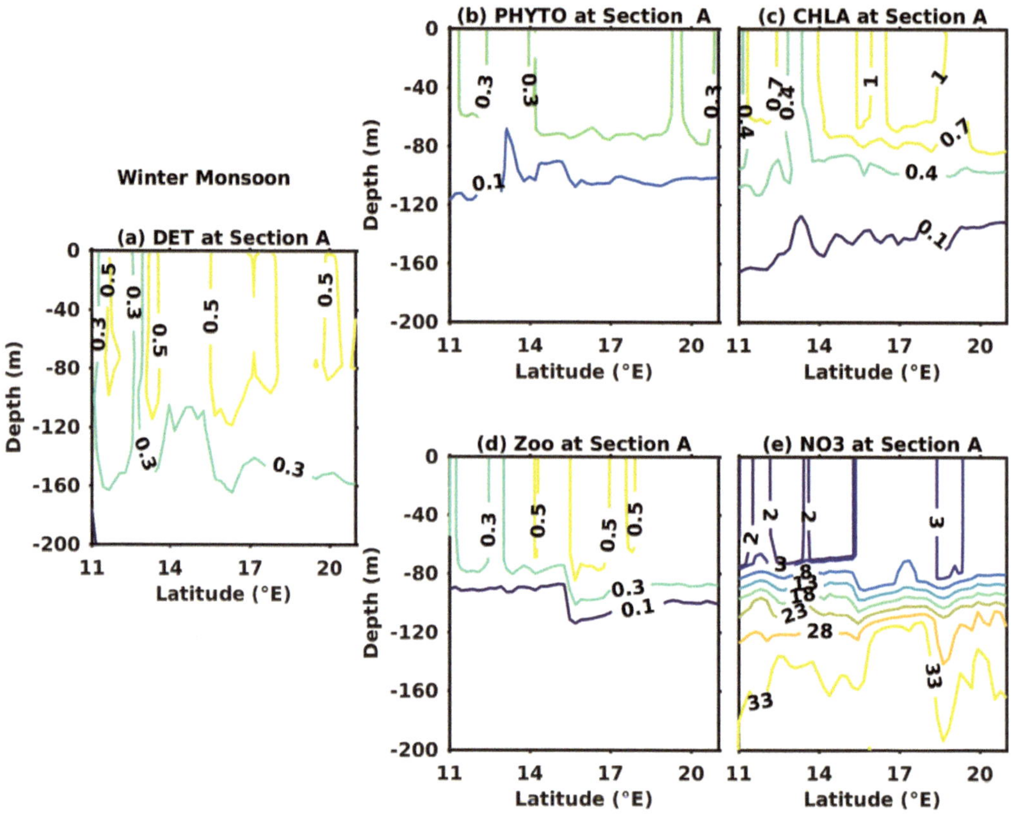

FIGURE 12.7 Vertical distribution of (a) detritus, (b) phytoplankton, (c) Chl-a, (d) zooplankton, and (e) nitrate during the winter monsoon along the central AS (64°E) at Section A.

with Chl-a and NO_3. Phytoplankton and zooplankton are high (up to 40 m) north of 12°N and 18°N with a maximum concentration >0.5 mMol N m^{-3}. The detrital concentration is high enough (>0.5 mMol N m^{-3}) north of 12°N up to 60 m depth, thus confirming a high concentration of zooplankton there. The MLD is 40–50 m deep along the section, while ILD is shallow (<150 m) south of 12°N and deep (>150 m) north of 12°N (Figure 12.10).

12.3.3.2 Winter Monsoon

Along 12°N–16°N, the salinity decreases (33–34 psu) due to freshening of water from adjoining rivers. The upper 40 m of the water column is devoid of oxygen except north of 12°N and 19°N, where the oxygen concentration is 200 mMol O_2 m^{-3} (Figure 12.12). The high Chl-a values (0.7 to 1 mg/m^3) are found along all the latitudes from surface to 60 m depth except between 14°N and 16°N, where values are comparatively less (0.4 mg/m^3). Convective mixing through northeasterly winds induces upwelling and deep mixing through vertical advection, resulting in high nitrate in the upper layers. Mixing is deep and more prominent north of 17°N. NO_3 is high along 13°N and north of 17°N. Deepening (70 m–80 m) of the nitracline occurs toward north of 17°N.

Phytoplankton and zooplankton concentrations show correspondence with each other. Phytoplankton growth is less compared to zooplankton grazing. South of 17°N, the phytoplankton concentration (0.3mMol N m^{-3}) is essentially restricted up to 40 m depth. North of 17°N, the abundance of phytoplankton is seen up to a greater depth (80 m). Zooplankton is at maximum (>0.9 mMol N m^{-3}) north of 12°N up to 40 m depth. The concentration of zooplankton greater than 0.5 mMol N m^{-3} remains up to 80 m depth, north of 17°N (Figure 12.13). Detritus concentration is high

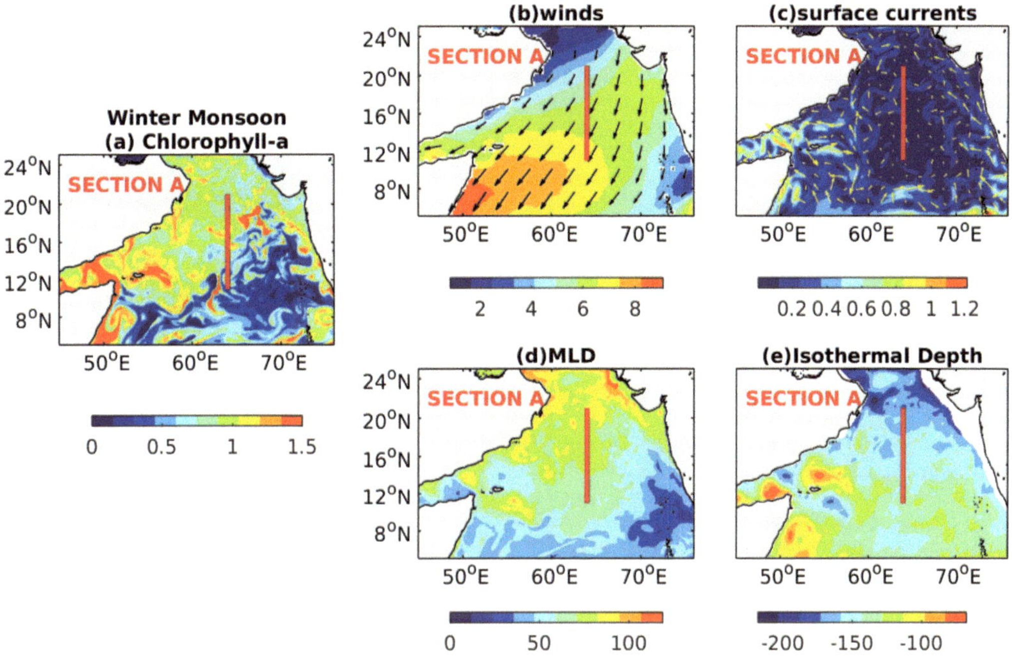

FIGURE 12.8 Spatial distribution of (a) Chl-a, (b) winds, (c) surface currents, (d) MLD, and (e) ILD during the winter monsoon along the central AS (64°E) at Section A.

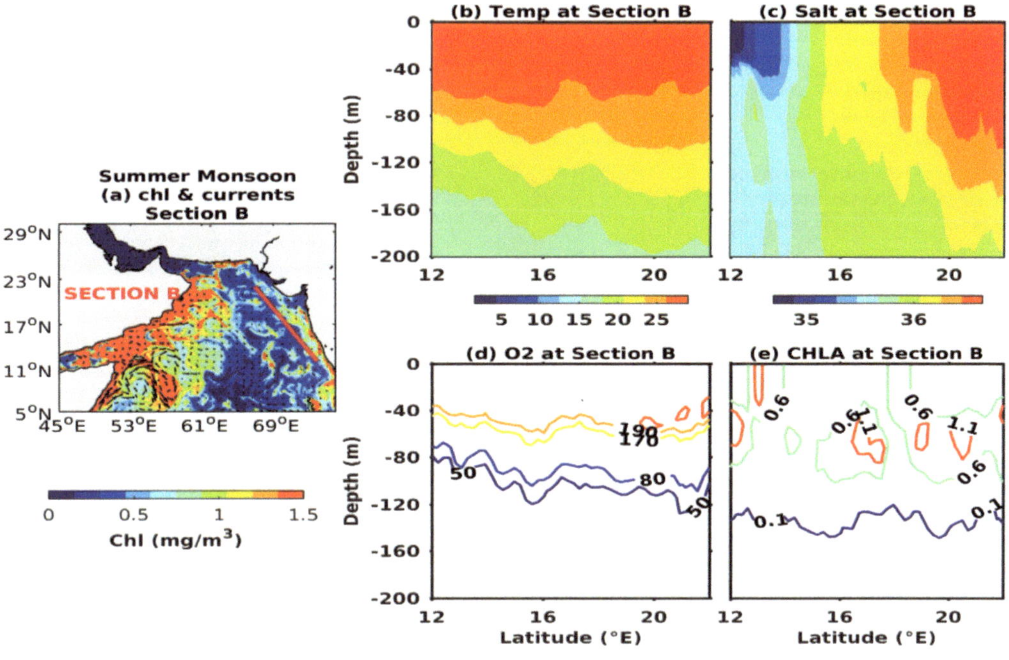

FIGURE 12.9 Variability of (a) surface Chl-a with surface currents and the vertical distribution of (b) oceanic temperature, (c) salinity, (d) oxygen, and (e) Chl-a during the summer monsoon along the eastern AS (74°E to 67°E and 12°N to 22°N) at Section B.

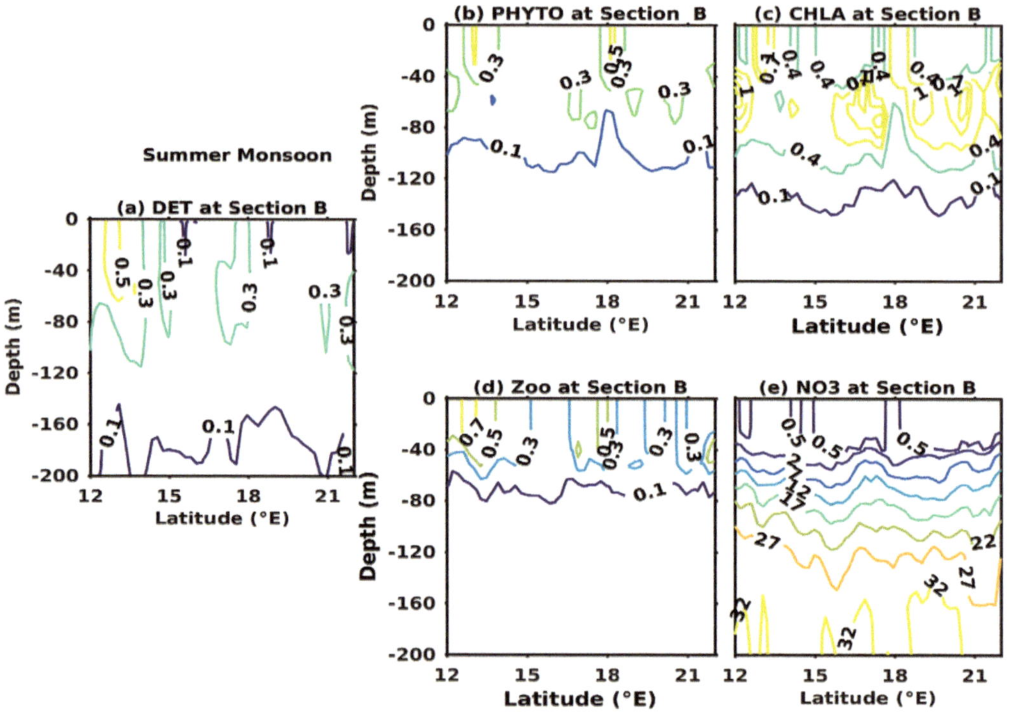

FIGURE 12.10 Vertical distribution of (a) detritus, (b) phytoplankton, (c) Chl-a, (d) zooplankton, and (e) nitrate during the summer monsoon along the eastern AS (74°E to 67°E and 12°N to 22°N) at Section B.

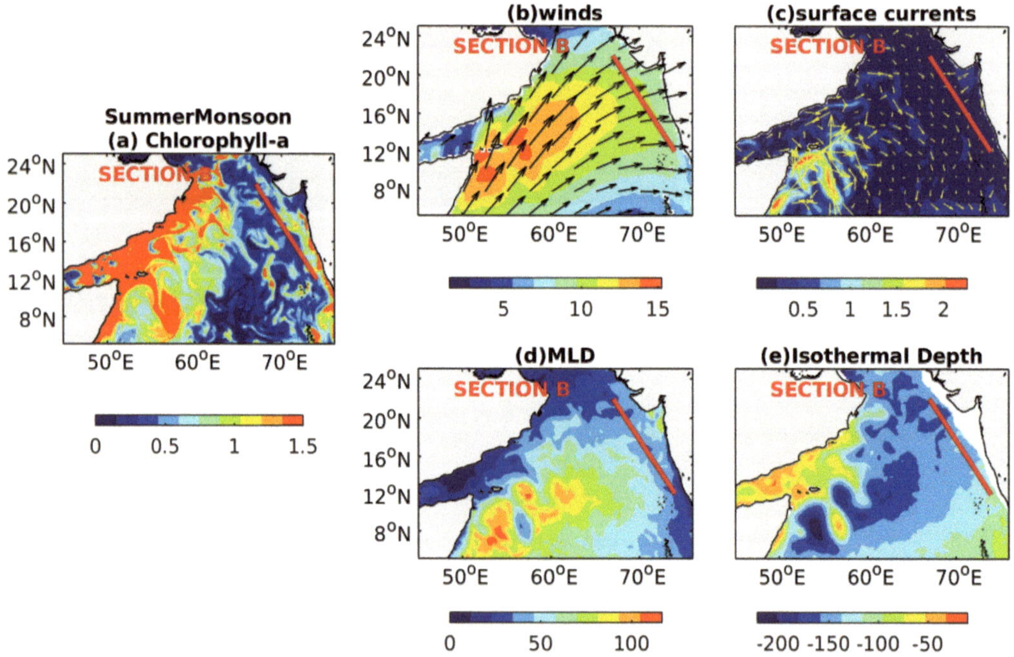

FIGURE 12.11 Distribution of (a) Chl-a, (b) winds, (c) surface currents, (d) MLD, and (e) ILD during the summer monsoon along the eastern AS (74°E to 67°E and 12°N to 22°N) at Section B.

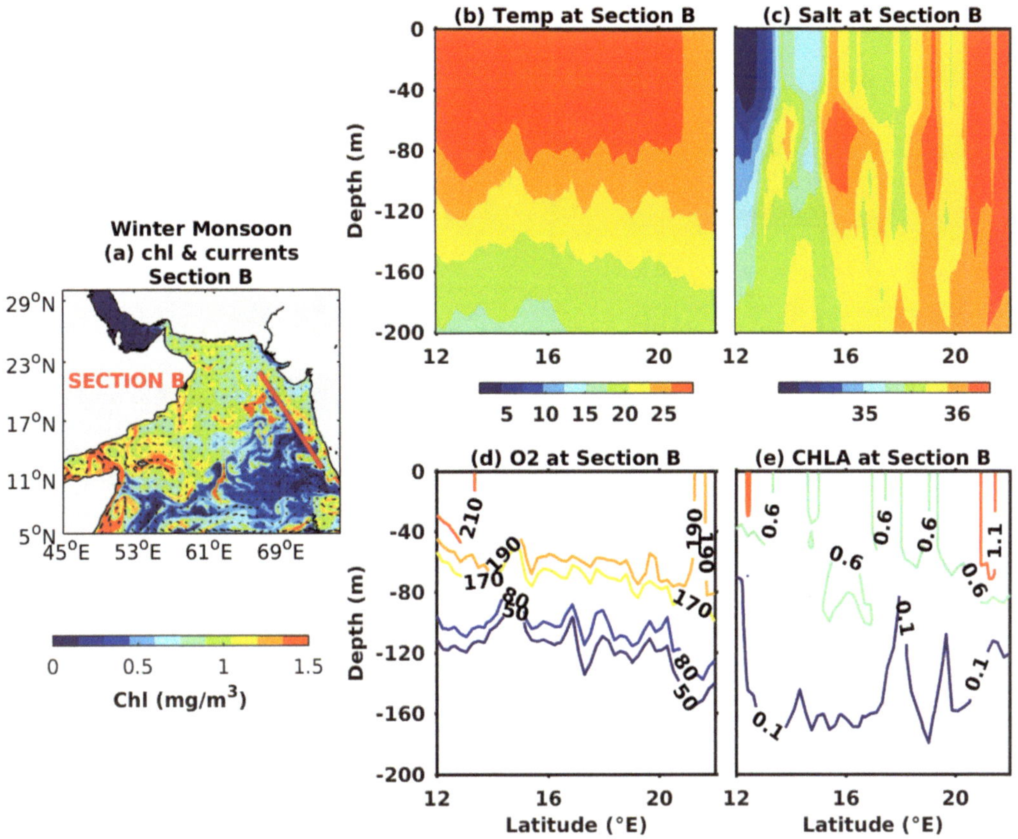

FIGURE 12.12 Variability of (a) surface Chl-a with surface currents and the vertical distribution of (b) oceanic temperature, (c) salinity, (d) oxygen, and (e) Chl-a during the winter monsoon along the eastern AS (74°E to 67°E and 12°N to 22°N) at Section B.

up to 100 m depth north of 12°N and 18°N. The MLD gets deeper (~90 m) toward the north, while ILD varies from 120 m to 150 m from south to north along Section B (Figure 12.14).

12.3.4 Western Arabian Sea

The western AS is a productive region, and the Chl-a concentration is high along the western coast as well as offshore regions. 'Section C' in the western AS is taken along 52°E–62°E and 12°N–22°N for studying the spatiotemporal variability of Chl-a and its related processes in the upper 200 m of the water column during summer and winter monsoon seasons.

12.3.4.1 Summer Monsoon

Somali current flows toward the north during the SWM. The high values of Chl-a in this season are induced by the coastal upwelling, horizontal advection, and upward Ekman pumping near the coast of Oman and Somalia that propagated offshore more than 500 km. The bloom in the off-Somalia region is entrained around the great whirl, which is an annually recurring anti-cyclonic pattern of the Somali current two-gyre system. The Chl-a concentration is high (ranging between 0.3 and 2 mg/m^3) in the upper 80 m of the water column. The SCM is found at 60 m depth north of 15°N with a concentration of 2 mg/m^3 (Figure 12.15). Oceanic temperature varies between 25°C and 26°C from the surface to 40 m depth. The salinity of the upper 40 m depth is in the range of 36 psu. The

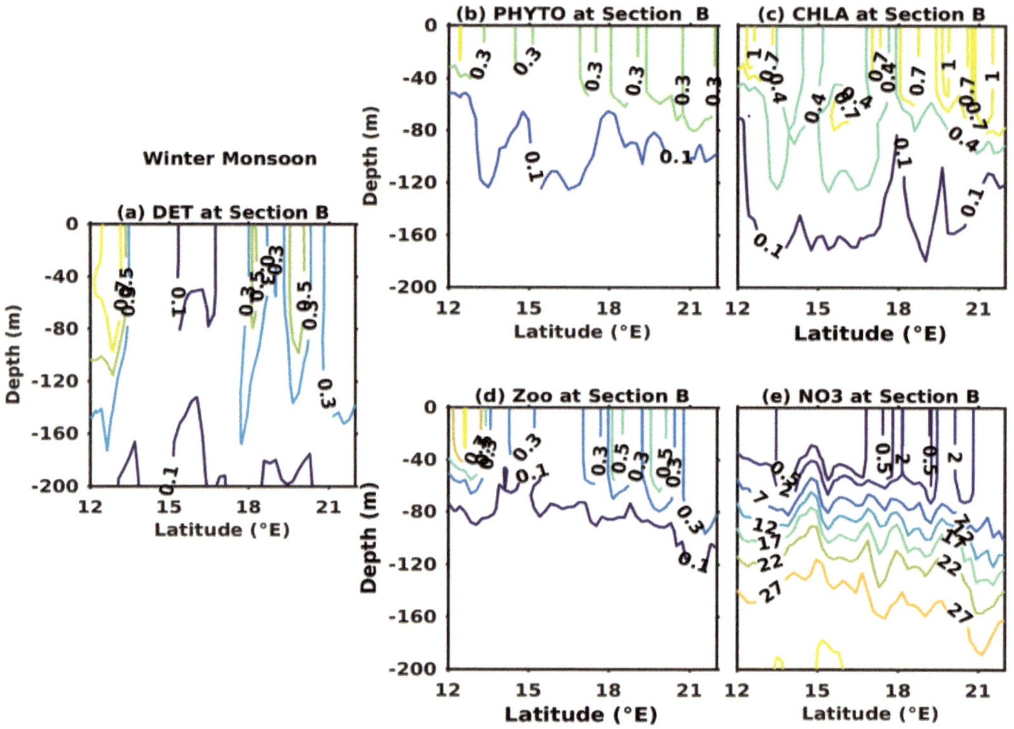

FIGURE 12.13 Vertical distribution of (a) detritus, (b) phytoplankton, (c) Chl-a, (d) zooplankton, and (e) nitrate during the winter monsoon along the eastern AS (74°E to 67°E and 12°N to 22°N) at Section B.

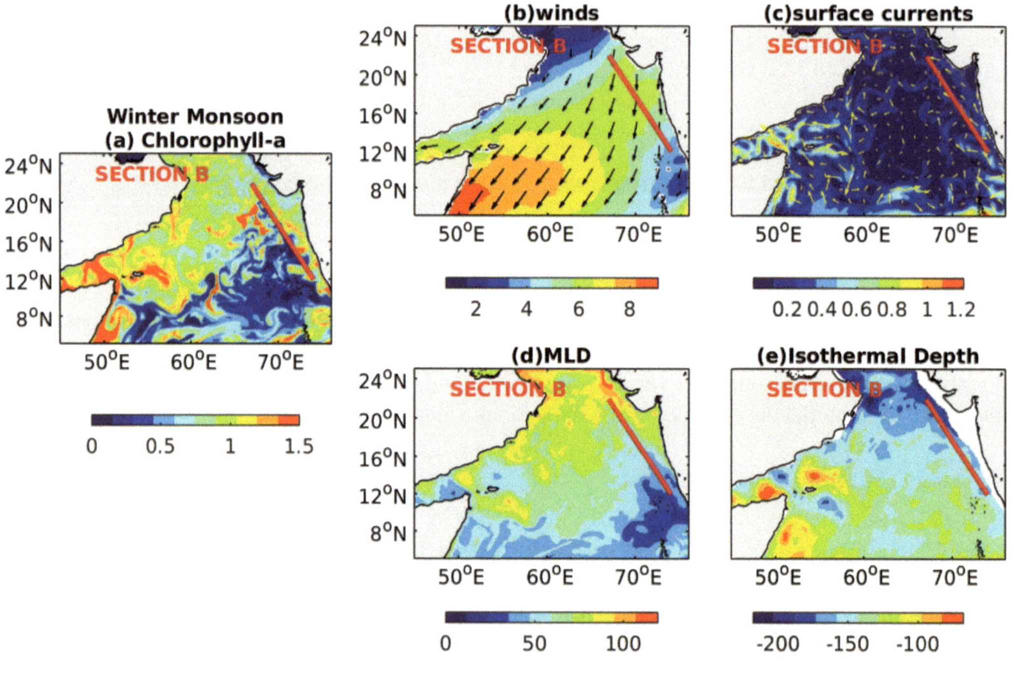

FIGURE 12.14 Spatial distribution of (a) Chl-a, (b) winds, (c) surface currents, (d) MLD, and (e) ILD during the winter monsoon along the eastern AS (74°E to 67°E and 12°N to 22°N) at Section B.

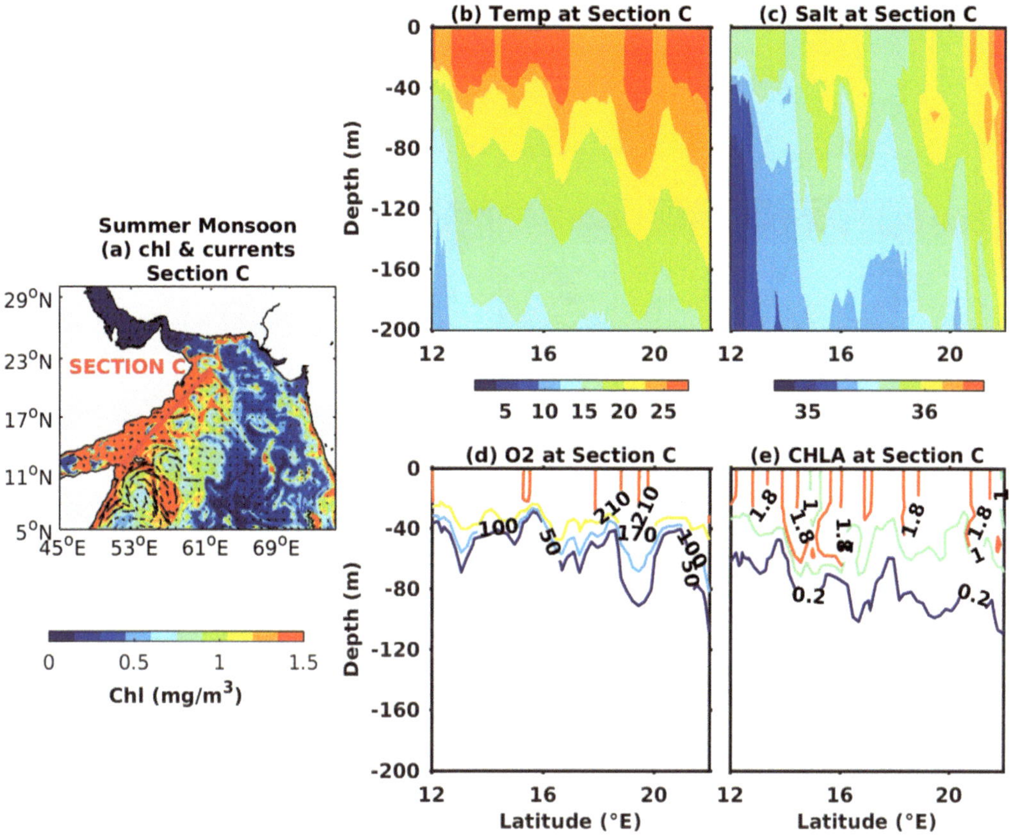

FIGURE 12.15 Variability of (a) surface Chl-a with surface currents and the vertical distribution of (b) oceanic temperature, (c) salinity, (d) oxygen, and (e) Chl-a during the summer monsoon along the western AS (52°E to 62°E and 12°N to 22°N) at Section C.

upper 30 m of the water column has a high concentration of O_2 with a concentration of 200 mMol O2 m^{-3} north of 15°N and 17°N (Figure 12.15). The upper 40 m of the water column also exhibits a high concentration of phytoplankton ranging between 0.7 and 0.5 mMol Nm^{-3}. The phytoplankton concentration decreases to less than 0.3 mMol Nm-3 below 60 m depth. The high phytoplankton concentration near the surface is attributed to high temperature (>22°C) and high nitrate level (>19 mMol N m^{-3}) during the southwest monsoon along the western AS (consistent with observations). The upper 20 m is devoid of zooplankton except along 17°N, where it is as high as 0.7 mMol N m-3. During the SWM season, the phytoplankton growth dominates over zooplankton grazing due to high productivity. Nitrate concentration is high in the upper 40 m of the water column. It ranges between 5 and 15 mMol N m^{-3} along Section C. The detritus concentration is also high (>1.5 mMol N m-3) in the upper 60 m along all the latitudes (Figure 12.16). The southwesterly winds are strong enough to erode the nitracline and nutrient-rich cold water in the upper surface (which drives phytoplankton growth). The forcing of upwelling is driven by the Somali jet (Figure 12.17).

The MLD is ~50 m along the section, but it is very deep toward the southwest direction of Section C. The ILD remains shallower up to 50 m depth along Section C. It shows that the thermocline shoaling is dominating for bloom in WAS compared to MLD deepening (Figure 12.17).

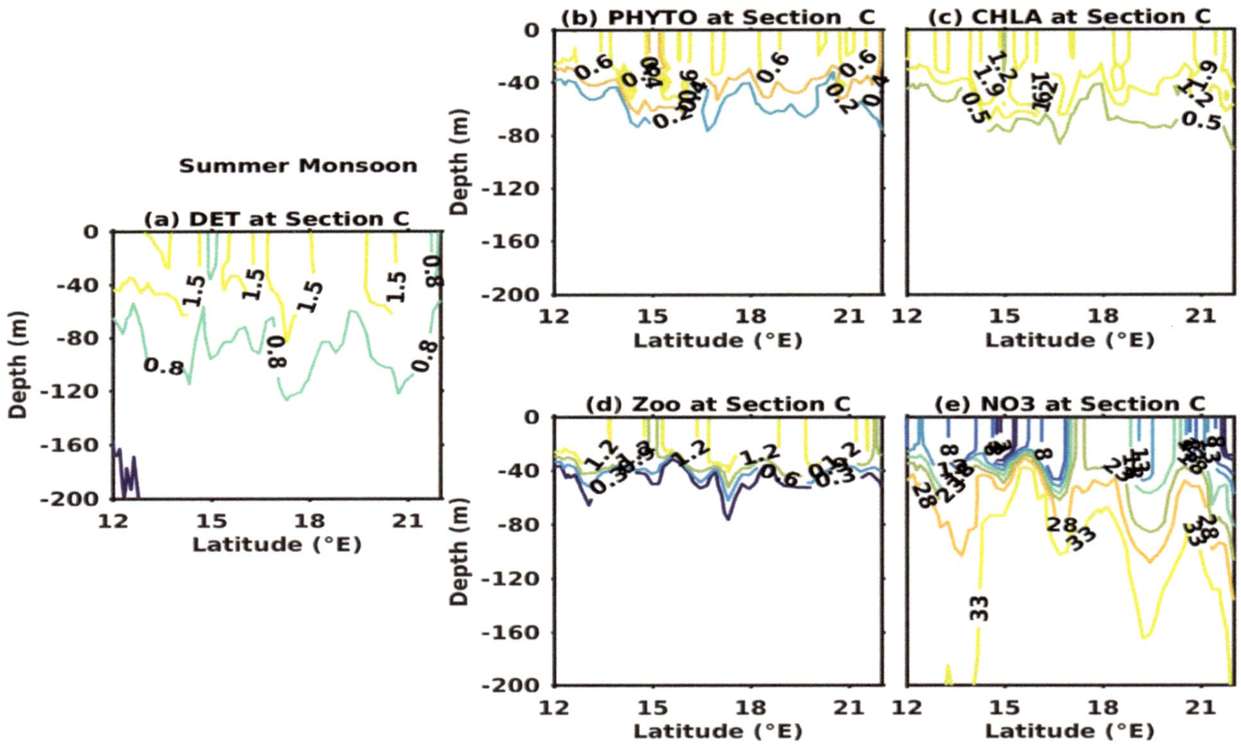

FIGURE 12.16 Vertical distribution of (a) detritus, (b) phytoplankton, (c) Chl-a, (d) zooplankton, and (e) nitrate during the summer monsoon along the western AS (52°E to 62°E and 12°N to 22°N) at Section C.

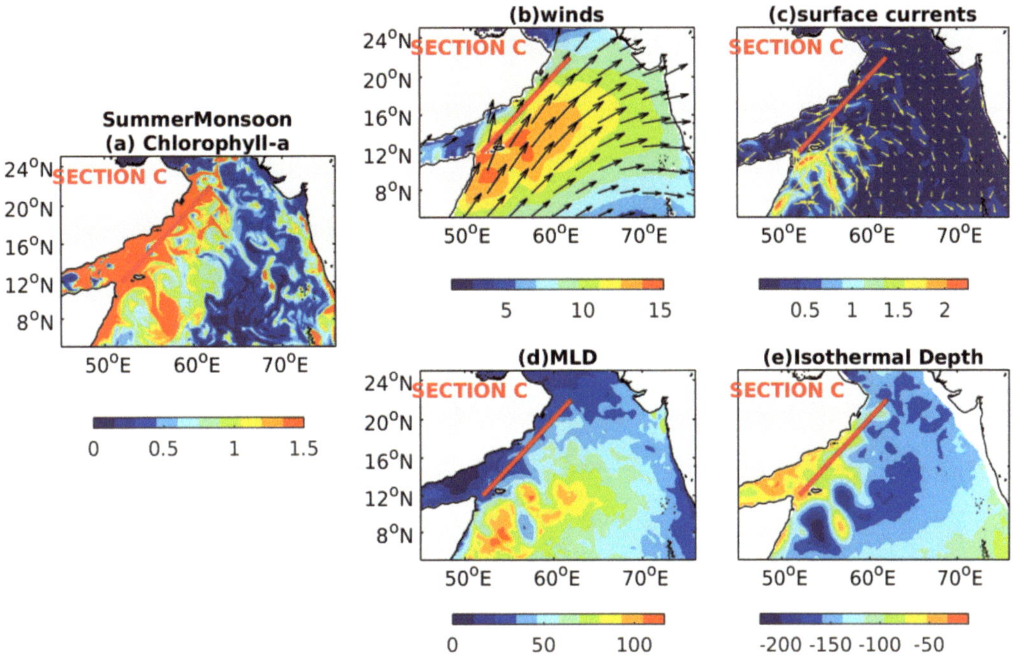

FIGURE 12.17 Distribution of (a) Chl-a, (b) winds, (c) surface currents, (d) MLD, and (e) ILD during the summer monsoon along the western AS (52°E to 62°E and 12°N to 22°N) at Section C.

12.3.4.2 Winter Monsoon

A temperature of nearly 25°C is obtained in the upper 40 m of the water column. The shoaling of the thermocline occurs along 14°N. The salinity of the upper 60 m increases from south to north, ranging 36 psu (Figure 12.18). High Chl-a is found along all the latitudes up to 80 m depth. The higher concentration of Chl-a, supported by elevated NO_3 (5–15 mMol N m^{-3} at 80 m) in the upper surface is due to convective mixing by northeasterly winds (Figures 12.19 and 12.20). The nitracline gets deeper from 60 m to 80 m toward north of 15°N.

Zooplankton grazing is more prominent than phytoplankton growth. The phytoplankton concentration remains approximately 0.4 mMol Nm^{-3} in the upper 40 m south of 18°N. The zooplankton concentration is high (>0.5 mMol Nm^{-3}) along all the latitudes. Moreover, it is found at deeper depths north of 15°N due to strong mixing. Detritus concentration is high south of 15°N. The oligotrophic condition is found in the southeastern AS. The MLD remains deeper (100 m) along this section. The ILD is shallow along 12°N and deep along 20°N (Figure 12.20).

12.4 CONCLUSIONS

The AS is a dynamic basin both in terms of physical and biogeochemical properties. In the present study, we tried to understand the dynamics of Chl-a variability on a seasonal timescale using numerical simulations as well as observational datasets. The seasonal variability of Chl-a and other biogeochemical parameters in different productive regions of AS was investigated, and the associated mechanisms are unraveled. The main outcomes of the study are as follows:

1. The basin average climatology showed strong seasonality of SST, net heat flux, winds, Chl-a, nitrate, and MLD in various regions of the AS.
2. In the central AS, during SWM, the high Chl-a bloom region occurred due to strong cyclonic wind-stress curl associated with the Findlater jet, which drives upward Ekman

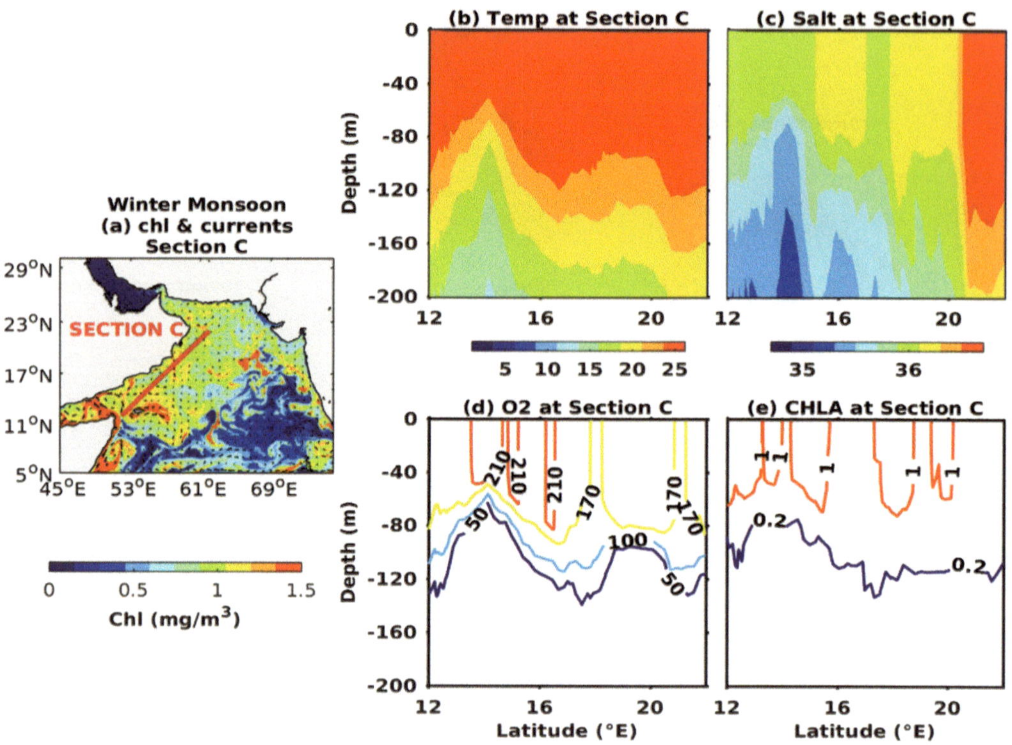

FIGURE 12.18 Variability of (a) surface Chl-a with surface currents and the vertical distribution of (b) oceanic temperature, (c) salinity, (d) oxygen, and (e) Chl-a during the winter monsoon along the western AS (52°E to 62°E and 12°N to 22°N) at Section C.

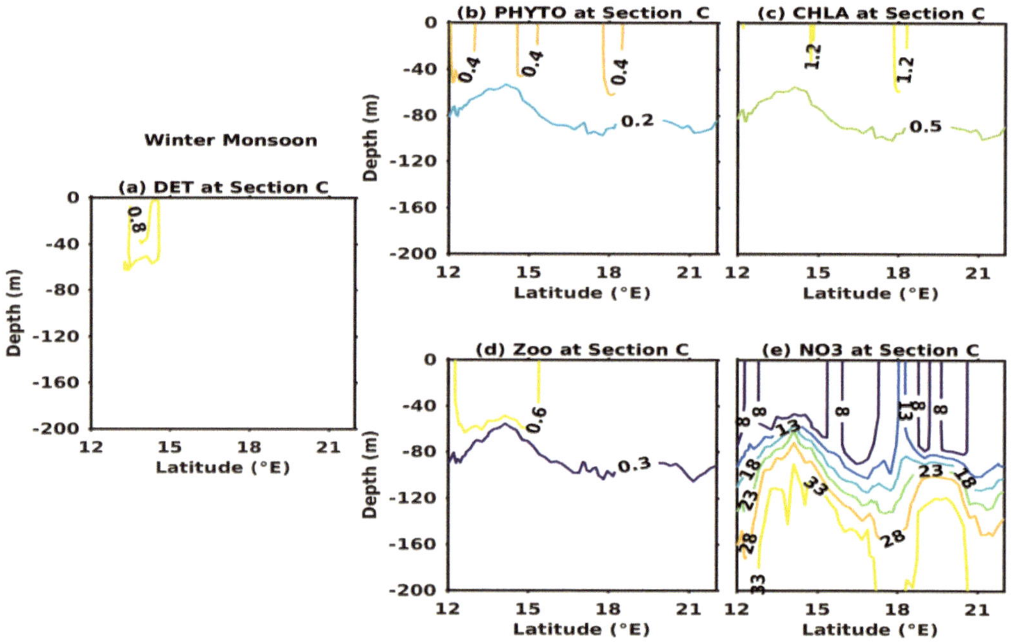

FIGURE 12.19 Vertical distribution of (a) detritus, (b) phytoplankton, (c) Chl-a, (d) zooplankton, and (e) nitrate during the winter monsoon along the western AS (52°E to 62°E and 12°N to 22°N) at Section C.

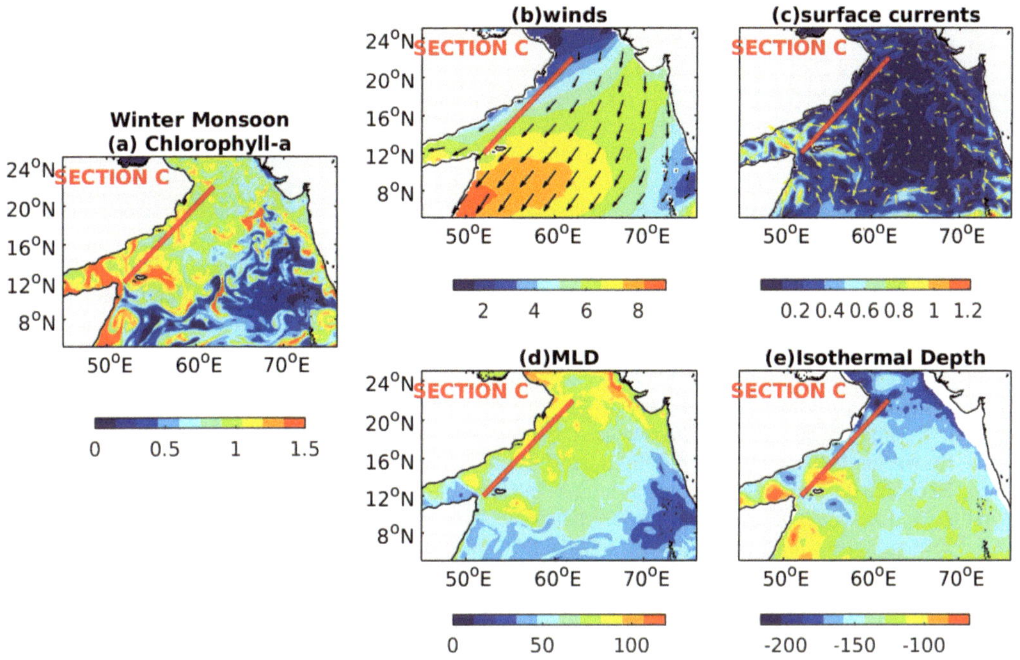

FIGURE 12.20 Distribution of (a) Chl-a, (b) winds, (c) surface currents, (d) MLD, and (e) ILD during the winter monsoon along the western AS (52°E to 62°E and 12°N to 22°N) at Section C.

pumping, upwelling, and SCM along 17°N and 64°E. During NEM, due to minimum temperature, maximum evaporation, and negative heat flux, evaporative heat loss drives convective mixing-induced upwelling. The MLD is shallow in SWM compared to NEM. In the western AS, very high Chl-a was observed during SWM due to the northward-flowing Somali current, coastal upwelling, horizontal advection, Ekman pumping near the coast of Oman and Somalia, and Great whirl. During NEM, upwelling occurs due to convective mixing and weaker eddy activity. Despite maximum nitrate, phytoplankton concentration is less due to excessive grazing by zooplankton during NEM. In the eastern AS, during SWM, elevated Chl-a was observed along 12°N to 15°N and 17°N to 19°N due to the propagation of Kelvin waves, which radiate westward propagating Rossby waves into the AS.

REFERENCES

Banse, K., and McClain, C. R. (1986). Winter blooms of phytoplankton in the Arabian Sea as observed by the Coastal Zone Color Scanner, Marine Ecology Progress Series, Vol. 34, pp. 201–211.

Brandes, J. A., Devol, A. H., Yoshinari, T., Jayakumar, D. A., and Naqvi, S. W. A. (1999). Isotopic composition of nitrate in the central Arabian Sea and eastern tropical North Pacific: A tracer for mixing and nitrogen cycles, Limnology and Oceanography, Vol. 43, pp. 1680–1689.

Brock, J. C., and McClain, C. R. (1992). Interannual variability in phytoplankton blooms observed in the Northwestern Arabian Sea during the Southwest Monsoon, Geophysical Research, Vol. 97 (C1), pp. 733–750.

Capone, D. G., Subramaniam, A., Montoya, J. P., Voss, M., Humborg, C., Johansen, A. M., Siefert, R. L., and Carpenter, E. J. (1998). An extensive bloom of the N2-fixing cyanobacterium Trichodesmium erythraeum in the central Arabian Sea, Marine Ecology Progress Series, Vol. 172, pp. 281–292.

Da Silva, A. M., Young, C. C., and Levitus, S. (1994). Atlas of Surface Marine Data, Vol. 1, Algorithms and Procedures. *NOAA Atlas NESDIS 6*, U.S. Department of Commerce, NOAA, NESDIS.

Di Lorenzo, E., Miller, A. J., Neilson, D. J., Cornuelle, B. D., and Moisan, J. R. (2004). Modelling observed California current mesoscale eddies and the ecosystem response, International Journal of Remote Sensing, Vol. 25 (7), pp. 1307–1312.

Egbert, G. D., and Erofeeva, S. Y. (2002). Efficient inverse modelling of Barotropic Ocean tides, Atmosphere and Ocean Technology, Vol. 19 (2), pp. 183–204.

Fasham, M. J. R., Ducklow, H. W., and McKelvie, S. M. (1990). A nitrogen-based model of plankton dynamics in the ocean mixed layer, Marine Research, Vol. 48, pp. 591–639.

Haidvogel, D., and Beckmann, A. (1999). *Numerical Ocean Modeling,* Environmental Science Management 2, Imperial College Press, London.

Hambright, K. D., Zohary, T., Easton, J., Azoulay, B., and Fishbein, T. (2001). Effects of zooplankton grazing and nutrients on the bloom-forming, N2-fixing cyanobacterium Aphanizomenon in Lake Kinneret, Journal of Plankton Research, Vol. 23 (2), pp. 165–174. https://doi.org/10.1093/plankt/23.2.165.

Hitchcock, G. L., Key E. L., and Masters, J. (2000). The fate of upwelled waters in the Great Whirl, August 1995, Deep Sea Research Part II, Topical Studies in Oceanography, Vol. 47 (7), pp. 1605–1621. https://doi.org/10.1016/S0967-0645(99)00156-3.

Latasa, M., and Bidigare, R. R. (1998). A comparison of phytoplankton populations of the Arabian Sea during the Spring Intermonsoon and Southwest Monsoon of 1995 as described by HPLC-analyzed pigments, Deep Sea Research Part II: Topical Studies in Oceanography, Vol. 45 (10–11), pp. 2133–2170.

Levitus, S., Locarnini, R. A., Boyer, T. P., Mishonov, A. V., Antonov, J. I., Garcia, H. E., Baranova, O. K., Zweng, M. M., Johnson, D. R., and Seidov, D. (2010). WOA 2009, NOAA Atlas NESDIS, Vol. 68, pp. 1–13.

Levy, M., Shankar, D., André,, J.-M., Shenoi, S. S. C., Durand, F., and Montégut, C. D. B. (2007). Basin-wide seasonal evolution of the Indian Ocean's phytoplankton blooms, Geophysical Research, Vol. 112, pp. 1–14.

Madhupratap, M., Kumar, S. P., Bhattathiri, P. M. A., Kumar, M. D., Raghukumar, S., Nair, K. K. C., and Ramaiah, N. (1996). Mechanism of the biological response to winter cooling in the northeastern Arabian Sea, Nature, Vol. 384 (6609), pp. 549–552.

Malanotte-Rizzoli, P., Hedstrom, K., Arango, H. G., and Haidvogel, D. B. (2000). Water mass pathways between the subtropical and tropical ocean in a climatological simulation of the North Atlantic Ocean circulation, Dynamics of Atmosphere and Oceans, Vol. 32, pp. 331–371.

McCreary Jr, J. P., Kohler, K. E., Hood, R. R., and Olson, D. B. (1996). A four-component ecosystem model of biological activity in the Arabian Sea, Progress in Oceanography, Vol. 37 (3–4), pp. 193–240.

Penven, P., Roy C., Lutjeharms, J. R. E., Colin de Verdiere, A., Johnson, A., Shillington, F., Freon, P., and Brundrit, G. (2001). A regional hydrodynamic model of the Southern Benguela, South African Journal of Science, Vol. 97, pp. 472–476.

Piketh, S. J., Tyson, P. D., and Steffen, W. (2000). Aeolian transport from southern Africa and iron fertilization of marine biota in the South Indian Ocean, South African Journal of Science, Vol. 96 (5), pp. 244–246.

Prasanna Kumar, S., Narvekar, J., Nuncio, M., Gauns, M., and Sardesai, S. (2009). What drives the biological productivity of the northern Indian Ocean? *Indian Ocean: Biogeochemistry and Ecological Variability,* AGU Book series, SIBER Monograph, pp. 35–58.

Rahaman, H., Ravichandran, M., Sengupta, D., Harrison, M. J., and Griffies, S. M. (2014). Development of a regional model for the North Indian Ocean, Ocean Modelling, Vol. 75, pp. 1–19.

Ravichandran, M., Girishkumar, M. S., and Riser, S. (2012). Observed variability of chlorophyll-a using Argo profiling floats in the southeastern Arabian Sea, Deep-Sea Research I, Vol. 65, pp. 15–25.

Sharada, M. K., Devasena, C. K., Swathi, P. S., Sundara Deepthi, M. V., Shelva Srinivasan, M. K., and Yajnik, K. S. (2015). Seasonal and interannual variability of marine ecosystem in the north Indian Ocean: A Model Evaluation Study, International Journal of Ocean and Climate Systems, Vol. 6 (1), pp. 19–34.

Shchepetkin, A., and McWilliams, J. (2005). The regional oceanic modeling system (ROMS): A split-explicit, free-surface, topography following-coordinate oceanic model, Ocean Modelling, Vol. 9, pp. 347–404.

Sil, S., and Chakraborty, A. (2011). Simulation of East India coastal features and validation with satellite altimetry and drifter climatology, Ocean and Climate Systems, Vol. 2 (4), pp. 279–289.

Smith, W. H. F., and Sandwell, D. T. (1997). Global sea floor topography from satellite altimetry and ship depth soundings, Science, Vol. 277 (5334), pp. 1956–1962.

Song, Y. T. and Haidvogel, D. B. (1994). A semi-implicit ocean circulation model using a generalized topography following coordinate system, Computational Physics, Vol. 115 (1), pp. 228–244.

Swathi, P. S., Sharada, M. K., and Yajnik, K. S. (2000). A coupled physical-biological-chemical model for the Indian Ocean, Earth System Science, Vol. 109 (4), pp. 503–537.

Veldhuis, M. J. W., Kraay, G. W., Bleijswijk, J. D. L. V., and Baars, M. A. (1997). Seasonal and spatial variability in phytoplankton biomass, productivity and growth in the northwestern Indian Ocean: The Southwest and Northeast Monsoon, 1992–1993, Deep Sea Research, Part I, Vol. 44 (3), pp. 425–449. https://doi.org/10.1016/S0967-0637(96)00116-1.

Wiggert, J. D., Jones, B. H. T. D. H., Weller, A., Marra, J., and Codispoti, L. A. (2000). The Northeast Monsoon's impact on mixing, phytoplankton biomass and nutrient cycling in the Arabian Sea, Deep Sea Research Part II: Topical Studies in Oceanography, Vol. 47 (7–8), pp. 1353–1385.

Wiggert, J. D., Murtugudde, R. G., and Christian, J. R. (2006). Annual ecosystem variability in the tropical Indian Ocean: Results of a coupled bio-physical ocean general circulation model, *Deep-Sea Research II,* Vol. 53, pp. 644–676.

Part V

Aquatic Ecosystem Management, Services, Threats, and Sustainability

13 Management of Aquatic Ecosystems and Aquatic Vegetation for Environmental Sustainability

Shilky, Subhashree Patra
Department of Environmental Sciences,
Central University of Jharkhand, Ranchi, India

*Purabi Saikia**
Department of Botany, Varanasi Hindu University, Varanasi, India

Amit Kumar
Department of Geography, Sikkim University, Gangtok, India

13.1 INTRODUCTION

Aquatic ecosystems are a critical element of the natural environment because they enable human activities like fishing, aquaculture, and recreation while also providing homes for a variety of flora and fauna and controlling water quality (Hassan et al., 2018). Aquatic vegetation makes positive contributions to water provision, pollution removal, flood control, and drought resistance (RNISR, 2020). The diversity of aquatic plant communities may have an impact on the functioning and provision of ecosystem services in wetland ecosystems (Englehardt and Ritchie, 2002; Kumari et al., 2020). Submerged aquatic vegetation fosters habitat for fauna, controls water flow, maintains sediment stability, and participates in biogeochemical cycling (Massicotte et al., 2015). Water habitats are essential for feeding, breeding, and migrating waterfowl as well as for fulfilling their preferred food and habitat needs (Havera, 1999). The base of the food chain is made up of both freshwater and marine macrophytes including plants and algae, which provide essential habitats for other aquatic species and act as a direct or indirect food source (Carpenter and Lodge, 1986). More than 70% of the oxygen in the world is produced by algae, which also serve as the basis of aquatic food chains (Smith et al., 2016). Macroalgae have served as a vegetable throughout Asia, particularly in Japan, where each person consumes 1.4 kg of macroalgae annually (Burting, 2003). In addition to providing nutrition, macroalgae are employed for their antibacterial, antiviral, antifouling, anti-inflammatory, cytotoxic, and antimitotic properties (Chen and Jiang, 2001). Important aquatic plant species including *Trapa natans* L., *Nelumbo nucifera* Gaertn., *Eleocharis dulcis* (Burm.f.) Trin. ex Hensch., *Ipomoea aquatica* Forssk., *Nasturtium officinale* W.T. Aiton, *Neptunia oleracea* Lour., *Colocasia esculenta* (L.) Schott, and some species of algae have been used as a wild stock or grown in flooded fields to fulfil nutritional demand, aquaculture, and animal fodder (Wersal and Madsen, 2012). Every component of a plant, including its stems, roots, rhizomes, tubers, seeds, and other parts, has been utilized as edibles, medicine, mulch, compost, and construction materials (Swapna

* Corresponding author.

DOI: 10.1201/9781003354000-18

et al., 2011). Some aquatic plant species, including the *N. nucifera*, are sacred and have religious utilities (Wersal and Madsen, 2012).

Water contamination, combined with limited water availability, has imposed a significant load on the environment (Ali et al., 2020). Water pollution is a significant, deadly, and uncontrolled global issue due to a lack of knowledge, awareness, and the strict application of eco-friendly norms, legislation, and financial resources (Eid et al., 2020). Water scarcity affects ~40% of the world's population as a result of climate change, fast urbanization, food demand, and unrestrained consumption of natural resources (Calzadilla et al., 2011). Population increase, industrialization, urbanization, and overuse of natural water resources contribute to deteriorating water quality (Enyoh et al., 2018). Global warming may exacerbate the imbalance between water supply and demand brought on by drought (Jung et al., 2010; van Loon et al., 2016). Unpredictable rainfall patterns and the resulting floods and droughts, inefficient water use, unrestricted groundwater extraction, water pollution, poor water quality due to open defecation, insufficient waste management laws, and interstate river disputes are all issues that must be addressed by the water management sector (Cronin et al., 2014). Untreated industrial and home wastewater can contain pesticides, oils, dyes, phenol, cyanides, toxic organic compounds, phosphorus, suspended particles, and heavy metals (HMs) (Pakdel and Peighambardoust, 2018). HMs can be found in large quantities in the environment (An et al., 2015) and the global pollution of HMs is attributed to economic and industrial activities such as metal processing, mining, geothermal energy plants, automobiles, paper, pesticide production, tanning, dyeing, and plating (Peligro et al., 2016). It is difficult to remove HMs from wastewater because they can take on a variety of chemical forms (Gall et al., 2015). Due to their inability to biodegrade, the majority of metals can easily move between trophic levels and accumulate in the biota over time (Zhu et al., 2016).

13.2 MAJOR CHALLENGES IN AQUATIC ECOSYSTEMS

The integrity and long-term health of aquatic ecosystems are in jeopardy due to significant challenges, which ultimately harm ecosystem services, biodiversity, and human well-being. Acid rain, industrial and domestic wastewater, HMs, pesticides, oil, and numerous other inorganic and organic toxic substances are the principal causes of water pollution (Verla et al., 2018). The overexploitation of aquatic resources seriously threatens the viability of aquatic ecosystems and the livelihoods of millions of people. Through overfishing, illegal fishing, and destructive fishing methods, fish populations can be reduced, food webs can be disrupted, and habitats can be damaged (Pauly and Zeller, 2016). One of the significant threats to aquatic ecosystems is climate change, which affects water temperature, precipitation, and acidity, contributing to sea-level rise and ocean acidification (IPCC, 2019). Corals, molluscs, and plankton, which produce calcium carbonate that is necessary for the construction of shells and skeletons, may experience decreased calcium carbonate availability due to ocean acidification (Doney et al., 2009). These modifications may affect aquatic animals and their habitats directly or indirectly, changing their range, abundance, and productivity (Poloczanska et al., 2016). The entire food chain may be affected by coral bleaching, fish migration, and changes in plankton ecosystems brought on by warming waters (Hoegh-Guldberg et al., 2018).

Non-native aquatic weeds, which humans frequently introduce through shipping, aquaculture, and recreational sailing, have the potential to harm aquatic environments on an ecological and financial level by displacing native species, altering food webs, and introducing diseases (Ricciardi et al., 2013). For instance, native fish and mollusc populations have decreased as a result of the invasion of quagga mussels (*Dreissena bugensis* Andrusov) and zebra mussels (*Dreissena polymorpha* Pallas) in freshwater ecosystems in North America and Europe (Nalepa and Clapp, 2018). Moreover, invasive fish species like the Asian carp (*Hypophthalmichthys* spp.) and Nile tilapia (*Oreochromis niloticus* L.) have had substantial negative ecological and economic effects on freshwater environments (Rahel and Olden, 2008). Invasive aquatic plant costs in the United States are estimated to be around $110 million per year, taking into account both control expenses and economic losses (Pimentel et al., 2005). It is possible to use eliminated weeds as a more practical and affordable bioenergy feedstock because they are a significant source of greenhouse gas emissions that are disposed of in landfills (Kaur et al., 2018). The dense infestations of non-native aquatic weeds have negatively impacted aquatic ecosystems, including usage of streams, drainage, fishing, water quality, fish and wildlife habitat, flood control, human and animal health, hydroelectric generation, irrigation, navigation, recreation, and eventually property values (Rockwell, 2003). *Pontederia crassipes* Mart., *Salvinia molesta* D. Mitch., *Pistia stratiotes* L., *Myriophyllum spicatum* L., *Typha latifolia* L., and *Ceratophyllum demersum* L. are the world's most dangerous invasive aquatic weeds (Space Applications Center, 2011). Their rapid expansion, numerous spread channels, and widespread dissemination may have detrimental effects on the environment and the economy. Mats of quickly multiplying water *P. crassipes* clog many reservoirs, pollute water supplies, infest rivers, dams, and drainage channels, and diminish the dissolved oxygen (DO) levels in supply water (Singh et al., 2024). They also endanger aquatic biodiversity by displacing native flora and fauna and causing irreversible changes to habitats (Kaur et al., 2018). *Hydrilla* is the aquatic invader that incurs the highest costs, drastically disrupting native plant communities and impacting fish populations too (Langeland, 1996). Largemouth bass (*Micropterus salmoides* Lacepède) and other fish species can benefit temporarily from *Hydrilla*, including increased juvenile survival (Tate et al., 2003). However, dense populations may harm other fish species, including important largemouth bass prey species like black crappies (*Pomoxis nigromaculatus* Lesueur) and sunfish species (*Lepomis* spp.) (Bonvechio and Bonvechio, 2006). A total of 15.3 Mha or 4.7% of the land in India is covered by wetlands, of which 1.32 Mha or 9% is infested with aquatic weeds in the post-monsoon season and 2.06 Mha or 14% in the pre-monsoon season (Space Applications Center, 2011). Aquatic weed management is a labor-intensive and costly process. In India, the total annual cost of weed treatment by humans is close to $490/ha/year, while the cost of physical weed control is close to $910/ha/year (Jayan and Sathyanathan, 2012). Nearly one million hectares of India's inland waterways are in danger due to the invasion of destructive aquatic weeds, of which the four most common are *P. crassipes*, *P. stratiotes*, *S. molesta*, and *Alternanthera philoxeroides* (Mart.) Griseb. (Space Applications Centre, 2011). *P. crassipes* invasion in the Harike wetland of Punjab, one of the Indian Ramsar sites is shown in Figure 13.1(a–e).

13.3 AQUATIC VEGETATION AND ITS ROLE IN ENVIRONMENT MANAGEMENT

The aquatic ecosystem is a cost-effective and resourceful phytoremediation technology for a vast contaminated area, as the aquatic plants absorb pollutants and HMs naturally (Pratas et al., 2014). Constructed/engineered artificial wetlands are created to treat and remediate specific types of wastewater, including agricultural, municipal, landfill leachate, stormwater, and industrial wastewater in a regulated setting throughout the world (Vymazal and Brezinova, 2015; Saeed et al., 2018; Wang et al., 2019). The built wetland with aquatic plants offers a very straightforward and affordable option for reducing water contamination and wastewater cleanup without affecting the resources of natural wetlands (Marchand et al., 2010; Rizzo et al., 2020). The best wetland plant for the phytoremediation of HMs from wastewater is *P. crassipes*, which can increase its biomass twofold in a matter of days when used in the built wetland technology-assisted phytoremediation of Cd, Pb, Cu, and Ar from industrial effluent (Rai, 2019). *P. crassipes* outperformed *P. stratiotes* and *Spirodela polyrhiza* (L.) Schleid. in terms of their ability to remediate Cu, Ni, Fe, Cd, Zn, and Cr in a 15-day experiment (Saha et al., 2016). The phytoremediation ability of *Cyperus alternifolius* R.Br., *Cynodon dactylon* (L.) Pers., and *T. latifolia* in a constructed wetland with refinery effluent was evaluated, and the results showed that the roots of the plants acquired the highest concentrations of Cr, Zn, Cd, Pb, and Fe, followed by the leaves and stem (Mustapha et al., 2017). Similarly, when *Typha domingensis* Pers. was grown in pots with municipal wastewater, the root and shoot tissues showed the greatest accumulation of Zn, Cd, Ni, Fe, and Mn during the first 48 hours of the study (Ali et al., 2020).

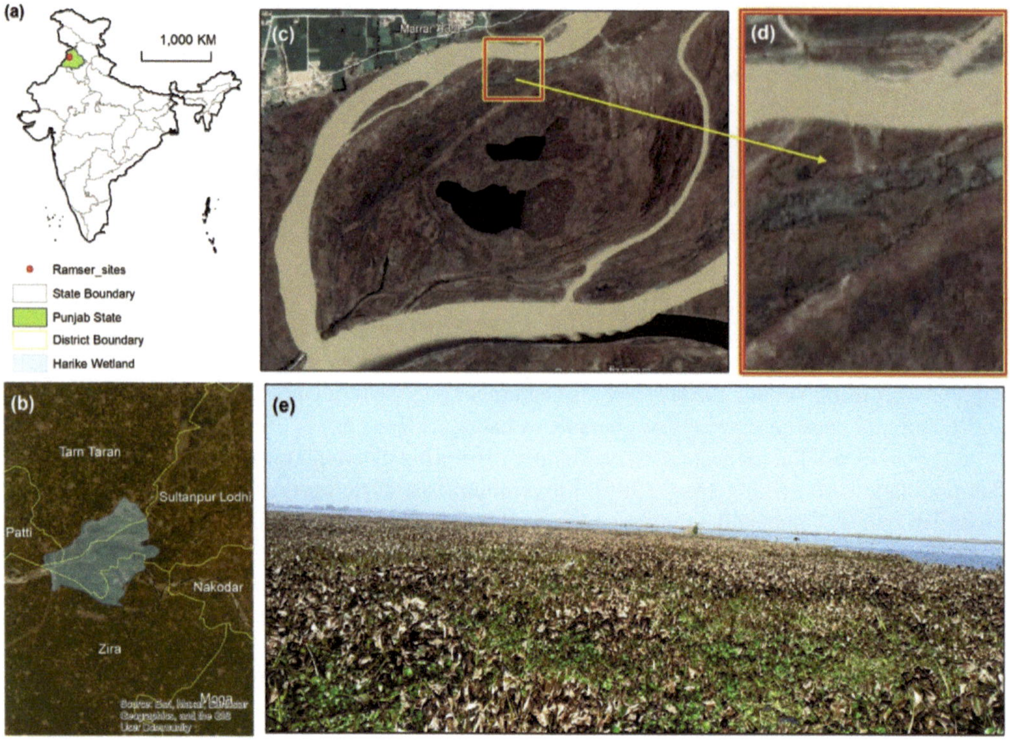

FIGURE 13.1 (a–e) Invasion of *Pontederia crassipes* Mart. in the Harike wetland of Punjab, India.

Numerous species of algae (*Chlorella vulgaris* Beij., *Phormidium* spp., *Spirulina platensis,* etc.) are known for their ability to remove HMs from water, which benefits the ecosystem (Feng et al., 2020).

In addition to removing contaminants and cleaning up contaminated waterways, aquatic vegetation also produces a significant amount of biomass that, if uncontrolled, contributes to the degradation of the water body and depletion of DO, making the management of aquatic biomass a serious issue (Muthusaravanan et al., 2018). By using a variety of techniques, including pyrolysis, gasification, hydrothermal liquefaction, and fermentation, waste aquatic plant biomass has emerged as a very affordable and environmentally friendly application for the production of biogas, biochar, and bioethanol (Awasthi et al., 2015). Biochar has received a lot of interest, primarily as an agricultural product for boosting soil fertility and crop productivity (Yang et al., 2021). It is also regarded as an effective method for reducing climate change, due to increased soil methane oxidation and carbon offset from reduced fossil fuel production, as well as decreased methane and nitric oxide (N_2O) emissions from the breakdown of biomass and other chemical fertilizers (Gonzalez-Cencerrado et al., 2020). Additionally, aquatic weed biomass can be used as a substitute feedstock for the production of biofuel to circumvent issues with agriculture and food security (Kaur et al., 2018). Because they are rich in carbohydrates, proteins, and fats, using them to produce biofuel would help reduce their growth in water bodies while also meeting current energy needs (Sarwar et al., 2002). Azolla may be used as a feedstock for the production of bioenergy and can be pyrolyzed to produce biochar with a high calorific value that can be used as a fuel or a soil amendment (Sivalingam et al., 2022). Furthermore, *P. crassipes* produces more biogas than *P. stratiotes*, and a pH range of 7.5–8.5 and a temperature of 35–40°C are ideal for biogas generation (Roy et al., 2021). Similarly, *Lemna minor* L. is used as a possible feedstock for making bioethanol due to its high cellulose and hemicellulose content (Abdul-Razak et al., 2020). Carbon content in microalgae biomass is ~50% on a dry weight basis, and *Nannochloropsis oculata* (Droop) Hibberd, *Phaeodactylum tricornutum*, and *C.*

TABLE 13.1
Aquatic Plants and Their Role in Ecosystem Management

Plant species	Life-form	Usage	Sources
Pontederia crassipes Mart.	Aquatic weed	Feed for livestock, improving soil fertility, sources of biofuel, etc.	Rawat et al. (2013); Njogu et al. (2015)
Lemna minor L.	Aquatic weed	Used as livestock feed	Njogu et al. (2015)
Pistia stratiotes L.	Aquatic weed	Source of biofuel	Rawat et al. (2013)
Phragmites australis (Cav.) Trin. ex Steud.	Aquatic weed	Effective in removing Pb from water	Khalid et al. (2017)
Hydrilla verticillata (L. f.) Royle	Aquatic weed	Phytoremediation of Cr and Fe	Najila and Anila (2022)
Ceratophyllum demersum L.	Bryophyte	Absorbs Cd	Kováčik et al. (2017)
Ulva lactuca L.	Algae	Ability to remove heavy metals such as Cd, Pb, and Ni from contaminated water	Zhang et al. (2021)
Chlorella vulgaris Beijerinck	Algae	Remove heavy metals such as Pb, Cd, and Cu from contaminated water	Perales-Vela et al. (2006)
Spirulina platensis Turpin Ex Gomont	Algae	Effective in removing heavy metals from water	Sen Gupta et al. (2011)
Microcystis aeruginosa Kützing	Algae	Removes heavy metals such as Cd and Pb from water	Abdel-Aty et al. (2013)
Gelidium amansii	Algae	Bioethanol production	Rajkumar et al. (2013)

vulgaris can grow quickly and duplicate their cells numerous times in a single day. They have the best chances of producing bioethanol because algae-derived carbohydrates can be converted into fermentable sugars (Rajkumar et al., 2013). Macroalgae such as *Gelidium amansii*, *Gracilaria* spp., and *Laminaria* spp. are suitable for bioethanol production (Tan et al., 2020) (Table 13.1).

13.4 TRADITIONAL TO ADVANCED REMOTE SENSING TECHNOLOGY FOR MONITORING AQUATIC ECOSYSTEMS

Monitoring of aquatic bodies for the initiation of the blue economy in a sustainable manner is one of the main objectives of several countries for the reduction and management of pollutant flow into aquatic ecosystems (Hafeez et al., 2018). Increasing anthropogenic activities and water contamination issues demand integrated and routine water monitoring processes for the efficient management of aquatic resources (Wang and Yang, 2019). Conventional methods were once thought to be the most reliable way to collect water samples using various vehicular transport systems at various depths for the detection and measurement of different physical and chemical parameters (Hafeez et al., 2018). Such approaches are accurate but need laboratory analysis by skilled experts, are time-consuming, and are geographically constricted (Cai et al., 2008). However, monitoring and regulating the environmental repercussions, and comprehending the mechanisms controlling their spatial distribution, need real-time observations of toxins and marine pollutants over a variety of spatial scales (Zielinski et al., 2009). Remote sensing applications are widely utilized by academicians, scientists, and resource managers for monitoring and managing aquatic ecosystems from a broad perspective (Klemas, 2011). In addition to photographic representation, remote sensing tools are able to access physical measurements of numerous aspects of aquatic ecosystems (Gutierres et al., 2016). Satellite and aerial tools are considered effective for tracking and mapping oil spills and simultaneously supplying data for models to detect pollutants over time and space (Zielinski et al., 2006). Initially, a coastal zone color scanner (CZCS) was used for monitoring ocean color, but later, with advanced tools and technologies, numerous spatial and spectral resolution satellite sensors such as geostationary ocean color imager (GOCI), moderate resolution imaging spectroradiometer (MODIS), ocean and land color imager (OLCI), visible infrared imager radiometer suite (VIIRS), Sentinel-2 multispectral instrument (MSI), and Landsat operational land imager (OLI) were developed that can obtain data on the color of water for monitoring and detecting pollutants from aquatic ecosystems (Hafeez et al., 2018). Thus, remote sensing tools play a vital role in the management and monitoring of aquatic resources, thereby contributing as a state-of-art technology to achieve the United Nations sustainable development goals (UN-SDGs).

13.5 POLICY INITIATIVES AND UN-SDG APPROACHES FOR SUSTAINABLE MANAGEMENT OF AQUATIC VEGETATION AND AQUATIC ECOSYSTEMS

Time-to-time monitoring and assessment of aquatic ecosystems are required to maintain and preserve aquatic habitats and related resources (Forio and Goethals, 2020). Moreover, further research and evaluations are necessary to identify the best management approaches and strategies to minimize or mitigate further water deterioration (De Pauw et al., 2006). Over 320 million tonnes of microplastic waste are produced in a year, and that amount is estimated to increase thrice by 2050 (Kumar et al., 2021). To protect aquatic ecosystems from plastic exposure and to monitor its harmful effects on aquatic life and people, several government agencies, research organizations, and universities have worked together on mitigation and management techniques (Jiang, 2018). UN-SDG 14 (Life below water) deals with the sustainable management of aquatic vegetation and aquatic ecosystems by emphasizing the need to minimize the quantity of marine plastic waste released from terrestrial activities in the ocean by 2025 (Walker, 2021). The use of plastic bags at the supermarket is charged for in Germany, Switzerland, the Netherlands, Norway, and Sweden (Luís and Spínola, 2010). Likewise, Portugal imposed a plastic bag fee in 2015, and the country's plastic use has decreased by 74% since then (Martinho et al., 2017). To prevent or control plastic and microplastic contamination from land-based and maritime sources, many agreements and international rules have been enacted (Chavarro, 2017). The International Treaty for the Prevention of Pollution from Ships (MARPOL) was established in 1973 and is regarded as the first treaty to address plastic limits, notably in Annex V, which addresses the explicit prohibition of dumping any type of plastic garbage (Kumar et al., 2021). To achieve the SDGs, the Bangladeshi government has recently placed a strong emphasis on promoting blue growth, where ocean-based resources play a crucial role (Vorosmarty et al., 2018). To effectively conserve marine biodiversity worldwide by 2030, the International Union for Conservation of Nature (IUCN) World Conservation Congress recommended safeguarding a minimum of 30% of each marine habitat and at least 30% of all the ocean for effective management and monitoring of aquatic diversity (IUCN, 2016). India has also taken several initiatives for the monitoring and protection of aquatic plants and ecosystems by introducing different acts and policies related to the conservation of aquatic fauna and flora in line with the UN-SDG 14. Among the significant policies implemented in India are the National Forest Policy, National Conservation Strategy and Policy Statement on Environment and Development, National Policy and Macro-level Action Plan on Biodiversity, National Biodiversity Action Plan (2009), National Agriculture Policy, National Water Policy, National Environment Policy (2006), and National Water Mission (Table 13.2). The Ministry of Earth Sciences (MoES) sanctioned a budget of INR 84.00 Crore from 2017 to 2020 to the Centre for Marine Living Resources and Ecology (CMLRE) for the implementation of a research and development program on Marine Living Resources (MLR) (India Science Wire, 2022). In addition to prohibitions and taxes, other practices like recycling and ecolabeling have also been used. Several countries have already started recycling, though different countries have different rates of recycling (Kumar et al., 2021).

TABLE 13.2
Various Indian Policies Related to Aquatic Vegetation and Aquatic Ecosystems

Policies	Objectives and action plans related to aquatic vegetation and aquatic ecosystems	Source
National Water Policy of India, 2012	Account for environmental requirements of the Himalayan areas, the aquatic ecosystem, wetlands, and embanked flood plains while planning	NWP, 2012
National Water Mission of India, 2008	Impact on aquatic vegetation and aquatic environment based on policy modification on climate change	NAPCC, 2008
National Conservation Strategy and Policy Statement on Environment and Development, 1992	Emphasizes the adversities caused by water pollution on aquatic vegetation and aquatic ecosystem	Nath, 1992
National Biodiversity Action Plan, 2008	Account for the protection and monitoring of aquatic vegetation and aquatic ecosystems in strategic action plan	NBAP, 2014

13.6 OPPORTUNITIES TO IMPROVE THE RESILIENCE OF AQUATIC ECOSYSTEMS

The vital aquatic habitats, including coral reefs, mangroves, and wetlands, must be managed and maintained by establishing buffer zones, marine protected areas, and conservation practices for aquatic ecosystems to function properly (Cinner et al., 2021). Promoting climate adaptation through actions like improving water management procedures, creating drought-resistant crops, and encouraging sustainable land use practices is crucial to increasing the resilience of aquatic ecosystems (Staudinger et al., 2021). Nature-based solutions (NbS) provide a potential method to increase the resilience of aquatic ecosystems by addressing these concerns through natural processes (Zari et al., 2019). One such method is wetland restoration because it provides critical ecological services like water filtering, carbon sequestration, and habitat for a diverse range of species (Rezaie et al., 2020). Wetland restoration can improve aquatic ecosystem resilience by boosting its ability to offset the effects of climate change and reducing community vulnerability to flooding and other natural disasters (Tornqvist et al., 2021). Similarly, floodplain restoration can help absorb and store excess water, lowering the risk of downstream floods (Keddy et al., 2009). Other NbS that can be used to improve the resilience of aquatic ecosystems, in addition to habitat restoration and green infrastructure, are the utilization of natural shorelines, the establishment of riparian buffers, and the restoration of mangrove forests (Barbier et al., 2011). For instance, oyster reef restoration can assist in improving the resilience of estuarine ecosystems by providing crucial habitat for a variety of aquatic species and also assist in improving water quality through filtering contaminants and minimizing sedimentation (Searles et al., 2022). Another option to improve the resilience of aquatic ecosystems is to use ecosystem-based fisheries management (EBFM), which is an approach to managing fish populations that considers the entire ecosystem rather than focusing simply on particular species (Hilborn, 2011). Furthermore, pursuing possibilities to lessen the vulnerabilities of destructive human activities, such as pollution, overfishing, habitat destruction, and invasive species, is also necessary to improve the resilience of aquatic ecosystems (Mcleod et al., 2019). Aquatic plants are underutilized resources that should be taken into consideration around the world to establish sustainable food and fiber systems (Hussain et al., 2020). On the other hand, the reliance on the biomass of *P. crassipes* would result in conflicts between user groups that want to spread and use biomass and those that want to restrict the plant species (Hill et al., 2011).

13.7 CHALLENGES IN THE MANAGEMENT OF AQUATIC ECOSYSTEMS AND AQUATIC VEGETATION

Despite their biological importance, aquatic environments suffer from a number of obstacles that make management difficult. For instance, pesticides such as "insecticides", "herbicides", and "fungicides" are widely used in agricultural areas throughout numerous countries, where they wash into aquatic environments and damage water resources (Mateo-Sagasta et al., 2017). However, it is challenging to control the flow of this contaminated water into rivers and lakes. Further, the challenges in environmental protection are due to a lack of national determination and inadequate facilities for wastewater treatment in industrial and agricultural sectors, often leading to discharge into surface water sources (Pirali Zefrehei et al., 2022). Moreover, transboundary water management poses significant issues (Riley, 2009), and without an effective plant invasion management regime for the entire ecological unit or landscape scale, there is an increased chance of conflicts and unsuccessful management interventions (Lim, 2014). A variety of alternative (non-operational) strategies have been tried with no success, particularly in the management of amphibious species. Examples include the use of hot foam in an attempt to eradicate *Crassula helmsii* in south England nature reserves (Bridge, 2005); the use of hydrogen peroxide and flame-throwers on *C. helmsii* and *Hydrocotyle ranunculoides*, which were also ineffective because only emergent parts were treated and submerged material remained unaffected (Leach and Dawson, 2000); and the use of liquid nitrogen, which produced similarly poor results despite being effective. These alternate strategies, for obvious reasons, provide little protection against aquatic invasive plants.

13.8 RECOMMENDATIONS AND FUTURE RESEARCH PROSPECTS

- Proper management of aquatic ecosystems is possible through the sustainable utilization of water and water-based resources.
- It is necessary to identify the diverse aquatic plants with better phytoremediation ability to minimize the problem of metal pollution and to utilize the biomass of noxious aquatic weeds for the production of value-added products like biogas, biofuel, bioethanol, etc. to convert waste into resources.
- The country's aquatic diversity and resources must be conserved through coordinated efforts incorporating new environmental program-based tactics and cutting-edge technological advancements.
- More research on deciphering biodiversity recommendations is needed to effectively comprehend public attitudes toward biodiversity and further incorporate the public in conservation options and decision-making processes.
- This greater participation may help in reducing the discrepancy between dispute and management of biodiversity in a sustainable way.

13.9 CONCLUSIONS

Aquatic ecosystems and aquatic vegetation provide a range of unique ecosystem goods and services from food, fodder, and water purification through phytoremediation to biofuel and bioethanol production. Additionally, aquatic vegetation is also responsible for water pollution and quality degradation due to the uncontrolled growth of invasive weeds in major wetlands of India and throughout the world. Management of the proliferation of aquatic weeds is one of the major challenges to maintaining water quality and quantity as well as protecting existing water resources in a sustainable manner. In prior years, the inland fishing industry and other land-based ecosystems have received higher priority for conservation and management than coastal and marine living resources. With the introduction of the sustainable economic concept and SDG targets, the government has strongly emphasized the utilization of marine and coastal resources. To achieve the desired goals of enhancing the protection, conservation, and monitoring of aquatic vegetation, the habitat of aquatic flora and fauna, and conserving biodiversity, rules with stringent enforcement mechanisms must be put in place. A suitable legislative framework is a crucial tool for the sustainability of the marine and coastal environment in achieving the SDG 14 targets. It will only be possible to achieve the objective of a sustainable economy by fostering a blue economy through the implementation of an integrated maritime policy. Therefore, establishing the proper legal framework is essential to carrying out the marine resources strategy efficiently.

REFERENCES

Abdel-Aty, A.M., Ammar, N.S., Ghafar, H.H.A., Ali, R.K., 2013. Biosorption of cadmium and lead from aqueous solution by fresh water alga *Anabaena sphaerica* biomass. *Journal of Advanced Research*, 4(4), pp.367–374.

Ali, S., Abbas, Z., Rizwan, M., Zaheer, I.E., Yavaş, İ., Ünay, A., Abdel-Daim, M.M., Bin-Jumah, M., Hasanuzzaman, M., Kalderis, D., 2020. Application of floating aquatic plants in phytoremediation of heavy metals polluted water: a review. *Sustainability*, 12(5), pp.1927.

An, B., Lee, C.G., Song, M.K., Ryu, J.C., Lee, S., Park, S.J., Zhao, D., Kim, S.B., Park, C., Lee, S.H., Hong, S.W., 2015. Applicability and toxicity evaluation of an adsorbent based on jujube for the removal of toxic heavy metals. *Reactive and Functional Polymers*, 93, pp.138–147.

Awasthi, P., Shrivastava, S., Kharkwal, A.C. and Varma, A., 2015. Biofuel from agricultural waste: a review. *Int J Curr Microbiol App Sci*, 4(1), 470–477.

Barbier, E.B., Hacker, S.D., Kennedy, C., Koch, E.W., Stier, A.C., Silliman, B.R., 2011. The value of estuarine and coastal ecosystem services. *Ecological Monographs*, 81(2), pp.169–193.

Bonvechio, K.I. and Bonvechio, T.F., 2006. Relationship between habitat and sport fish populations over a 20-year period at West Lake Tohopekaliga, Florida. *North American Journal of Fisheries Management*, 26(1), 124–133.

Bridge, T., 2005. Controlling New Zealand pygmyweed Crassula helmsii using hot foam, herbicide and by burying at Old Moor RSPB Reserve, South Yorkshire, England. *Conservation Evidence*, 2, pp.33–34.

Burting, J. 2003. Macroalgae as a vegetable in Asia, *Journal of Applied Phycology*, 15(5), 427–434.

Cai, L., Liu, P., Zhi, C., 2008. Discussion on remote sensing based on water quality monitoring methods. *Geomatics & Spatial Information Technology*, 31(4), pp.68–73.

Calzadilla, A., Rehdanz, K., Tol, R.S., 2011. Water scarcity and the impact of improved irrigation management: a computable general equilibrium analysis. *Agricultural Economics*, 42(3), pp.305–323.

Carpenter, S.R. and Lodge, D.M. 1986. "Effects of submersed macrophytes on ecosystem processes", *Aquatic Botany*, 26(4), 341–370.

Chavarro, J.M., 2017. Common concern of humankind and its implications in international environmental law. In: *Globalization and Common Responsibilities of States*, Feyter, K.D. (ed.), Routledge Taylor and Francis Group, New York, pp.337–351.

Chen, F. and Jiang, Y. 2001. "Bioactive compounds from marine macroalgae", *Journal of Applied Phycology*, 13(4), 223–226.

Cinner, J.E., Barnes, M.L., Gurney, G.G., Lockie, S. and Rojas, C., 2021. Markets and the crowding out of conservation-relevant behavior. *Conservation Biology,* 35(3), 816–823.

Cronin, A.A., Prakash, A., Priya, S., Coates, S., 2014. Water in India: situation and prospects. *Water Policy*, 16(3), pp.425-441.

de Pauw, N., Gabriels, W., Goethals, P.L., 2006. River monitoring and assessment methods based on macroinvertebrates. In: *Biological Monitoring of Rivers: Applications and Perspectives*, Ziglio, G., Flaim, G., Siligardi, M. (eds.), John Wiley and Son, Ltd., Chichester, pp.113–134.

Doney, S.C., Fabry, V.J., Feely, R.A. and Kleypas, J.A. 2009. "Ocean acidification: the other CO2 problem", *Annual Review of Marine Science*, 1, 169–192.

Eid, E.M., Galal, T.M., Sewelam, N.A., Talha, N.I., Abdallah, S.M., 2020. Phytoremediation of heavy metals by four aquatic macrophytes and their potential use as contamination indicators: a comparative assessment. *Environmental Science Pollution Research*, 27, pp.12138–12151.

Engelhardt, K.A., Ritchie, M.E., 2002. The effect of aquatic plant species richness on wetland ecosystem processes. *Ecology*, 83(10), pp.2911–2924.

Enyoh, C.E., Verla, A.W., Egejuru, N.J., 2018. pH variations and chemometric assessment of borehole water in Orji, Owerri Imo State, Nigeria. *International Journal of Environmental and Analytical Chemistry*, 5, pp.1–9.

Feng, M., Zhao, Y., Chen, X., Lu, Q., 2020. Biosorption of cadmium and copper from aqueous solutions by *Spirulina platensis*: equilibrium and kinetic studies. *Journal of Ocean University of China*, 19(3), pp.607–613.

Forio, M.A.E., Goethals, P.L., 2020. An integrated approach of multi-community monitoring and assessment of aquatic ecosystems to support sustainable development. *Sustainability*, 12(14), p.5603.

Gall, J.E., Boyd, R.S., Rajakaruna, N., 2015. Transfer of heavy metals through terrestrial food webs: a review. *Environmental Monitoring and Assessment*, 187(4), pp.1–21.

González-Cencerrado, A., Ranz, J.P., Jiménez, M.T.L.F., Gajardo, B.R., 2020. Assessing the environmental benefit of a new fertilizer based on activated biochar applied to cereal crops. *Science of The Total Environment*, 711, pp.134668.

Gutierres, F., Teodoro, A.C., Reis, E., Neto, C., Costa, J.C., 2016. Remote sensing technologies for the assessment of marine and coastal ecosystems. In: *Seafloor Mapping along Continental Shelves: Research and Techniques for Visualizing Benthic Environments*, Finkl, C.W., Makowski, C. (eds.), Springer International Publishing, pp.69–104.

Hassan, S.M., Rahman, R.A., Kamaruddin, R.H. and Madlul, N.S., 2018. Effect of exposure of African catfish (*Clarias batrachus*) to magnetic field on water properties and egg hatching. *Borneo Journal of Marine Science and Aquaculture* (BJoMSA), 2, 54–59.

Havera, S.P. 1999. Waterfowl of Illinois: status and management. Champaign: Illinois Natural History Survey.

Hilborn, R., 2011. Future directions in ecosystem-based fisheries management: a personal perspective. *Fisheries Research*, 108(2–3), pp.235–239.

Hill, W.R., Rinchard, J. and Czesny, S., 2011. Light, nutrients and the fatty acid composition of stream periphyton. *Freshwater Biology*, 56(9), 1825–1836.

Hoegh-Guldberg, O., Jacob, D., Taylor, M., Guo, Y., Bindi, M., Brown, S., and others 2018. "Impacts of 1.5°C of global warming on natural and human systems" in Global Warming of 1.5°C: An IPCC Special Report. Intergovernmental Panel on Climate Change (IPCC). Available at: https://www.ipcc.ch/sr15/ (accessed 30 July 2024).

Hussain, S., Jõudu, I., Bhat, R., 2020. Dietary fiber from underutilized plant resources—A positive approach for valorization of fruit and vegetable wastes. *Sustainability*, 12(13), p.5401.

India Science Wire., 2022. Initiatives to conserve marine resources. Available at URL. https://vigyanprasar.gov.in/isw/Initiatives-to-conserve-marine-resources.html. Accessed on: 06 April 2023.

IPCC 2019. IPCC Special Report on the Ocean and Cryosphere in a Changing Climate. Available at: https://www.ipcc.ch/srocc (accessed 30 July 2024).

IUCN., 2016. Increasing marine protected area coverage for effective marine biodiversity conservation. Available at URL. https://portals.iucn.org/library/sites/library/files/resrecfiles/WCC_2016_RES_050_EN.pdf. Accessed on: 06 April 2023.

Jayan, P.R. and Sathyanathan, N., 2012. Aquatic weed classification, environmental effects and the management technologies for its effective control in Kerala, India. *International Journal of Agricultural and Biological Engineering*, 5(1), 76–91.

Jiang, J.Q., 2018. Occurrence of microplastics and its pollution in the environment: a review. *Sustainable Production and Consumption*, 13, pp.16–23.

Jung, M., Reichstein, M., Ciais, P., Seneviratne, S.I., Sheffield, J., Goulden, M.L., et al., 2010. Recent decline in the global land evapotranspiration trend due to limited moisture supply. *Nature*, 467(7318), pp.951–954.

Kaur, M., Kumar, M., Sachdeva, S., Puri, S.K., 2018. Aquatic weeds as the next-generation feedstock for sustainable bioenergy production. *Bioresource Technology*, 251, pp.390–402.

Keddy, P.A., Fraser, L.H., Solomeshch, A.I., Junk, W.J., Campbell, D.R., Arroyo, M.T., Alho, C.J., 2009. Wet and wonderful: the world's largest wetlands are conservation priorities. *BioScience*, 59(1), pp.39–51.

Khalid, S., Shahid, M., Niazi, N. K., Murtaza, B., Bibi, I., Dumat, C., ... Ali, N., 2017. A comparison of technologies for remediation of heavy metal contaminated soils. *Journal of Geochemical Exploration*, 182, pp.247–268.

Klemas, V., 2011. Remote sensing techniques for studying coastal ecosystems: an overview. *Journal of Coastal Research*, 27(1), pp.2–17.

Kováčik, J., Babula, P., Hedbavny, J., 2017. Comparison of vascular and non-vascular aquatic plant as indicators of cadmium toxicity. *Chemosphere*, 180, pp.86–92.

Kumar, R., Verma, A., Shome, A., Sinha, R., Sinha, S., Jha, P.K., ... Vara Prasad, P.V., 2021. Impacts of plastic pollution on ecosystem services, sustainable development goals, and need to focus on circular economy and policy interventions. *Sustainability*, 13(17), pp.9963.

Kumari, R., Shukla, S.K., Parmar, K., Bordoloi, N., Kumar, A., Saikia, P., 2020. Wetland conservation and restoration for ecosystem services and halt biodiversity loss: an Indian Perspective. In: *Restoration of Wetland Ecosystem: A Trajectory towards a Sustainable Environment*, Upadhyay, A.K., Singh, R., Singh, D.P. (eds.), Chapter 6, Springer Nature Singapore Pte Ltd., pp.75-85. https://doi.org/10.1007/978-981-13-7665-8_6

Langeland, K.A., 1996. Hydrilla verticillata (LF) Royle (Hydrocharitaceae), "the perfect aquatic weed". Castanea, 293–304.

Leach, J., Dawson, H., 2000. Is resistance futile? The battle against *Crassula helmsii. Journal of Practical Ecology and Conservation*, 4(1), pp.7–17.

Lim, M., 2014. Is water different from biodiversity? Governance criteria for the effective management of transboundary resources. *Review of European, Comparative & International Environmental Law*, 23(1), pp.96–110.

Luís, I.P., Spínola, H., 2010. The influence of a voluntary fee in the consumption of plastic bags on supermarkets from Madeira Island (Portugal). *Journal of Environmental Planning and Management*, 53(7), pp.883–889.

Marchand, L., Mench, M., Jacob, D.L., Otte, M.L., 2010. Metal and metalloid removal in constructed wetlands, with emphasis on the importance of plants and standardized measurements: a review. *Environmental Pollution*, 158(12), pp.3447–3461.

Martinho, G., Balaia, N., Pires, A., 2017. The Portuguese plastic carrier bag tax: the effects on consumers' behavior. *Waste Management*, 61, pp.3–12.

Mateo-Sagasta, J., Zadeh, S.M., Turral, H., Burke, J., 2017. Water pollution from agriculture: a global review. Executive summary.

Mcleod, E., Anthony, K.R., Mumby, P.J., Maynard, J.A., Beeden, R., Graham, N.A., Heron, S.F., Hoegh-Guldberg, O., Jupiter, S.D., MacGowan, P., Mangubhai, S., Marshall, N.A., Marshall, P.A., McClanahan, T.R., McLeod, K.L., Nyström, M., Obura, D.O., Parker, B.A., Possingham, H.P., Salm, R.V., Tamelander, J., 2019. The future of resilience-based management in coral reef ecosystems. *Journal of Environmental Management*, 233, pp.291–301.

Mustapha, H.I.A., van Bruggen, J.J., Lens, P.N.L., 2017. Fate of heavy metals in vertical subsurface flow constructed wetlands treating secondary treated petroleum refinery wastewater in Kaduna, Nigeria. *International Journal of Phytoremediation*, 20, pp.44–53.

Muthusaravanan, S., Sivarajasekar, N., Vivek, J.S., Paramasivan, T., Naushad, M., Prakashmaran, J., Gayathri, V. and Al-Duaij, O.K., 2018. Phytoremediation of heavy metals: mechanisms, methods and enhancements. *Environmental chemistry letters*, 16, 1339–1359.

Najila, N., Anila, G., 2022. Heavy metal absorption and phytoremediation capacity of macrophytes of Polachira wetland of Kollam District, Kerala, India. *Research Journal of Chemistry and Environment*, 26(1), pp.90–96.

Nalepa, G. and Clapp, D.W., 2018. Fanconi anaemia and cancer: an intricate relationship. *Nature Reviews Cancer*, 18(3), 168–185.

Nath, K., 1992. National Conservation Strategy and Policy Statement on Environment and Development. Government of India, Ministry of Environment & Forest. Available at URL: https://moef.gov.in/wp-content/uploads/2017/07/introduction-csps.pdf. Accessed on: 15 April 2023.

National Action Plan on Climate Change (NAPCC)., 2008. National Water Mission. Government of India Ministry of Water Resources. Comprehensive Mission Document, vol. 2, New Delhi. Available at URL. https://nwm.gov.in/sites/default/files/nwm28756944786.pdf. Accessed on: 14 April 2023.

National Biodiversity Action Plan., 2014. Ministry of Environment, Forests, and Climate Change, Government of India. Available at URL. https://www.cbd.int/doc/world/in/in-nbsap-v3-en.pdf. Accessed on: 16 April 2023.

National Water Policy (NWP)., 2012. Government of India, Ministry of Water Resources. Available at URL. https://nwm.gov.in/sites/default/files/national%20water%20policy%202012_0.pdf. Accessed on: 15 April 2023.

Njogu, P., Kinyua, R., Muthoni, P., Nemoto, Y., 2015. Biogas production using water hyacinth (*Eichhornia crassipes*) for electricity generation in Kenya. *Energy and Power Engineering*, 7, pp.209–216.

Pakdel, P.M., Peighambardoust, S.J., 2018. A review on acrylic-based hydrogels and their applications in wastewater treatment. *Journal of Environmental Management*, 217, pp.123–143.

Pauly, D. and Zeller, D. 2016. "Catch reconstructions reveal that global marine fisheries catches are higher than reported and declining", Nature Communications, 7, Article 10244. doi: 10.1038/ncomms10244.

Peligro, F.R., Pavlovic, I., Rojas, R. Barriga, C., 2016. Removal of heavy metals from simulated wastewater by in situ formation of layered double hydroxides. *Chemical Engineering Journal*, 306, pp.1035–1040.

Perales-Vela, H.V., Peña-Castro, J.M., Cañizares-Villanueva, R.O., Brown-Bojórquez, F., 2006. Heavy metal detoxification in eukaryotic microalgae. *Chemosphere*, 64(1), pp.1–10.

Pimentel, D., Zuniga, R., Morrison, D., 2005. Update on the environmental and economic costs associated with alien-invasive species in the United States. *Ecological Economics*, 52(3), pp.273–288.

Pirali zefrehei, A.R., Kolahi, M., Fisher, J., 2022. Ecological-environmental challenges and restoration of aquatic ecosystems of the Middle-Eastern. *Scientific Reports*, 12(1), p.17229.

Poloczanska, E.S., Brown, C.J., Sydeman, W.J., Kearney, S., MacKenzie, B.A., Schoeman, D.S., and Moore, P.J. 2016. "Global imprint of climate change on marine life", *Nature Climate Change*, 6(3), 283–290. doi: 10.1038/nclimate2922.

Pratas, J., Paulo, C., Favas, P.J., Venkatachalam, P., 2014. Potential of aquatic plants for phytofiltration of uranium-contaminated waters in laboratory conditions. *Ecological Engineering*, 69, pp.170–176.

Rahel, F.J. and Olden, J.D., 2008. Assessing the effects of climate change on aquatic invasive species. *Conservation biology*, 22(3), 521–533.

Rai, P.K., 2019. Heavy metals/metalloids remediation from wastewater using free-floating macrophytes of a natural wetland. *Environmental Technology & Innovation*, 15, pp.100393.

Rajkumar, R., Yaakob, Z., Takriff, M.S., 2013. Potential of the micro and macro algae for biofuel production: A brief review. *Bioresources*, 9(1), pp.1606–1633.

Rawat, I., Kumar, R.R., Mutanda, T. and Bux, F., 2013. Biodiesel from microalgae: a critical evaluation from laboratory to large scale production. *Applied Energy*, 103, 444–467.

Rezaie, A.M., Loerzel, J., Ferreira, C.M., 2020. Valuing natural habitats for enhancing coastal resilience: wetlands reduce property damage from storm surge and sea level rise. *PLoS One*, 15(1), p.e0226275.

Ricciardi, A., Hoopes, M.F., Marchetti, M.P. and Lockwood, J.L. 2013. "Progress toward understanding the ecological impacts of non-native invasive species", *Ecological Applications*, 23(3), 567–572. doi: 10.1890/12-1411.1.

Riley, S., 2009. Preventing transboundary harm from invasive alien species. *Review of European Community & International Environmental Law*, 18(2), pp.198–210.

Rizzo, A., Bresciani, R., Martinuzzi, N., Masi, F., 2020. Online monitoring of a long-term full-scale constructed wetland for the treatment of winery wastewater in Italy. *Applied Sciences*, 10(2), p.555.

Rockwell, W.H., 2003. Summary of a survey of the literature on the economic impact of aquatic weeds. Report to the Aquatic Ecosystem Restoration Foundation, 18 p. http://www.aquatics.org/pubs/economic_impact.pdf.

Roy, U.K., Radu, T. and Wagner, J.L., 2021. Carbon-negative biomethane fuel production: Integrating anaerobic digestion with algae-assisted biogas purification and hydrothermal carbonisation of digestate. *Biomass and Bioenergy*, 148, p.106029.

Royal Netherlands Institute for Sea Research, 2020. River plants counter both flooding and drought to protect biodiversity. *ScienceDaily*. Retrieved April 16, 2023, from www.sciencedaily.com/releases/2020/07/200717101013.htm

Saeed, T., Muntaha, S., Rashid, M., Sun, G., Hasnat, A., 2018. Industrial wastewater treatment in constructed wetlands packed with construction materials and agricultural by-products. *Journal of Cleaner Production*, 189, pp.442–453.

Saha, M., Sarkar, S., Sarkar, B., Sharma, B.K., Bhattacharjee, S. and Tribedi, P.,2016. Microbial siderophores and their potential applications: a review. *Environmental Science and Pollution Research*, 23, 3984–3999.

Saha, P., Shinde, O., Sarkar, S., 2017. Phytoremediation of industrial mines wastewater using water hyacinth. *International Journal of Phytoremediation*, 19(1), pp.87–96.

Searles, A.R., Gipson, E.E., Walters, L.J., Cook, G.S., 2022. Oyster reef restoration facilitates the recovery of macroinvertebrate abundance, diversity, and composition in estuarine communities. *Scientific Reports*, 12(1), pp.1–15.

Sen Gupta, S., Chatterjee, S., Hazra, S., 2011. Biosorption of heavy metals by cyanobacteria. *Indian Journal of Microbiology*, 51(4), pp.438–450.

Singh, C.K., Sodhi, K.K. and Singh, D.K., 2024. Understanding the bacterial community structure associated with the Eichhornia crassipes rootzone. *Molecular Biology Reports*, 51(1), p.35.

Smith, K.L., Huffard, C.L., Sherman, A.D. and Ruhl, H.A., 2016. Decadal change in sediment community oxygen consumption in the abyssal northeast Pacific. *Aquatic geochemistry*, 22, 401–417.

Space Applications Centre (SAC)., 2011. Space Applications Centre (SAC) National Wetland Atlas SAC Indian Space Research Organisation, Ahmedabad.

Staudinger, M.D., Lynch, A.J., Gaichas, S.K., Fox, M.G., Gibson-Reinemer, D., Langan, J.A., Teffer, A.K., Thackeray, S.J., Winfield, I.J. 2021. How does climate change affect emergent properties of aquatic ecosystems? *Fisheries*, 46(9), pp.423–441.

Swapna, M.M., Prakashkumar, R., Anoop, K.P., Manju, C.N., Rajith, N.P., 2011. A review on the medicinal and edible aspects of aquatic and wetland plants of India. *Journal of Medicinal Plants Research*, 5, pp.7163–7176.

Tan, L., Zhong, J., Jin, Y.L., Sun, Z.Y., Tang, Y.Q. and Kida, K., 2020. Production of bioethanol from unwashed-pretreated rapeseed straw at high solid loading. *Bioresource Technology*, 303, p.122949.

Tate, W.B., Allen, M.S., Myers, R.A., Nagid, E.J. and Estes, J.R., 2003. Relation of age-0 largemouth bass abundance to hydrilla coverage and water level at Lochloosa and Orange Lakes, Florida. *North American Journal of Fisheries Management*, 23(1), 251–257.

Törnqvist, T.E., Cahoon, D.R., Morris, J.T., Day, J.W., 2021. Coastal wetland resilience, accelerated sea-level rise, and the importance of timescale. *AGU Advances*, 2(1), p.e2020AV000334.

van Loon, A.F., Gleeson, T., Clark, J., van Dijk, A.I.J.M., Stahl, K., Hannaford, J., et al., 2016. Drought in the anthropocene. *Nature Geoscience*, 9(2), pp.89–91.

Verla, A.W., Verla, E.N., Amaobi, C.E., Enyoh, C.E., 2018. Water pollution scenario at River Uramurukwa flowing through Owerri Metropolis, Imo State, Nigeria. *International Journal of Scientific Research*, 3, pp.40–46.

Vorosmarty, C.J., Osuna, V.R., Cak, A.D., Bhaduri, A., Bunn, S.E., Corsi, F., ... Uhlenbrook, S., 2018. Ecosystem-based water security and the Sustainable Development Goals (SDGs). *Ecohydrology & Hydrobiology*, 18(4), pp.317–333.

Vymazal, J., Březinová, T., 2015. The use of constructed wetlands for removal of pesticides from agricultural runoff and drainage: a review. *Environment International*, 75, pp.11–20.

Walker, T.R., 2021. Micro plastics and the UN sustainable development goals. *Current Opinion in Green and Sustainable Chemistry*, 30, pp.100497.

Wang, X., Yang, W., 2019. Water quality monitoring and evaluation using remote sensing techniques in China: a systematic review. *Ecosystem Health and Sustainability*, 5(1), pp.47–56.

Wang, Y., Meng, D., Fei, L., Dong, Q., Wang, Z., 2019. A novel phytoextraction strategy based on harvesting the dead leaves: cadmium distribution and chelator regulations among leaves of tall fescue. *Science of the Total Environment*, 650, pp.3041–3047.

Wersal, R.M., Madsen, J.D., 2012. Aquatic plants, their uses and risks: a review of the global status of aquatic plants. FAO: Rome, Italy, p. 97.

Yang, L., Zhang, Y., Kang, S., Wang, Z. and Wu, C., 2021. Microplastics in soil: A review on methods, occurrence, sources, and potential risk. *Science of the Total Environment*, 780, p.146546.

Zari, M.P., Kiddle, G.L., Blaschke, P., Gawler, S., Loubser, D., 2019. Utilising nature-based solutions to increase resilience in Pacific Ocean Cities. *Ecosystem Services*, 38, p.100968.

Zhang, P., Kong, X., Bakker, E. S., Xu, J., & Zhang, M. 2021. Temperature affects carbon and nitrogen-stable isotopic signatures of aquatic plants. *Aquatic Sciences*, 83, 1–11.

Zhu, C., Tian, H., Cheng, K., Liu, K., Wang, K., Hua, S., Gao, J., Zhou, J., 2016. Potentials of whole process control of heavy metals emissions from coal-fired power plants in China. *Journal of Cleaner Production*, 114, pp.343–351.

Zielinski, O., Busch, J.A., Cembella, A.D., Daly, K.L., Engelbrektsson, J., Hannides, A.K., Schmidt, H., 2009. Detecting marine hazardous substances and organisms: sensors for pollutants, toxins, and pathogens. *Ocean Science*, 5(3), pp.329–349.

Zielinski, O., Hengstermann, T., Robbe, N., 2006. Detection of oil spills by airborne sensors. In: *Marine Surface Films: Chemical Characteristics, Influence on Air-Sea Interactions and Remote Sensing*, Gade, M., Hühnerfuss, H., Korenowski, G.M. (eds.), Springer, Berlin, Heidelberg, pp.255–271.

14 Potential Fishing Zones Estimation along the Coast of India Using Geospatial Technology

*Md. Wasim, Smrutisikha Mohanty, Prem C. Pandey**
Department of Life Sciences, School of Natural Sciences, Shiv Nadar Institution of Eminence, Greater Noida, Uttar Pradesh, India

Arvind Chandra Pandey
Department of Geoinformatics, School of Natural Resource Management, Central University of Jharkhand, Brambe, Ranchi, Jharkhand, India

14.1 INTRODUCTION

Over 30 million people live along the Indian coastline, and their lives are linked to the ocean in one way or another. Coastal fishing is a significant part of India's national economy. Nearly seven million people live directly on the coast over 8100 km, dependent on fishing for their livelihood (Chakraborty et al., 2019). Based on the oceanographic conditions of the water, there is the possibility to predict the abundance of fish in that area (Mursyidin et al., 2015). The most important parameters used to determine fish availability and the abundance of pelagic species are changes in sea surface chlorophyll-a (Chl-a) concentrations and sea surface temperature (SST) (Fitrianah et al., 2016; Karuppasamy et al., 2020) as well as help in predicting the potential fishing zone (PFZ) (Guidetti et al., 2010; Nammalwar et al., 2013; Fitrianah et al., 2015). The concentration of surface chlorophyll is a proxy for the biomass of phytoplankton and is used to determine the habitat of fish stocks (O'Reilly et al., 1998; Dan Teknik et al., 2015; Karuppasamy et al., 2020). PFZ is the zone where fish congregate due to large amounts of food. The availability of food is greater in this zone compared to the open sea, due to optimal SST and the high concentration of surface Chl-a occurrence at a given time (Giri et al., 2016; Harahap et al., 2020). PFZs offer several advantages for fishermen, such as reducing search time and saving valuable fuel and human effort, thus improving the profitability and socio-economic status of fishermen (Chakraborty et al., 2019).

Geospatial technology allows the monitoring of changes in marine ecosystems around the world (Nurdin et al., 2013) because it provides a means to view and evaluate physical and biological variability, as well as to quantify them visually and systematically throughout the region (O'Reilly et al., 1998). In recent decades, geospatial technology has played a dominant role in analyzing Chl-a and SST to estimate PFZs (Devi et al., 2015; Harahap et al., 2020; Suhadha and Asriningrum, 2020). The remotely sensed SST and Chl-a data can be retrieved respectively from NOAA-AVHRR, Oceansat-2, MODIS products, and the ESSO-Indian National Center for Ocean Information Services (INCOIS) for forecasting and mapping fishing zones (Xingwei and Baide, 1988; Solanki et al., 2005a; Solanki et al., 2005b; Nair et al., 2022). INCOIS, Hyderabad, India, provides PFZ advisories all along India's coastline, based on the fact that ocean processes that enhance biological production

* Corresponding author.

 DOI: 10.1201/9781003354000-19

leave traces on ocean surface parameters that can be monitored by satellite remote sensing techniques and used to predict fish availability (https://incois.gov.in/MarineFisheries/PfzAdvisory). The advisories provide a georeferenced map highlighting areas likely to contain large schools of fish.

Initially, methods of identifying PFZ by satellite data were developed by detecting the SST gradients through ocean features like fronts, eddying, and upwelling (Lasker et al., 1981; Laurs et al., 1984; Maul et al., 1984; Xingwei and Baide, 1988). Setiawan et al. (2022) used the Moderate Resolution Imaging Spectroradiometer (MODIS)-based SST and Chl-a products to map the distribution pattern of PFZ based on seasonal variability in the Bangka Strait waters using remote sensing and geographical information system (GIS) techniques (Setiawan et al., 2022). The most PFZ points were found in transition season I (March–May) (636 PFZ points), whereas the fewest PFZ points were found in transition season II (September–November) (total of 219 PFZ points). Chakraborty et al. (2019) used the AVHRR SST and MODIS-AQUA Chl-a products for modeling marine ecosystems on a regional scale for short-term prediction of satellite-aided operational fishery advisories. The average length of the potential fishing zone observed off Gujarat is 27.80 $\pm$ 7.2 km (33.07 $\pm$ 3.2 km), whereas off Andhra Pradesh, it is 28.27 $\pm$ 10.9 km (52.48 $\pm$ 8.7 km) from 2010–2016 using satellite model data. Karuppasamy et al. (2020) used the MODIS-AQUA SST and Chl-a products to identify the PFZ zone for the Thoothukudi coast (Bay of Bengal) from 2013 to 2017, for the month of June. PFZ was classified into three classes: High PFZ, Medium PFZ, and Low PFZ, and the highest PFZ zone was found for 2015 in Sippikulam and Vembar, whereas low PFZ was found for 2017. There is an urgent need to disseminate PFZ data to fishermen for the actual area for fish harvesting to improve the economy and thus benefit the fishing community.

PFZ estimation using geospatial technologies requires the collection of imagery to identify regions with suitable conditions for fishing activities in and around coastal zones. Several factors play a crucial role in PFZ estimations, such as sea surface temperature, Chl-a, ocean color and turbidity, sea surface height (SSH), sea surface winds and currents along with seasonal and interannual variability in the region. Moreover, considering these factors and applying machine-learning methods, data mining on remotely sensed data is one of the foremost requirements for PFZ estimation in the coastal region. Apart from these, vessel monitoring is taken into account for their movement and harvesting from the most suitable places for fish using their habitat estimation. Remotely sensed data are employed to map marine and coastal habitats such as seagrass beds, coral reefs, mangroves, and rocky shores to understand the movement of fish. The aforementioned habitats serve as important breeding, nursery, and feeding grounds for many fish species, making them key fishing potential zones, where SST, SSH, Chl-a, colored dissolved organic matter (CDOM), turbidity, and ocean currents are considered and PFZs are estimated.

The present study aims to demonstrate the effectiveness of remote sensing technology in identifying PFZs. MODIS-AQUA-based SST and Chl-a products have been utilized in the study to identify PFZ along the Indian coastline for three different seasons, i.e., summer season (April–June 2022), rainy season (July–September 2022), and winter season (November 2022–January 2023).

14.2 STUDY AREA

The study area chosen for the present study includes the coastline of India (see Figure 14.1). Coastal areas in India refer to a region along the country's extensive coastline stretching more than 7517 km (4671 miles). The coastline of India is located along the Bay of Bengal (BOB), the Arabian Sea of India, and the Indian Ocean. The Indian mainland coast includes nine coastal states and two union territories having 66 coastal districts. According to the CPDAC (2017), the morphology of the coast has 97 significant estuaries, 34 lagoons, 11% rocky coastline, 36% muddy flats, and 43% sandy beaches. Thirteen major ports, 46 fishing harbors, and 187 minor ports are located in the area. The coastal regions of the Indian region are also an important economic hub and have a large number of ports and fishing industries. The eastern coastal plain of India is a 100–130 km-wide stretch of land lying between the Eastern Ghat Ranges and the Bay of Bengal, whereas the western coastal

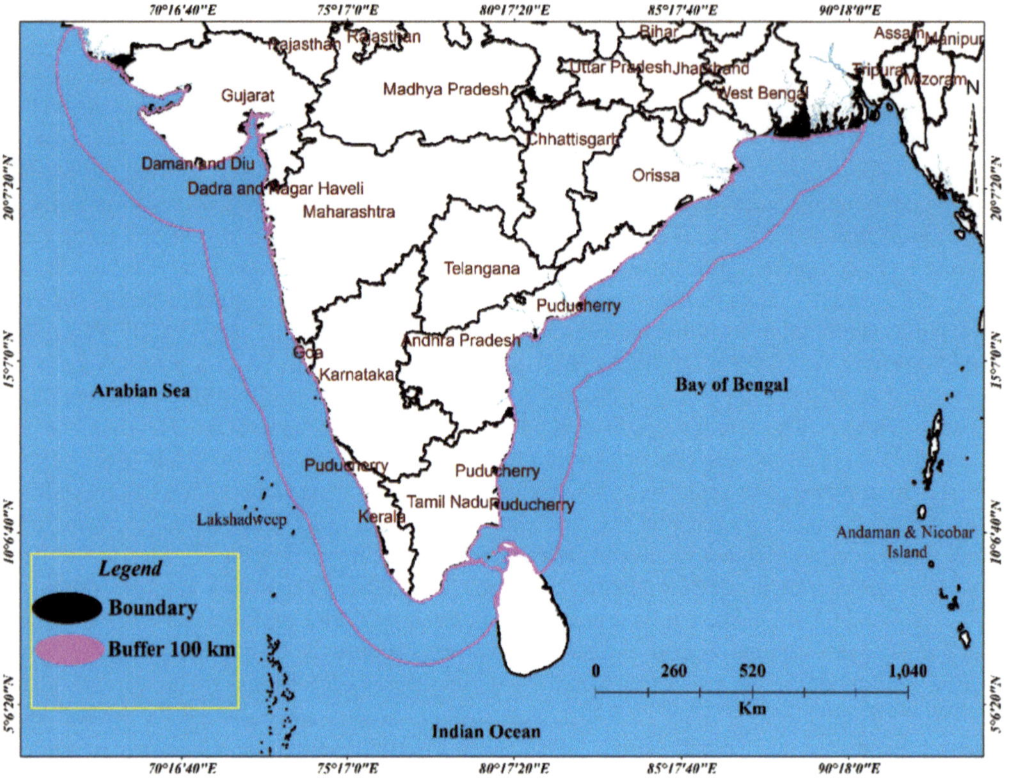

FIGURE 14.1 Map showing the study area of the present study.

lowland is a relatively narrow strip of land stretching 50–100 km between the Western Ghat and the Arabian Sea (Vaidyanadhan, 1991). According to the INCOIS-PFZ-advisory (2023), a total of 432 PFZ advisories and 586 fish landing centers have been identified along the Indian coast. The major problem with the fishing zone is the changing variation due to the inflow of river water, temperature variation, and overall production of phytoplankton. These factors control the abundance of fish in the coastal zones of any country. Moreover, for optimal SST, a higher concentration of Chl-a is required for the breeding and flourishing of fish.

14.3 MATERIALS AND METHODS

In the present study, remotely sensed MODIS-AQUA SST and Chl-a level-3 products were downloaded from the website (https://oceancolor.gsfc.nasa.gov/data/aqua) and validated from the INCOIS-PFZ-advisory (2023). The data for marine fish landings in India for the year 2022 was downloaded from FCSA (2022). The selected dataset along with the image acquisition dates and resolution are listed in Table 14.1. The MODIS-AQUA remotely sensed SST (MODIS_L3m_SST), Chl-a (MODIS_L3m_CHL) product, was projected using the reprojection tool in the SeaDAS 7.5.3 Software and was exported to GeoTIFF format. A 100 km buffer along the coastline of India has been created using the Buffer (Analysis) tool in Arc GIS 10.3 software to observe the variation in SST, Chl-a and to identify PFZ. The average SST and Chl-a data for three different seasons, i.e., summer season (April–June 2022), rainy season (July–September 2022), and winter season (November 2022–January 2023) were processed using the raster calculator tool, and outputs were clipped from the Area of Interest (AOI) using the extract by mask tool in Arc GIS 10.3 software. SST and Chl-a products were reclassified using Reclassify tools (Spatial Analyst), and each factor

TABLE 14.1
Data Used for Potential Fishing Zone Study

Data	Spatial resolution	Time period	Purpose	Source
AQUA-MODIS (SST product)	4 km *4 km	Summer season (April–June 2022); rainy season (July–September 2022); winter season (November 2022–January 2023)	Mapping of sea surface temperature variation along the Indian coastline	NASA (https://oceancolor.gsfc.nasa.gov/l3)
MODIS-AQUA (Chl-a product)	4 km *4 km	Summer season (April–June 2022); rainy season (July–September 2022); winter season (November 2022–January 2023)	Mapping of chlorophyll-a variation along the Indian coastline	NASA (https://oceancolor.gsfc.nasa.gov/l3/)

was given a weight of 50% by using the weighted sum tool in Arc GIS 10.3. The weighted sum was performed to obtain PFZ. With the usability of the raster to multipoint tool in Arc GIS 10.3 software, the raster image of SST and Chl-a was converted to points to observe the correlation between these two parameters, and the correlation was plotted in R-studio. The flowchart of the methodology adopted for the present study is shown in Figure 14.2.

14.4 RESULTS AND DISCUSSIONS

This section deals with the outcomes of the present study to explore the PFZs along the coastal zones of India. It outlines the estimation of SST, Chl-a map, and observes the correlation between SST and chlorophyll-a for three different seasons. The study aims to analyze the optimal SST with a high Chl-a concentration that leads to a good fisheries zone. A high Chl-a concentration indicates more phytoplankton growth, which is the major food source for fish.

14.4.1 Sea Surface Temperature

Mapping of SST (see Figure 14.3) has been done along the coastline of India with a buffer area of 100 km^2 to observe the variation in the SST during different seasons. The highest SST (~31.44 °C) was observed along the southeastern coast of Tamil Nadu during the summer season (April–June, 2022), and >28.52 °C to <31.44 °C was observed along the coast of Karnataka, Goa, and Maharashtra. A trend of continuous variation in temperature (increasing–decreasing) was observed along the coast of Tamil Nadu, Andhra Pradesh, and Puducherry, whereas lower SST (~24.23 °C) was observed along the coast of Odisha, West Bengal, Gulf of Kutch, and Gulf of Khambhat (Gujarat) and the southern coast, compared to other regions, showing temperature range from ~28.52 °C to ~31.44 °C during the summer season. High SST >28.52 °C to 31.44 °C was observed along the eastern coast of India, including Tamil Nadu, Andhra Pradesh, and Puducherry, Odisha, and West Bengal, whereas low SST (24.23 °C to <28.52 °C) was seen along the coast of Kerala, Karnataka, Goa, Maharashtra, and the southern coast of India. However, SST variation (moderate to low) was observed along

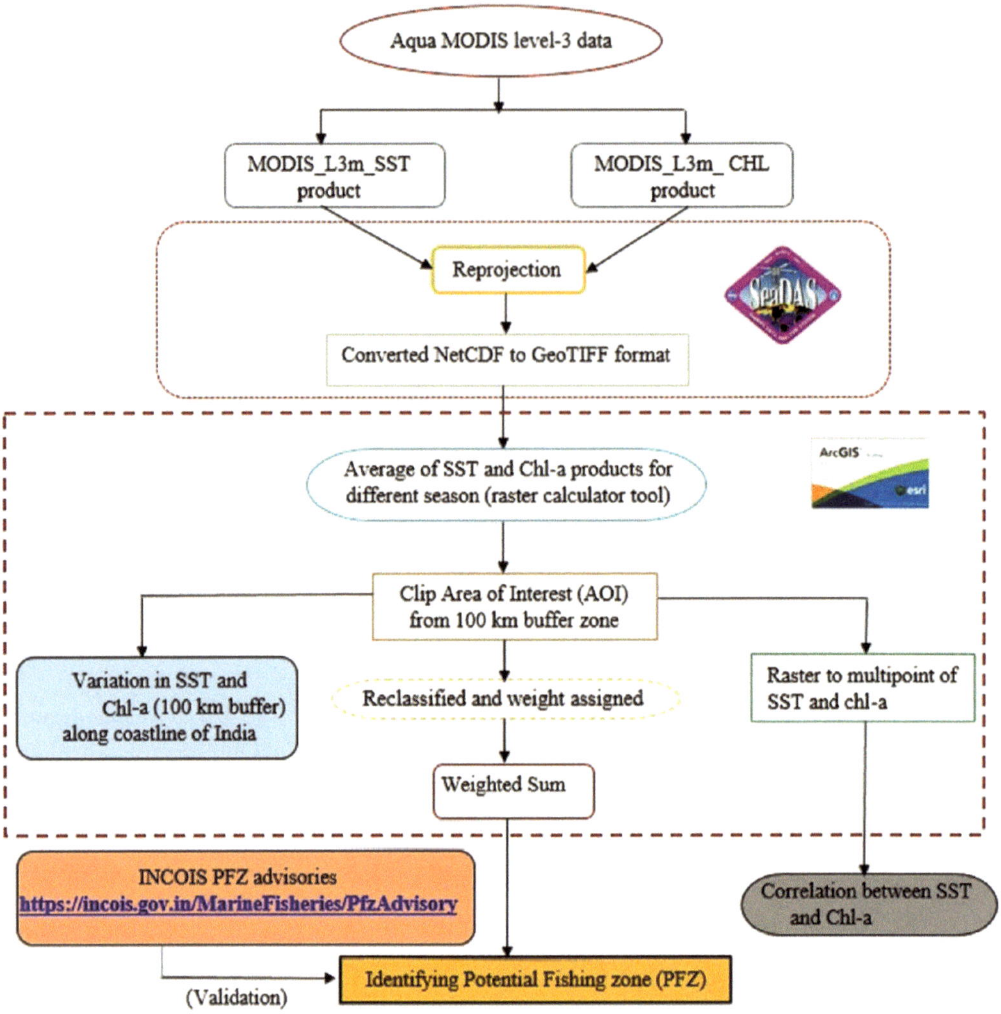

FIGURE 14.2 Flowchart of the methodology adopted for the present study.

the Gulf of Kutch and the Gulf of Khambhat (Gujarat) during the rainy season (July–September). During the winter season (November 2022–January 2023), high SST (>28.52 °C to 31.44 °C) was seen along the coast of Kerala, Karnataka, Goa, Maharashtra, and the southern coast of India, whereas continuous variation in temperature (increasing–decreasing) was observed along the coast of Tamil Nadu, Andhra Pradesh, and Puducherry. However, low SST (~24.24 °C) was observed along the coast of Odisha, West Bengal, the Gulf of Kutch, and the Gulf of Khambhat (Gujarat) during the winter season.

14.4.2 Chlorophyll-a

Chlorophyll-a mapping (see Figure 14.4) has been done along the coastline of India with a buffer area of 100 km^2 to observe the variation in the concentration of Chl-a during different seasons. The concentration of surface chlorophyll is a proxy for the biomass of phytoplankton and is used to determine the habitat of fish stocks (O'Reilly et al., 1998; Nurdin et al., 2013; Dan Teknik et al.,

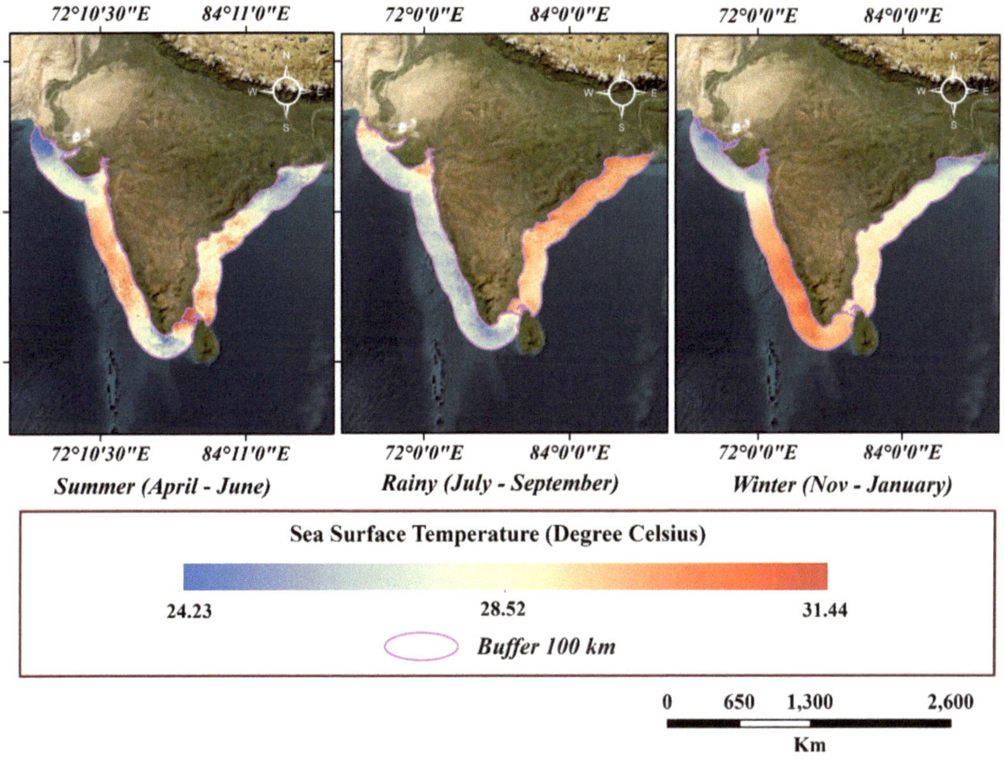

FIGURE 14.3 Spatial variation map of SST during the summer, rainy, and winter seasons.

2015; Karuppasamy et al., 2020). High concentrations of Chl-a (20.5 mg m^{-3}) were seen along the coast of Kerala, the Gulf of Khambhat (Gujarat), and the southern coast of India, while moderate to high concentrations (6.26 mg m^{-3} to 20.5 mg m^{-3}) of Chl-a were observed along the coastline of Odisha, West Bengal, and some parts of Andhra Pradesh, Tamil Nadu, and the Gulf of Kutch (Gujarat). However, a low concentration of Chl-a (~0.167 mg m^{-3}) was observed in the periphery of the eastern and western coasts of India and some parts of Tamil Nadu during the summer season (April–June 2022). In the rainy season, a variation in Chl-a concentration from 6.35 mg m^{-3} to 20.5 mg m^{-3} was observed along the western coastline of India, including Kerala, Karnataka, Goa, Maharashtra, and the Gulf of Khambhat (Gujarat). A moderate concentration (0.167 to ~6.35 mg m^{-3}) was reported in the Gulf of Kutch, some parts of Tamil Nadu and Andhra Pradesh, and a minute portion of Odisha and West Bengal. The periphery of the eastern and western coasts of India showed a low concentration of Chl-a (0.167 mg m^{-3}) in the months of July–September. During the winter season, high concentrations of Chl-a (6.35 mg m^{-3} to 20.5 mg m^{-3}) were observed along the coastline of the Gulf of Khambhat, Gulf of Kutch, Gujarat, Maharashtra, Karnataka, Goa, Odisha, West Bengal, and some parts of Kerala, whereas the coastline of Tamil Nadu and Andhra Pradesh showed moderate concentrations (~6.35 mg m^{-3}) of chlorophyll-a. However, the periphery of the eastern, western, and southern coasts showed a low concentration (0.167 mg m^{-3}) of Chl-a from November 2022 to January 2023.

14.4.3 Correlation between SST and Chlorophyll-a

The relationship between SST and Chl-a (see Figure 14.5) was calculated for three different seasons, , summer season (Apri–lJune, 2022), rainy season (July–September, 2022), and winter season

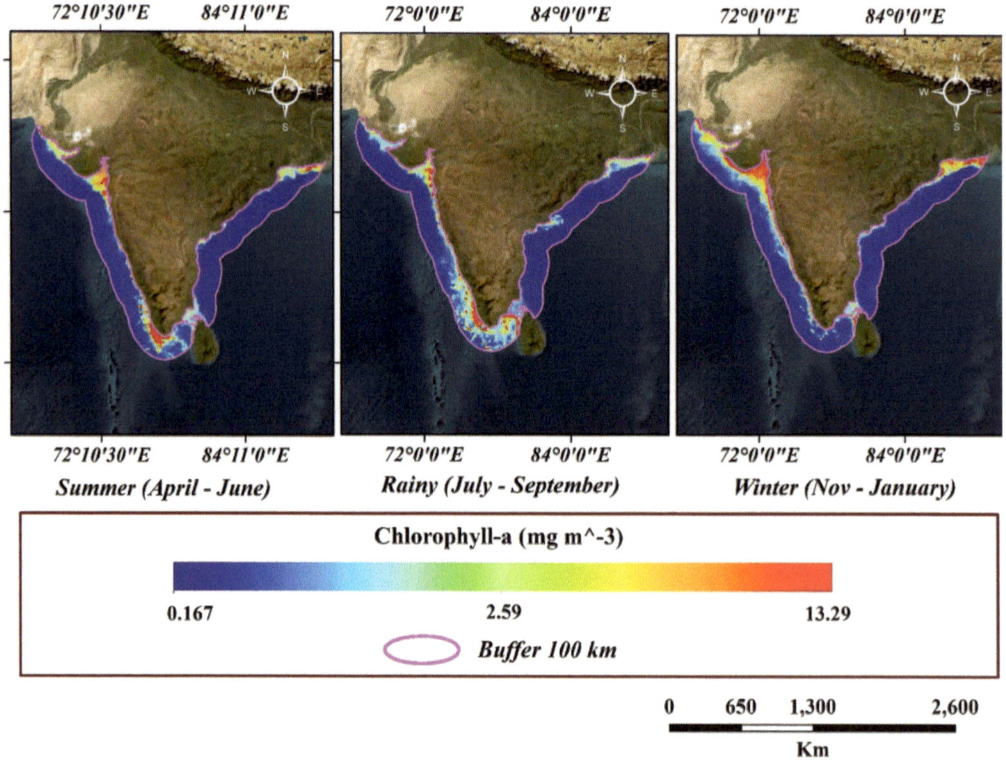

FIGURE 14.4 Spatial variation map of chlorophyll-a during the summer, rainy, and winter seasons.

(November 2022–January 2023) (see Figure 14.5a, b, c). We found that between SST and Chl-a there was a negative correlation (inversely), with the coefficient of correlation at ($R^2 = 0.401$, $P < 0.001$) for the summer season, ($R^2 = 0.302$, $P < 0.001$) for the rainy season, and ($R^2 = 0.615$, $P < 0.001$) for the winter season, which means if SST increases, the Chl-a concentrations will decrease, and vice versa (Solanki et al., 2001; Yoder et al., 2002; Navarro and Ruiz, 2006; Radiarta and Saitoh, 2008). These phenomena are due to the upwelling process. In the event of upwelling, the bottom layer of cold, nutrient-rich water and rich nutrients will rise to the surface to encourage phytoplankton growth (Zainuddin, 2007; Sukresno et al., 2008). Results indicate that optimal SST and high Chl-a concentrations are required for determining the PFZ. Nurdin et al. (2013) reported that SST contributes approximately 29.4% to Chl-a variation. However, about 71.6% of the Chl-a variation is influenced by other factors, which may include rainfall, wind, current, eddy, and salinity.

14.4.4 Potential Fishing Zone Mapping

The most important parameters used to determine fish availability and the abundance of pelagic species are changes in Chl-a concentrations and SST (Fitrianah et al., 2016; Karuppasamy et al., 2020), which help to predict the PFZ (Guidetti et al., 2010; Nammalwar et al., 2013; Fitrianah et al., 2015). High PFZ is categorized in the area having SST ranges between 24°C and 27°C, whereas SST ranges between 27°C and 30°C are categorized as moderate PFZ. However, low PFZ is considered in an area having a temperature range of less than 24°C and more than 30°C (Muhammad et al., 2022). The PFZ (see Figure 14.6) was estimated based on satellite-derived data for SST and Chl-a along the coastline of India with a buffer area of 100 km^2 during different seasons. During the summer season, high PFZ was observed along the southern part of Kerala and Tamil Nadu, Odisha

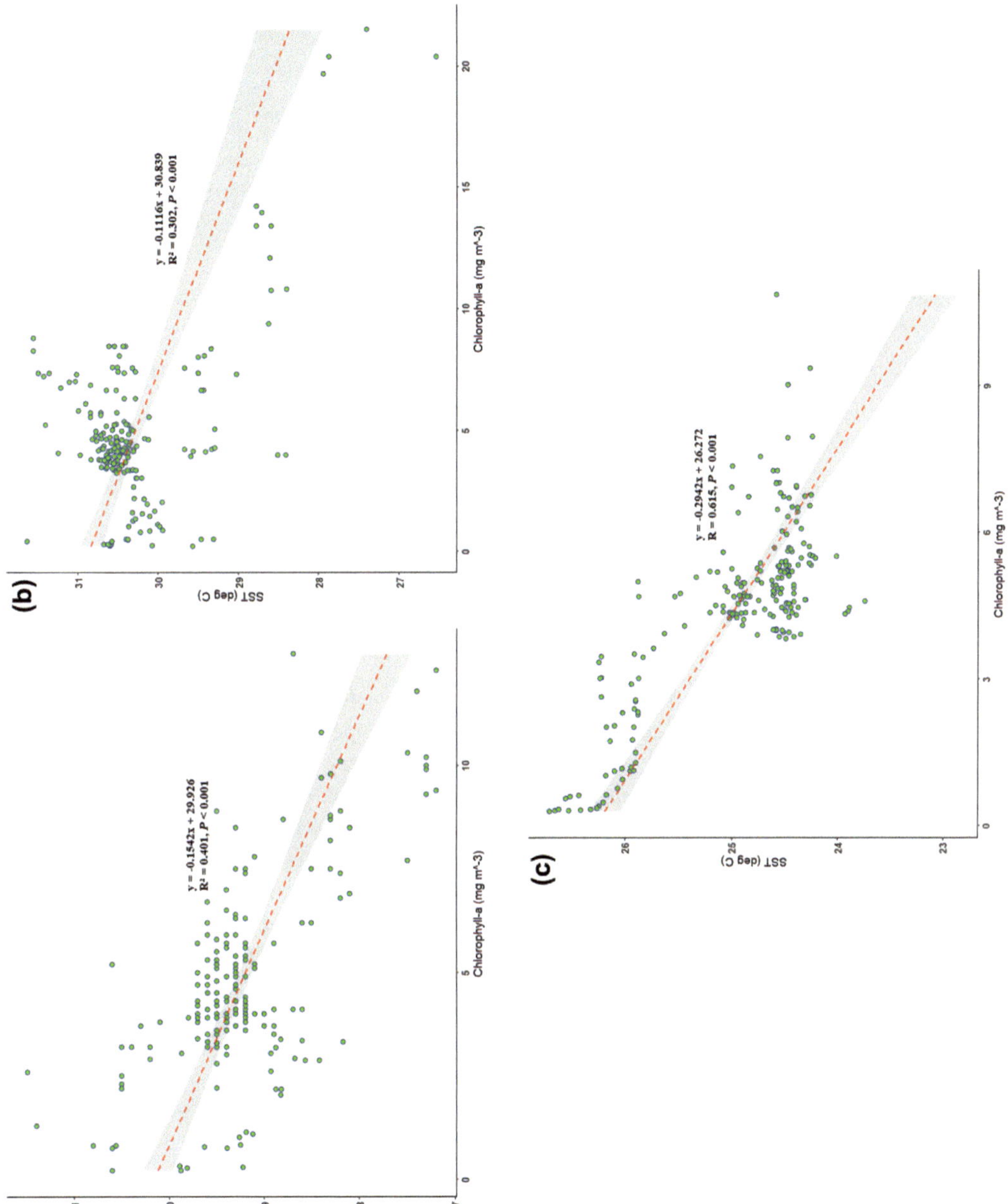

FIGURE 14.5 Correlation between sea surface temperature and chlorophyll-a during (a) the summer, (b) rainy, and (c) winter seasons.

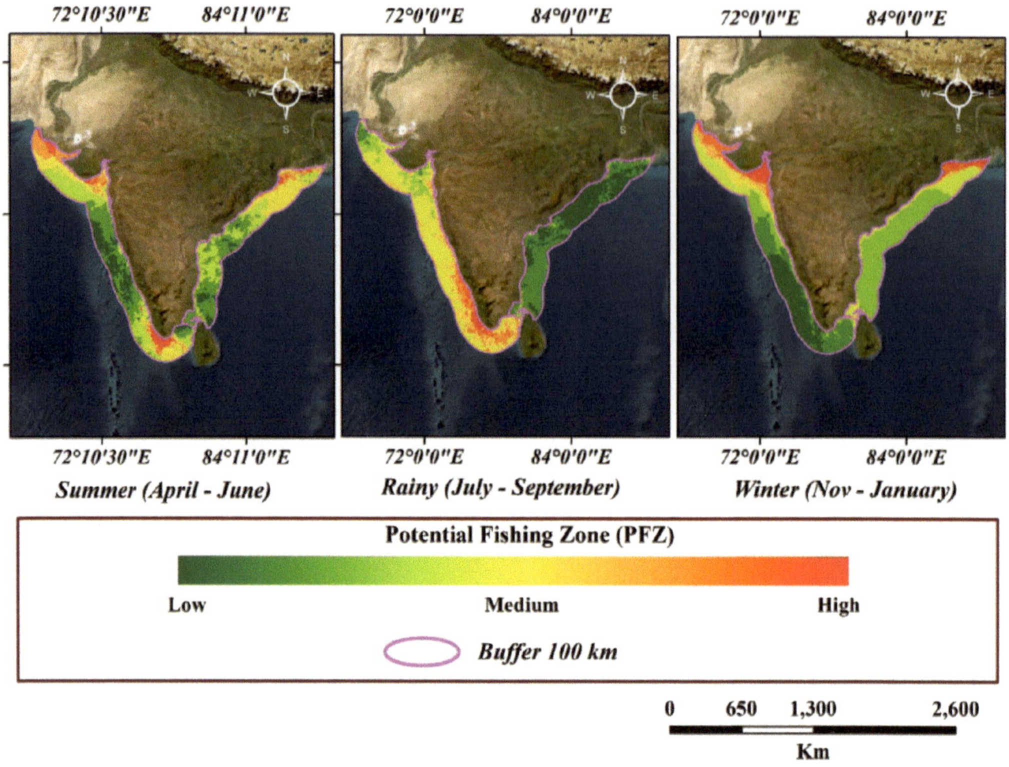

FIGURE 14.6 Probable potential fishing zone map generated in the present study along the Indian coastline.

(Puri, Kendrapada, Bhadrak), and West Bengal (south and north 24 Parganas), whereas a moderate zone was seen in the Gulf of Khambhat, the Gulf of Kutch, the periphery of the southern coast, and the coastline of Odisha and West Bengal. The remaining periphery of the eastern and western coasts showed low PFZ. In the rainy season, the coastline of Kerala, Goa, and parts of Karnataka, including Dakshina Kannada, showed high PFZ, whereas moderate PFZ was observed along the coastline of Karnataka, Goa, Maharashtra, Gujarat, the Gulf of Khambhat, and the southern part of Tamil Nadu. However, the periphery of the western coast showed low-to-moderate variation in PFZ, and low PFZ was observed in the Gulf of Kutch and the eastern coast of India. The coast of West Bengal, Odisha, including Balasore, Bhadrak, Kendrapara, and Jagatsinghpur, and the Gulf of Khambhat showed high PFZ during the winter season, whereas high-to-moderate variation was observed in the Gulf of Kutch. The moderate zone comprised the coast of Maharashtra, Goa, and Tamil Nadu, including Ramanathapuram, Pudukkottai, Thanjavur, Nagapattinam, and parts of Odisha, including Puri and Ganjam. The periphery of the eastern, western, and southern coasts of India showed low PFZ. The results of the present study were found to be similar to web maps developed by INCOIS (INCOIS-PFZ-advisory, 2023).

14.4.5 Marine Fish Landing Sites vs Fishing Potential Zones

Since its inception in 1947, the Central Marine Fisheries Research Institute of the Indian Council of Agricultural Research (ICAR) has been assigned different tasks, from the collection of scientific data to the estimation of marine fish landing sites, and has provided information for healthy as well as diverse fish stock monitoring (CMFRI-ICAR, 2023). This information assists planners in making conclusive decisions for policy by looking at the harvest patterns. Figure 14.7a, b, c represents

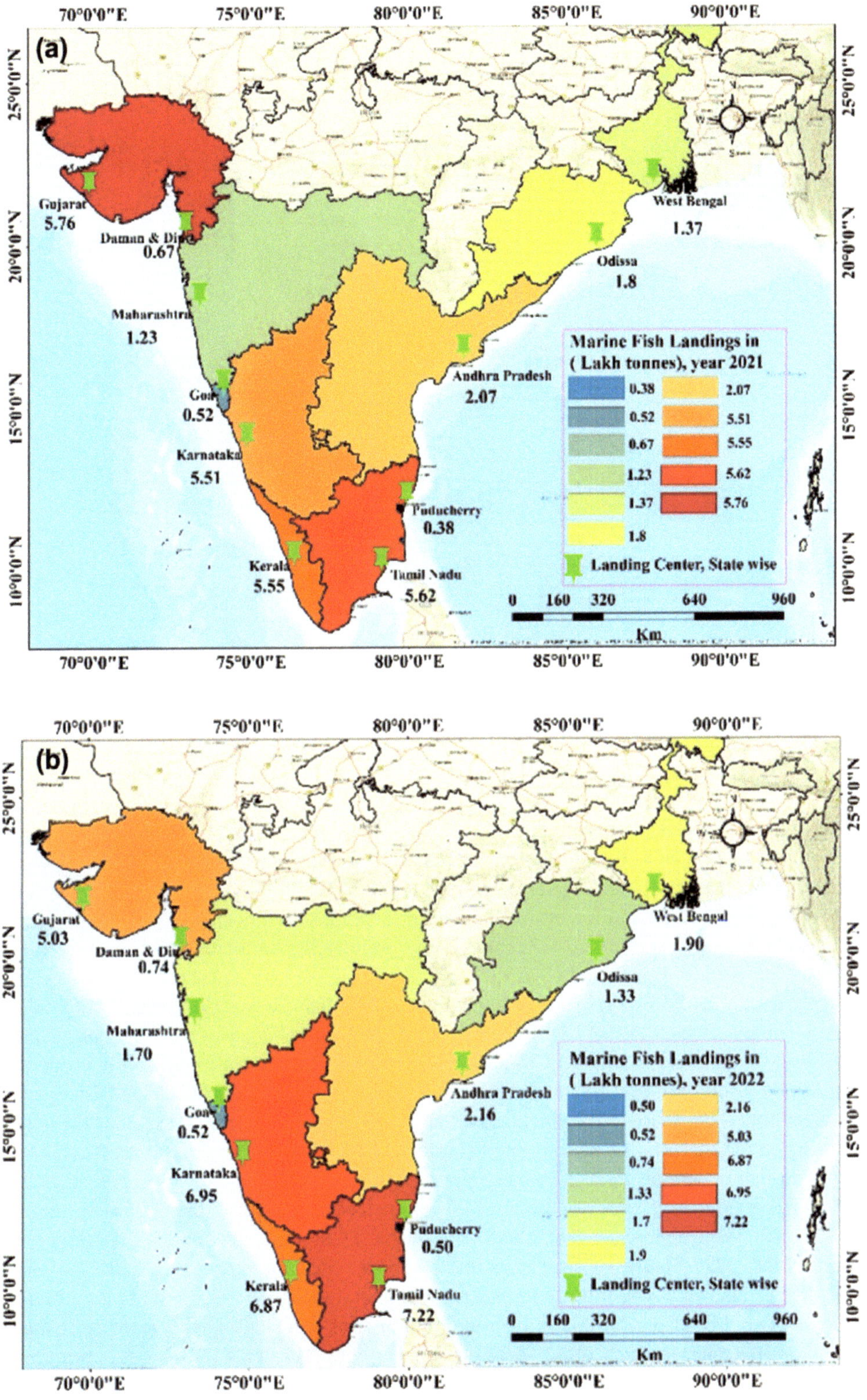

FIGURE 14.7 Map showing the marine fish landings (lakhs tonnes) information in India (a) for the year 2021, (b) for the year 2022, and (c) change in marine fish landings (lakh tonnes) in India from year 2021 to 2022.

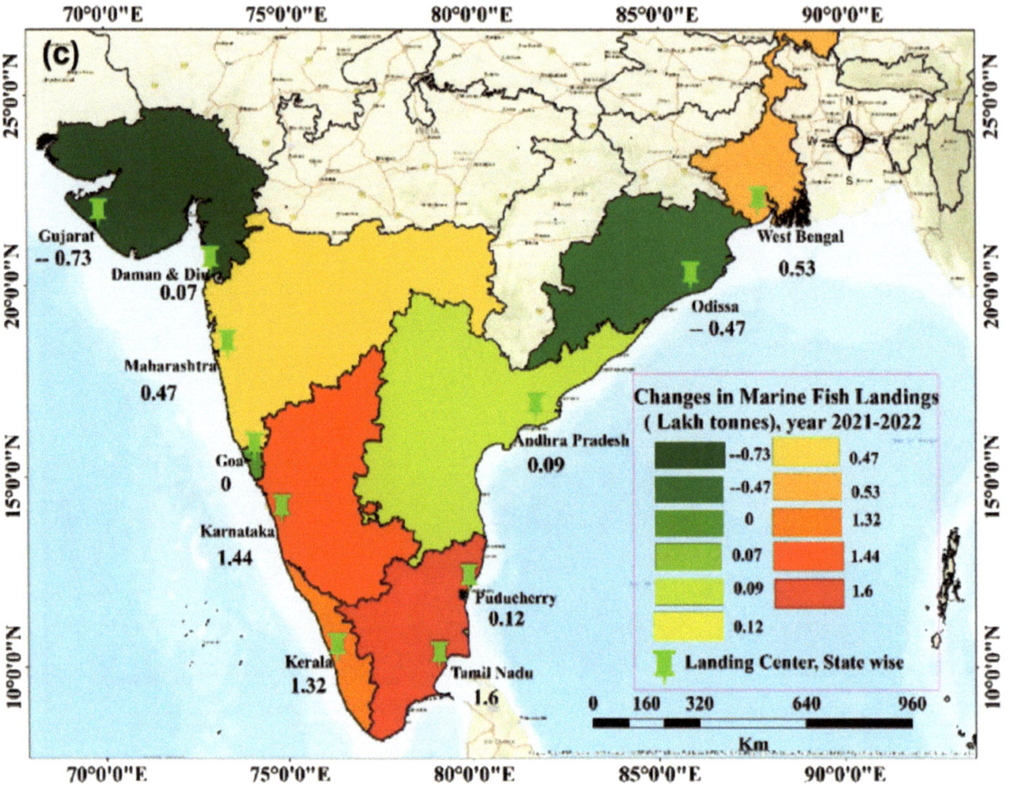

FIGURE 14.7 Contd.

marine fish landing (lakh tonnes) centers of India for the years 2021 and 2022, respectively. The harbor of fish was found to be highest for the year 2021, on the coast of Gujarat (~5.76 lakh tonnes) and lowest in Puducherry (~0.38 lakh tonnes), while in 2022, the fish capture was highest in Tamil Nadu (~7.22 lakh tonnes) and lowest in Puducherry (~0.5 lakh tonnes). The change in marine fish landing sites from the year 2021–2022 is shown in Figure 14.7c. An increase in the collection of fish catch was reported in Tamil Nadu (~1.6 lakh tonnes), followed by Karnataka (~1.44 lakh tonnes) and Kerala (~1.32 lakh tonnes), whereas a decrease in fish capture was observed in Gujarat (~−0.73 lakh tonnes) and Odisha (~−0.47 lakh tonnes). Goa shows no change in fish harvesting. The optimal SST and high concentration of Chl-a (Figures 14.3 and 14.4) were found on the coast of Tamil Nadu, Kerala (southernmost tip), and Karnataka, observed for summer (April–June 2022) and rainy (July–September 2022) seasons, which indicates a good fishing zone as presented in Figure 14.6. According to the marine fish landing map (Figure 14.7b) for the year 2022, these states show an increase in fish collection in lakh tonnes. However, the winter season (November 2022–January 2023) shows good fishing sites (see Figure 14.6) on the coast of Gujarat, the Gulf of Khambhat, the Gulf of Kutch, West Bengal, and some parts of Odisha. This output may be the reason for good fish harvesting for 2022 (Figure 14.7b). The estimated marine fish landings along the coast of mainland India during 2022 were 3.49 million tonnes (showing an increase of 14.53% compared to the landings in 2021) (CMFRI-ICAR, 2023). Fishing landing sites representing the years 2021 and 2022, along with differences during these years, have been demonstrated in Table 14.2. Similar trends and patterns of fishing sites were observed by the outcomes with respect to the marine fishing sites developed by FCSA.

TABLE 14.2
Marine Fish Landings (lakhs tonnes) Statewise along with Changes for the Years 2021 and 2022

Coastal state	Year 2021*	Year 2022**	Changes
Gujarat	5.76	5.03	–0.73
Daman and Diu	0.67	0.74	0.07
Maharashtra	1.23	1.7	0.47
Goa	0.52	0.52	0
Karnataka	5.51	6.95	1.44
Kerala	5.55	6.87	1.32
West Bengal	1.37	1.9	0.53
Odisha	1.8	1.33	–0.47
Andhra Pradesh	2.07	2.16	0.09
Puducherry	0.38	0.5	0.12
Tamil Nadu	5.62	7.22	1.6

* Year 2021, 3.05 million tonnes of fish were caught from the landing center, **whereas 3.49 million tonnes of fish were harvested in the year 2022 (FCSA).

14.5 CONCLUSION

The outcomes of the present research focus on providing PFZs while determining the relationship between SST and Chl-a concentration and its impact on fish diversity along the coastline of India with a buffer zone of 100 sq. km. The SST between ~24.23 °C and <28.52 °C with Chl-a concentration of >6.53 mg m-3 to 20.5 mg m-3 is found to be suitable for high PFZ, whereas the moderate zone has a Chl-a concentration of >0.167 mg m-3 to ~6.53 mg m-3 and SST of ~28.52 °C to <31.44 °C. However, the low PFZ is identified as having an SST of ~31.44 °C and Chl-a concentration of ~01.67 mg m-3. A similar result for the categorization of PFZ was obtained by Muhammad et al. (2022). This is also in accordance with the fishing landing sites provided by FCSA for years 2021 and 2022 (Year 2021, 3.05 million tonnes of fish were harvested from the landing center, whereas 3.49 million tonnes of fish were harvested in the year 2022 [FCSA, 2022]).

On the sea surface, upwelling can be indicated by lower SST and higher Chl-a concentration since the lifting of the water mass delivers cold water rich in nutrients needed for phytoplankton growth. It is often followed by an increase in fishing productivity as a result of high concentrations of Chl-a during the upwelling period (Sachoemar Suhendar et al., 2010; Friedland et al., 2012; Sachoemar et al., 2012). An optimal temperature range between 25°C and 29°C is required for the breeding and releasing of eggs for fish (Iqbal et al., 2021; NTAU, 2023). A good correlation between SST and Chl-a was observed during the winter season ($R^2 = 0.615$, $P < 0.001$) as compared to the summer and rainy seasons. The low correlation ($R^2 = 0.302$, $P < 0.001$) observed in the rainy season may be due to high runoff, fluctuation in water levels, and irregular rainfall patterns (Patrick, 2016). Thus, from the above study, it is concluded that the winter season is more suitable for fish catching. The MODIS-AQUA product of SST and Chl-a, along with the weighted sum technique, is quite effective for the identification of PFZs along the coastline of India. By integrating several criteria such as SST, SSH, habitat mapping, turbidity, and CDOM for PFZ, and leveraging remote sensing technology, fisheries managers and stakeholders can effectively identify and prioritize fishing potential zones, support sustainable fisheries management, and enhance socio-economic development in coastal and marine regions. Identifying the potential fishing zone using remote sensing techniques has proved to be a valuable tool in enhancing the fish catch and revenue for fishermen and reducing the search time, fuel costs, and human efforts. Future studies may include availability of near real-time data, which could help delineate potential fishing zones along the coastal tracts of any region daily using a given technique. It is concluded that PFZ provides a platform for the region's rich biodiversity and serves as an ecological indicator of marine biodiversity.

ACKNOWLEDGEMENT

The authors are grateful to the Shiv Nadar Institution of Eminence for help and support.

REFERENCES

Chakraborty, K., Maity, S., Lotliker, A.A., Samanta, A., Ghosh, J., Masuluri, N.K., Swetha, N., Bright, R.P., 2019. Modelling of marine ecosystem in regional scale for short term prediction of satellite-aided operational fishery advisories. *Journal of Operational Oceanography* 12, S157–S175.

CMFRI-ICAR, 2023. Fishery resources assessment, economics & extension division. Available online https://www.cmfri.org.in/ accessed March 2023.

CPDAC, 2017. Report of sub-committee of Coastal Protection and Development Advisory Committee (CPDAC) on Coastal Data Collection, Compilation, and Publication (2017), CPDAC Secretariat: CWC, New Delhi.

Dan Teknik, M., Nurdin, S., Mustapha, M.A., Lihan, T., Abd Ghaffar, M., 2015. Determination of potential fishing grounds of Rastrelliger kanagurta using satellite remote sensing and GIS technique. *Sains Malaysiana* 44, 225–232.

Devi, G.K., Ganasri, B., Dwarakish, G., 2015. Applications of remote sensing in satellite oceanography: A review. *Aquatic Procedia* 4, 579–584.

FCSA, 2022. Fish landings in India 2022. Published by CMFRI India. Available online http://eprints.cmfri.org.in/16883/1/Fish%20Landings%20in%20India_2022.pdf accessed March 2023.

Fitrianah, D., Fahmi, H., Hidayanto, A.N., Arymurthy, A.M., 2016. A data mining based approach for determining the potential fishing zones. *International Journal of Information and Education Technology* 6, 187–191.

Fitrianah, D., Hidayanto, A.N., Gaol, J.L., Fahmi, H., Arymurthy, A.M., 2015. A spatio-temporal data-mining approach for identification of potential fishing zones based on oceanographic characteristics in the Eastern Indian Ocean. *IEEE Journal of Selected Topics in Applied Earth Observations and Remote Sensing* 9, 3720–3728.

Friedland, K.D., Stock, C., Drinkwater, K.F., Link, J.S., Leaf, R.T., Shank, B.V., Rose, J.M., Pilskaln, C.H., Fogarty, M.J., 2012. Pathways between primary production and fisheries yields of large marine ecosystems. *PloS One* 7, e28945.

Giri, S., Manna, S., Chanda, A., Chowdhury, A., Mukhopadhyay, A., Chakraborty, S., Hazra, S., 2016. Implementing a spatial model to derive potential fishing zones in the Northern Bay of Bengal lying adjacent to West Bengal Coast, India. *Journal of the Indian Society of Remote Sensing* 44, 59–66.

Guidetti, P., Bussotti, S., Pizzolante, F., Ciccolella, A., 2010. Assessing the potential of an artisanal fishing co-management in the Marine Protected Area of Torre Guaceto (southern Adriatic Sea, SE Italy). *Fisheries Research* 101, 180–187.

Harahap, S.A., Syamsuddin, M.L., Purba, N.P., 2020. Range of sea surface temperature and chlorophyll-α values based on mackerel catches in the northern waters of West Java, Indonesia. *AACL Bioflux* 13, 2265–2272.

INCOIS-PFZ-advisory, 2023. Potential Fishing Zone (PFZ) advisory. Available online https://incois.gov.in/MarineFisheries/PfzAdvisory accessed March 2023.

Iqbal, S., Atique, U., Abbas, F., Ahmad, S., & Haider, M. S. (2021). Impact of Temperature Variations on Breeding Behavior of Cirrhinus mrigala during Induced Spawning. *Pakistan Journal of Zoology,* 1–6. DOI: 10.17582/journal.pjz/20200510140530

Karuppasamy, S., Ashitha, T., Padmanaban, R., Shamsudeen, M., Silva, J., 2020. A remote sensing approach to monitor potential fishing zone associated with sea surface temperature and chlorophyll concentration. *Indian Journal of Geo Marine Sciences*, 49(6),1025–1030.

Lasker, R., Peláez, J., Laurs, R.M., 1981. The use of satellite infrared imagery for describing ocean processes in relation to spawning of the northern anchovy (Engraulis mordax). *Remote Sensing of Environment* 11, 439–453.

Laurs, R.M., Fiedler, P.C., Montgomery, D.R., 1984. Albacore tuna catch distributions relative to environmental features observed from satellites. *Deep Sea Research Part A. Oceanographic Research Papers* 31, 1085–1099.

Maul, G., Williams, F., Roffer, M., Sousa, F., 1984. Remotely sensed oceanographic patterns and variability of bluefin tuna catch in the Gulf of Mexico. *Oceanologica Acta* 7, 469–479.

Muhammad, M., Priana, A.W., Affan, J.M., Haridhi, H.A., Irwan, I., Yuni, S.M., Setiawan, I., 2022. Mapping potential fishing zones based on sea surface temperature and Chlorophyll-A in the Waters of Aceh Besar, Indonesia. E3S Web of Conferences. EDP Sciences, p. 02002.

Mursyidin, M., Munadi, K., Muchlisin, Z., 2015. Prediksi zona tangkapan ikan menggunakan citra klorofil-a dan citra suhu permukaan laut satelit aqua modis di perairan Pulo Aceh. *Jurnal Rekayasa Elektrika* 11, 176–182.

Nair, P.G., Joseph, S., Pillai, N., Abdulla, M.H.A., 2022. Seasonality in recurrent Potential Fishing Zones along India's Southwest coast and its relationship to the prevailing hydrographic settings. *Regional Studies in Marine Science* 51, 102191.

Nammalwar, P., Satheesh, S., Ramesh, R., 2013. Applications of remote sensing in the validations of potential fishing zones (PFZ) along the coast of North Tamil Nadu, India. *Indian Journal of Geo-Marine Sciences* 42, 283–292.

Navarro, G., Ruiz, J., 2006. Spatial and temporal variability of phytoplankton in the Gulf of Cádiz through remote sensing images. *Deep Sea Research Part II: Topical Studies in Oceanography* 53, 1241–1260.

NTAU, 2023. Fisheries, TNAU agritech portal. Available online agritech.tnau.ac.in/fishery/fish_cul_ornamental.html accessed April 2023.

Nurdin, S., Mustapha, M., Lihan, T., 2013. The relationship between sea surface temperature and chlorophyll-a concentration in fisheries aggregation area in the archipelagic waters of Spermonde using satellite images. AIP Conference Proceedings. American Institute of Physics, pp. 466–472.

O'Reilly, J.E., Maritorena, S., Mitchell, B.G., Siegel, D.A., Carder, K.L., Garver, S.A., Kahru, M., McClain, C., 1998. Ocean color chlorophyll algorithms for SeaWiFS. *Journal of Geophysical Research: Oceans* 103, 24937–24953.

Patrick, A., 2016. Influence of rainfall and water level on inland fisheries production: A review. *Archives of Applied Science Research,* 8, 44–51.

Radiarta, I.N., Saitoh, S.-I., 2008. Satellite-derived measurements of spatial and temporal chlorophyll-a variability in Funka Bay, southwestern Hokkaido, Japan. *Estuarine, Coastal and Shelf Science* 79, 400–408.

Sachoemar, S.I., Yanagi, T., Aliah, R.S., 2012. Variability of sea surface chlorophyll-a, temperature and fish catch within Indonesian region revealed by satellite data. *Marine Research in Indonesia* 37, 75–87.

Sachoemar Suhendar, I., Tetsuo, Y., Nani, H., Muhamad, S., Fanny, M., 2010. Seasonal variability of sea surface chlorophyll-a and abundance of pelagic fish in Lampung Bay, Southern Coastal Area of Sumatra, Indonesia. *Coastal Marine Science* 34, 82–90.

Setiawan, A., Agustriani, F., Ningsih, E.N., Ulqodry, T.Z., 2022. Distribution pattern of potential fishing zones in the Bangka Strait waters: an application of the remote sensing technique. *The Egyptian Journal of Remote Sensing and Space Science* 25, 257–265.

Solanlki, HU., Dwivedi, R., Nayak, S., Jadeja, J., Thakar, D., Dave, H., Patel, M., 2001. Application of Ocean Colour Monitor chlorophyll and AVHRR SST for fishery forecast: Preliminary validation results off Gujarat coast, northwest coast of India. *Indian Journal of Marine Sciences,* 30(3), 132–138.

Solanki, H., Dwivedi, R., Nayak, S., Naik, S., John, M., Somvanshi, V., 2005a. Cover: application of remotely sensed closely coupled biological and physical processes for marine fishery resources exploration. *International Journal of Remote Sensing* 26, 2029–2034.

Solanki, H., Mankodi, P., Nayak, S., Somvanshi, V., 2005b. Evaluation of remote-sensing-based potential fishing zones (PFZs) forecast methodology. *Continental Shelf Research* 25, 2163–2173.

Suhadha, A.G., Asriningrum, W., 2020. Potential fishing zones estimation based on approach of area matching between thermal front and mesotrophic area. *Jurnal Ilmu Dan Teknologi Kelautan Tropis* 12, 565–581.

Sukresno, B., Suniada, K.I., Riset, B., 2008. Observasi pengaruh ENSO terhadap produktifitas primer dan potensi perikanan dengan menggunakan data satelit di Laut Banda. *Globe* 10, 97–107.

Vaidyanadhan, R., 1991. Quaternary deltas of India. Geological Society of India, 1–291.

Xingwei, S., Baide, X., 1988. Quick reporting state of fishery and sea on the east China Sea and the Yellow Sea with NOAA. International Geoscience and Remote Sensing Symposium, 'Remote Sensing: Moving Toward the 21st Century'. IEEE, pp. 1405–1408.

Yoder, J.A., Schollaert, S.E., O'Reilly, J.E., 2002. Climatological phytoplankton chlorophyll and sea surface temperature patterns in continental shelf and slope waters off the northeast US coast. *Limnology and Oceanography* 47, 672–682.

Zainuddin, M., 2007. Pemetaan daerah potensial penangkapan ikan kembung lelaki (Rastrelliger kanagurta) di perairan kabupaten Bantaeng, Sulawesi Selatan (Mapping of potential fishing grounds of Rastrelliger kanagurta in Bantaeng waters, South Sulawesi). *Jurnal Sains dan Teknologi* 7, 57–64.

15 Impact of Climate Change and Land Use Land Cover Dynamics on Wetland Ecosystem in Punjab Using Geospatial Technology

Mohit Arora, Sashikanta Sahoo, Brijendra Pateriya*
Punjab Remote Sensing Centre, PAU Campus, Ludhiana, Punjab, India

Isha Swati
Punjab Agricultural University, Ludhiana, Punjab, India

Parag Roy
Central University of Karnataka, Kadaganchi, Karnataka, India

15.1 INTRODUCTION

Wetland is land where the water table is at, near, or above the land surface, and it can be salt water or fresh water depending on the climate or ecosystem. It can be either permanent or seasonal. Wetlands are among the most biologically diverse, complex, most productive, and sensitive ecosystems on Earth. Wetlands are the most productive ecosystem types as, through photosynthesis, they can actively isolate and accumulate carbon as plant biomass or organic matter in soil. Wetlands encompass a large and heterogeneous spectrum of aquatic habitats and are found at the interface between terrestrial ecosystems (such as forests and grasslands) and waters (such as rivers, lakes, estuaries, and oceans), strongly dependent on the water cycle. Wetlands are a major feature of the landscape in almost all parts of the Earth (Singh et al., 2020) and act as critical habitats for a variety of plants, fish, shellfish, and other wildlife (Klemas, 2011). According to the Cowardin system, there are five major wetland types: (i) marine, (ii) estuarine, (iii) lacustrine, (iv) palustrine, and (v) riverine. Compared to all terrestrial ecosystems, wetlands have the highest density of carbon, which enables wetlands to play an important role in global biogeochemical carbon cycles and climate change (Kayranli et al., 2010). Wetlands play an important role in the ecological system, covering 6% of the Earth's surface (~1280 million ha) and containing about 12% of the global carbon pool (IPCC, 1996; Sahagian and Melack, 1998; Finlayson and Davidson, 1999; Ferrati et al., 2005; Junk et al., 2013; Barros and Albernaz, 2014; Singh et al., 2020). There are more than 2400 Ramsar sites in the world, covering about 2.5 million square kilometers. A Ramsar site is an intergovernmental environmental treaty under the Ramsar Convention, also known as "The Convention on Wetlands" established on February 2, 1971, in Iran by UNESCO, which came into action on December 21, 1975, for wetlands of international importance. The Ramsar Convention on Wetlands has long recognized the need to develop techniques that can fill gaps in baseline inventory. These Ramsar wetlands are important

* Corresponding author.

DOI: 10.1201/9781003354000-20

for supporting the great diversity of ecosystems, hydrology, environment, and mammalian species, and they serve as recreational spots for nature lovers, migratory bird watchers, researchers, and others. According to the Ramsar Convention on Wetlands, Article 1, wetlands are defined as areas of marsh, fen, peatland, or water, whether natural or artificial, permanent or temporary, with water that is static or flowing, fresh, brackish, or salt, including areas of marine water the depth of which at low tide does not exceed 6 meters. Wetlands can be identified by the following characteristics: (i) the wetlands should at least periodically support predominantly hydrophytes, (ii) the substrate is predominantly in undrained hydric soil, and (iii) the substrate is saturated with water or covered by shallow water at some time during the growing season of each year.

Despite their vital function, many wetlands are under threat and being destroyed worldwide due to climate change and anthropogenic activities (Sivakumar and Ghosh, 2016; Kundapura et al., 2019; Singh et al., 2020). There are many different factors, biotic and abiotic, that influence the function of wetlands. Wetland degradation and loss are a worldwide phenomenon, progressing faster than in other ecosystems, not only because of the complex relationships and feedback between climate and wetlands but also because of the interdependent effects of non-climatic and climatic drivers, especially in coastal and low-lying areas. Climate change has been identified as the major threat to wetlands. Climate change refers to any change in climate conditions over any period of time. It presents a great challenge to conserving wetlands and will make future efforts to restore and manage wetlands more complex (Erwin, 2009). It causes shifts in species composition, habitat destruction, and degradation in existing wetlands worldwide (Hulme, 2005; Erwin, 2009). Loss of carbon stored in soil, changes in soil structure, and more frequent drying or flooding are the effects of climate change on wetlands. Climate change affects wetlands on a bigger scale and poses a major threat to its animal and plant species. The impact of climate change on inland wetland ecosystems largely depends on the rate and magnitude of change in two critical climate drivers: temperature and water availability from precipitation and runoff. Additionally, sea level and storm surges will be important drivers in coastal and low-lying wetlands. Climate change can affect wetlands through the direct and indirect effects of rising temperature, changes in rainfall intensity and frequency, extreme climatic events such as drought, flooding, and the frequency of storms. The increasing temperature and the amount of increasing carbon dioxide in the atmosphere reduce rainfall. Wetlands capture CO_2 from the atmosphere, making them nature's own solution to the climate emergency. Hurricanes, forestry practices, sea level change, and certain agricultural practices can increase erosion or sedimentation. Drastic climate change causes significant damage to agriculture, human life, and its surrounding ecosystem. Some human modification such as fitting pipes and building channels can alter wetland types, elevation, or hydrology. The flow of water into wetlands is reduced by human activities such as groundwater or surface water withdrawal. Many human activities, converting wetlands into agricultural and residential areas, pollution, and urbanization impact the climate. Wasteland's waterlogged conditions cause inefficient decomposition that exceeds the rate of production and results in an anoxic state in which vast amounts of carbon accumulate in wetlands, making them a sink of carbon. Over the past three decades, many researchers have been looking at the ecological and hydrological impacts resulting from climate change worldwide (Jain et al., 2008; Erwin, 2009). Tropical wetlands are considered highly vulnerable to climate change, and degradation seems to be progressing faster than other ecosystems. The climate change impacts on wetland ecosystems result not only in the loss of biodiversity but also in the loss of sources of economic benefits. Anthropogenic impacts have also been more noteworthy than natural impacts. Anthropogenic activities include population growth, urbanization, agricultural expansion, infrastructure development, etc. Climate change, together with human pressure, has changed ecosystems more rapidly and extensively in the last five decades (Kumar and Chopra, 2009; Sahoo et al., 2021). Many wetlands have disappeared and been converted to industrial, agricultural, and residential use.

Wetlands play a crucial role in maintaining the overall health of our environment. Wetlands help to improve the water quality by removing excess nutrients, such as nitrogen and phosphorus, which can be harmful to aquatic ecosystems if they are added in large amounts. They can help reduce the risk of downstream flooding by slowing down the flow of water and releasing it gradually over time. During periods of flooding, they trap suspended solids and nutrient loads. Wetlands accumulate and store large amounts of organic matter in the form of vegetation and soil, preventing it from being released into the atmosphere as carbon dioxide, which helps to reduce greenhouse gas emissions. Wetlands also stabilize and prevent erosion caused by wind and water. They also provide for outdoor activities such as bird watching, fishing, boating, and nature photography, and attract tourists and contribute to local economies through recreational and educational activities.

Wetlands in India are distributed in different geographical regions, and the area coverage is about 58.2 million hectares (ha), accounting for nearly 4.75% of the total geographical area. They are important repositories of aquatic biodiversity (Singh et al., 2020). India falls in the tropical region, and tropical wetlands are considered highly vulnerable to climate change, and degradation is progressing faster than in other ecosystems. India has 54 sites designated as wetlands of international importance (Ramsar sites). The wetlands of Punjab state are not stable, and climatic changes are causing them to deteriorate. The National Wetland Atlas (2011) mapped 1381 wetlands in Punjab state, covering 86,283 ha which accounted for 1.71% of the total geographical area of the region (SAC, 2011). There are about 40 (8 natural and 32 man-made) significant wetlands in Punjab state (Singh et al., 2020). Wetland management requires an understanding of the scientific aspects of wetlands, prevention or reduction of additional stress, balanced with legal, institutional, and economic realities that can reduce the ability of wetlands to change. The Ramsar site wetlands in Punjab are Harike wetland (1990), Ropar wetland (2002), Kanjli wetland (2002), Keshopur-Miani community reserve (2020), Beas conservation reserve (2020), and Nangal wildlife sanctuary (2020). There are many inherent problems, both natural and man-made, faced by wetlands, and many wetlands are under threat from developmental activities and population pressures. The main climatic change factors responsible for wetland degradation in the Punjab region are the decreasing flow in perennial rivers, less rainfall in the monsoon season, and increasing temperatures. The objective of this study is mainly focused on climate change impacts and land use and land cover (LULC) dynamics on the Harike and Keshopur wetlands in the Punjab region, India. It provides insights into trophic state changes, growth and spread of aquatic vegetation, and the impact of natural and anthropogenic sources of pollution. Wetlands are highly dependent on water levels. Harike is one of the largest wetlands in northern India, and Keshopur, which lies in the former flood plains of the rivers Ravi and Beas. This study will help explain the environmental aspects and protection of the biodiversity of wetland ecosystems for future sustainability.

15.2 STUDY AREA

Punjab is located in the northern state of India and has a total area of 50,362 sq. km. The extension of Punjab state is from 29°18'0"N to 32°19'12"N latitude and 73°33'0"E to 76°30'0"E longitude. Punjab is divided into 23 districts and shares a boundary with Jammu and Kashmir in the north, Himachal Pradesh in the north-east and east, Haryana in the south-east and south, Rajasthan in the south and south-west, and shares the international boundary with Pakistan on the western side. There are three perennial rivers, which are Satluj, Beas, and Ravi, which along with their tributaries drain the state. The Satluj River, which is the longest river in Punjab, is a major source of water supply, drinking, washing, bathing, etc. It originates from Manasarover Lake in Tibet and enters Punjab near Nangal, moves on to plains at Ropar, passes through districts SBS Nagar, Ludhiana, Jalandhar, Kapurthala, Moga, Firozpur, and reaches Harike wetland (Ramsar Convention) before crossing over to Pakistan (Sharma et al., 2018). Punjab is one of the important food grain-producing states of India and its wetlands act as sources of irrigation, hydropower generation, and shelters for varieties of birds, such as permanent and migratory birds. This study is mainly concerned with the two major sites: Harike wetland, which has a total area coverage of 222 sq. km and is situated in the Tarn Taran district, and Keshopur wetland, which has a total area coverage of 7.1 sq. km and is situated in the Gurdaspur district (Figure 15.1). Harike wetland is also known as "Hari-Ke-Pattan" and together

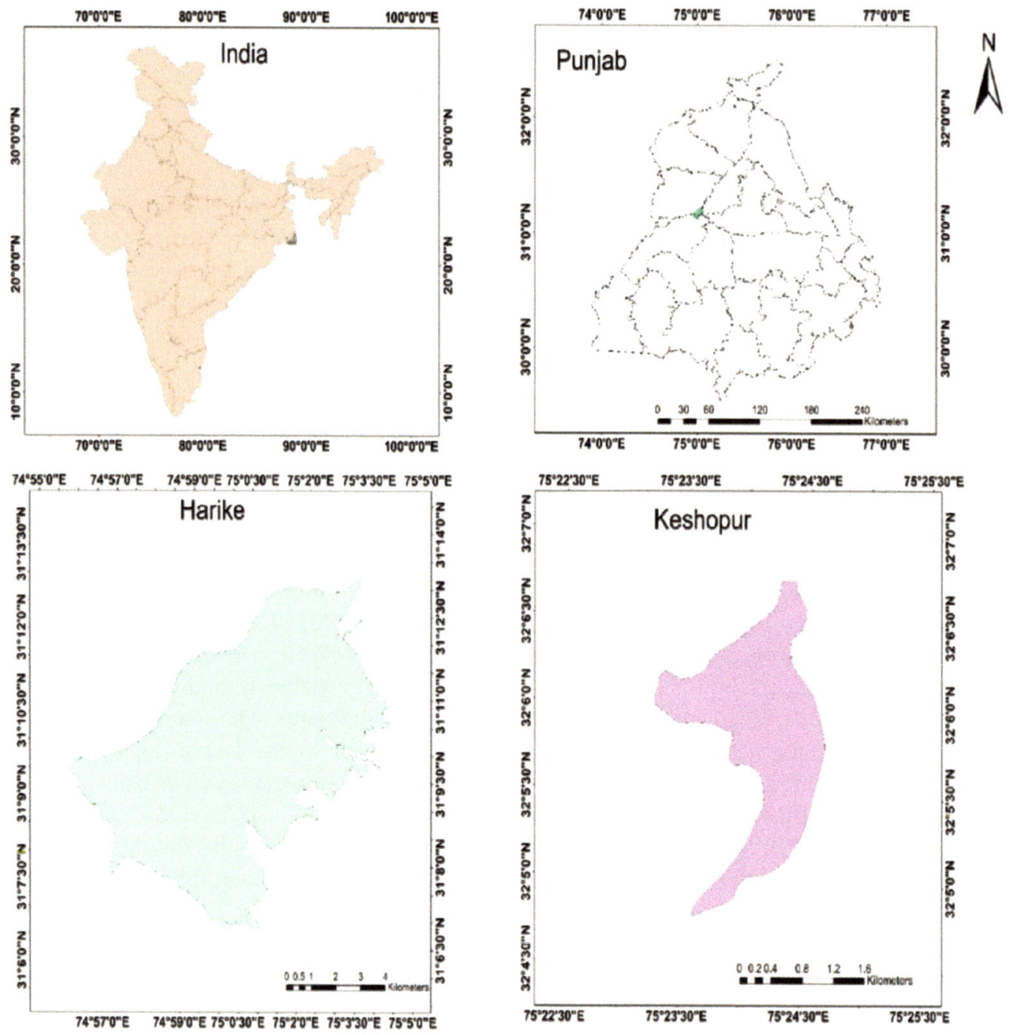

FIGURE 15.1 Study aea map of Harike and Keshopur wetlands over Punjab state.

with the Harike lake is the largest wetland of north India. It is located between latitudes 31°05'15"N to 31°40'15"N and longitudes 74°55'30"E to 75°07'30"E, and lies at a distance of 57 km from the historic city of Amritsar and about 150 km from the capital of Punjab, which is Chandigarh. The headworks is located at the downstream confluence of the Beas and Sutlej rivers just south of Harike village. The rich biodiversity of the wetland plays a vital role in maintaining the hydrological balance in the catchment with its vast fauna of waterfowls, including a number of globally threatened species, responsible for the recognition accorded to this wetland in 1990 by the Ramsar Convention, as one of the Ramsar sites in India, for conservation, development, and preservation of the ecosystem. Keshopur wetland is also known as the Keshopur-Miani community reserve or *chhmab* in the local Punjabi dialect. It is located between latitudes 32°04'23"N to 32°06'33"N and longitudes 75°21'6"E to 75°24'26"E, and lies at a distance of 5 km from the city of Gurdaspur. Keshopur community reserve is a dynamic freshwater ecosystem. It is the first-ever notified community reserve of India under wildlife. It lies in the former flood plains of the rivers Ravi and Beas. It is a dynamic freshwater ecosystem and consists of a mosaic of natural marshes, aquaculture ponds, and agricultural

TABLE 15.1
Landsat Satellite Data Information during Both Seasons (Pre- and Post-monsoon)

Seasons	Satellite Data Id	Date of Acquired Data	Path/ Row	Cloud Cover (%)
Pre-Monsoon	LC08_L2SP_148038_20200224_20200822_02_T1	24 February 2020	148/38	<10
	LT05_L2SP_148038_20100503_20200824_02_T1	3 May 2010	148/38	<10
	LE07_L2SP_148038_20000515_20200918_02_T1	15 May 2010	148/38	<10
Post-Monsoon	LC08_L2SP_148038_20201122_20210315_02_T1	22 November 2020	148/38	<10
	LT05_L2SP_148038_20101127_20200823_02_T1	27 November 2010	148/38	<10
	LE07_L2SP_148038_20001022_20200918_02_T1	22 October 2000	148/38	<10

wetlands where crops such as lotus and chestnut are cultivated. Keshopur-Miani community reserve is the first ever to witness 4,500 birds but by 2016 it was around 25,000 birds.

15.3 MATERIALS AND METHODOLOGY

This study used Landsat series Landsat 5, Landsat 7, and Landsat 8 for the study of spatial and temporal patterns of LULC dynamics over Harike and Keshopur wetlands in Punjab during the period from 2000 to 2020 for both seasons (pre- and post-monsoon). The Landsat program is the longest-running enterprise for the acquisition of satellite imagery of the Earth. It is a joint mission of the National Aeronautics and Space Administration (NASA) and the United States Geological Survey (USGS) program. The first Landsat 1 was launched in 1972, and now the most recent, Landsat 9, was launched on 27 September 2021. Landsat satellites have acquired millions of images, and the images have been archived in the United States and can be viewed from the USGS "Earth Explorer" website. It has a spatial resolution of 30 m and a temporal resolution of 16 days. Landsat 5 and Landsat 7 data contain 7 bands whereas the Landsat 8 contains 11 bands (Table 15.1).

15.3.1 Landsat Series Data

15.3.1.1 Landsat 5 Thematic Mapper

Landsat 5 is a low Earth orbit satellite which was launched on 1 March 1984 to collect the imagery of the surface of the Earth. Landsat 5 carries the "Thematic Mapper" and "Multi Spectral Scanner". It ran successfully for 29 years and was officially decommissioned on 5 June 2013. Landsat 4–5 Thematic Mapper consists of seven spectral bands with a spatial resolution of 30 m for bands 1 to 5 and 7. Band 6, which is a thermal infrared, has a spatial resolution of 120 m but is resampled to 30 m per pixel. The approximate scene size is 170 km north-south by 183 km east-west.

15.3.1.2 Landsat 7 Enhanced Thematic Mapper Plus

Landsat 7 Enhanced Thematic Mapper Plus (ETM+) is the seventh satellite of the Landsat program, and it was launched on 15 April 1999. It was officially decommissioned on 7 February 2017. Landsat 7 ETM+ images consist of eight spectral bands with a spatial resolution of 30 m for bands 1 to 7. The resolution for band 8, which is panchromatic, is 15 m. All bands can collect one of two gain settings, high or low for increased radiometric sensitivity and dynamic range. Band 6 collects both high and low gains for all scenes. Its approximate scene size is 170 km north-south by 183 km east-west.

TABLE 15.2
Spectral Bands and Wavelength Ranges of Landsat 5, Landsat 7 and Landsat 8 and Their Uses

Sr. No.	Satellite Data	Spectral Bands	Wavelength Range (nm)	Use Area
1.	Landsat 8	Band 1	450-520/433-453	Aerosol/Coastal
2.	Landsat 5, 7, and 8	Band 2	450-515	Blue
3.	Landsat 5, 7, and 8	Band 3	525-600	Green
4.	Landsat 5, 7, and 8	Band 4	630-680	Red
5.	Landsat 5, 7, and 8	Band 5	760-900/845-885	NIR
6.	Landsat 5, 7, and 8	Band 6	1550-1750/1560-1660	SWIR-1
7.	Landsat 5, 7, and 8	Band 7	2100-2350/2100-2300	SWIR-2
8.	Landsat 8	Band 8	500-680	Panchromatic
9.	Landsat 8	Band 9	1360-1390	Cirrus
10.	Landsat 5, 7, and 8	Band 10	1031-1236/1060-1120	Thermal-1
11.	Landsat 8	Band 11	1150-1250	Thermal-2

15.3.1.3 Landsat 8 Operational Land Imager and Thermal InfraRed Sensor

Landsat 8 was launched on 11 February 2013 as the eighth satellite of the Landsat program. It uses two sensor payloads, the Operational Land Imager (OLI) and the Thermal InfraRed Sensor (TIRS). Respectively, these two instruments collect image data for nine shortwave bands and two longwave thermal bands. It consists of nine spectral bands with a spatial resolution of 30 m for bands 1 to 7 and 9. New band 1, which is ultra-blue, is useful for coastal and aerosol studies. Band 9 is useful for cirrus cloud detection. The resolution of band 8, which is panchromatic, is 15 m. Thermal bands 10 and 11 are useful in providing more accurate surface temperatures and are collected at 100 m. All Landsat series data have been taken from the USGS site. All collected data have been pre- and post-processed through ERDAS Imagine, ENVI Classic software, and ARC GIS used for mapping and statistical analysis (Figure 15.2). The Landsat 5, Landsat 7 and Landsat 8 level–2 imagery have been used for change analysis of LULC over the Harike and Keshopur wetlands.

15.3.2 Estimating Land Surface Temperature from Landsat Series Data

For the retrieval of emissivity-corrected land surface temperature (LST) maps, the following steps are followed: First, the digital numbers (DNs) of thermal bands are converted to spectral radiance by the following expression (Chen et al., 2006):

$$L\lambda = [(L_{max} - L_{min})/(QCAL_{max} - QCAL_{min})]^* (QCAL - QCAL_{min}) + L_{min} \quad (15.1)$$

where $L\lambda$ is the spectral radiance and λ is the wavelength, $QCAL_{min}$ and $QCAL_{max}$ are the minimum and maximum DN of the images respectively, QCAL = DN, and L_{max} and L_{min} are the spectral radiances for the thermal band at $QCAL_{max}$ and $QCAL_{min}$, respectively (given in the header file of the images).

Then, the brightness temperature of the field of view of the satellite under the assumption of uniform emissivity is estimated from the spectral radiance ($L\lambda$) using the following expression (Weng, Lub and Schubringa, 2004; Chen et al., 2006; Pan, 2015):

Brightness temperature, TB (K):

$$TB = K_2 / \ln[(K_1 / L\lambda) + 1] \quad (15.2)$$

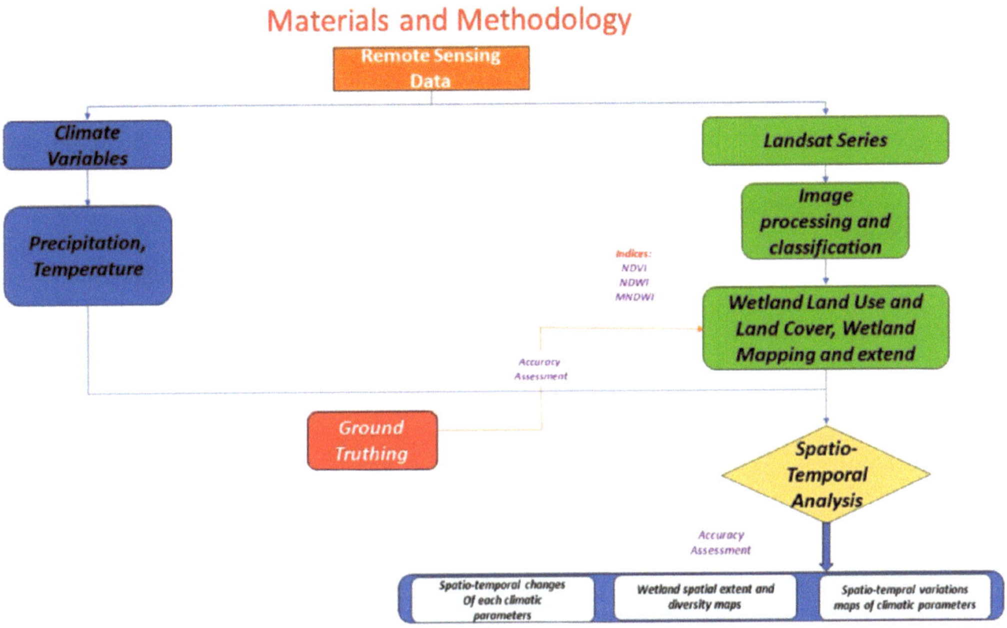

FIGURE 15.2 Flowchart of materials and methodology.

where K_1 (K) and K_2 (Wm–2) are pre-launch calibration constants for Landsat 5, 7, and 8 sensors. Finally, the LST (K) can be derived from the brightness temperature (K) by the following formula (Weng, Lub, and Schubringa, 2004):

$$LST\ (°C) = TB/[1+W(TB/p)*\ln(\varepsilon)] - 273.15 \quad (15.3)$$

where W is the wavelength of emitted radiance; $p = h \times c/\sigma = 1.438 \times 10{-}2$ mK (σ is the Boltzmann constant, h is the Planck constant, and c is the velocity of light).

$$\varepsilon = 0.004*Pv + 0.986 \quad (15.4)$$

$$Pv = [(NDVI - NDVI_{min})/(NDVI_{max} - NDVI_{min})] \quad (15.5)$$

where Pv is the fractional vegetation or proportion of vegetation.

15.3.3 Estimating Rainfall Data from Global Precipitation Measurement

The precipitation data were taken from the Global Precipitation Measurement (GPM). GPM is a joint mission of NASA and the Japanese Aerospace Exploration Agency (JAXA) for environmental research. The name of the satellite is the GPM Core Observatory satellite; it was launched on 27 February 2014, and it carries advanced instruments for precipitation measurements from space. Its mission is to observe global rainfall and snowfall. It is the extended form of the Tropical Rainfall Measuring Mission (TRMM), which was launched on 27 November 1997 and was in operation until 2015. TRMM was also a joint mission of NASA and JAXA and its mission was for environmental research to monitor and study tropical and subtropical rainfall and storms. It provided rain rate in mm/hr and had a spatial resolution of 0.25 degrees in latitude and longitude, covering areas from 50N to 50S. Its temporal resolution was 3 hours. GPM Core Observatory satellite continues the sampling of TRMM but extends its observations by covering the higher latitudes, encompassing the entire globe from the Antarctic Circle to the Arctic Circle. GPM provides rain rates in mm/hr with a spatial resolution of 0.1 degree * 0.1 degree, which is roughly 10 km * 10 km and its temporal resolution is 30 minutes. GPM uses two instruments: the GPM Microwave Imager (GMI) and the Dual-Frequency Precipitation Radar (DPR). The GMI uses 13 different microwave channels to observe energy from different types of precipitation for estimating heavy to light rainfall and for detecting snowfall. The Dual-Frequency Precipitation Radar uses two bands, Ku and Ka, which provide three-dimensional measurements of precipitation structure and characteristics. The Ka-band frequency scans across a region of 78 miles (125 km) and is nested within the wider scan of the Ku-band frequency of 158 miles (254 km). All the collected data has been processed using MATLAB 2018b software. The rainfall anomaly has been derived from each monthly dataset, and the average value of rainfall for Harike and Keshopur wetlands over the baseline period (January 2000 to December 2020) has been calculated in ArcGIS 10.4 software.

15.4 RESULTS AND DISCUSSION

The results show that Harike and Keshopur wetlands have experienced significant changes due to the influences of natural and human activities inside the sanctuary area. Landsat datasets have been taken at intervals of 10 years: 2000, 2010, and 2020 for the assessment of LULC dynamics and LST, while GPM data have been used from 2000 to 2022 to observe the monthly variations of rainfall over both study areas. This study used Landsat 5, Landsat 7, and Landsat 8 data for mapping and monitoring wetland extent and LST both spatially and temporally (Figures 15.3 and 15.5).

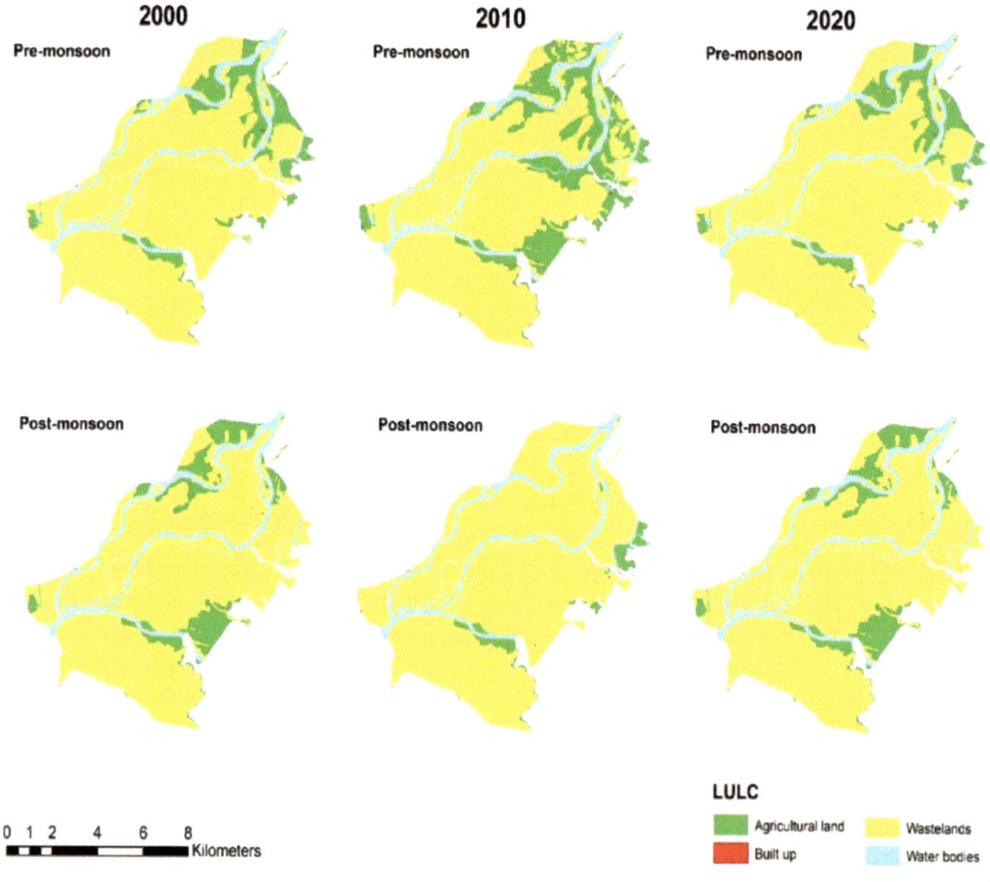

FIGURE 15.3 Land use and land cover dynamics over Harike wetland during the period from 2000 to 2020.

Monitoring wetland changes spatially and temporally is not feasible by ground measurements due to the high cost and limited observations. We classified Landsat images into four different classes including (i) agricultural land, in which we have added aquatic vegetation and cropland, (ii) built-up areas in which we have added helmeted and dispersed houses, (iii) water bodies, in which we have included lakes and rivers, and (iv) wastelands, in which we have included shrub lands, algal bloom, sandy areas, and waterlogged areas.

15.4.1 Dynamics of Harike Wetlands during Pre- and Post-Monsoon

Harike wetland has experienced significant changes due to the influences of natural and human activities inside the sanctuary area during the pre- and post-monsoon periods from 2000 to 2020. Figure 15.3 illustrates the land use/land cover dynamics of Harike wetland. In the year 2000 pre-monsoon season, Harike wetland comprised 18.05 sq. km of agriculture, 0.003 sq. km of built-up, 6.95 sq. km of water bodies, and 49.02 sq. km of wastelands, whereas in the 2010 pre-monsoon season it comprised 18.02 sq. km of agriculture, 0.02 sq. km of built-up, 6.94 sq. km of water bodies, and 51.32 sq. km of wastelands. In 2020, it comprised 19.44 sq. km of agriculture, 0.04 sq. km of built-up, 6.94 sq. km of water bodies, and 49.01 sq. km of wastelands (Figure 15.4). After analyzing these datasets, we found that there is a net increase in agricultural land, built-up areas, and wastelands. Similarly, in the year 2000 post-monsoon season, it comprised 18.81 sq. km of agriculture, 0.003 sq. km of built-up, 6.94 sq. km of water bodies, and 58.78 sq. km of wastelands, whereas in the year 2010, it comprised 18.89 sq. km of agriculture, 0.02 sq. km of built-up, 6.94 sq. km of water bodies, and 64.69 sq. km of wastelands. In the year 2020, it comprised 25.82 sq. km of agriculture, 0.04 sq. km of built-up, 6.94 sq. km of water bodies, and 49.02 sq. km of wastelands. After analyzing these datasets we found that there is a net increase in agricultural land, built-up areas, and wastelands. Harike wetland is the largest wetland of northern India, and its land is very fertile because it is situated along the river Sutlej. The expansion of cultivated lands and built-up areas can be seen in this wetland. Encroachment inside this wetland for cultivation and grazing activities, as well as an increasing number of built-up areas, are considered the major obstacles in the management of the wetland.

15.4.2 Dynamics of Keshopur Wetlands during Pre- and Post-Monsoon

Significant changes have been experienced by Keshopur wetland due to the influence of natural and human activities during the pre- and post-monsoon periods from 2000 to 2020. Figure 15.5 illustrates the land use and land cover of the Keshopur wetland ecosystem. In the year 2000 pre-monsoon season, Keshopur wetland comprised 3.03 sq. km of agriculture, 0.63 sq. km of water bodies, and 1.57 sq. km of wastelands, whereas in the 2010 pre-monsoon season, it comprised 2.99 sq. km of agriculture, 0.67 sq. km of water bodies, and 0.92 sq. km of wastelands. In 2020, it comprised 2.98 sq. km of agriculture, 0.009 sq. km of built-up areas, 0.86 sq. km of water bodies, and 0.90 sq. km of wastelands (Figure 15.6). After analyzing these datasets, we found that there is a net decrease in agricultural land and wastelands, and an increase in built-up areas and water bodies. Similarly, in the year 2000 post-monsoon, it comprised 3.03 sq. km of agriculture, 0.63 sq. km of water bodies, and 1.34 sq. km of wastelands, whereas in the year 2010, it comprised 3.03 sq. km of agriculture, 0.54 sq. km of water bodies, and 0.92 sq. km of wastelands. In the year 2020, it comprised 3.02 sq. km of agriculture, 0.009 sq. km of built-up areas, 0.84 sq. km of water bodies, and 0.51 sq. km of wastelands. After analyzing these datasets we found that there is a net decrease in agricultural land and wastelands, and an increase in built-up areas and water bodies. The agricultural land has decreased due to issues with soil, social atmosphere, overgrazing, land surface temperature variations, and land degradation due to soil erosion, converting it into wastelands, and built-up areas. The built-up areas are increasing, and there is an increase in water bodies due to the adverse effect of rainfall. Keshopur wetland lies in the former flood plains of the rivers Ravi and Beas.

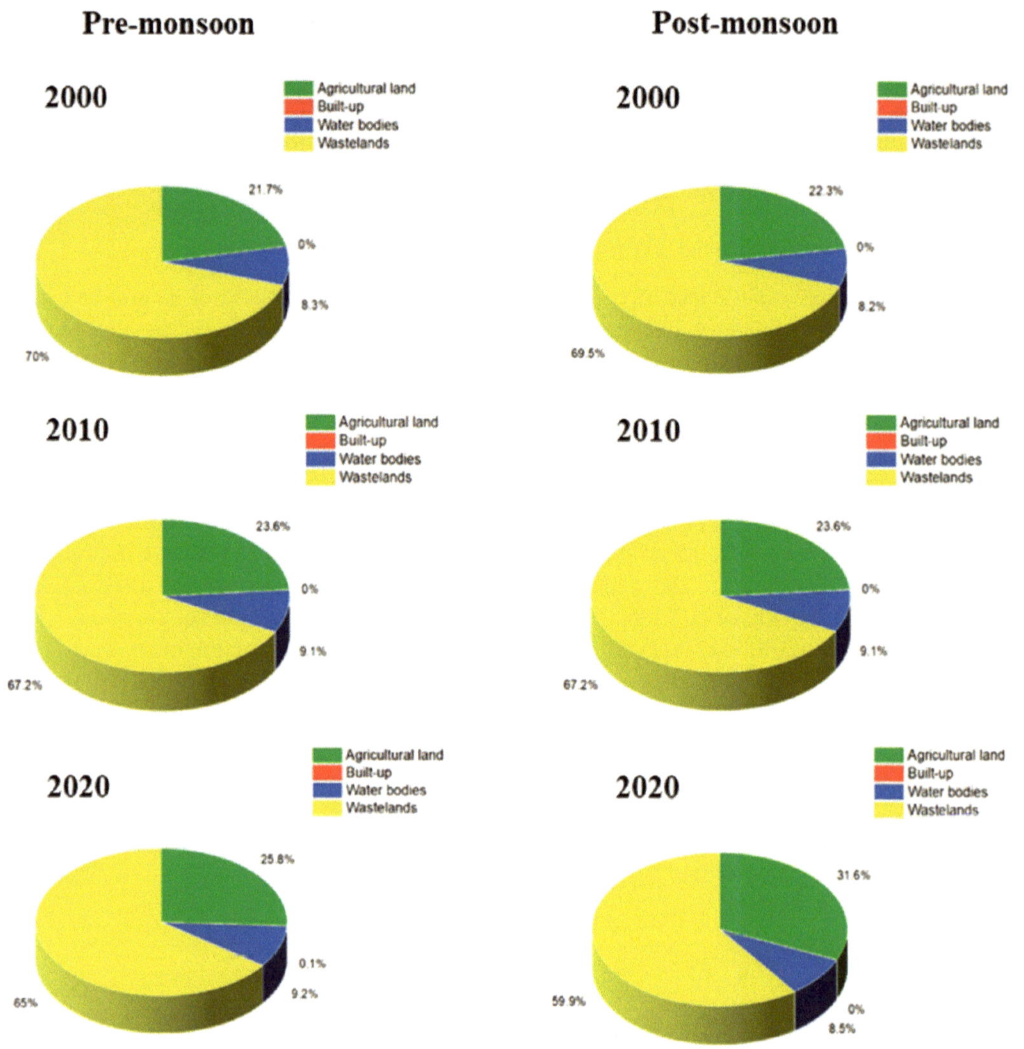

FIGURE 15.4 Pie chart of Harike wetland during the years 2000 to 2020 for both pre- and post-monsoon season.

15.4.3 Assessment of Land Surface Temperature over Harike and Keshopur Wetlands

The spatial pattern of land surface temperature for both seasons (pre- and post-monsoon) has been derived from Landsat 5, Landsat 7, and Landsat 8 data to assess its spatial and temporal variations over both wetlands (Figures 15.7 and 15.8). LST maps have been prepared to delineate the location and intensity of temperature. The results show that the LST pattern over the study region varies considerably. The maximum LST value has been recorded as 53°C in Harike wetland and 49°C in Keshopur wetland. Harike wetland has experienced higher land surface temperature compared with Keshopur wetland during the period from 2000 to 2020. Both study areas experienced high LST in the year 2010. This can be because of global changes, pollution, climatic effects, and man-made activities. Remote sensing data collection and efficient monitoring of LST are fundamental to minimizing the impact on natural resources. Wetlands are among the most biologically diverse ecosystems on Earth and a wide variety of plant and animal species, including endangered ones,

TABLE 15.3
Total Area Coverage of Different Classes over Harike Wetland for Both Seasons (Pre- and Post- Monsoon) during 2000 to 2020

		Area of LULC Class (sq. km)			
Season	**Year**	**Agriculture Land**	**Built-up**	**Water Bodies**	**Wastelands**
Pre-monsoon	2000	18.0511	0.00367	6.94877	58.3314
	2010	18.029	0.02146	6.94884	51.3297
	2020	19.4411	0.04084	6.94877	49.0175
Post-monsoon	2000	18.8174	0.00367	6.94877	58.7844
	2010	18.8968	0.02146	6.94877	51.6976
	2020	25.8223	0.04084	6.94877	49.0235

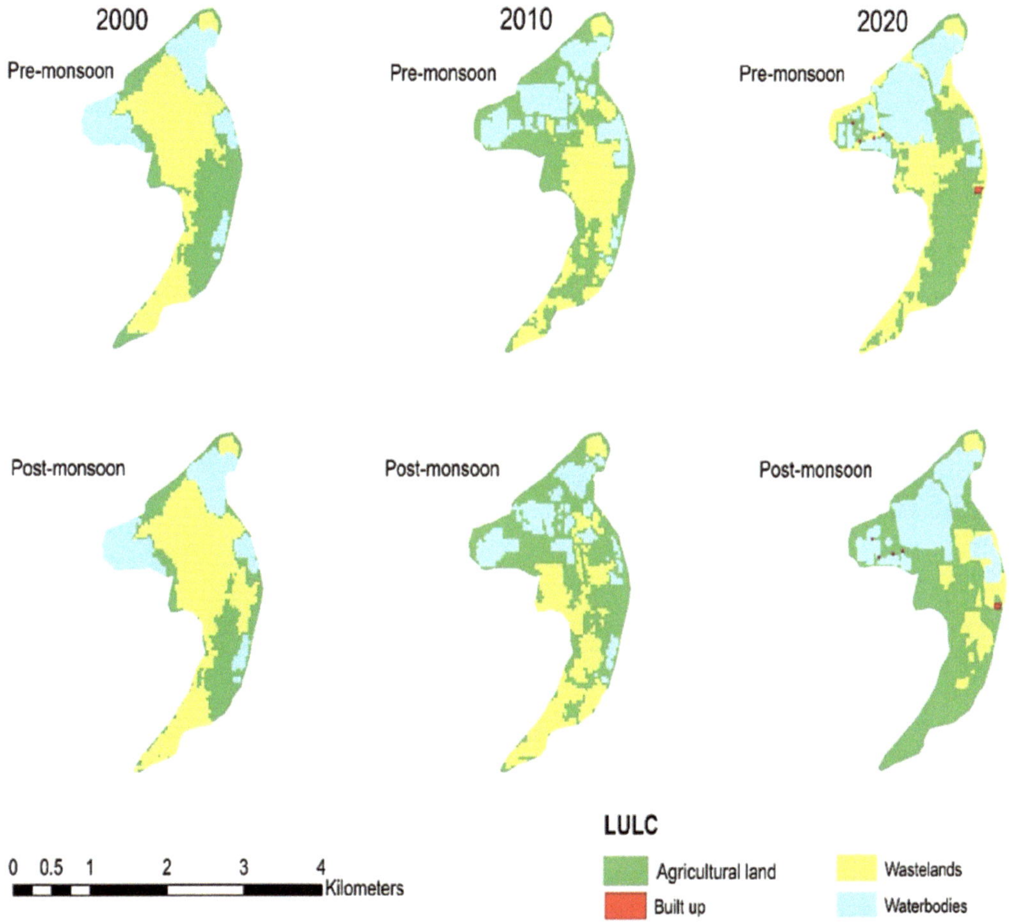

FIGURE 15.5 Land use and land cover dynamics over Keshopur wetland during the period from 2000 to 2020.

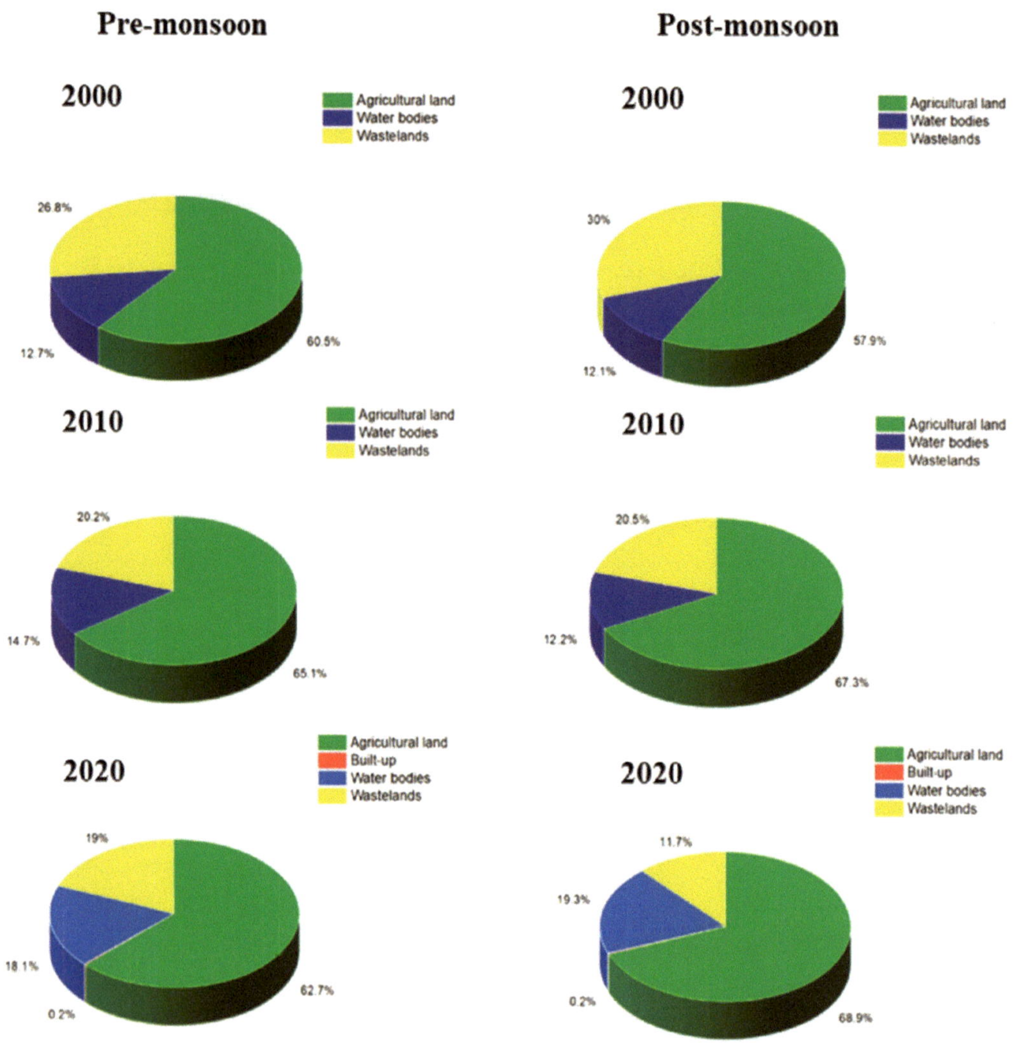

FIGURE 15.6 Pie chart of Harike wetland during the years 2000 to 2020 for both pre- and post-monsoon season.

shelter in and get their habitats from wetlands. The resultant LST maps could be utilized by government officials and resource managers effectively for damage assessment.

15.4.4 Assessment of Rainfall over Harike and Keshopur Wetlands

The trends of monthly rainfall have been derived from Global Precipitation Measurement data for Harike and Keshopur wetlands. The monthly rainfall variations during the period from 2000 to 2022 provide information on the magnitude, intensity, and duration of rainfall experienced by the Harike and Keshopur wetlands. The rainfall occurring is not always the same among years for all regions. Monthly rainfall data have been used to create annual cumulative rainfall. The annual trends of rainfall at Harike exhibit fluctuating patterns for each year, with rainfall values of 317.57 mm in 2000, 327.53 mm in 2001, 327.53 mm in 2002, 253.62 mm in 2003, 121.14 mm in 2004, 263.60 mm in 2005, 282.35 mm in 2006, 182.80 mm in 2007, 359.46 mm in 2008, 220.35 mm in 2009, 394.16 mm in 2010, 296.20 mm in 2011, 254.36 mm in 2012, 343.91 mm in 2013, 263.56 mm in 2014, 320.87 mm in

TABLE 15.4

Total Area Coverage of Different cCasses over Keshopur Wetland for Both Seasons (Pre- and Post-Monsoon) during 2000 to 2020

		Area of LULC Class (sq. km)			
Season	**Year**	**Agriculture Land**	**Built-up**	**Water Bodies**	**Wastelands**
Pre-monsoon	2000	3.03466	NA	0.63566	1.57014
	2010	2.99478	NA	0.67447	0.92844
	2020	2.98871	0.00968	0.86315	0.90673
Post-monsoon	2000	3.03428	NA	0.63566	1.34251
	2010	3.03429	NA	0.54855	0.92571
	2020	3.02201	0.00968	0.84472	0.51141

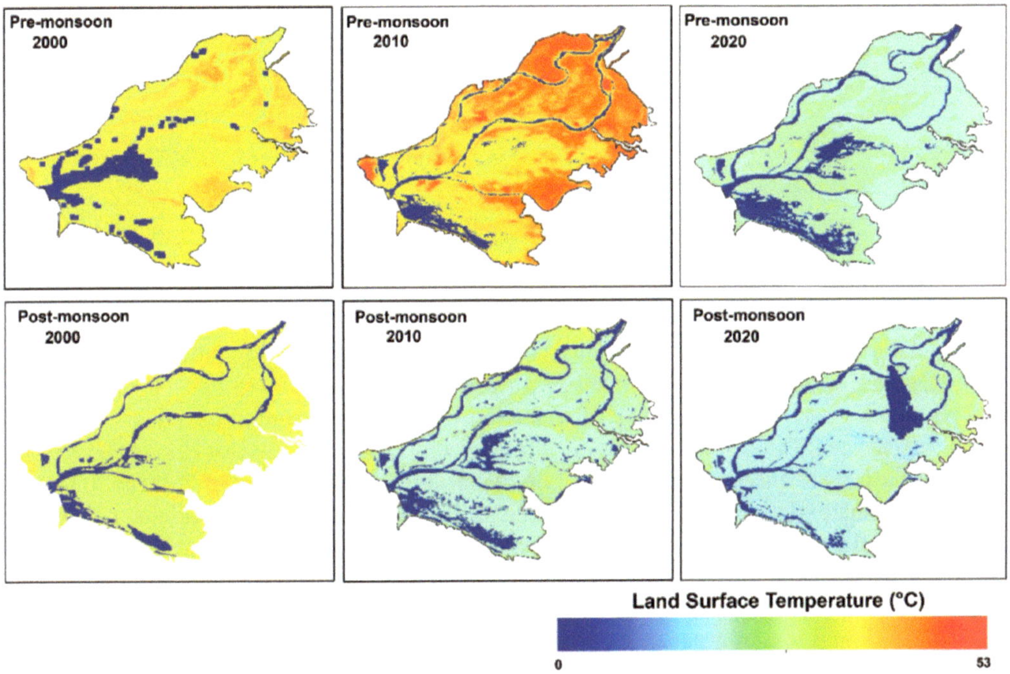

FIGURE 15.7 Spatial variations of land surface temperature over Harike wetland during the period from 2000 to 2020.

2015, 269.14 mm in 2016, 552.53 mm in 2017, 384.75 mm in 2018, 335.16 mm in 2019, 461.32 mm in 2020, 436.65 mm in 2021, and 508.40 mm in 2022 respectively. The results indicated that the Harike wetland has received the maximum cumulative rainfall of up to 552.53 mm in 2017, with a mean rainfall of 325.09 mm during the period from 2000 to 2022 (Figure 15.9a). Similarly, the annual trends of rainfall at Keshopur wetland exhibit fluctuating patterns for each year, with rainfall values of 348.97 mm in 2000, 743.72 mm in 2001, 743.72 mm in 2002, 565.80 mm in 2003, 235.10 mm in 2004, 225.05 mm in 2005, 380.90 mm in 2006, 306.75 mm in 2007, 430.32 mm in 2008, 322.94 mm in 2009, 413.24 mm in 2010, 303.40 mm in 2011, 308.62 mm in 2012, 618.14 mm in 2013, 382.90 mm in 2014, 682.20 mm in 2015, 472.98 mm in 2016, 438.73 mm in 2017, 382.10 mm in 2018, 526.95 mm in 2019, 406.80 mm in 2020, 934.63 mm in 2021, and 348.00 mm in 2022 respectively. The results

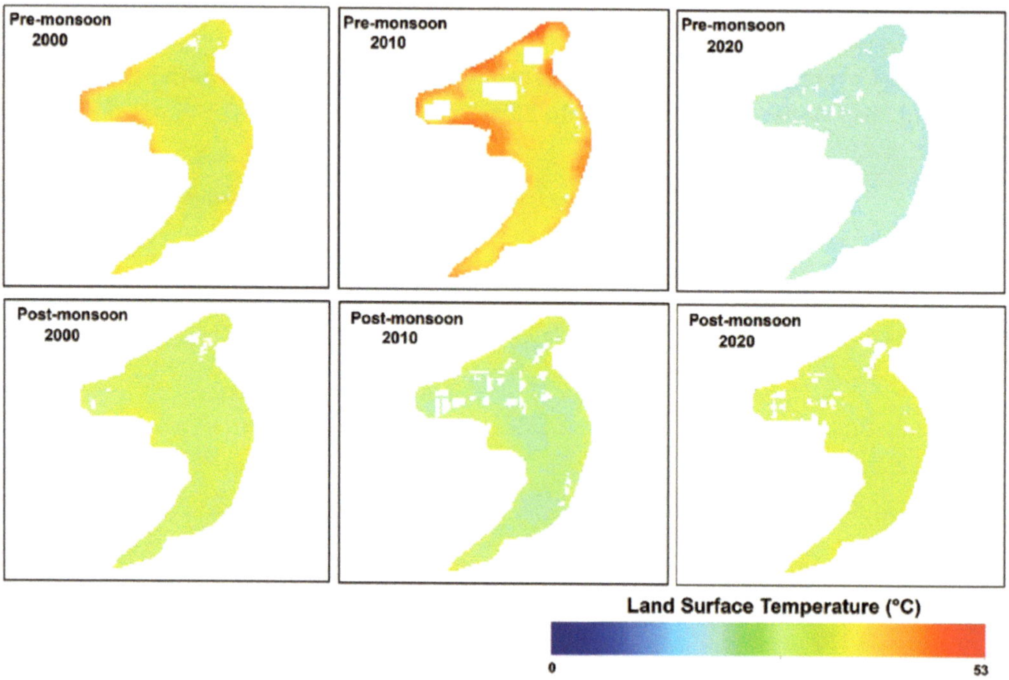

FIGURE 15.8 Spatial variations of land surface temperature over Keshopur wetland during the period from 2000 to 2020.

indicated that the Keshopur wetland received the maximum cumulative rainfall of up to 934.63 mm in 2021, with a mean rainfall of 457.48 mm during the period from 2000 to 2022 (Figure 15.10b).

The monthly rainfall anomaly has been derived from monthly rainfall data and mean rainfall over the study area. The spatial distribution of rainfall anomalies indicates that Harike and Keshopur wetlands have experienced unprecedented rainfall variations during the years 2000 to 2022, which affect the climate, flora, and fauna of that region. The rainfall anomaly represents deviations of annual rainfall from long-run averages, which threaten the flora and fauna of that region. A positive anomaly indicates that the observed rainfall was more than the reference or base value while the negative anomaly indicates that the observed rainfall was lower than the reference value. The spatial and temporal variations of water quality parameters are associated with climatic variables, rainfall, and high runoff (Arora et al., 2022). Monitoring and understanding the water cycle and water conditions is paramount to researchers, scientists, and policymakers for sustainable management.

Wetlands are increasingly isolated and disconnected, making adjustments to rapid climate change through animal and plant dispersal very difficult. Thus, climate change clearly represents an additional, significant threat to wetlands, which will interact with existing human-caused stresses. Wetlands are of greater importance in our ecological cycle and also provide food, water aquifers, and groundwater recharge, protect us from flood and storms, and control the soil erosion and climate stability. An ecosystem is a dynamic complex of biotic communities and the non-living environment which interacts as a functional unit. It affects both the non-living and the biotic components of the ecosystems and their whole structure and functioning. In this chapter, we have tried to explain how the rainfall has changed in the wetlands, how much rainfall has occurred, and whether the rainfall was good or bad. We have also shown the extent of the wetlands and how wetlands are affected due to climate change.

The integration of remote sensing and GIS techniques has been widely used by many researchers for mapping and monitoring wetland ecosystems. Chopra et al. (2001) has worked on the mapping,

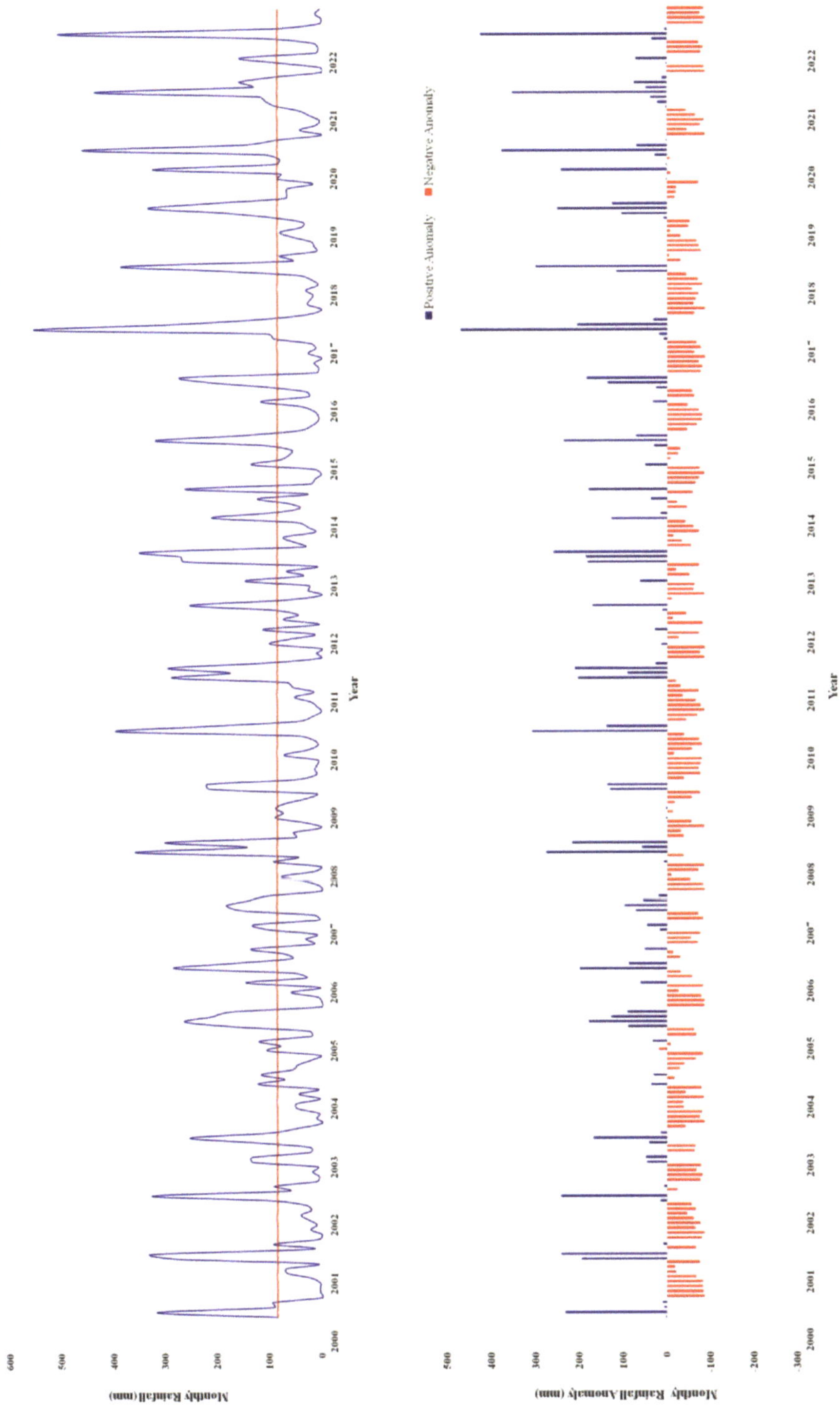

FIGURE 15.9 (a) Monthly rainfall variations and (b) rainfall anomalies over Harike wetland during the period from 2000 to 2022.

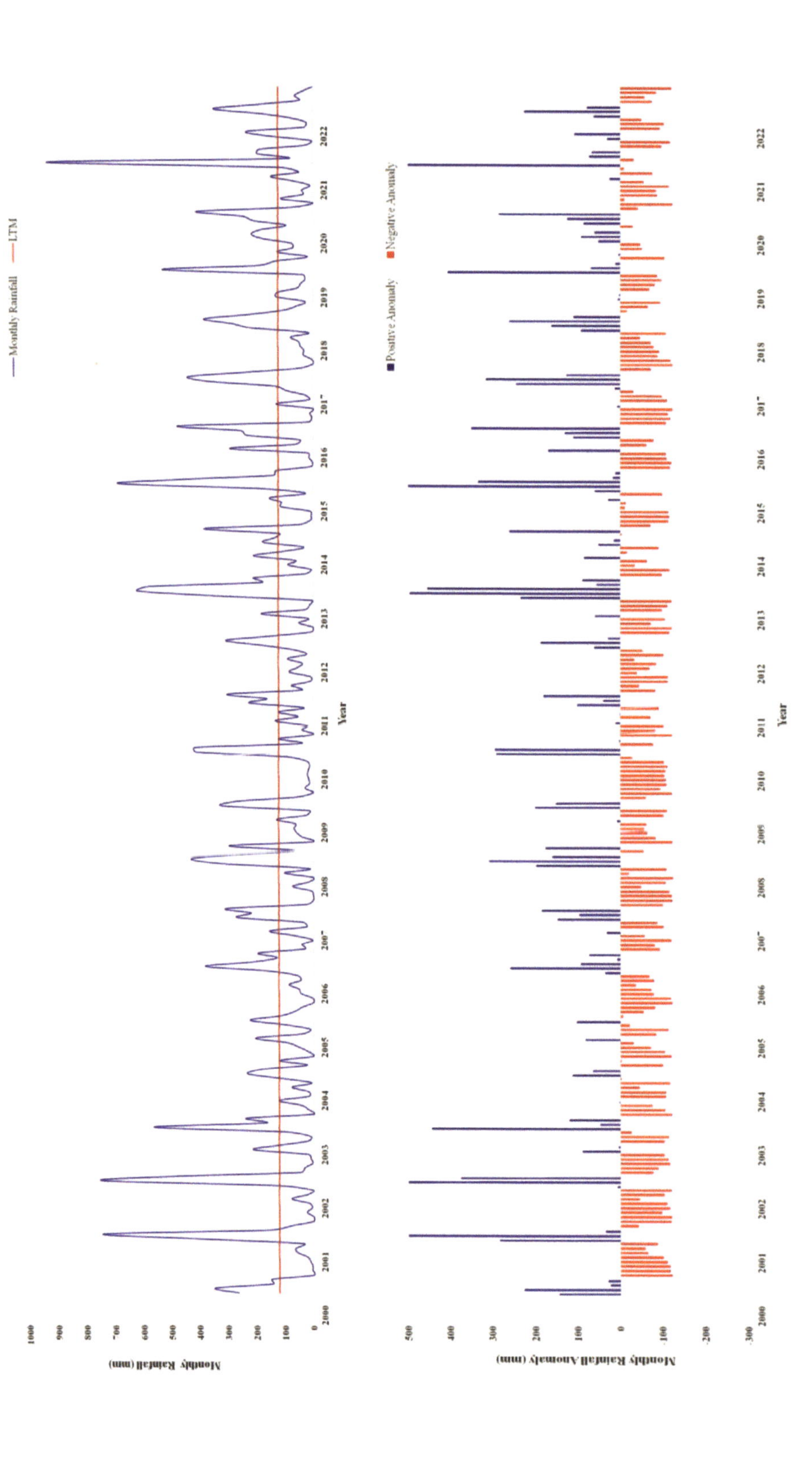

FIGURE 15.10 (a) Monthly rainfall variations and (b) rainfall anomalies over Keshopur wetland during the period from 2000 to 2022.

monitoring, and conservation of the Harike wetland ecosystem, Punjab, India, through Indian Remote Sensing Satellite (IRS) remote sensing. IRS-1A LISS-II data were used to study the land use and land cover map of the wetland zone during post-monsoon (October 1991), spring (February 1992), and pre-monsoon (May 1992) seasons and multi-season data. This paper showed land use/land cover of the Harike wetland ecosystem, water spread and aquatic vegetation, turbidity, water quality of the lake, and threats to the Harike wetland. They concluded that the growth of weeds, the inflow of industrial effluents, and the addition of agricultural waste products and fertilizers in the pond area have resulted not only in shrinkage in the lake but have also affected the habitat for the ecosystem and the survival of birds. Jain et al., 2008 has studied the impact of the declining trend of flow on the Harike wetland. The methodology used for this study involved the creation of a database, processing of field data, and analysis of rainfall-runoff. He concluded that the wetland of Harike is reducing due to various reasons such as silting and the decreasing trend in flow in upstream catchment. Kumar et al. (2018) has worked on Punjab wetlands to study their gaps in policies and threats to wetlands. The classification of wetlands was based on their size. Indian Space Research Organization (ISRO) officials divided the wetlands into two categories based on size: those of 2.25 hectares and greater are included in larger wetlands, while those less than 2.25 hectares are included in smaller wetlands. This paper has also discussed wetland management policies and their lack of implementation. It was concluded that the management and conservation of medium and small-sized wetlands should be the same as for large-sized wetlands. Kumar et al. (2020) have worked on mapping and monitoring selected wetlands of Punjab, India, using geospatial techniques over a span of 28 years from 1990/1991 to 2018. The datasets used here are Landsat 5, Landsat 7, Landsat 8, and IRS LISS-IV. They used different indices such as NDVI, NDWI, TWI, and MNDWI for wetland studies. These indices were used for LULC dynamics in wetlands and surface water dynamics. This study concluded that wetlands located in hilly areas for the purpose of hydroelectric power generation had little fluctuation in water surface while Harike wetland, which is located in the plains, suffered major fluctuations in the extent of water spread. Singh et al. (2020) conducted a remote sensing and GIS-based analysis of temporal land use/land cover and water quality changes in the Harike wetland ecosystem, Punjab, India, during 2006 to 2014 and 2014 to 2018. The wetland acts as a sink for pollutants, and this paper concluded that agricultural land, forest, and built-up areas increased while wasteland, water bodies, and wetlands decreased during the period 2006 to 2018.

15.5 CONCLUSION

This study concludes that wetland changes over the study area are a major concern due to long-term climatic changes and human activities. The significant reduction of highly biodiverse wetland areas must be addressed. The integration of remote sensing and GIS techniques is useful for the monitoring of wetland ecosystems. Landsat imagery has been used for the mapping of wetland dynamics during the pre- and post-monsoon season. The results showed that a significant change has been observed in the Harike and Keshopur wetlands and many wetlands have been converted into agricultural lands and built-up areas. Additionally, there is also land degradation due to soil erosion. Intense human activities have resulted in more changes in the wetlands. This study indicates that wetland monitoring is essential for its ecosystem management and for reducing greater impact in the future. This will help policymakers implement new strategic policies for the wetlands.

ACKNOWLEDGEMENTS

The authors are grateful to the USGS and NASA for providing Landsat Imagery and Global Precipitation Measurement (GPM) rainfall data in the public domain. The authors express thanks to scientists and researchers from Punjab Remote Sensing Centre, Ludhiana, for their valuable suggestions and support during in-house paper review.

REFERENCES

Arora, M., Mudaliar, A., and Pateriya, B. (2022). Assessment and monitoring of optically active water quality parameters on wetland ecosystems based on remote sensing approach: A case study on Harike and Keshopur wetland over Punjab Region, India. *Engineering Proceeding*, 27, 84. https://doi.org/10.3390/ecsa-9-13361.

Barros, D.F., and Albernaz, A.L.M. (2014). Possible impacts of climate change on wetlands and its biota in the Brazilian Amazon. *Brazilian Journal of Biology*, 74(4), 810–820.

Chen, X.L., Zhao, H.M., Li, P.X., and Yin, Z.Y. (2006). Remote sensing image-based analysis of the relationship between urban heat island and land use/cover changes. *Remote Sensing of Environment*, 104(2), 133–146.

Chopra, R., Verma, V.K., and Sharma, P.K. (2001). Mapping, monitoring and conservation of Harike wetland ecosystem, Punjab, India, through remote sensing. *International Journal of Remote Sensing*, 22(1), 89–98.

Erwin, K.L. (2009). Wetlands and global climate change: The role of wetland restoration in a changing world. *Wetlands Ecology and Management*, 17(1), 71–84. http://dx.doi.org/ 10.1007/ s11273-008-9119-1.

Ferrati, R., Canziani, G.A., and Moreno, D.R. (2005). Estero dellbera: Hydrometeorological and hydrological characterization. *Ecological Modelling*, 186, 3–15.

Finlayson, C.M., and Davidson, N.C., summary report. In Finlayson, C.M, and Spiers, A.G. (eds). (1999). Global review of wetland resources and priorities for wetland inventory, supervising scientist report 144, Canberra, 1–13.

Hulme, P.E. (2005). Adapting to climate change: Is there scope for ecological management in the face of a global threat? *Journal of Applied Ecology*, 42, 784–794.

IPCC (International Panel on Climate Change). (1996). Climate change 1996—impacts, adaptations and mitigation of climate change: Scientific technical analysis. Contribution of working group II to the second assessment report of the IPCC. Cambridge University Press, Cambridge.

Jain, S.K., Sarkar, A., and Garg, V. (2008). Impact of declining trend of flow on Harike Wetland, India. *Water Resources Management*, 22, 409–421. https://doi.org/10.1007/s11269-007-9169-9

Junk, W.J., An, S.Q., Finlayson, C.M., Gopal, B., Kvet, J., Mitchell, S.A., Mitsch, W.J., and Robarts, R.D. (2013). Current state of knowledge regarding the world's wetlands and their future under global climate change: A synthesis. *Aquatic Science,* 75, 151–167.

Kayranli, B., Scholz, M., Mustafa, A., and Hedmark, Å. (2010). Carbon storage and fluxes within freshwater wetlands: A critical review. *Wetlands*, 30, 111–124.

Klemas, V. (2011). Remote sensing of wetlands: Case studies comparing practical techniques. *Journal of Coastal Research*, 27(3), 418–427.

Kumar, G., & Singh, K. K. (2020). Mapping and monitoring the selected wetlands of Punjab, India, using geospatial techniques. *Journal of the Indian Society of Remote Sensing*, 48(4), 615–625.

Kumar, V., and Chopra, A.K. (2009). Impact of climate change on biodiversity of India with special reference to Himalayan region-An overview. *Journal of Applied and Natural Science*, 1(1), 117–122.

Kumar, G., and Kaur, A. (2018). Status of wetlands in Punjab: A review on policy frameworks. *Asian Journal of Multidimensional Research (AJMR)*, 7(10), 169–177.

Kumar, G., and Singh, K.K. (2020). Mapping and monitoring the selected wetlands of Punjab, India, using geospatial techniques. *Journal of the Indian Society of Remote Sensing*, 48(4), 615–625.

Kundpura, S., Kommoju, R., and Verma, I. (2019). Assessment of changes in wetland storage in Gurupura river basin of Karnatka, India, using Remote Sensing and GIS techniques. In: Rathinasamy, M., Chandramouli, S., Phanindra, K.B.V.N., Mahesh, U. (Eds.), *Water Resources and Environment Engineering II.* © Springer Nature Singapore, Pte Ltd., pp. 57–68. https://doi.org/10.1007/978-98.

Pan, J., Durand, M., Sandells, M., Lemmetyinen, J., Kim, E.J., Pulliainen, J., Kontu, A. and Derksen, C., (2015). Differences between the HUT snow emission model and MEMLS and their effects on brightness temperature simulation. *IEEE Transactions on Geoscience and Remote Sensing,* 54(4), 2001–2019.

SAC. (2011). National wetland atlas. http://saconenvis.nic.in/publication%5CNWIA_National_atlas.pdf.

Sahagian, D., and Melack, J. (1998). Global wetland distribution and functional characterization: Trace gases and the hydrologic cycle. IGBP Report 46.

Sahoo, S., Majumder, A., Swain, S., Gareema, Pateriya, B., and Al-Ansari, N. (2022). Analysis of decadal land use changes and its impacts on urban heat island (UHI) using remote sensing-based approach: A smart city perspective. *Sustainability*, 14(19), 11892.

Sahoo S, Swain S, Goswami A, Sharma R, Pateriya, B. (2021). Assessment of trends and multi-decadal changes in groundwater level in parts of the Malwa region, Punjab India. *Groundwater Sustainable Development,* 14:100644. https://doi.org/10.1016/j.gsd.2021.100644

Sharma, C., Jindal, R., Singh, U.B., and Ahluwalia, A.S. (2018). Assessment of water quality of river Satluj, Punjab, India. *Journal of Sustainable Water Resources Management*, 4(4), 809. https://doi.org/10.1007/s40899-017-0173-9.

Singh, S., Bhardwaj, A., and Verma, V.K. (2020). Remote sensing and GIS based analysis of temporal land use/land cover and water quality changes in Harike wetland ecosystem, Punjab, India. *Journal of Environmental Management*, 262. https://doi.org/10.1016/j.jenvman.2020.110355.

Sivakumar, R., and Ghosh, S. (2016). Wetland spatial dynamics and mitigation study: An integrated remote sensing and GIS approach. *Natural Hazards*, 80, 975–995.

Weng, Q., Lu, D., & Schubring, J. (2004). Estimation of land surface temperature–vegetation abundance relationship for urban heat island studies. *Remote sensing of Environment*, 89(4), 467–483.

16 Coral Reefs Threat Assessment and Comparative Analysis of Neil and Agatti Island, India, Using Multispectral Approach

*Md. Wasim, Smrutisikha Mohanty, Prem C. Pandey**
Department of Life Sciences, School of Natural Sciences, Shiv Nadar Institution of Eminence, Greater Noida, Uttar Pradesh, India

Basheer Ahammed KK
Universitas Udayana (UNUD), Denpasar, Bali, Indonesia

Arvind Chandra Pandey
Department of Geoinformatics, School of Natural Resource Management, Central University of Jharkhand, Brambe Ranchi,Jharkhand, India

16.1 INTRODUCTION

Coral reefs, often referred to as the "marine rainforest" of the sea (Gajendran, 2007) constitute high-productivity marine communities characterized by rich biodiversity (Richmond, 1993). About 30%–40% of all fish species on earth are located in coral reefs, indicating their crucial role in upholding the balance of marine resource productivity and serving as an indicator of the health of the marine environment (Hourigan et al., 1988; Hein et al., 2015). Owing to both natural forces and anthropogenic impacts, coral reef ecosystems have undergone rapid transformation at both global and regional scales in recent decades. Disturbingly, over 60% of the world's coral reefs are currently facing significant threats (Burke et al., 2011; Wilkinson, 2008). On a regional scale, coral reef degradation is exacerbated by human activities including overexploitation, coastal development, destructive fishing practices, and the impact of pollution and runoff (Burke et al., 2011; Mora, 2008). Due to the increasing trend of global warming with the concomitant rise in atmospheric carbon dioxide coupled with ocean acidification, degradation and bleaching of coral reef ecosystems are prevalent (Baker et al., 2008; Guldberg et al., 2007). About 55% of the world's coral reefs are affected by overfishing and destructive fishing (Bruno and Selig, 2007; Burke et al., 2011). Coral bleaching is a phenomenon in which corals, under stress, undergo a discoloration as a result of expelling symbiotic algae residing in their tissues. This process can have a significant impact on the morphology and mortality of the corals (Loya et al., 2001; Smith et al., 2018). Excessive tourism activities are also a triggering factor for coral degradation (Huges, 2003). Tourism causes direct damage to coral reefs through activities such as coral breakage by divers, caused by SCUBA tanks or other equipment, and the generation of sediment clouds. Additionally, indirect damage is inflicted by tourism facilities through factors like sewage discharge, runoff, sedimentation, coastal alteration, and the

* Corresponding author.

 DOI: 10.1201/9781003354000-21

disposal of anthropogenic waste onto reef areas, including boating-related impacts (Barker et al., 2004; Dinsdale et al., 2004; Hawkins et al., 1993).

Coral reefs are being affected by climate change as they are highly sensitive to thermal stress due to low adaptation to variations in temperature (Berstein et al., 2007). Researchers have shown that an increase in sea surface temperature (SST) by 1°C to 3°C causes an increase in acidification, resulting in a reduced ability of coral to form calcium carbonate (Pandolfi et al., 2011). The increasing concentration of CO_2 causes difficulty in making calcium carbonate shells for marine organisms resulting in higher mortality and frequent bleaching (Bernstein et al., 2007). The intensity of coral bleaching is directly proportional to sea surface temperature; it escalates with rising temperature and diminishes as temperature decreases (Safaie et al., 2018; Hoegh et al., 2007).

Coral reefs throughout the Indian subcontinent have encountered 29 widespread bleaching events since 1989, as reported on www.reefbase.org. In the years 1998, 2002, 2005, and 2010 coral bleaching was observed around Andaman Island (Krishnan et al., 2011; Pet-Soede et al., 2000; Rajasuriya et al., 1999; Ravindran et al., 1998; Roy et al., 2006, 2014).

Substantial recovery of the coral reef in the Indian Ocean was observed post-1998 (Baker et al., 2008). However, in the year 2010, the Andaman Islands experienced a severe episode of mass bleaching, attributed to a rapid increase in SST (Krishnan et al., 2011). Early records of 30%–40% coral bleaching in Lakshadweep, the Gulf of Mannar, and the Gulf of Kachchh indicated that 60%–90% mortality of coral reefs occurred due to El Nino effects (Arthur, 2000; Arthur et al., 2005).

The study can also be compared with the status of other coral reef ecosystems across the world. Water temperatures recorded in the northern part of the Great Barrier Reef (GBR) during the 2015–2016 austral summer surpassed both the empirical bleaching thresholds and the historical climate record. This matched observations of widespread and severe bleaching in the northern sector of the GBR (GBRMPA, 2016a, 2016b). In the summer, a rapid increase in the average ocean temperature in the tropical Atlantic and the Caribbean led to the most severe coral-bleaching event, with approximately 80% of coral experiencing bleaching and more than 40% succumbing to mortality during that period (Kennedy and Lindsey, 2010).

Physical factors such as salinity, temperature, and light play a pivotal role in the survival, growth, and photosynthesis of corals (Ferrier-Pagès et al., 1999; Baird and Hughes, 2000; Ferrier-Pagès et al., 2007; Chow et al., 2009). Research indicates that changes in salinity by ±2 psu from normal levels can lead to a decrease in coral growth rates (Ferrier-Pagès et al., 1999). The rapid fluctuation of salinity levels may induce coral mortality (Sakai et al., 1989; Jokiel et al., 1993) by affecting the photosynthesis of zooxanthellae, resulting in reduced energy transferred to corals (Muthiga and Szmant, 1987; Manzell and Lirman, 2003).

Solar radiation, either in the form of ultraviolet (UV) or photosynthetically active radiation (PAR), can lead to damage the exposed corals, and in certain instances, coral bleaching occurs even in the absence of elevated temperatures (Fitt et al., 2001; Brown et al., 2002). Sudden increases in radiation exposure have also been associated with bleaching events (Hoegh-Guldberg and Smith, 1989). Laboratory experiments have demonstrated that elevated radiation, in conjunction with increased temperature, induces additional bleaching beyond the impact of temperature alone (Anthony et al., 2007). This phenomenon is further substantiated by molecular biomarker studies conducted by Downs et al. (2000).

Satellite-based sensors are commonly employed for coral health monitoring due to its capability to offer multi-temporal datasets for periodic assessment of coral reefs (Goodman et al., 2013; Mumby and Harborne, 1999). Remote sensing and GIS technology help in scientific investigation, management, as well as mapping and classification of coral classes using various classification techniques. Roelfsema et al. (2013) used multispectral Thematic Mapper (TM) satellite data for mapping the spatial distribution of microalgae in coral reefs. Changes in the coral reef in Hurghada, Egypt, were detected using Landsat images of different generations, i.e. Landsat 5 (TM), Landsat 7 (ETM), and Landsat 8 (OLI) (El-Askary et al., 2014).

In the present study, an attempt was made for temporal mapping (2002 and 2020) of coral reefs in the Neil Island of Andaman-Nicobar and Agatti Island of Lakshadweep using Landsat 7 ETM and Landsat 8 OLI satellite images. The depth invariant bottom index technique has been applied to identify the submerged and heterogeneous coral reef classes. Digital image processing and supervised classification techniques were applied for classifying the various coral reefs along with the impact of climatic and marine parameters viz., SST, PAR, salinity, sea level rise (SLR), turbidity, surface algal bloom (SAB), and colored dissolved organic matter (CDOM) on coral bleaching.

16.2 STUDY AREA

The Andaman-Nicobar Islands, spanning an area of 6,408 km^2, are situated within latitudes 13° 41' 19''' to 10° 30' 14" N and longitudes 93° 01' 56" to 92° 30' 37" E in the Bay of Bengal with a maximum elevation of 732 m (2,401.57 ft). In contrast, the Lakshadweep islands (32.62 km^2) are located within latitudes 10° 19' 40" to 11° 38' 24" N and longitudes 72° 47' 04" to 72° 08' 49" E in the Arabian Sea. The selected two islands for study (Figure 16.1) are Neil Island (11° 49' 48" N and 93° 2' 24" E, Area: 13.7 km^2), an island of Andaman located in Ritchie's Archipelago, and Agatti Island (10° 30' 0" N and 73° 0' 0" E, Area: 3.22 km^2), situated on a coral atoll called Agatti Atoll in the Union Territory of Lakshadweep. Neil Island, renamed Shaheed Dweep in December 2018, is a comparatively flat island best known for its unparalleled biodiversity and rich coral reefs. The average summer and winter temperatures on this island are 30.2°C and 23.0°C, respectively. Agatti Island is situated on the eastern arc of the coral shoal, measuring just over 5.6 km in length and approximately 1000 m wide at the broadest point. The island features a remarkably beautiful lagoon on its western side, inhabited by a diverse array of multi-colored fishes and coral reefs. The average

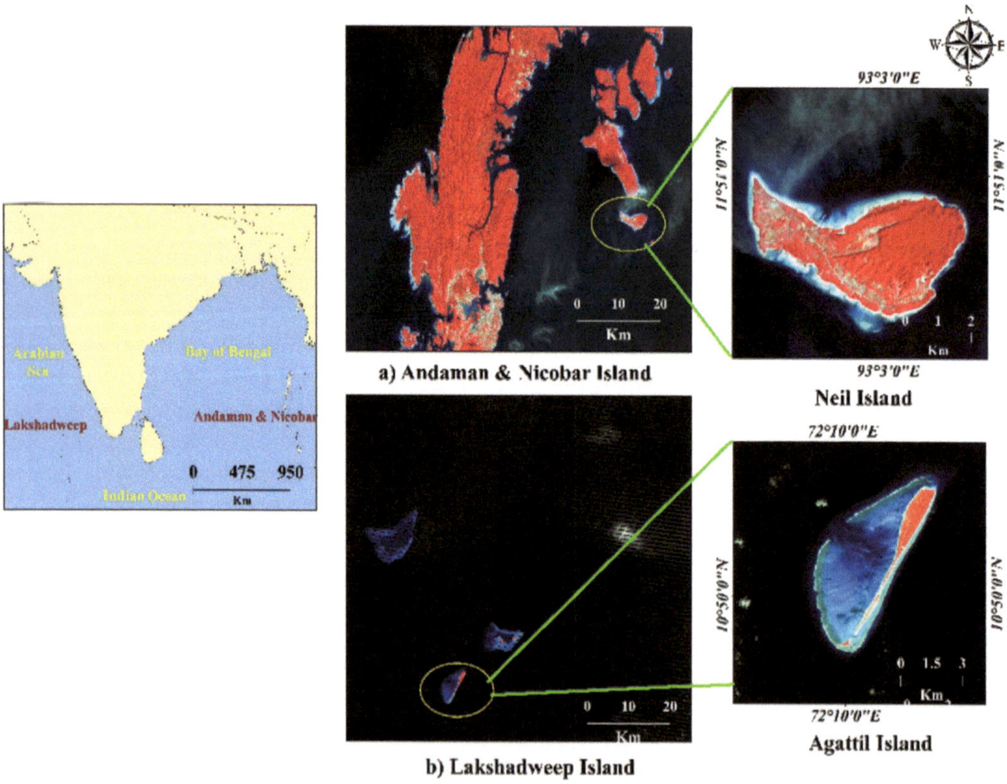

FIGURE 16.1 Location map of the study area, Agatti Island and Neil Island, along with a false-color composite image of Landsat 7.

summer and winter temperatures on this island are 32.0°C and 28.0°C, and humidity ranges from 70% to 76% for most of the year. According to the census of India 2011, the total population of Neil Island was 3040 with a population density of 222 persons per km^2, whereas for Agatti Island the total population was 7700 with a population density of 2386 persons per km^2.

16.3 MATERIALS AND METHODS

In this study, the utilization of Landsat 7 (ETM) and Landsat 8 (OLI/TIRS) satellite data with a spatial resolution of 30 m provided a robust foundation for mapping and monitoring coral reef habitats on selected islands. The choice of these specific satellite sensors allows for the acquisition of multispectral imagery, enabling researchers to discern key environmental features and characteristics of coral ecosystems. The spectral bands captured by Landsat 7 and 8 sensors are particularly well suited for differentiating various benthic cover types and assessing the health of coral reefs. Landsat satellite imageries were downloaded from the website (www.earthexplorer.usgs.gov) of the United States Geological Survey (USGS), whereas MODIS-based SST and PAR products were downloaded from Giovanni (https://giovanni.gsfc.nasa.gov/giovanni/), and sea level rise and salinity data were downloaded from the Asia-Pacific Data Research Centre (http://apdrc.soest.hawaii.edu/). The detail of the dataset used in the present study is illustrated in Table 16.1.

The depth invariant bottom index technique was applied for the identification of the benthic type submerged and heterogeneous coral reef classes on both islands. This method usually transforms image radiance/reflectance data into a relative index value using suitable pairs of visible bands and is used for the elimination of water column effects that arise due to absorption and scattering through the water column. Before applying the DII technique, the land part is extracted because this technique applies to the water part only. To distinguish between land and water areas, threshold values were determined using the Near-Infrared (NIR) band. The blue band (0.490 μm), green band (0.560 μm), and NIR band (0.842 μm) have been used for the present research objectives. The deglint process was applied to resolve the problem related to sun glint and low clouds. To improve the image quality, haze removal and atmospheric correction along with the application of the crisp filter were performed. These procedures serve to sharpen the image and enhance the contrast between bright

TABLE 16.1
Data Used for the Present Study

S. No.	Data Used	Source	Spatial Resolution	Purpose	Time Period
1.	Landsat 7 (ETM+)	USGS	30 m	Coral classification, Turbidity, SABI, CDOM for year 2002	2002-02-07 for (Neil Island), 2002-02-15 for (Agatti Island)
2.	Landsat 8 (OLI/TIRS)	USGS	30 m	Coral classification, Turbidity, SABI, CDOM for the year 2020	2020-02-01 for Neil Island, 2020-02-16 for Agatti Island
3.	MODIS (SST product)	GIOVANNI	4 km	SST variation	April, May, June (2002–2020)
4.	MODIS (PAR product)	GIOVANNI	4 km	PAR variation	April, May, June (2002–2020)
5.	Sea level rise data	Asia-Pacific Data Research Centre	Globally by 1°	To observe sea level rise from 2002 to 2020	April, May, June (2002–2020)
6.	Salinity	Asia-Pacific Data Research Centre	Globally by 1°	To observe salinity from 2002 to 2020	April, May, June (2002–2020)

and dark regions, effectively highlighting distinct features of the coral reefs. The coral habitat on both islands was classified using a supervised classification technique with a maximum likelihood classifier. Signatures were meticulously obtained based on the reflectance pattern of various classes. The algorithm was applied for computing the Surface Algal Bloom Index (SABI), CDOM, and turbidity. Finally, the net bleaching for both islands during the period from 2002 to 2020 was determined using the image differencing technique. The pivot tool was used for creating the confusion matrix in ArcGIS 10.3 and further statistical analysis was done in MS Excel software. A 10 km buffer zone was created using Buffer Analysis in ArcGIS 10.3 around the islands to observe the impact of SST, PAR, salinity, and SLR from the years 2002-2020, and further trend analysis was done to observe the variation of these parameters and their relationship with coral reef temporal variation. The overall flowchart of the methodology adopted for the study is shown in Figure 16.2.

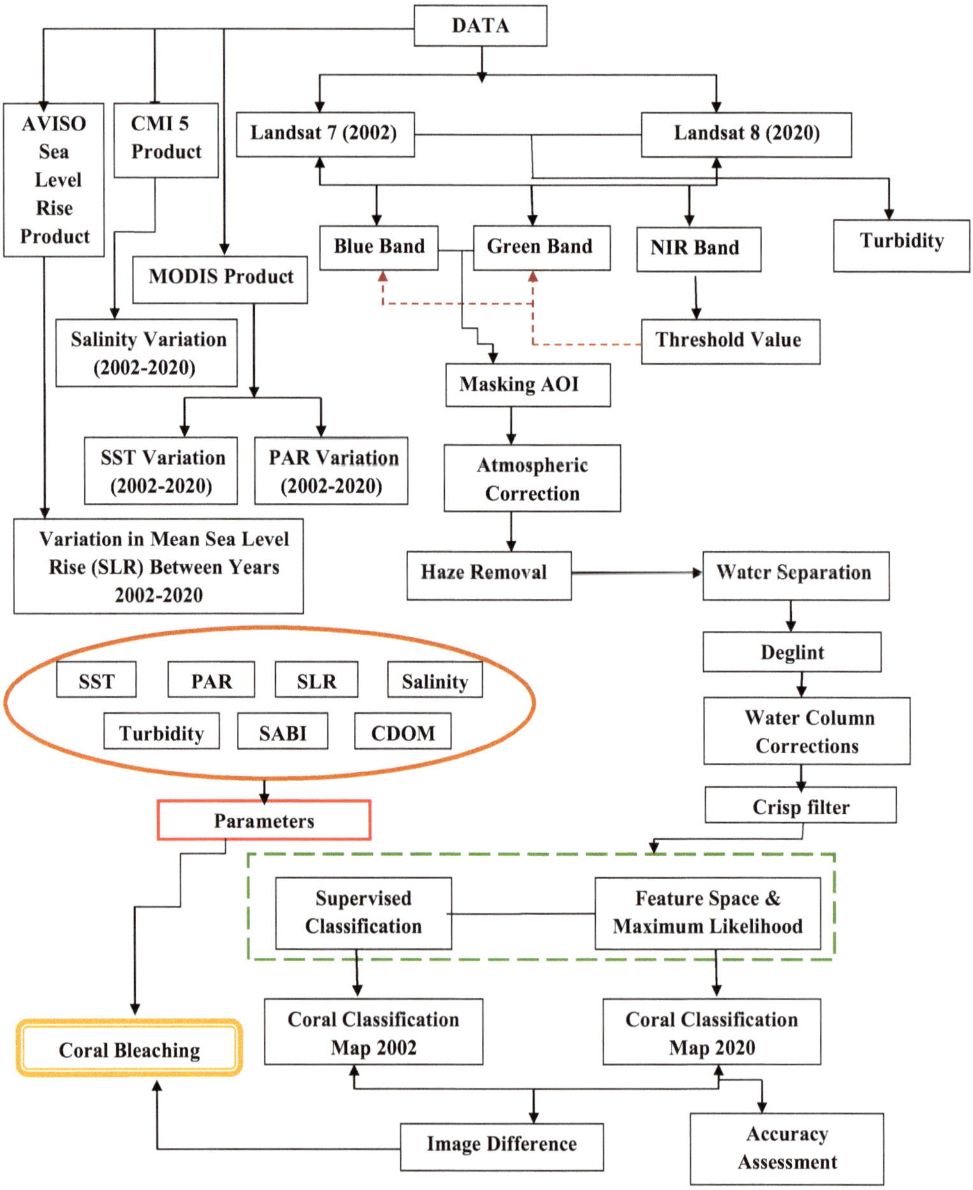

FIGURE 16.2 Flowchart of the methodology adopted in the study.

16.3.1 Image Processing

Image processing was conducted through several stages including haze removal, atmospheric correction, and crisp filter application on satellite images of the selected islands of Andaman-Nicobar and Lakshadweep. Image enhancement involved the correction of the water column by using the DII (Dark Object Subtraction) transformation.

16.3.1.1 Image Subsetting and Masking

Subset or masking is a method employed in the Region of Interest (ROI) selection technique to reinforce geospatial analysis and concentrate on the study area. The ROI was masked by creating a simple polygon around the region of interest using the Extract by Mask tool in ArcGIS 10.3.

16.3.1.2 Atmospheric Correction

Atmospheric correction was executed to address issues related to the scattering and absorption phenomena caused by atmospheric molecules. Additionally, this correction aimed to eliminate the angular effect that could potentially reduce the quality of information in satellite imagery data by up to 10%, depending on the spectral channel (Che and Price, 1992; Ishizaka et al., 1992). In this study, atmospheric correction was applied to Landsat 7 (ETM+) and Landsat 8 (OLI/TIRS) using R-Studio software. Equation 16.1 was utilized to address the problem as given below:

$$pA' = Mp*Q_{cal} + A_p \tag{16.1}$$
$$\text{Sun angle correction} = pA'/\sin_{(z)}$$

In the given equations:

pA' stands for the TOA (Top of Atmosphere) planetary reflection, without correction for the sun angle.

M_p is the band-specified multiplicative rescaling factor from the metadata.

Ap is the band-specified additive rescaling factor derived from the quantized and calibrated standard value pixel (Q_{cal}) in the imagery, and z denotes the sun elevation angle sourced from metadata.

16.3.1.3 Haze Removal

Haze removal is an algorithm designed to address the interference caused by the light scattering phenomenon through haze particles. Its purpose is to enhance the visual quality of images by mitigating the impact of haze. In this study, haze removal was executed utilizing the Haze Reduction tool in ERDAS Imagine 2014, with the primary objective of improving image contrast.

16.3.2 Water Separations

Before applying the deglint and depth invariant index technique, it is important to separate land from water, because these apply to water parts only. To distinguish water from land, the NIR band was employed, leveraging the characteristic of water appearing dark and land appearing bright in this spectral range. A threshold value of 2100 for Neil Island and 1900 for Agatti Island was established. This separation was achieved by drawing a line with the cross-section line tool in ArcGIS 10.3, crossing from land (bright areas) into the water (dark areas).

16.3.3 Deglint

The deglint process was implemented on the water component to rectify the radiometric contribution arising from sun glint and low clouds (Hedley et al., 2005). This process has been performed for the radiometric correction of blue and green bands. To execute the deglint process, a linear regression was established between the NIR band and visible bands (i.e. green and blue bands). This was achieved by converting the raster data into points using the raster-to-point tool in ArcGIS 10.3. The stability of the linear regression was ensured by selecting a sample area of image pixels. The equation that has been used for removing sun glint and low clouds is shown below, as proposed by Hedley et al. (2005).

$$R_i' = R_i - b_i\,(R_{NIR} - Min_{NIR}) \tag{16.2}$$

where:
R_i' is the deglint pixel in band i,
R_i is the reflectance from visible bands (blue and green bands),
b_i is the regression slope
R_{NIR} is the NIR band value and Min_{NIR} is the minimum NIR value of the sample, which indicates the pixel brightness of NIR with no sun glint.

The use of the minimum NIR pixel is generally preferred over the maximum NIR pixel as it is less susceptible to problematic outliers.

16.3.4 Applying Depth Invariant Bottom Index

The depth invariant indices transformation method (DII) is used for the elimination of water column effects that arise due to absorption and scattering through the water column. This method was proposed by Lyzenga (1978) and was designed to discriminate various bottom types using multispectral image data. The initial assumption underlying this approach is that the radiance ratios of two distinct benthic covers remain independent of water depth, as long as the attenuation coefficients are consistent across the selected bands. Prior to applying water column correction, a calculation of the attenuation coefficient was conducted through statistical analysis. This method is specifically employed in waters with high transparency, and its efficacy is contingent on the wavelength. The blue band (0.490 μm) and green band (0.560 μm) were chosen for this algorithm due to their enhanced penetration effect in water. This choice facilitates the identification of submerged coral reefs in both shallow and deep regions.

$$DII = \ln(L_i) - [(ki / k_j)\ln(L_j)] \tag{16.3}$$

$$k_i / k_j = a + \sqrt{a^2 + 1}$$

$$a = \frac{\sigma_{jj} - \sigma_{ii}}{2\sigma_{ij}}$$

$$\sigma_{ij} = \overline{x_i x_j} - \left(\overline{x_i} * \overline{x_j}\right)$$

where
L_i and L_j represent the reflectance values of bands i and j,
k_i / k_j is the ratio of the attenuation coefficient of band i and j,
σ_{ii} = is the variance of the band i,
σ_{jj} = is the variance of the band j,
σ_{ij} = is the covariance of the band *i* and *j*
$\overline{x_i}$ = is the mean value of band i,
$\overline{x_j}$ = is the mean value of band j, and
$\overline{x_i x_j}$ = is the mean value of the product of bands *i* and *j*.

16.3.5 Crisp Filter

The crisp filter was employed for the spatial enhancement of the image. This filter enhances the overall luminance of the input image while increasing fine detail (edge) within a multispectral image without affecting broader-scale features. The crisp filter was applied using the crisp filter tool under the spatial section in ERDAS Imagine 2014. This process aimed to sharpen the edges, aiding in the differentiation of various spectral signatures and facilitating the identification of different coral classes.

16.3.6 Supervised Classification

In the final stage, supervised classification was conducted using the maximum likelihood classifier in ERDAS Imagine software. The process involved the utilization of training data, and relevant signatures were meticulously collected based on the reflectance patterns of various coral classes. The signature editor tool was employed to define distinct classes of coral reefs for the years 2002 and 2020.

16.3.7 Turbidity

Turbidity arises when the presence of the suspended particles becomes conspicuous. The mapping of turbidity was conducted using the following equation (Ritchie et al., 2003):

$$\text{Turbidity} = 380.32{*}R_{rs}\ (\text{red band}) - 1.7826 \tag{16.4}$$

The R^2 value for Equation 16.4 is 0.84, with a significance level of $\alpha < 0.05$.

16.3.8 Surface Algal Bloom Index

The surface algal bloom index (SABI) is a four-band algorithm developed for mapping the spatial distribution of surface and near-surface algae bloom. SABI mapping was carried out using the equation (Alawadi, 2010):

$$\text{SABI} = (B5 - B4)/(B2 + B3) \tag{16.5}$$

where

B2, B3, B4, and B5 represent the reflectance values of blue, green, red, and NIR bands of Landsat imagery.

16.3.9 Colored Dissolved Organic Matter

Colored dissolved organic matter (CDOM) is the optically measurable component of dissolved organic matter found in environments and is a complex mixture of hundreds and thousands of individual, unique organic matter molecules, primarily leached from decaying detritus and organic matter (Cobal, 2007). CDOM mapping was conducted using the equation as follows (Chen et al., 2017):

$$\text{aCDOM}^{(440)} = 40.75e^{-2.463x} \tag{16.6}$$
$$x = Rrs(B3)/Rrs\ (B4)$$

where

B3 and B4 represent the reflectance value of the green and red bands of Landsat imagery.

16.4 RESULTS AND DISCUSSION

16.4.1 Sun Glint Removals

Sun glint occurs in an image when the sensor's viewing angle is equal to the angle of reflection of the direct solar beam. Sun glint contamination results in anomalously bright pixels that dominate sensor reflection and can severely degrade the quality of an image in the optical image data. In order to solve the sun glint problem, Hedley et al. (2005) proposed Equation 16.2, where the slope and R^2 value were obtained by using the correlation between NIR and visible bands to determine the corrected pixel values.

The regression was performed between the visible bands (blue, green) and the NIR band to solve the sun glint problem for Neil Island, Andaman-Nicobar Island, and Agatti Island, Lakshadweep. With the help of the regression slope (bi) values, as given in Tables 16.2 and 16.3 for Landsat 7 and Landsat 8, and with the help of Equation 16.2 given by Hedley et al. (2005), the glint problem, arising from the radiometric contributions from sun glint and low clouds, was successfully addressed for both the islands.

Sun glint problems have been solved for Neil Island and Agatti Island, as shown in Figures 16.3 and 16.4. In images (a, b) and (e, f) radiometric contributions from sun glint and low clouds have been observed in blue and green bands, whereas in images (c, d) and (g, h), the problem arising due sun glint in blue and green bands has been solved using Equation 16.2.

TABLE 16.2
The Results of the Linear Regressions between the Visible Bands and the Corresponding NIR Band for Neil Island

	Landsat 7		Landsat 8	
	NIR band		NIR band	
	Blue band	Green band	Blue band	Green band
Slope (bi)	0.687	0.490	0.7403	0.698
R^2	0.901	0.889	0.886	0.882

TABLE 16.3
The Results of the Linear Regressions between the Visible Bands and the Corresponding NIR Band for Agatti Island

	Landsat 7		Landsat 8	
	NIR band		NIR band	
	Blue band	Green band	Blue band	Green band
Slope(bi)	0.7901	0.6055	0.7375	0.763
R^2	0.8714	0.8694	0.816	0.892

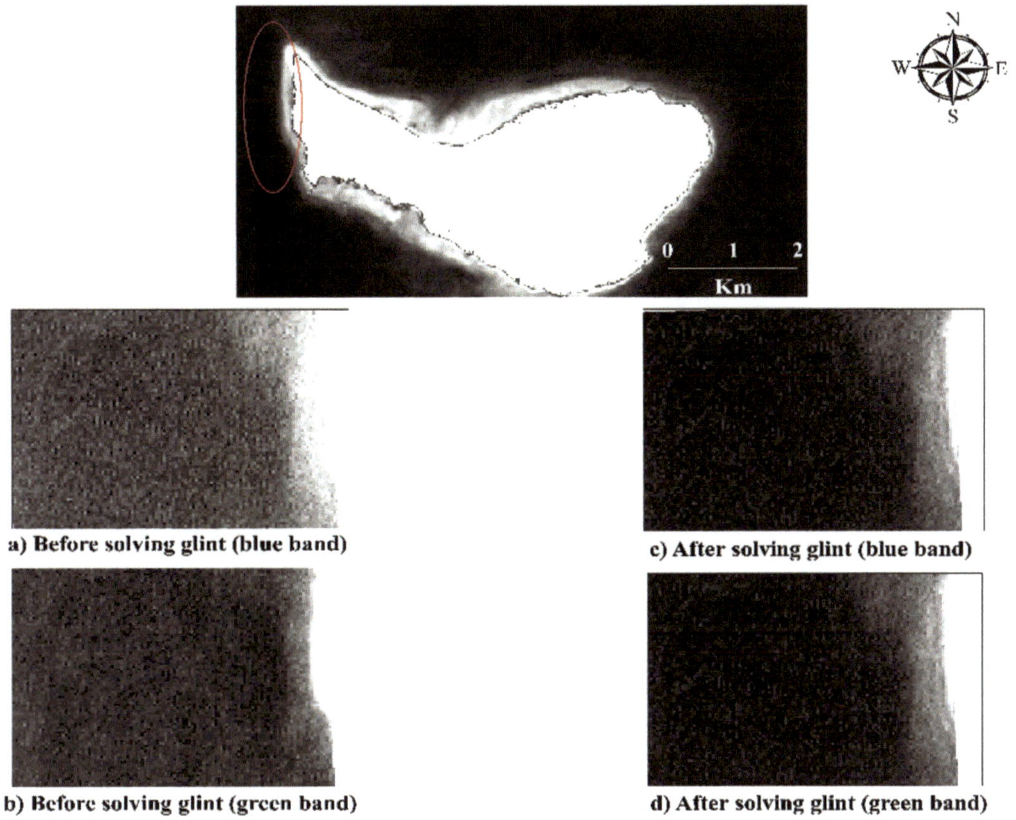

FIGURE 16.3 Zoom image (a, b) before solving the sun glint problem in blue and green bands and image (c, d) after solving the sun glint problem in blue and green bands for Neil Island.

16.4.2 Coral Reef Characteristics

By employing the Feature Space and Maximum Likelihood Classifier, supervised classification was performed for Neil Island (Figure 16.5) and Agatti Island (Figure 16.6) for the years 2002 and 2020. Coral reef classes observed in Neil Island include reef flat, reefward front, sand over reef, and mud over reef, whereas for Agatti Island various coral classes like reefward front, reef crest, reef flat, sand over reef, along with the lagoon (deep lagoon, shallow lagoon), and seagrass have been observed. The change in the area of coral classes for Neil Island and Agatti Island from the year 2002 to 2020 is shown in Figures 16.7 and 16.8.

Both the islands have shown some classes viz. reef flat, reefward front, and sand across the reef as common. Reef flat has shown a decrement of 55.63% and 62.07% from the year 2002 to 2020, mostly observed in the northern parts of Agatti Island and the western part across Neil Island, respectively. Reefward front was observed mostly in the northern parts of Agatti Island in contrast to a 31.58% decrement over Neil Island, as observed in the north-western and south-eastern parts. Sand across the reef exhibited a decrease of 38.58% (in southern parts) over Agatti Island, contrary to a 44.36% increase (in northern parts) over Neil Island. Mud over the reef in Neil Island increased by 57.28%, as observed in northern and south-western parts, whereas seagrass, deep lagoon, and shallow lagoon exhibited an increase of 48.54%, 33.68%, and 60.66%, mostly observed in the western, south-central, and south-western periphery in Agatti Island from the year 2002 to 2020.

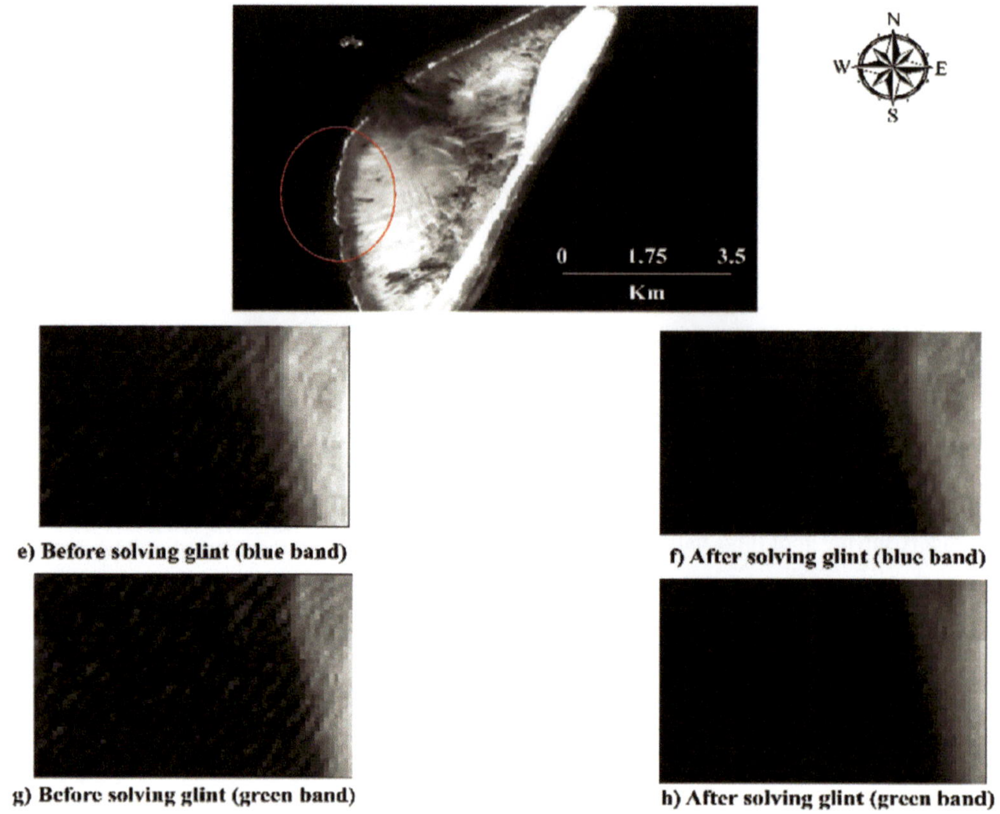

FIGURE 16.4 Zoom image (e, f) before solving the sun glint problem in blue and green bands and image (g, h) after solving the sun glint problem in blue and green bands for Agatti Island.

16.4.3 Accuracy Assessment

Random points were systematically chosen across the classified map and validated by exporting them in KML format using the Google Earth Pro platform. For the years 2002 and 2020, Neil Island had a total of 349 and 325 random points respectively, while Agatti Island had 512 and 535 random points for the corresponding years. Accuracy assessments for classified images of both islands were conducted for the years 2002 and 2020. The overall accuracy of classified images for Neil Island in the years 2002 and 2020 was 85.3% and 84.6% with a Kappa coefficient of 0.816 and 0.806, respectively. Similarly, the overall accuracy of classified images of Agatti Island for the years 2002 and 2020 was 85.0% and 85.1% with a Kappa coefficient of 0.828 and 0.814, respectively. User accuracy signifies the likelihood that a pixel classified into a specific category accurately represents that category on the ground for a classified image. On the other hand, the producer accuracy indicates the quality of the map classification in the classified image.

16.4.4 Parameterization of Different Factors and Their Impact on the Coral Reef Ecosystem

16.4.4.1 Sea Surface Temperature and Photosynthetically Active Radiation

The intensity of coral bleaching exhibits a direct proportionality to SST; it increases with rising temperature decline (Safaie et al., 2018; Hoegh et al., 2007). Excessive increase in SST causes thermal stress conditions on the coral reef ecosystem and leads to the phenomenon of bleaching. A 10 km

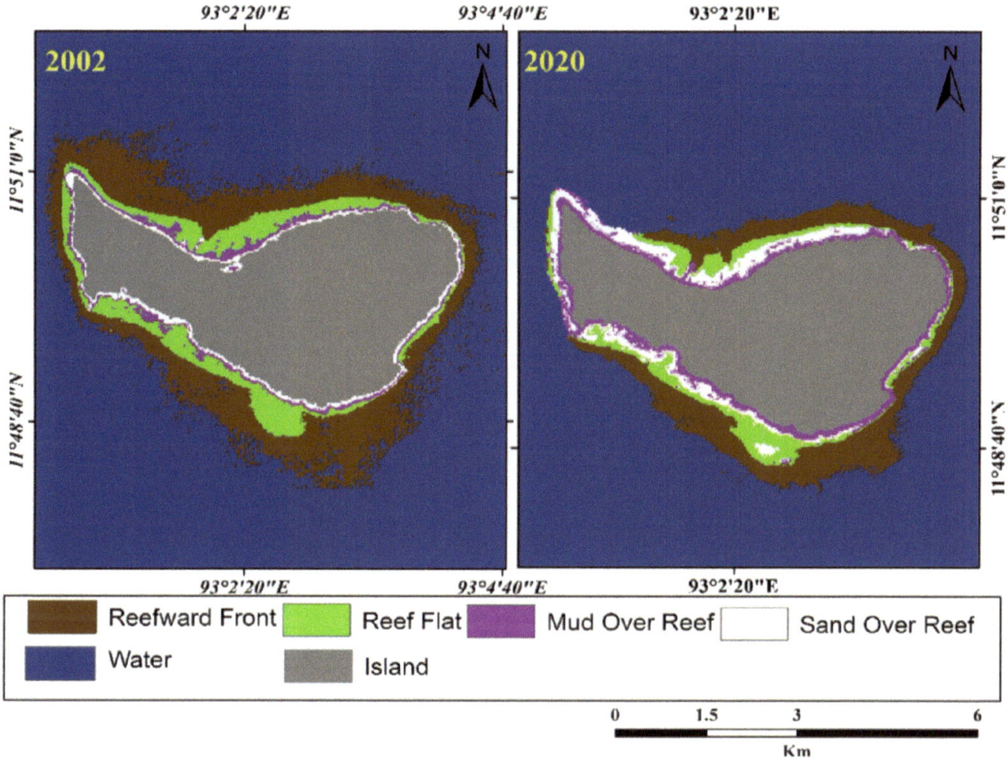

FIGURE 16.5 Coral reef map of Neil Island.

buffer was created along the boundary of the islands to observe the impact of SST and PAR from the year 2002 to 2020 during summer (April–June) on the coral reef ecosystem.

In spite of having an increasing-decreasing behavior of SST, both islands, viz. Neil and Agatti, have shown an overall spiked value for SST with an increasing trend (red line). The year 2010 for Neil Island showed a major spike in SST value (~32.7°C), which could be attributed to a major El-Nino event. These graphs (Figure 16.9) proved the adverse impact of SST on the coral reef ecosystem during different time periods, which may be the major factor for coral bleaching during the years 2002–2020.

Solar radiation, whether in the form of UV or PAR, has been documented to induce damage to exposed corals. In certain instances, coral bleaching occurs due to radiation exposure, even in the absence of elevated temperatures (Fitt et al., 2001; Brown et al., 2002). Additionally, abrupt surges in the amount of radiation received by corals have been associated with bleaching events (Guldberg and Smith, 1989). An increasing-decreasing pattern in PAR (Figure 16.10) was observed within the 10 km of buffer, with an overall increasing trend (red line) observed from the years 2002 to 2020.

High SST along with high PAR (SST >30°C and PAR >51 Einstein/m^{-2} day^{-1}) was observed over Neil Island during April for the years 2004 (SST ~31.06, PAR ~55.22), 2005 (SST ~31.64, PAR ~53.09), 2007 (SST ~30.61, PAR ~52.66), 2009 (SST ~30.18, PAR ~51.22), 2010 (SST ~31.72, PAR ~55.05), 2012 (SST ~30.82, PAR ~52.85), 2017 (SST ~30.36, PAR ~51.43), and 2018 (SST ~30.76, PAR ~53.11). Similar observation was obtained over Agatti Island during April for the years 2003 (SST ~30.46, PAR ~54.90), 2004 (SST ~30.70, PAR ~53.45), 2005 (SST ~30.61, PAR ~52.88), 2007 (SST ~30.04, PAR ~55.23), 2010 (SST ~30.48, PAR ~51.26), 2012 (SST ~30.29, PAR ~53.44), 2018 (SST ~30.26, PAR ~53.37), and for May 2013 (SST ~30.05, PAR ~51.61), 2012 (SST ~30.29, PAR ~54.51), and 2014 (SST ~30.25, PAR ~50.63).

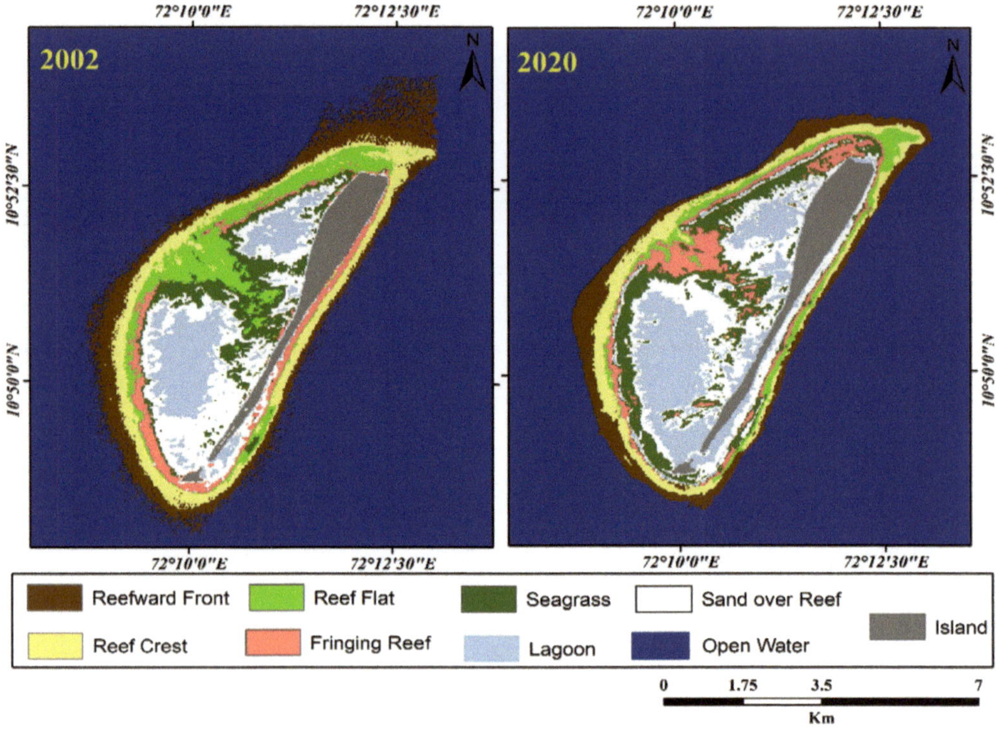

FIGURE 16.6 Coral reef map of Agatti Island.

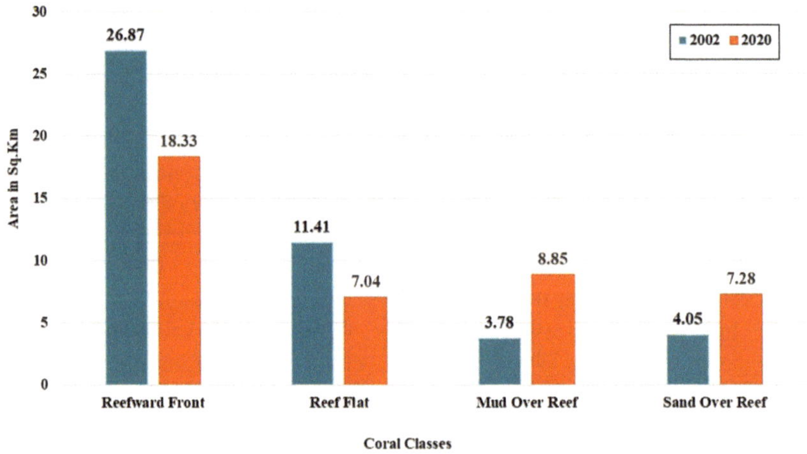

FIGURE 16.7 Change in the area of coral classes on Neil Island.

Abnormally high PAR (>51 Einstein/m^{-2} day$^{-1)}$ along with high SST (>30°C) affects the photochemical pathway known as photoinhibition (Brown and Dunne, 2002), possibly comprising the major factor responsible for deteriorating coral health by damaging coral tissue and causing the bleaching of coral reefs during successive years on these islands.

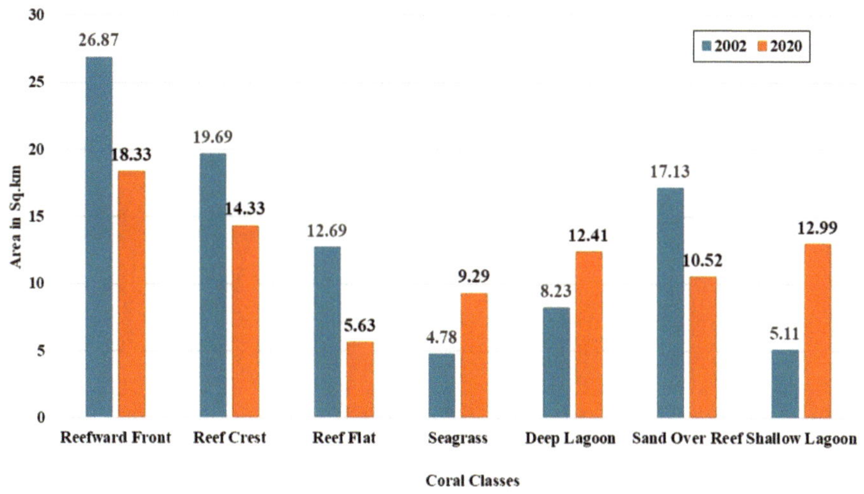

FIGURE 16.8 Change in the area of coral classes on Agatti Island.

16.4.4.2 Impact of Turbidity and Sea Level Rise

Turbidity mapping was performed for both islands in order to estimate the sediment influx in water and its impact on the coral reef ecosystem. The intensity of sunlight was reduced by total suspended sediment, resulting in a reduction of photosynthesis. An increase in water temperature in the vicinity of the island also results in heat absorbed by the suspended sediment (ASTM, 2000). The maps depict a higher turbidity zone shown in red and a lower turbidity zone in orange (Figure 16.11). In the year 2020, turbidity increased for Neil Island and Agatti Island, as indicated by the circled areas. This increase in turbidity became a governing factor contributing to coral health damage and subsequent coral bleaching in both Islands.

Sedimentary processes responsible for encroachment in important aspects, viz. photosynthesis and food intake for corals, are generally triggered by sea level rise, which then induces turbidity issues (Zweifler et al., 2021; Bessell-Browne et al., 2017; Macdonald, 2015). A 10 km buffer was created to observe the impact of sea level rise on the Neil and Agatti islands (Figure 16.12). Both islands show an overall spiked value for SLR with an increasing trend (red trend line). The rising sea level could be a potential factor contributing to the increased turbidity observed in the year 2020, which could possibly be an important reason for coral bleaching on both islands.

16.4.4.3 Algal Bloom and CDOM

Food preparation and intake by corals are hindered by harmful algal blooms (HABs) which block sunlight required for photosynthesis (Bauman et al., 2010). Both islands have witnessed an increase in the concentration of algal bloom for the year 2020, as shown in brown with the circle (Figure 16.13) possibly due to marine pollution and anthropogenic activities, contributing to coral bleaching in both the islands.

More coral bleach susceptibility and possible related diseases were observed in clear water due to the rich concentration of CDOM (West and Salm, 2003). A considerable increase in CDOM concentration was observed in the year 2020 along the coast of both islands (as shown in the circle) (Figure 16.14) possibly due to the rise in the level of pollutants, indicating more risk to the coral ecosystem due to the hindrance to sunlight required for photosynthesis.

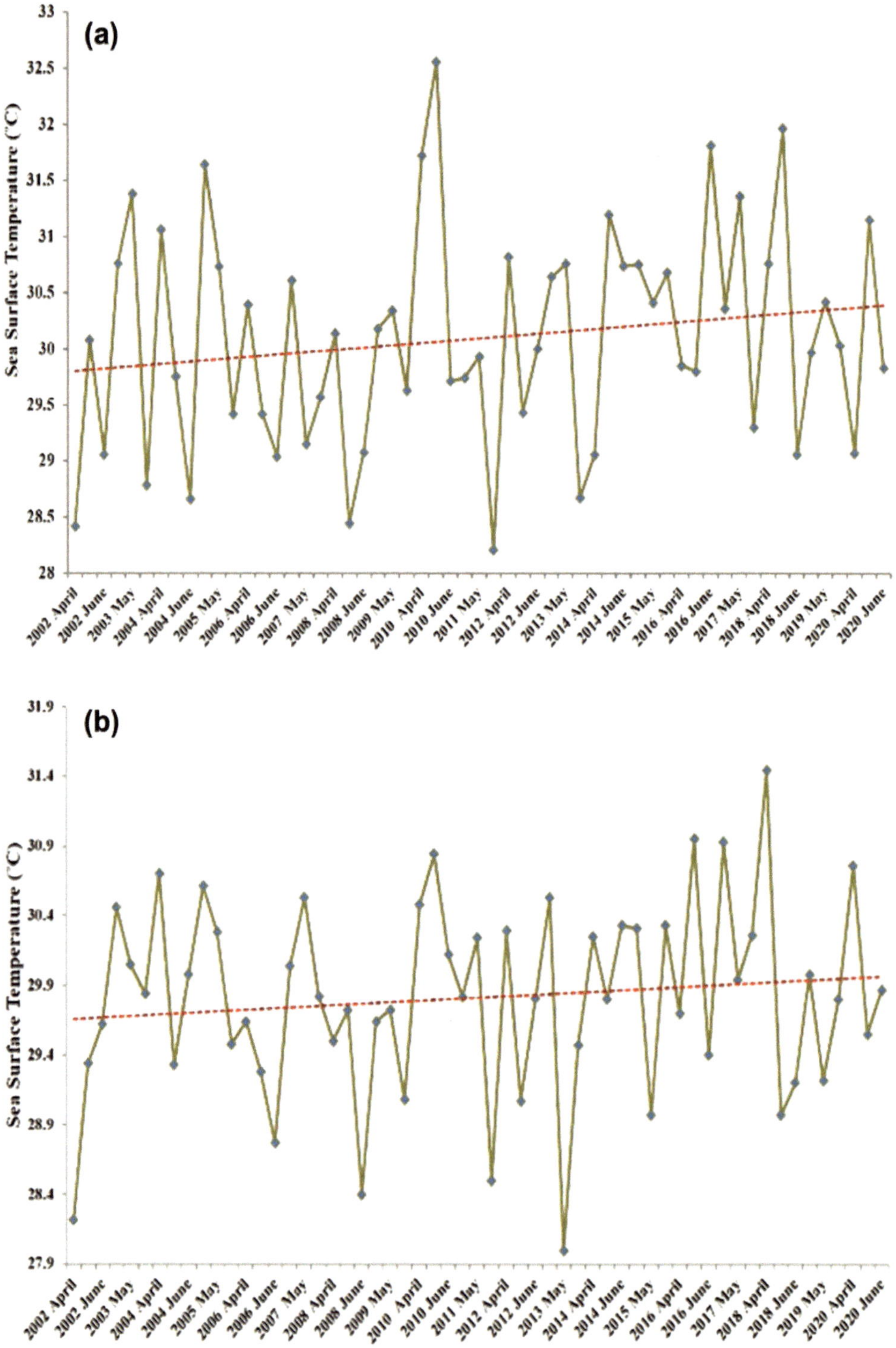

FIGURE 16.9 Trend of sea surface temperature (2002–2020) during summer: (a) Neil Island, (b) Agatti Island.

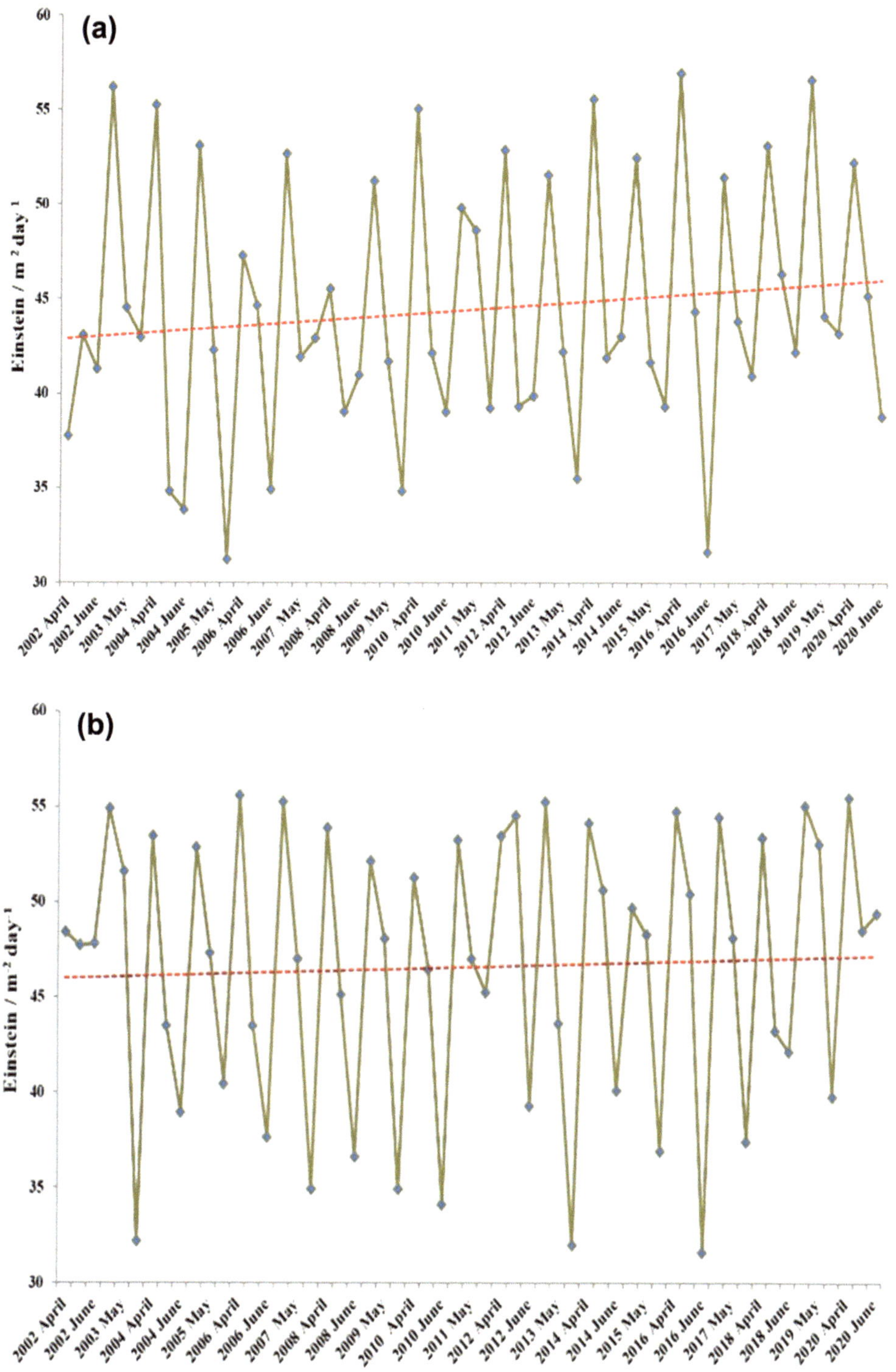

FIGURE 16.10 Trend of photosynthetically active radiation (2002–2020) during summer: (a) Neil Island, (b) Agatti Island.

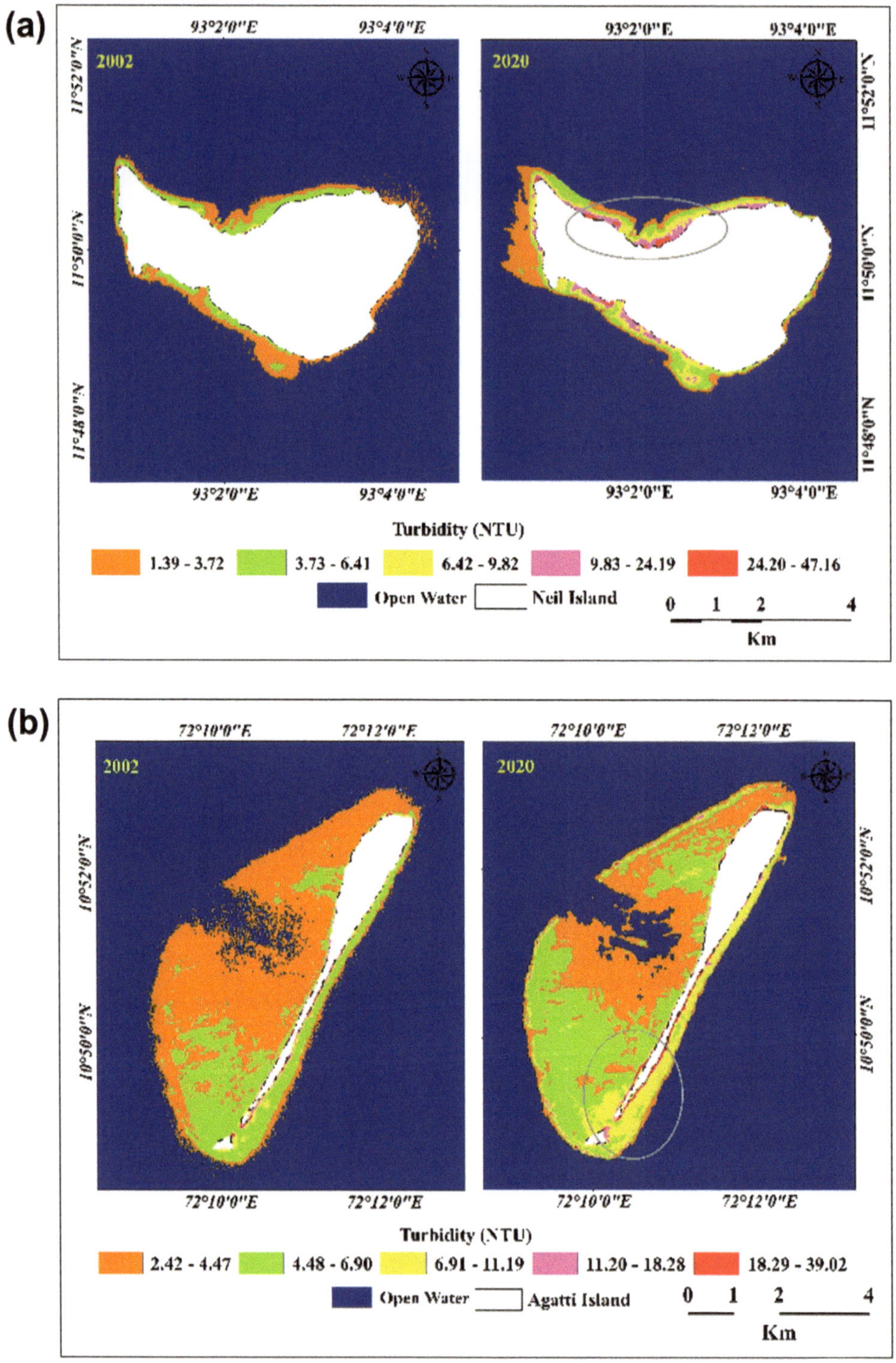

FIGURE 16.11 Turbidity maps: (a) Neil Island, (b) Agatti Island.

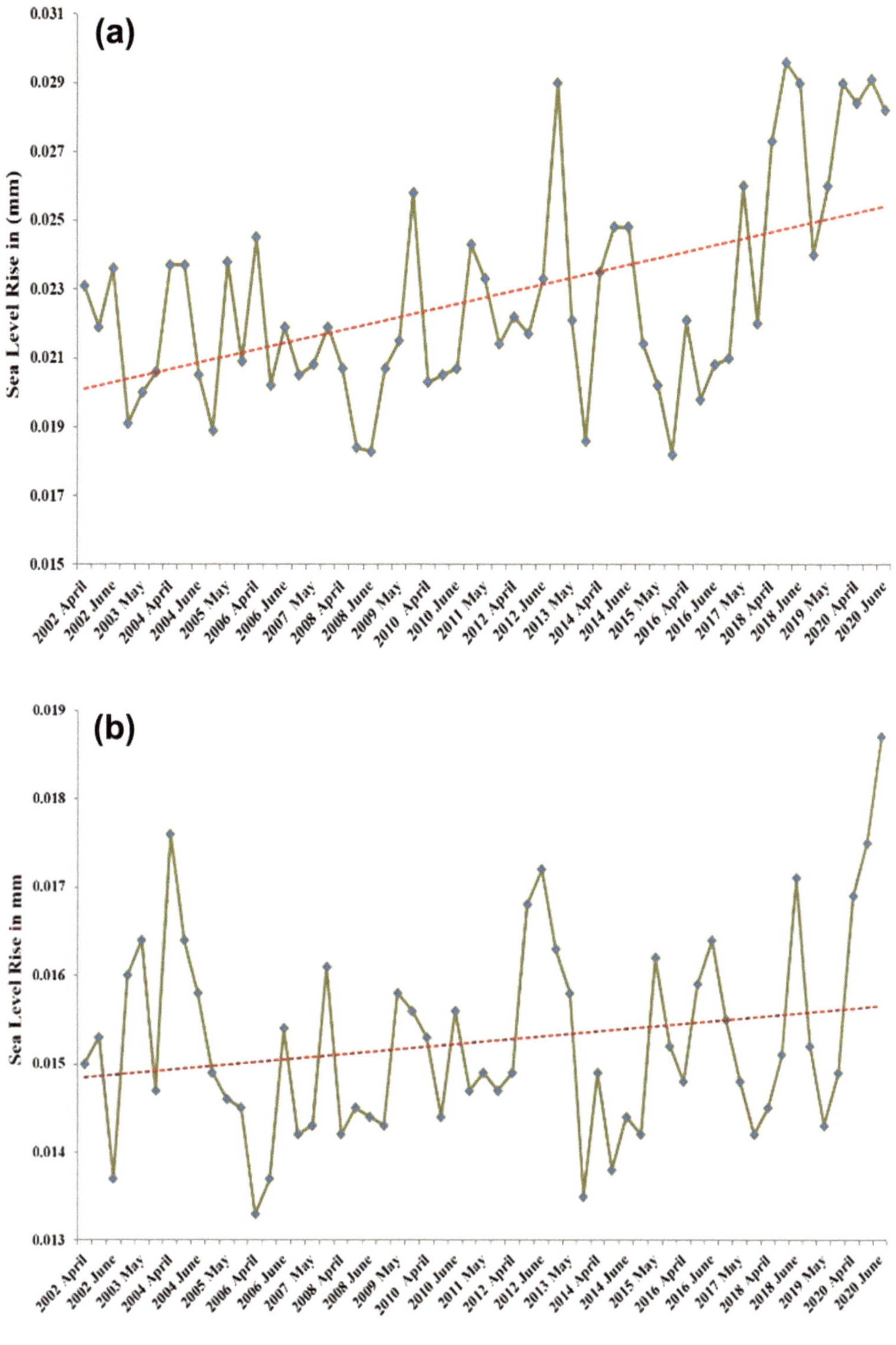

FIGURE 16.12 Trend of sea level rise (2002–2020) during summer: (a) Neil Island, (b) Agatti Island.

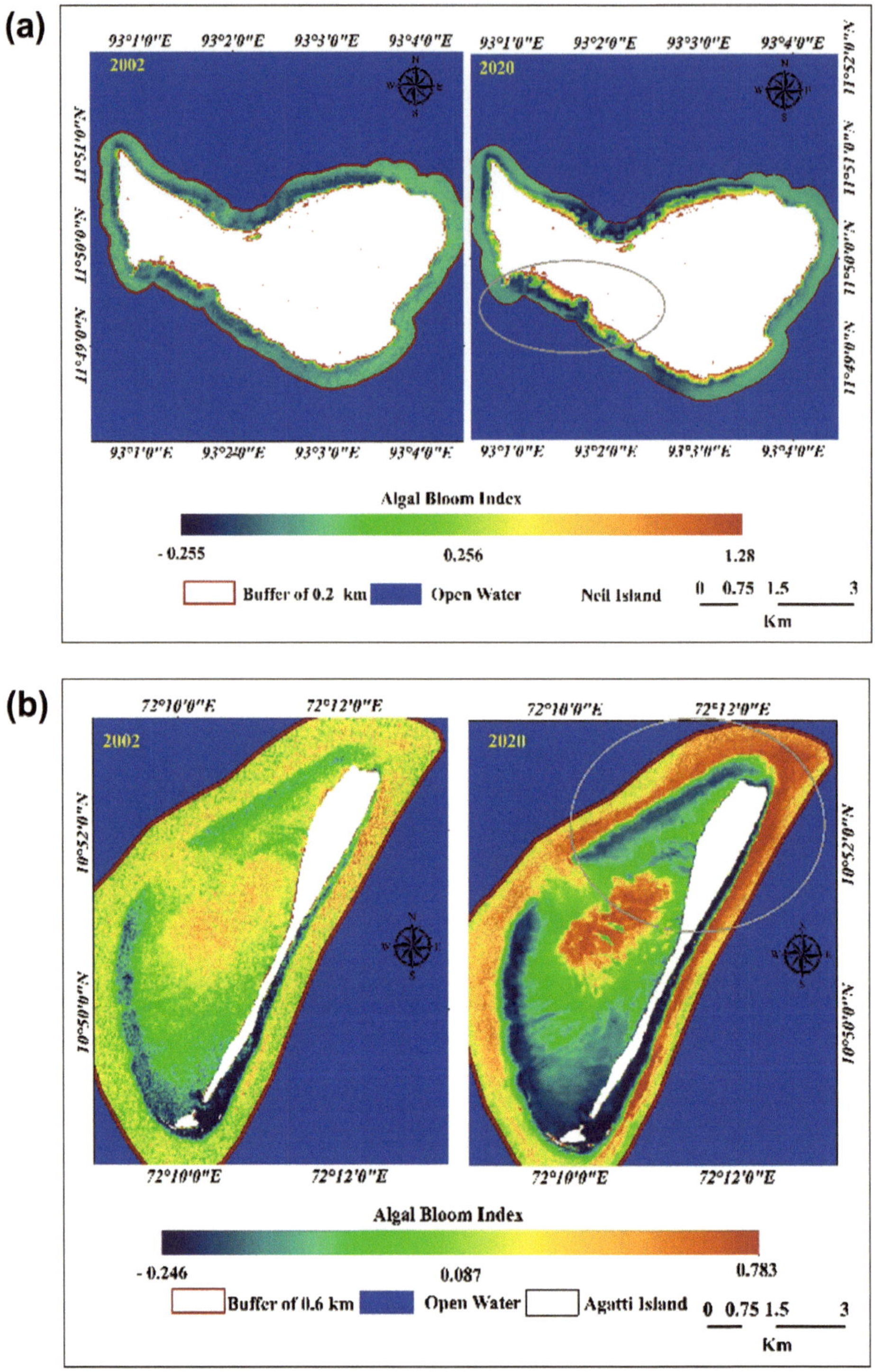

FIGURE 16.13 Algal bloom map: (a) Neil Island, (b) Agatti Island.

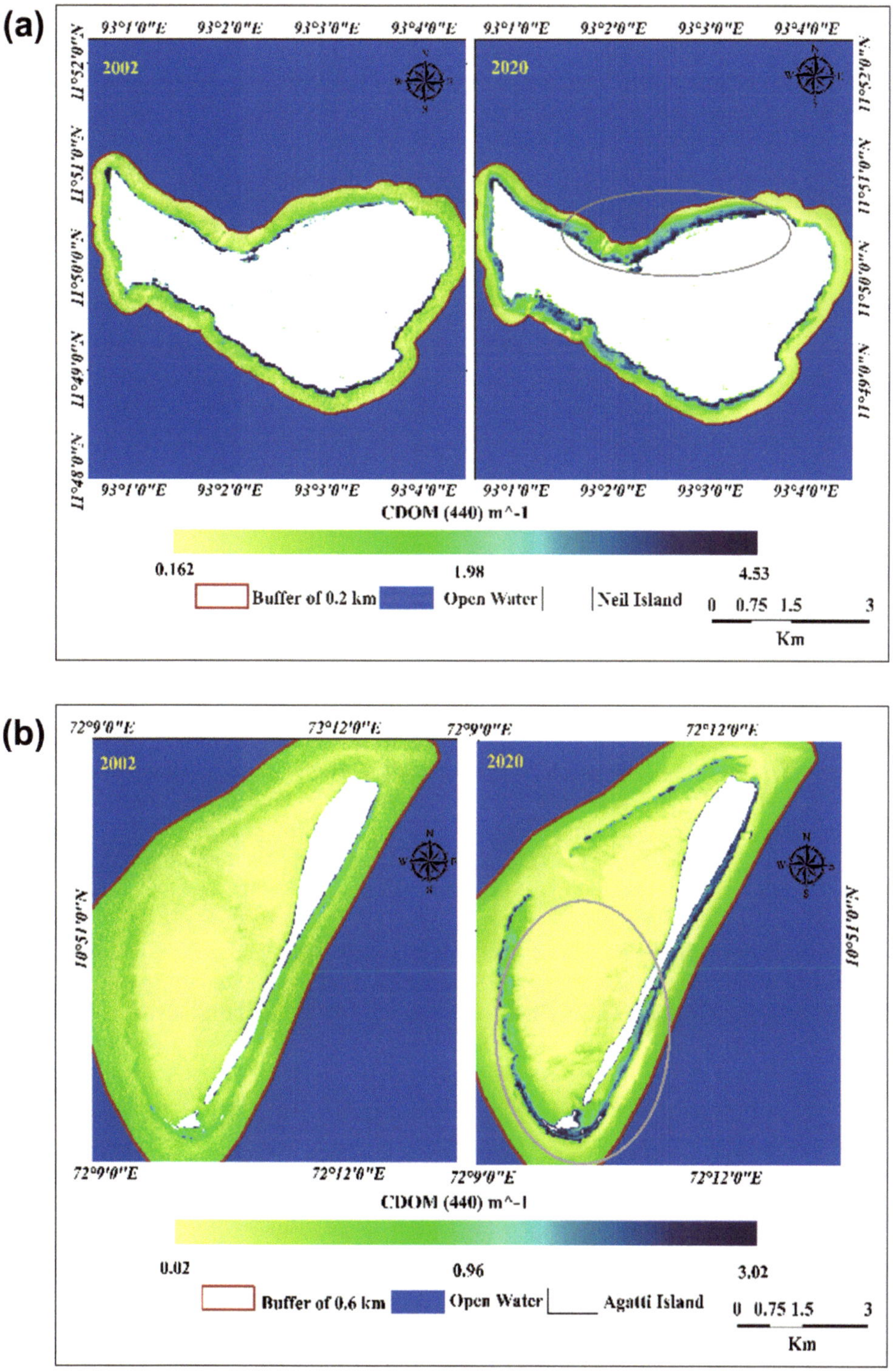

FIGURE 16.14 CDOM map: (a) Neil Island (b) Agatti Island.

16.4.4.4 Salinity

The growth rates of coral decrease when there is a change in salinity by ±2 psu from normal levels was observed by Ferrier et al. (1999). Additionally, rapid fluctuations in salinity have been linked to coral mortality (Sakai et al., 1989; Jokiel et al., 1993).

To observe the variation in salinity and its impact on coral health from the years 2002 to 2020, a buffer of 10 km was created along the boundary of the island. An overall increasing trend of salinity (Figure 16.15) was observed along both islands, which could be attributed to the increasing evaporation against precipitation, rendering a potential indicator for the deterioration of coral health.

16.4.5 Coral Bleaching and Mortality

Various coral reef classes were observed along both islands, in which some classes exhibited a decrease and some an increase. In Neil Island, the coverage of the reef flat and reefward front was more extensive in 2002 compared to 2020. Conversely, sand over reef and mud over reef covered a larger area in 2020. Similarly, for Agatti Island, reefward front, reef crest, reef flat, and sand over reef exhibited a larger area in 2002, while the area covered by deep lagoon, seagrass, and shallow lagoon was higher in 2020. A decrease in the area of coral classes indicates coral degradation and coral bleaching, possibly due to the combined effects of climatic changes, tourism activities, overfishing, disease, aquatic invasive species, sedimentation, variation in sea surface temperature, variation in PAR, thermal stress, acidification, variation in salinity, sea level rise, etc. On the contrary, some coral classes exhibited an area increase during 2020, possibly reflecting their adaptation to climatic conditions or the less adverse impact of anthropogenic activities in their region of dominance, leading to no effect of excessive bleaching on these coral reef classes.

Image difference technique was used to map the coral bleaching area from 2002 to 2020 (Figure 16.16) which indicated that an area of 11.48 sq. km (Neil Island) and 20.23 sq. km (Agatti Island) falls under coral bleaching. The estimation of the coral mortality rate and percentage of area under bleaching was done by using standard Equations 16.7 and 16.8 (Wasim et al., 2021).

$$\text{Coral mortality } \% = (\text{bleached area/total coral class area}) * 100 \quad (16.7)$$

$$\text{Coral mortality rate} = (\text{bleached area/t}) \quad (16.8)$$

t = difference between the latest and oldest observation years

The results of the present study have been compared with earlier studies on coral classification using hyperspectral data for Agatti Island (Velloth et al., 2014). Another study witnessed coral reef bleaching from May to June in the year 2010, due to variations in sea surface temperature (Vinot et al., 2011). A great intensity of mass bleaching of coral was also observed in Andaman Island during the year 2010 compared to 1998, owing to the rapid increase in sea surface temperature (Baker et al., 2008).

The result of coral mortality was juxtaposed with the findings of Sreenath et al. (2015). According to their study, the reef around Agatti Atoll exhibited an average live coral cover of 48.6%, while the recorded average dead coral coverage was 21.6%, with a significant portion of the deceased corals being covered by algae. This gives clear evidence that bleaching occurred from 2002 to 2020 on Agatti Island, as demonstrated in the present study. Over Neil Island, the result of the present study has been compared with the studies of coral reef mapping in the South Andaman Islands by Dharanirajan et al. (2011), who opined that the impact of the tsunami in Andaman Island during 2004 severely affected the outer coral reef classes, whereas Majumdar et al. (2019) indicated similar effects in the inland area of Andaman Island. Anthropogenic influences like touching coral with hands, standing on corals, snorkeling, etc., may have triggered coral bleaching and degradation with a subsequent decrease in select coral reef areas in Neil Island from 2002 to 2020. The findings on coral mortality in Neil Island were compared with a previous study conducted by Sarkar et al. (2013). signifying that the percentage of bleached corals (including partially bleached, fully

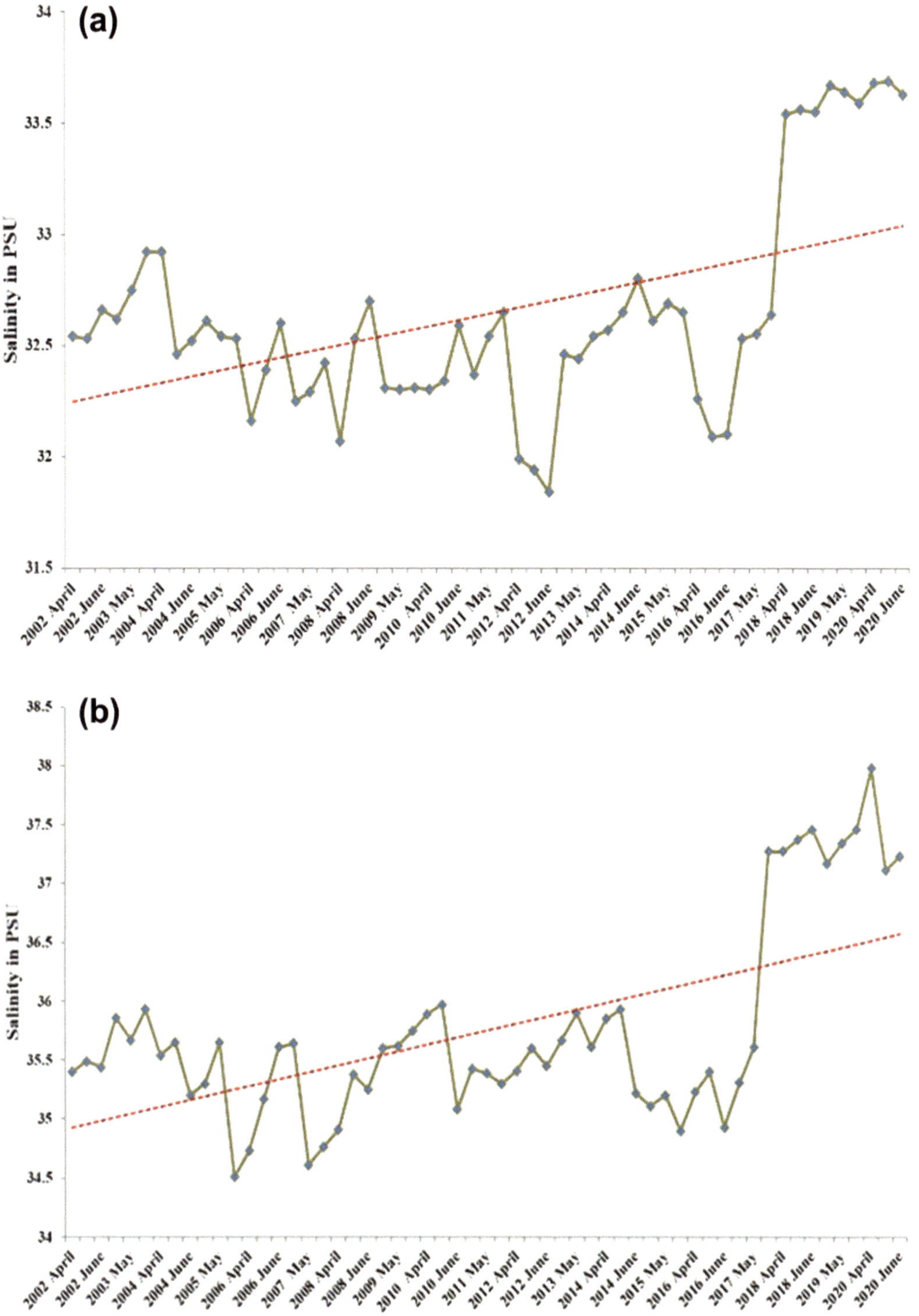

FIGURE 16.15 Trend of salinity (2002–2020) during summer: (a) Neil Island, (b) Agatti Island.

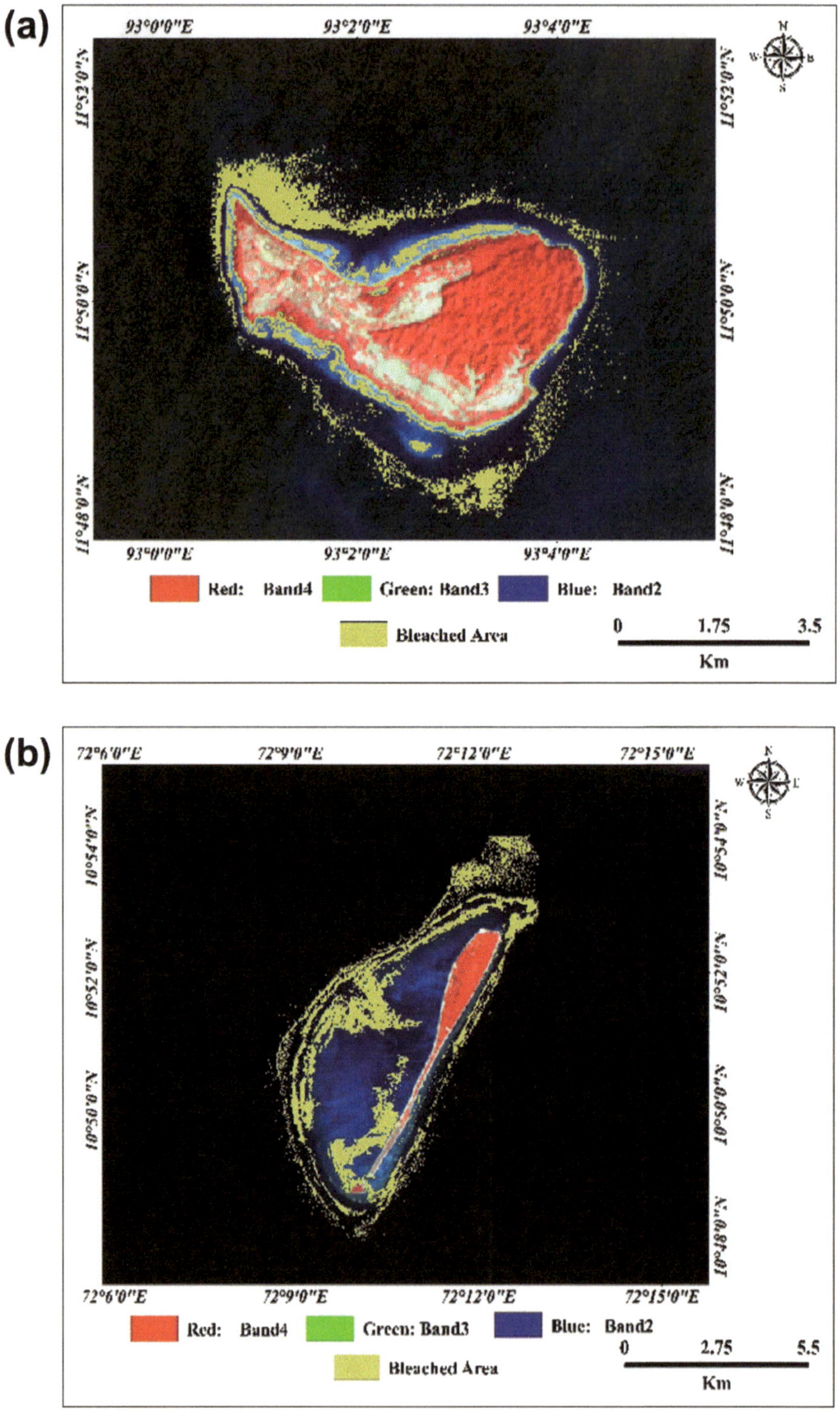

FIGURE 16.16 Map showing the coral bleaching of (a) Neil Island and (b) Agatti Island, along with a false-color composite (FCC) of Landsat 7.

bleached but not dead, and dead corals) as a component of total estimated data (excluding sand and algae) was highest at Ross Island (88.7%), followed by Havelock (86.2%), North Bay (84.1%), Chidiyatapu (82.4%), Neil (77.0%), and Jolly Buoy (43.3%). This corroborates that bleaching has occurred in Neil Island as per the statistics presented in our study.

16.5 CONCLUSION

The present study demonstrated satellite-based coral reef mapping using a multispectral data set by applying the depth invariant bottom type index with a supervised classification technique. The bottom type map was improved by applying water column correction technique through the integration of the texture layer in the supervised classification, resulting in improved coral classification. Temporal mapping indicated that the area of different types of coral reefs decreased on both islands due to coral degradation or bleaching. Long-term coral reef monitoring is important to understand the perceived impact of climate conditions and marine environmental variables, in addition to existing tourism, marine pollution, overfishing, and other anthropogenic impacts on the coral reef ecosystem. Major parameters/factors that affected mass coral bleaching in the study area include SST, SSLR, PAR, salinity, CDOM, turbidity, and SABI. PAR of 50 $E/m^2/day$ with a minimum SST of 30°C causes photoinhibition, which leads to an adverse impact on coral tissue on both islands. Algal bloom, CDOM, and turbidity for the year 2020 proved to be a major source of threat to the coral ecosystem on both islands, possibly reflecting anthropogenic activities in the form of excessive tourism and discharge of marine pollutants. Sea level rise observed within the 10 km buffer zone indicated a negative impact on coral health and proved to be a major source of the rise in turbidity on both islands. The study proved that the remote sensing technique is quite effective in monitoring coral reefs and influencing factors that result in the degradation of the coral ecosystem.

ACKNOWLEDGEMENTS

The authors thank the United States Geological Survey (USGS) for providing Landsat data. They also thank "Giovanni and the Asia Pacific platform" for providing products used for parameterization in the present study. The authors are grateful to SNIOE for help and support.

REFERENCES

Alawadi, F. (2010). Detection of surface algal blooms using the newly developed algorithm surface algal bloom index (SABI). *Proceedings of The International Society for Optical Engineering*, 7825, 1–14.

Anthony, K.R.N., Connolly, S.R., Hoegh-Guldberg, O. (2007). Bleaching, energetics, and coral mortality risk: Effects of temperature, light, and sediment regime. *Limnology and Oceanography*, 52, 716–726.

Arthur, R. (2000). Coral bleaching and mortality in three Indian reef regions during an El Niño southern oscillation event. *Current Science,* 79, 1723–1729.

Arthur, R., Done, T.J., Marsh, H. (2005). Benthic recovery four years after an El Niño-induced coral mass mortality in the Lakshadweep atolls. *Current Science,* 89(4), 694–699.

Baird, A.H., Hughes, T.P. (2000). Competitive dominance by tabular corals: an experimental analysis of recruitment and survival of understorey assemblages. *Journal of Experimental Marine Biology and Ecology,* 251, 117–132.

Baker, A.C., Glynn, P.W., Riegl, B. (2008). Climate change and coral reef bleaching: An ecological assessment of longterm impacts, recovery trends and future outlook. *Estuarine, Coastal and Shelf Science*, 80, 435–471.

Barker, N.H.L., Roberts, C.M. (2004). Scuba diver behaviour and the management of diving impacts on coral reefs. *Biological Conservation*, 120, 481–489. http://dio.org/10.1016/j.biocon.2004.03.021.

Bauman, A.G., Burt, J.A., Feary, D.A., Marquis, E., Usseglio, P. (2010). Tropical harmful algal blooms: An emerging threat to coral reef communities? *Marine Pollution Bulletin,* 60, 2117–2122. https://doi.org/10.1016/j.marpolbul.2010.08.015.

Bessell-Browne, P., Negri, A. P., Fisher, R., Clode, P. L., Duckworth, A., & Jones, R. (2017). Impacts of turbidity on corals: The relative importance of light limitation and suspended sediments. *Marine Pollution Bulletin*, 117(1–2), 161–170

Brown, B., Dunne, R., Goodson, M., Douglas, A. (2002). Experience shapes the susceptibility of a reef coral tobleaching. *Coral Reefs*, 21, 119–126.

Bruno, J.F., Selig, E.R. (2007). Regional decline of coral cover in the Indo-Pacific: Timing, extent, and sub regional comparisons. *PLoS One,* 2, e711.

Burke, L., Reytar, K., Spalding, M., Perry, A. (2011). Reefs at risk revisited – World resources Institute, Washington DC, pp. 130.

Che, N., Price, J. (1992). Survey of radiometric calibration results and methods for visible and near infrared channels of NOAA-7,-9, and-11 AVHRRs. *Remote Sensing of Environment,* 41, 19–27.

Chen, J., Zhu, W., Tian, Y.Q., Yu, Q. (2017). Estimation of colored dissolved organic matter from Landsat-8 Imagery for complex Inland water; A case study of Lake Huron. *IEEE Transcations on Geoscience and Remote Sensing,* 99, 1–12. https://doi.org/10.1109/TGRS.2016.2638828.

Chow, A.M., Ferrier-Pagès, C., Khalouei, S., Reynaud, S., Brown, I.R. (2009). Increased light intensity induces heat shock protein Hsp60 in coral species. *Cell Stress Chaperon,* 14, 469–476. https://doi.org/10.1007/s12192-009-0100-6.

Coble, P. (2007). Marine optical biogeochemistry: The chemistry of ocean color. *Chemical Reviews*, 107(2), 402–418. https://doi.org/10.1021/cr050350+.

Dharanirajan, K., Ramanujam, N., Ramesh, D., Gurugnanam, B. (2011). Coral reef mapping in south Andaman Islands using Remote sensing and GIS. *Asian Journal of Geoinformatics*, 11(1), 1–7.

Dinsdale, E.A., Harriott, V.J. (2004). Assessing anchor damage on coral reefs: A case study in selection of environmental indicators. *Environmental Management*, 33, 126–139. https://doi.org/10.1007/s00267-003-3056-9.

Downs, C.A., Mueller, E., Phillips, S., Fauth, J.E., Woodley, C.M. (2000). A molecular biomarker system for assessing the health of coral (Montastraea faveolata) during heat stress. *Marine Biotechnology,* 2, 533–544. https://doi.org/10.1007/s101260000038, PMID: 14961177.

El-Askary, H., Abd El-Mawla, S., Li J., El-Hattab, M., El-Raey, M. (2014). Change detection of coral reef habitat using Landsat-5 TM, Landsat 7 ETM+, and Landsat 8 OLI data in the Red Sea (Hurghada, Egypt). *International Journal of Remote Sensing,* 35, 2327–2346.

Ferrier-Pagès, C., Gattuso, J.P., Jaubert, J. (1999). Effect of small variations in salinity on the rates of photosynthesis and respiration of the zooxanthellate coral *Stylopora pistillata. Marine Ecology-Progress Series,* 181, 309–314.

Ferrier-Pagès, C., Richard, C., Forcioli, D., Allemand, D., Pichon, M., Shick, J.M. (2007). Effects of temperature and UV radiation increases on the photosynthetic efficiency in four scleractinian coral species. *Biological Bulletin,* 213, 76–87.

Fitt, W., Brown, B., Warner, M., Dunne, R. (2001). Coral bleaching: Interpretation of thermal tolerance limits and thermal thresholds in tropical corals. *Coral Reefs*, 20, 51–65. https://doi.org/10.1007/s003380100146.

Gajendran, N. (2007). Endangered coral island: A case study of Dahlak in Red Sea, Indian. *Journal of Science and Technology*, 1(1), 31–33. http://www.indjst.org.

GBRMPA. (2016a). GD-2016-0040-BleachingStatus-AsAt13-June-2016 [Internet]. GBRMPA; [cited, 2016]. Available from: http://www.gbrmpa.gov.au/__data/assets/pdf_file/0009/246384/GD-2016-0040-BleachingStatus-AsAt13-June-2016.pdf.

GBRMPA. (2016b). Interim report: 2016 coral bleaching event on the Great Barrier Reef. Available from: http://hdl.handle.net/11017/3044.

Goodman, J.A., Purkis, S.J., Phinn, S.R. (2013). *Coral reef remote sensing: A guide for mapping, monitoring and management.* Springer Science & Business Media, pp. 446.

Hawkins, J.P., Roberts, C.M. (1993). Can Egypt's coral reefs support ambitious plans for diving tourism? Proceedings of the 7th International Coral Reef Symposium, Vol. 2, Guam, 22–27 June 1992, 1007–1013.

Hedley, J.D., Harborne, A.R., Mumby, P.J. (2005). Simple and robust removal of sun glint for mapping shallow-water benthos. *International Journal of Remote Sensing*, 26(10), 2107–2112.

Hein, M.Y., Lamb, J.B., Scott, C., Willis, B.L. (2015). Assessing baseline levels of coral health in a newly established marine protected area in a global scuba diving hotspot. *Marine Environmental Research,* 103, 56–65.

Hoegh-Guldberg, O., Mumby, P.J., Hooten, A.J., Steneck, R.S., Greenfield, P., Gomez, E., Harvell, C.D., Sale, P.F., Edwards, A.J., Caldeira, K., Knowlton, N., Eakin, C.M., Iglesias-Prieto, R., Muthiga, N., Bradbury, R.H., Dubi, A., Hatziolos, M.E. (2007). Coral reefs under rapid climate change and ocean acidification. *Science,* 318, 1737–1742.

Hoegh-Guldberg, O., Smith, G.J. (1989). The effect of sudden changes in temperature, light and salinity on the population density and export of zooxanthellae from the reef corals Stylophora pistillata Esper and Seriatopora hystrix Dana. *Journal of Experimental Marine Biology and Ecology,* 129, 279–303.

Hourigan, T.F., Tricas, T.C., Reese, E.S. (1988). Coral reef fishes as indicators of environmental stress in coral reefs. *Marine Organisms as Indicators,* 1988, 107–135.

Hughes, T.P. (2003). Climate change, human impacts, and the resilience of coral reefs. *Science*, 301, 929–933. https://doi.org/10.1126/science.1085046.

Ishizaka, J., Fukushima, H., Kishino, M., Saino, T., Takahashi, M. (1992). Phytoplankton pigment distribution in regional upwelling around the Izu Peninsula detected by coastal zone color scanner in May 1982. *Journal of Oceanography,* 48, 305–327.

Jokiel, P.L., Hunter, C.L., Taguchi, S., Watarai, L. (1993). Ecological impact of a fresh-water "reef kill" in Kaneohe Bay, Oahu, Hawaii. *Coral Reefs,* 12, 177–184.

Kennedy, C., and Lindsey, R. (2010). Coral bleaching could be as severe as 2005 event. *NOAA Science and Information for a climate Smart Nation, Climate.gov.* Retrieved from: https://www.climate.gov/news-features/featured-images/coral-bleaching-could-be-severe-2005-event#:~:text=In%202005%2C%20many%20areas%20of,percent%20later%20died%20from%20disease.

Krishnan, P., Dam Roy, S., George, G., Srivastava, R.C., Anand, A., Murugesan, S., Kaliyamoorthy, M., Vikas, N., Soundararajan, R. (2011). Elevated sea surface temperature during May 2010 induces mass bleaching of corals in the Andaman. *Current Science*, 100, 111–117.

Loya, Y., Sakai, K., Yamazato, K., Nakano, Y., Sambali, H., van Woesik, R. (2001). Coral bleaching: The winners and the losers. *Ecology Letters,* 4, 122–131. https://doi.org/10.1046/j.1461-0248.2001.00203.x.

Lyzenga, D.R. (1978). Passive remote sensing techniques for mapping water depth and bottom features. *Applied Optics*, 17(3), 379–383.

Macdonald, R. K. (2015). Turbidity and light attenuation in coastal waters of the Great Barrier Reef. *Doctoral dissertation*, James Cook University.

Majumdar, D., Chakraborty, S., Saha, S., Datta, D. (2019). Geospatial analysis of the effects of tsunami on coral and mangrove ecosystems of Mayabunder in Andaman Island, India. *Brazillian Journal of Oceanography*, 67, ISSN 1679-8759. http://dx.doi.org/10.1590/s1679-87592019025806715.

Manzell, D., Lirman, D. (2003). The photosynthetic resilience of Porites furcata to salinity disturbance. *Coral Reefs,* 22, 537–540.

Mora, C. (2008). A clear human footprint in the coral reefs of the Caribbean. *Proceedings of the Royal Society B: Biological Sciences,* 275, 767–773.

Mumby, P.J., Harborne, A.R. 1999. Development of a systematic classification scheme of marine habitats to facilitate regional management and mapping of Caribbean coral reefs. *Biological Conservation,* 88, 155–163.

Muthiga, N.A., Szmant, A.M. (1987). The effects of salinity stress on the rates of aerobic respiration and photosynthesis in the hermatypic coral Siderastrea siderea. *Biological Bulletin,* 173, 539–551.

Pandolfi, J.M., Connolly, S.R., Marshall, D.J., Cohen, A.L. (2011). Projecting coral reef futures under global warming and ocean acidification. *Science,* 333, 418–422.

Pet-Soede, L., Wafar, M., Venkataraman, K., Rajan, P.T., Wilhelmsson, D. (2000). The status of the coral reefs of India following the bleaching event of 1998, in: Souter, D., Obura, D., Linden, O. (Eds.), *Coral Reef Degradation in the Indian Ocean*, status report 2000, CORDIO, SAREC Marine science Program, Stockholm University, Sweden, pp. 69–74.

Rajasuriya, A., Maizan, H.M., Subramaniam, B.R., Jason, R. (1999). Coral reef ecosystems in South Asia. Coral reef degradation in the Indian Ocean: Status reports and project presentations 1999, pp. 10–23.

Ravindran, J., Raghukumar, C., Raghukumar, S. (1998). Disease and stress-induced mortality of corals in Indian reefs and observations on bleaching of corals in the Andamans. *Current Science*, 76, 233–236.

Richmond, R.H. (1993). Coral reefs: Present problems and future concerns resulting from anthropogenic disturbance. *American Zoologist,* 33, 524–536.

Ritchie, C.J., Zimba, V.P., Everitt, H.J. (2003). Remote sensing techniques to assess water quality Photogramm. *Engineering Remote Sensing,* 69, 695–704.

Roelfsema, C., Phinn, S., Jupiter, S., Comley, J., Albert, S. (2013). Mapping coral reefs at reef to reef-system scales, 10s–1000s km2, using object-based image analysis. *International Journal of Remote Sensing*, 34, 6367–6388.

Roy, S.D., George, G., Soundararajan, R., Krishnan, P., Murugesan, S., Kaliyamoorthy, M. (2014). Status of coral reefs in Andaman after two major catastrophic events—tsunami of December 2004 and bleaching of May 2005. *Ecology Environment and Conservation*, 20, 539–544.

Roy, S.D., Krishnan, P., George, G. (2006). Assessment of coral reef health using satellite data, in: Srivastava, R.C., Ganesh Kumar, B., Jayakumar, V., Domodaran, T., Ravisankar, N., Jeyakumar, S., George, G., Tomar, S. (Eds.), *Annual Report 2005–06*. Central Agricultural Research Institute, Port Blair, pp. 73–74.

Safaie, A., Silbiger, J.N., McCblanahan, T.R., Pawlak, G., Barshis, D.J., Hench, J.L., Rogers, J.S., Williams, G.J., Davis, K.A. (2018). High frequency temperature variability reduces the risk of coral bleaching. *Nature Communications*, 9, 1671.

Sakai, K., Snidvongs, A., Nishihara, M. (1989). A mapping of a coral- based, non-reefal community at Khang Khao Island, in ner part of the Gulf of Thailand: Interspecific competition and community structure. *Galaxea, Journal of the Coral Reef Studies*, 8, 185–216.

Sarkar, S., Ghosh, A.K. (2013). Coral bleaching a nemesis for the Andaman reefs: Building an improved conservation paradigm. *Ocean & Coastal Management,* 71, 153–162. https://doi.org/10.1016/j.ocecoaman.2012.09.010.

Sreenath, K.R., Jasmine, S., George, R.M., Lakshmanan, R., Koya, M., Kingsley, J. (2015). Community structure and spatial pattern in hard coral diversity of Agatti Island, Lakshadweep, India. *Indian Journal of Fisheries,* 62(3), 35–44.

Stuart-Smith, R.D., Brown, C.J., Ceccarelli, D.M., Edgar, G.J. (2018). Ecosystem restructuring along the Great Barrier Reef following mass coral bleaching. *Nature,* 560, 92–96.

Velloth, S., Mupparthy, R.S., Nayak, S. (2014). Coupled correction and classification of hyperspectral imagery for mapping coral reefs of Agatti and Flat Island, India. *International Journal of Remote Sensing,* 35, 5544–5561. https://doi.org/10.1080/01431161.2014.926410.

Vinoth, R., Mohan, G., Kumar, T.T.A., Thangaraddjou, T. (2011). Coral reefs bleaching at Agatti Island of Lakshadweep Atolls, India. *Journal of Ocean University China*, ISSN 1672-5182. https://doi.org/10.1007/s11802-012-1822-0.

Wasim, M., Pandey, A.C., Kumar, A., Dwivedi, C.S. (2021). Spatio-temporal mapping to investigate coral bleaching in Andaman and Nicobar Islands, India using geoinformatics. *Journal of the Indian Society of Remote Sensing*. https://doi.org/10.1007/s12524-021-01345-2.

West, J.M., Salm, R.V. (2003). Resistance and resilience to coral bleaching: Implications for coral reef conservation and management. *Conservation Biology,* 17, 956–967.

Wilkinson, C. (2008). Status of coral reefs of the world: 2008 Global coral reef monitoring network and reef and Rainforest Research Center. Townsville, Australia, pp. 296.

Zweifler, A., O'leary, M., Morgan, K., & Browne, N. K. (2021). Turbid coral reefs: past, present and future—a review. *Diversity*, 13(6), 251.

17 Plastic Pollution is a Serious Menace to Ecosystem Health with Special Reference to Aquatic Ecosystems and its Associated Challenges, Opportunities, and Mitigations

*Anannya Dahiya, Dasputre Yash Sanjay Kumar, Suthar Shubham Kumar, Ashutosh K. Pandey**
School of Agricultural Science, Sharda University, Noida, Uttar Pradesh, India

Chongtham Allaylay Devi
School of Agriculture, Galgotias University, Greater Noida, Uttar Pradesh, India

17.1 INTRODUCTION

The ubiquitous presence and levels of plastic pollution have become a major cause of concern to Earth's ecological processes. Anthropogenic activities significantly contribute to plastic pollution. Plastic pollution levels are equivalent to natural geophysical processes, influencing even biodiversity and ecosystem functions. Plastics, originating in the early 1900s, rapidly evolved into a ubiquitous menace to both the environment and public health over subsequent decades (Sala and Castellani, 2019). In contemporary times, the global output of plastic resins and synthetic fibers has skyrocketed, surpassing 450 million tons per year, thus aggravating the grave challenges of solid waste and litter management. Previous studies indicate an alarming surge in plastic waste generation, with estimates projecting a monumental 26 billion tons by 2050, underscoring the urgent need for proactive intervention (de Sousa, 2021).

The United Nations indicates that the severity of the issue has enshrined plastic waste mitigation as a cornerstone of its 2030 Agenda for Sustainable Development, underscoring its critical importance in global efforts to foster a more sustainable future (https://www.un.org/en/observances/zero-waste-day). Addressing the multifaceted ramifications of plastic pollution demands collaborative action on a global scale, encompassing policy reforms, technological innovations, and widespread public awareness initiatives (Yan et al., 2024). A disquieting reality emerges as over 60% of all plastics ever produced find their way into landfills, natural habitats, and water bodies, perpetuating a cycle of environmental issues (Tekman et al., 2022). This pervasive pollution not only tarnishes landscapes but also poses a grave threat to marine ecosystems, with an alarming 8 million

* Corresponding author.

DOI: 10.1201/9781003354000-22

tons of plastic waste infiltrating our oceans annually (Schmaltz et al., 2020), (https://www.iucn.org/resources/issues-brief/marine-plastic-pollution).

The repercussions of inadequate solid waste management extend far beyond environmental concerns, manifesting in approximately 30 diseases linked to pollution (Kwun Omang et al., 2021). These health impacts disproportionately affect vulnerable populations in developing nations, leading to the premature demise of hundreds of thousands to a million individuals annually (Kwun Omang et al., 2021). Moreover, the adverse impact on marine biodiversity is palpable, with an escalating number of species falling victim to entanglement, ingestion, and other hazards associated with plastic pollution. From marine mammals to seabirds and fish, the toll of plastic waste on wildlife populations is undeniable, threatening the delicate balance of ocean ecosystems (Schmaltz et al., 2020). This concerning trend underscores the urgent need for concerted global efforts to address and mitigate the multifaceted challenges posed by plastic waste. Collaboration among governments, industries, communities, and environmental organizations is imperative to develop sustainable solutions and enact policies that promote responsible consumption, waste reduction, and circular economies (Kibria et al., 2023).

When plastics are thrown into the environment, they break down due to mechanical interventions (ploughing, drying and rewetting, freezing and warming), chemical reactions (oxidation and hydrolysis), or microbial/enzymatic actions into smaller particles known as microplastics (MPs) and nanoplastics (NPs) (Garcia-Muñoz et al., 2023). As a result of widespread use and extremely slow rates of degradation, plastic pollution and its management have emerged as a pressing imperative, meriting attention on the global scale.

Nanoplastics, defined as plastic particles with sizes smaller than 1 μm, have emerged as a significant environmental and health concern in recent years (Lai et al., 2022). Resulting from the breakdown and fragmentation of larger plastic debris, nanoplastics pose unique challenges due to their small size and potential widespread distribution in ecosystems (Somashekharappa and Rajendrachari, 2022). These minute particles have been detected in various environmental compartments, including water bodies and the atmosphere, raising concerns about their impact on ecological balance and human health. Research into nanoplastics is essential to comprehensively understand their sources, distribution, behavior, and potential risks to both the environment and human well-being. By examining the existing body of knowledge and conducting further investigations, scientists aim to elucidate the complex interactions between nanoplastics and living organisms, assess their toxicity, and develop strategies to mitigate their adverse effects.Nanoplastics in agricultural systems have emerged as a significant environmental concern. Micro and nanoplastics (MNPs) in agricultural soil ecosystems pose risks to soil health, plant growth, and the overall ecosystem. The pervasive nature of NPs is becoming increasingly prevalent in agroecosystems (Moeck et al., 2022). The presence of MNPs in soils is raising concerns about their impact on biota and the environment (Maddela et al., 2023). The potential risks to soil physicochemical properties, microbiological aspects, and plant growth illustrate the complex interactions of NPs in agricultural environments (Verma et al., 2023) (Figure 17.1). Additionally, the translocation of nanoplastics into crop grains has been studied, raising questions about food safety and the potential entry of NPs into the food chain (Jiang et al., 2022).

The present study aims to set the stage for delving into the need for continued research on plastic pollution for the benefit of environmental sustainability and human health. Understanding their sources, behavior, and impacts is crucial for effective mitigation strategies and safeguarding ecosystem and human health. Especially nanoplastics, being smaller and more numerous than microplastics, can potentially pose greater health risks to humans. Nanoplastics can enter the human body more easily and may have adverse biological effects, particularly in the gastrointestinal system.

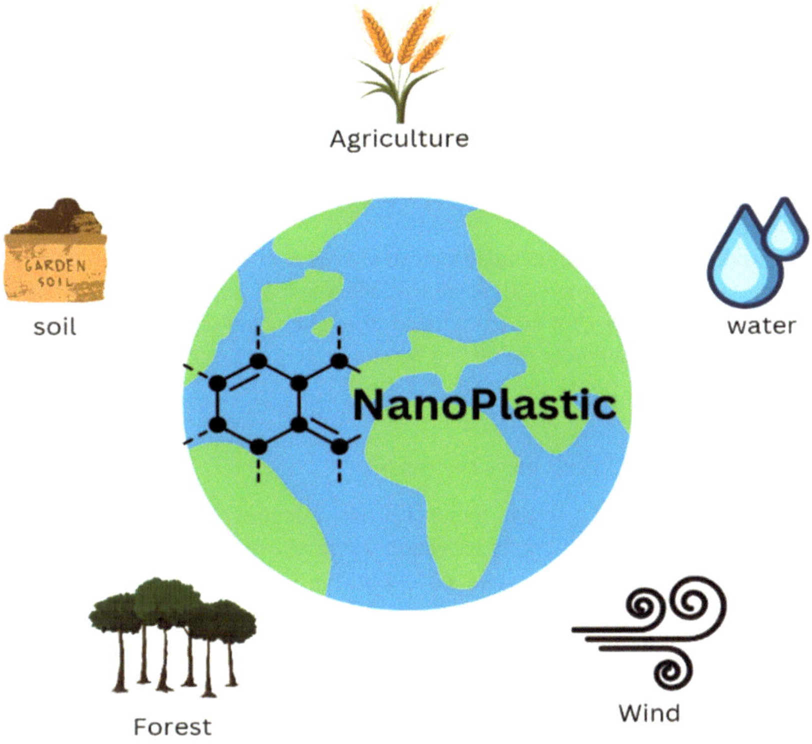

FIGURE 17.1 Microplastics and nanoplastics in the environment can have varying effects across different domains.

17.2 IMPACT OF PLASTICS ON THE ENVIRONMENT

17.2.1 Airborne NPs and Their Impact

It is important to remember that a variety of human activities are the cause of the existence of NPs in the environment. For example, numerous MP particles are released into the atmosphere when larger plastic objects, such as bottles and bags, break down due to mechanisms including photodegradation and mechanical breakdown (Bhat et al., 2023). Furthermore, the amount of plastic in the air is increased by synthetic fabrics, which are often used in apparel and household products and release microfibers when washed and worn (Yee et al., 2021). Moreover, this ubiquitous contamination is exacerbated by the atmospheric deposition of MPs, which originate from land-based sources such as agricultural practices and urban runoff.

Once in the air, these particles may be carried by wind currents over great distances and can even spread around the world through precipitation, thus infecting even far-off places with plastic pollution (Facciolà et al., 2021). Comprehending the many origins and intricate routes of dispersal of airborne NPs highlights the pressing requirement for all-encompassing approaches to alleviate their detrimental effects on the environment and human well-being.

17.2.1.1 Potential Impacts on Air Quality and Ecosystems

Exposure to MNPs through the air, water, and food has a direct influence on human health. It also has an indirect impact on food security due to its impact on ecosystems (Kögel et al., 2020). Indeed, airborne NPs have a significant negative influence on the environment. These microscopic particles can aggregate after deposition in a variety of environmental compartments, such as soil and water bodies, where they pose serious threats to species and ecosystems.

When marine species consume NPs in aquatic habitats, the resultant bioaccumulation and biomagnification down the food chain can have a significant effect on whole ecosystems (Tuuri et al., 2023). Furthermore, the presence of NPs in soil can disrupt microbial activity and soil health, which can impact plant development and nutrient cycling. Furthermore, airborne NPs may operate as carriers of dangerous substances like heavy metals and persistent organic pollutants, worsening their effects on the environment.

In order to address the environmental effects of airborne NPs, coordinated efforts must be made to minimize plastic pollution at its source, put in place efficient waste management procedures, and create cutting-edge technology for environmental plastic removal and rehabilitation. The biodiversity that depends on ecosystems for living can be preserved and protected by reducing the environmental impact of NPs.

17.2.2 Impact of Plastic Pollution on the Soil

MNPs also affect the soil and it is also found that they affect agricultural soil. A variety of channels, such as sewage sludge, compost, or irrigation water, allow nanoplastics to build up in soil and contaminate both natural ecosystems and agricultural areas. Their presence disturbs soil microbial populations, which are vital to plant health and nutrient cycling, impacting the functioning of the ecosystem as a whole (Zhou et al., 2023). Crop development and production are impacted by nanoplastics' effects on soil structure, soil organic matter, and pH, organic carbon storage, and electrical conductivity (Verma et al., 2023). Additionally, they could absorb and transfer dangerous substances like pesticides or heavy metals, worsening soil contamination.

17.2.2.1 Impact on Soil Microbial Diversity

NPs in soil negatively affects microbial biomass and activity due to their small size and surface area, inhibiting microbial growth (Pérez-Reverón et al., 2023). Plastics, including MNPs, can act as both stressors and resources for soil microbes, depending on various factors such as plastic characteristics and microbial community composition (Gao et al., 2023). MP pollution in soil can lead to mixed effects on plant growth and yield, with biodegradable plastics posing a greater risk (Iqbal et al., 2020). Plastics can also have a positive impact on soil microbial activity by serving as an external carbon source for microbes (Gaoet al., 2023).

Furthermore, the introduction of MPs into soil environments presents a complex scenario for plant productivity (Lin et al., 2020). The intricate interplay between plastic pollution and soil microbial communities underscores the multifaceted nature of this environmental concern. Therefore, understanding the dynamics of plastic pollution in soil requires considering the intricate relationships between plastic properties, soil conditions, and microbial activity. Consequently, research efforts continue to unravel the mechanisms through which plastics impact soil ecosystems, aiming to mitigate potential ecological repercussions and devise sustainable management strategies (Iqbal et al., 2020). Moreover, the effects of microplastics, particularly low density ployethlene, on bacterial community dynamics highlight the need for comprehensive assessments of plastic pollution's consequences on soil biota (Omidoyin et al., 2023). As the understanding of the interactions between microplastics and soil microbiota evolves, it becomes increasingly apparent that addressing this environmental challenge demands interdisciplinary approaches and concerted global efforts.

17.2.2.2 Effects of Plastic Pollution on Microbial Biodiversity in Soil

MNPs exert intricate influences on microbial communities in agricultural soils, shaping essential ecological processes and soil fertility. These minute plastic particles can disrupt soil nutrient cycling by altering soil enzyme activities and microbial functions (Omidoyin et al., 2023). While MPs might induce shifts in microbial populations and soil properties, NPs could pose detrimental effects on soil organisms and ecosystem dynamics (Verma et al., 2023). The pervasive presence of MNPs in agricultural soils underscores the urgency of understanding their long-term impacts on soil microbial diversity and function, vital for sustaining crop productivity and ecosystem resilience (Verma et al., 2023). Studies suggest that microplastic exposure in soil can accelerate bacterial turnover rates, underscoring the complex interplay between plastics and soil microbiota (Omidoyin et al., 2023). However, unraveling the precise mechanisms underlying nanoplastic-mediated changes in microbial community composition remains a critical research frontier. Developing effective mitigation strategies necessitates comprehensive investigations to safeguard soil health and ecosystem integrity in the face of plastic pollution. Further interdisciplinary research is imperative to address this pressing environmental challenge and promote sustainable agricultural practices.

17.2.2.3 Long-Term Effects on Soil Health

NPs have been found to accumulate in soil environments at a rapid rate, potentially leading to serious consequences for ecosystems and human health (Zhou et al., 2023). The extremely small size and high surface area of nanoplastics can exacerbate their impact on soil ecosystems. The small size and increased surface area of NPs intensify their interaction with soil organisms and biogeochemical processes. Soil microbiota may suffer, leading to compromised soil fertility and nutrient cycling (Pérez-Reverón et al., 2023). The fate of NPs in soil is influenced by physical properties rather than chemical structures, highlighting the complexity of their interactions with soil health (Pathan et al., 2020). While chemical composition influences plastic degradation, the fate of NPs in soil is primarily determined by physical properties. Factors such as size, shape, and surface charge govern their behavior and persistence in soil matrices (Pathan et al., 2020). Nanoplastics can potentially leach toxic substances into soil and water, posing long-term risks to soil health and overall environmental quality (https://www.unep.org/news-and-stories/story/plastic-planet-how-tiny-plastic-particles-are-polluting-our-soil).

NPs can infiltrate agricultural soils through various pathways, including irrigation with contaminated water, the application of plastic mulches, and the breakdown of larger plastic debris (Pérez-Reverón et al., 2023). As a result, NPs accumulate in soil matrices and pose a potential threat to soil health and agricultural productivity (Pérez-Reverón et al., 2023). The high surface area-to-volume ratio of NPs enhances their reactivity with soil components, potentially leading to changes in soil aggregation, porosity, and compaction. NPs can disrupt microbial communities, affecting soil enzymatic activities, nitrogen fixation, and carbon sequestration processes. These alterations may have cascading effects on soil fertility and crop productivity (Astner et al., 2023). NPs can interact with plant roots and translocate within plant tissues, raising concerns about their potential impacts on crop growth, nutrient uptake, and food quality (Moeck et al., 2023). Studies have shown that NP exposure can induce physiological changes in plants, including oxidative stress, altered gene expression, and reduced photosynthetic efficiency (Moeck et al., 2023).

17.2.3 Impact of Plastic Pollution on Aquatic Ecosystems: Special Emphasis

The collection of various contaminants and microbes on the surface of microplastics worsens the impact. There is no doubt that the universality of MNPs in aquatic environments has impacted food chains and webs as well as ecosystem processes (Zhao et al., 2023). Because of their many uses and resistance to degradation, NPs (diameter <100 nm) are widely distributed in the environment and may be found in soil, water, and the atmosphere (Liu et al., 2022). Similarly, humic substances (humic acid, fulvic acid, humin, etc.) and extracellular polymeric substances (DNA, proteins, carbohydrates, etc.) released by organisms and degradation products of organic matter, respectively, make up vast amounts of naturally occurring organic matter in the environment. Aquatic species can be harmed by MNPs and the pollutants they transport at the tissue, cellular, and molecular levels. The adverse effects of pollutants and MNPs on aquatic life are scant. Amobonye et al. (2021) suggested that tiny plastic particles emit harmful plastic additives and have the potential to absorb other substances. As a result, they act as sinks for different hazardous substances, increasing their bioavailability, toxicity, and mobility (Figure 17.2). Moreover, after being consumed by lesser creatures, MNPs may pose a trophic transfer risk to humans and other higher animals.

Ecosystems are seriously threatened by MNPs, which are produced by a variety of industrial processes and the environmental degradation of bigger plastic particles (Geremia et al., 2023). Because of their tiny size and resilience to degradation, nanoplastics survive once they are introduced into ecosystems. They mostly infiltrate aquatic habitats as byproducts of particle emissions. Their tenacity causes them to accumulate in soils and bodies of water (Musa et al., 2024). Ecosystems face a variety of hazards posed by nanoplastics, such as altered microbial populations, perturbations to nutrient cycles, and detrimental impacts on the growth and development of aquatic creatures.

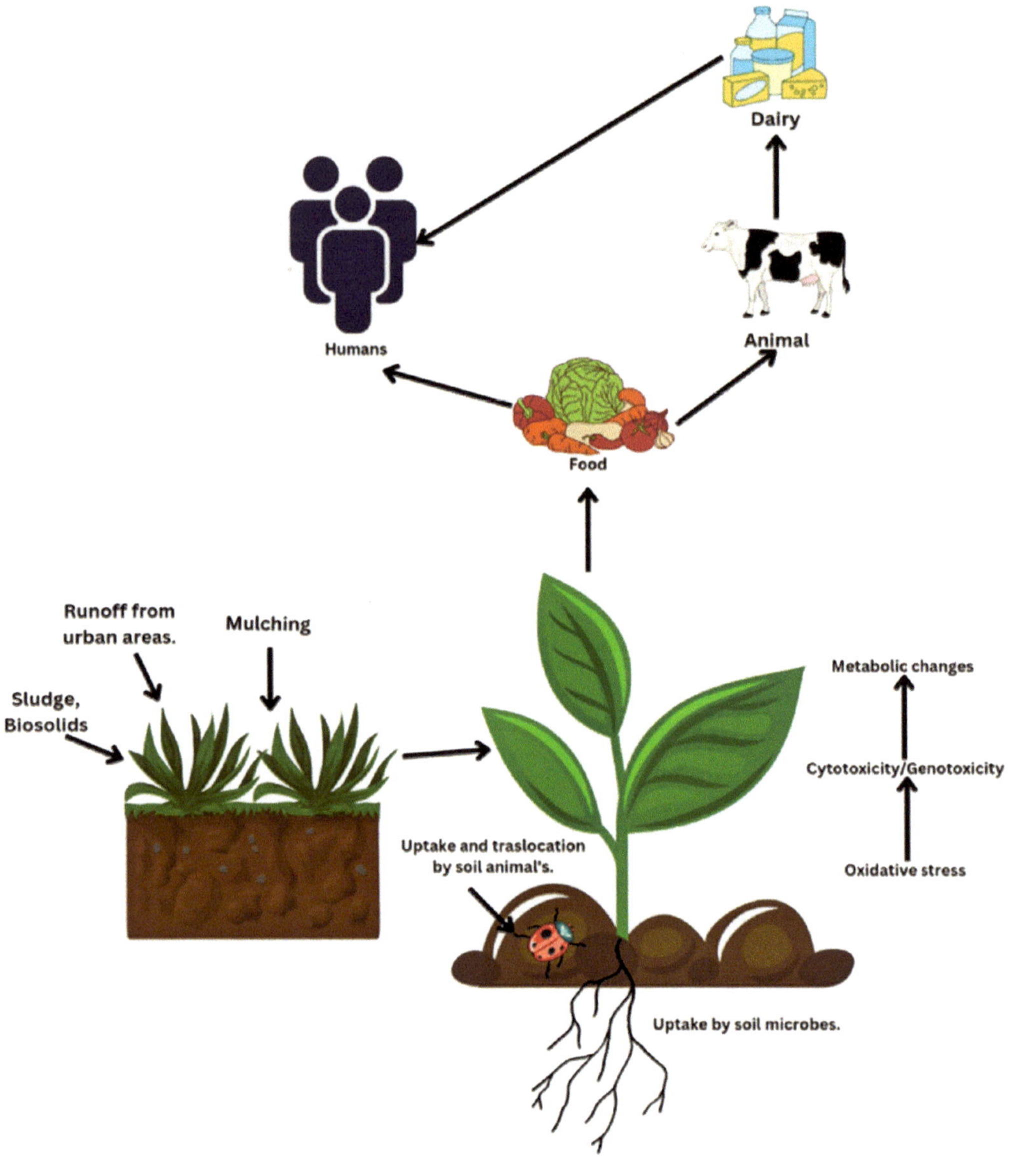

FIGURE 17.2 Nanomaterial's flow and mobility within agricultural and food systems.

Moreover, nanoplastics have the potential to worsen the problem of plastic pollution, adding to an already dire environmental situation.

17.2.3.1 Impact of Plastic Pollution on the Marine Ecosystem

Marine ecosystems may be impacted by the release of dissolved organic carbon caused by the photodegradation of nanoplastics in marine habitats (Geremia et al., 2023). Furthermore, these contaminants have the ability to penetrate the food chain, which might endanger human health and lead to a decline in biodiversity. Therefore, reducing the harmful impacts of nanoplastics on the environment and human health requires addressing their presence and influence in ecosystems. Because of their extensive distribution in aquatic environments and their interactions with living things, MNPs together represent serious hazards to the environment and human health. Particles of plastic smaller than 100 nm are known as NPs, and they differ from bigger particles of plastic in that they have a higher surface area and may be more harmful. NP particles are frequently found throughout aquatic habitats because they are either purposefully generated for different industrial uses or result from the breakdown of bigger plastic waste (Geremia et al., 2023).

By interfering with aquatic creatures' metabolism, reproduction, and mortality, NPs can have a negative impact on them. Furthermore, it is possible for NPs to build up in the tissues of terrestrial plants, which upsets terrestrial ecosystems. Although the intricate relationships between NPs and ecosystems are currently poorly understood (Amobonye et al., 2021), earlier studies indicate that they may change nutrient cycling and microbial populations, which might have a domino effect on ecosystem health. In order to reduce the detrimental impacts of nanoplastics on ecosystems and human health, it is necessary to conduct multidisciplinary research and implement efficient mitigation techniques.

17.2.3.2 Impact of Nanoplastics on Aquatic Systems

A significant worldwide problem is the existence of tiny plastic contaminants in the aquatic environment (Bulannga et al., 2024). MNPs build up in aquatic ecosystems as part of bigger plastic waste floating in the water, as a result of land-based operations, abrasion of fishing gear, and paint used in shipping. Studies show that MPs may be found in both the world's most remote areas and metropolitan coastal settings (Trevisan et al., 2022). Although most aquatic ecosystems naturally tend to dilute pollution to some level, serious contamination causes changes in the community's flora and fauna (Bashir et al., 2020).

The substantial threat that NPs represent to biodiversity, ecosystem health, and ultimately the well-being of humans in aquatic environments emphasizes the urgent need for mitigation measures and more study into the effects of these materials. NPs have a lot to do with the co-contaminants of the aquatic ecosystem. As we know, due to increasing water pollution, other contaminants like pesticides and heavy metals (metals with densities greater than 5 g/cm^3 and atomic numbers greater than 20) (Saidon et al., 2024), are also found in the aquatic ecosystem (Bashir et al., 2020).

So, once taken up, these pollutants can travel across various aquatic ecosystems more easily by attaching themselves to NPs. This process raises the potential damage to aquatic creatures by extending the persistence of pollutants in the environment and widening their dispersion range. Additionally, pollutants may be concentrated and delivered to organisms through the use of nanoplastics as carriers. This can amplify the harmful effects of contaminants in the food chain (Trevisan et al., 2022). As a result, the relationship between co-contaminants and NPs presents a complex risk to the well-being and stability of aquatic ecosystems, underscoring the pressing need for a thorough investigation and proactive management techniques to address this expanding environmental issue.

17.2.3.3 Evaluation of Nanoplastic Effects on Aquatic Microorganisms

The effects of NPs on aquatic microorganisms are a topic of increasing interest and investigation. Research has demonstrated that nanoplastics can negatively impact aquatic microorganisms in a number of ways. First of all, hazardous substances and other contaminants that stick to their surfaces might be transported by nanoplastics. These contaminants can be released through the ingestion of nanoplastics by microbes, leading to toxicity and interference with normal cellular processes. Furthermore, by offering fresh surfaces for adherence and colonization, nanoplastics have the potential to modify microbial communities and upset the delicate equilibrium of microbial populations (Shi et al., 2023) that is necessary for ecosystem health. It has been proposed that MNPs' capacity to penetrate biological barriers and hence boost their effective uptake by organisms can be enhanced by interactions with environmental entities in the aquatic environment, thereby increasing their bio-invasiveness (Zhang et al., 2022). As a result, it is possible to anticipate the significant buildup of plastic pieces in the aquatic environment, which will result in a concerning MNP burden. They travel through different water levels with the flow of the water, depositing in the mud at the bottom or floating in the water, whereupon they are consumed by creatures and pose harm to the environment.

Furthermore, ingesting NPs can cause physical harm to microbes because the particles can build up in the digestive system and result in obstructions or abrasions (Trevisan et al., 2022). The bacteria may be less able to metabolize nutrients effectively as a result of this barrier, which might result in decreased feeding efficiency and even famine. As a result, the reduced amount of nutrients they receive might impair their ability to develop and reproduce, which will worsen the effect on the population dynamics in aquatic environments.

Furthermore, bacteria may experience stress reactions in response to NPs, which might impair their general tolerance to environmental stresses and physiological processes. Aquatic organisms are seriously threatened by the presence of NPs in aquatic habitats (Mofijur et al., 2021; Zhang et al., 2022), which may have a domino impact on entire ecosystems.

17.2.3.4 Consequences for Fish and Other Marine Organisms

When we talk about MNPs in the marine environment, we're talking about their original condition prior to leakage into the ecosystem. These sources might be primary or secondary (Chen et al., 2023).

Nanoplastics can bioaccumulate in aquatic settings, which has significant effects on fish species' growth and well-being. Various aquatic creatures, including fish, as well as birds and mammals, including humans, can consume and accumulate microplastics (Barría et al., 2020). When nanoplastics find their way into bodies of water, fish and other aquatic life easily consume them, which has a series of negative consequences.

MPs have demonstrated the capacity to adsorb and concentrate harmful substances found in the environment, including metals (such as Ni, Cu, Zn, and Pb) and persistent organic pollutants (POPs) such as dichlorodiphenyltrichloroethane (DDT), polychlorinated biphenyls (PCBs), and polybrominated diphenyl ethers (PBDEs) (Barría et al., 2020). This buildup may result in bodily harm, including organ failure and tissue damage, which would impair regular physiological processes. When marine species swallow NPs, they may experience oxidative stress, immune system malfunction, physiological disturbances, and physical injury (Soares et al., 2020). Moreover, hazardous contaminants can be absorbed and transported by NPs, making aquatic environments even more hazardous. Because of their larger surface area, which allows them to pass through biological barriers, enter membranes, and be absorbed into tissues and organs, NPs are more harmful than MPs (Liu et al., 2022). Particularly at sizes less than 100 nm, the accumulation of NPs has been reported in the digestive systems, brain, heart, liver, yolk sac, gonads, and gills of both vertebrate and invertebrate species (Trevisan et al., 2022).

Furthermore, aquatic organisms may experience stunted growth and development as a result of NP interference with metabolic processes. Species exposed to NP pollution have been shown to exhibit reproductive abnormalities, such as lower fertility and diminished egg viability. When consumed in large enough quantities, NPs can cause death, upsetting the delicate balance of freshwater food webs and having a domino effect on ecosystem dynamics. When marine life consumes NPs, the effects can be more complex than just bodily injury and physiological disturbance. Moreover, immune system dysfunction brought on by NPs in aquatic habitats might weaken marine life's defenses against infections and make them more vulnerable to illness.

17.2.3.5 Impact of NPs on Freshwater

NPs, which are particles smaller than 1 μm, are found in freshwater ecosystems through a number of routes (Kukkola et al., 2021). These include the discharge of wastewater from domestic and industrial sources, agricultural runoff that contains fertilizers and pesticides that contain nanoplastics, and atmospheric deposition, in which particles of nanoplastics suspended in the air settle on water bodies.

Plastics are thought to originate in freshwater and travel through terrestrial habitats before ending up in the seas (Strungaru et al., 2019). After being introduced, NPs have a variety of interactions with aquatic environments that affect the health of aquatic species, the composition of the sediment, and the quality of the water. Their long-term persistence worsens their environmental impact since they keep building up in freshwater environments, which continuously threatens the resilience and functionality of ecosystems. Because of their tiny size, these particles are quite robust and may stay in the environment for a long time. Since they can build up in sediments, float on the water's surface, or be consumed by aquatic life, nanoplastics represent a serious long-term danger to aquatic ecosystems, eventually upsetting ecological balance and biodiversity.

17.2.3.6 Biodegradable Microplastics

Just because biodegradable polymers break down quickly in the environment, they have become a popular ecological substitute (González-Pleiter et al., 2019). However, although often touted as a solution to plastic pollution, they can undergo degradation processes that result in the formation of secondary NPs (Tong et al., 2022). This transformation may worsen the burden of plastic pollution in freshwater ecosystems, compounding the challenges already posed by NPs. As these secondary NPs enter the environment, they contribute to the accumulation of MPs in water bodies, further impacting aquatic life and ecosystem health. The complex interactions between biodegradable plastics and environmental conditions can influence the rate and extent of secondary NPs formation, highlighting the need for comprehensive assessments of their environmental fate and impact. Understanding the mechanisms driving this secondary release is essential for developing effective strategies to mitigate the adverse effects of NPs pollution on freshwater ecosystems (Tong et al., 2022).

17.2.4 Impact on Human Health

When discussing the possible health impacts of airborne NPs, it is important to emphasize the consequences of human exposure to these ubiquitous contaminants (Luo et al., 2024). Breathing in airborne NPs poses serious dangers to respiratory health since these tiny particles can enter the respiratory system through the alveoli, pass through biological barriers, and find their way into the bloodstream, where they might accumulate in key organs or cause systemic health problems.

NPs have the potential to cause respiratory disorders including asthma, bronchitis, fibrosis, and pneumothorax to worsen, cause lung inflammation, and hasten the onset of more serious respiratory ailments when breathed in (Rodrigues et al., 2022). Moreover, research on the long-term health impacts of chronic exposure to airborne NPs is still ongoing, so additional studies are required to fully understand their impact.

Furthermore, there is concern that NPs can pass through biological barriers. In order to limit exposure and protect human health, addressing the health dangers associated with airborne NPs involves multidisciplinary efforts, including strict air quality legislation, cutting-edge pollution control technology, and public awareness campaigns.

17.3 RISK ASSESSMENT AND MITIGATION STRATEGIES

Because of the serious risks associated with NPs due to both human health and the environment, a careful risk assessment and practical mitigation techniques are required (Cunningham et al., 2023).

Research indicates that our understanding of the effects of NPs is limited. Therefore, extensive research is necessary to clarify the toxicity processes of NPs in order to address this challenge (Catarino et al., 2023).

Risk assessment models need to conduct a thorough analysis of several factors in order to estimate the dangers related to NPs. This entails giving careful thought to exposure pathways, which include air, water, and food chains, as well as direct and indirect exposure routes (Noventa et al., 2021). The evaluation should also explore the complex mechanisms of bioaccumulation, looking at how NPs build up in species over time and may even extend up the food chain, posing threats to the environment and human health (Wang et al., 2021). Furthermore, a thorough investigation of the possible health effects of NPs is necessary, including both immediate effects like oxidative stress and inflammation as well as long-term effects like carcinogenicity and reproductive toxicity (Noventa et al., 2021).

A diverse approach is required for mitigation methods intended to address the dangers associated with NPs. One important component is concentrating efforts on cutting the production of NPs by creating and utilizing cutting-edge manufacturing techniques that reduce the number of NPs particles produced. Furthermore, it is imperative to improve waste management systems, highlighting the need for effective recycling strategies and appropriate disposal approaches to avert environmental contamination with NPs. To further reduce the number of NPs released into the environment, strict laws governing the usage and disposal of plastic products are essential (Cunningham et al., 2023).

These rules ought to include things like outright prohibitions on specific plastic items, the adoption of extended producer responsibility programs, and the tight enforcement of pollution control guidelines. Stakeholders may successfully limit the hazards presented by NPs to human health and the environment by implementing rigorous regulatory frameworks, improving waste management practices, and reducing the production of NPs through a holistic strategy.

Transitioning to bio-based plastics offers a promising avenue for reducing the environmental footprint of plastic production and consumption (Okeke et al., 2023). These alternatives, derived from renewable sources such as plants or microorganisms, exhibit biodegradability and lower persistence in the environment compared to traditional plastics. By shifting toward bio-based plastics, industries and consumers can decrease their reliance on fossil fuel-derived plastics, thereby mitigating the generation of NPs pollution in diverse ecosystems. Furthermore, the adoption of bio-based alternatives aligns with sustainability goals and promotes a circular economy model by closing the loop on plastic waste (Okeke et al., 2023).

Embracing sustainable agricultural methods not only diminishes reliance on plastic-based materials but also fosters a healthier relationship between farming and the environment. By implementing agroecological approaches such as crop rotation, organic farming, and integrated pest management, farmers can curtail the need for plastic inputs while promoting soil health and biodiversity. Additionally, adopting precision agriculture techniques and utilizing biodegradable mulching materials offer effective alternatives to conventional plastic-based practices, thereby mitigating the release of NPs into agricultural ecosystems. Moreover, promoting agroforestry systems and incorporating cover crops can enhance soil structure, water retention, and nutrient cycling, further reducing the environmental impact of NPs. Collaborative efforts among policymakers, researchers, and agricultural stakeholders are vital for advancing sustainable farming practices and safeguarding agricultural landscapes from NP contamination (Okeke et al., 2023).

17.4 LEGISLATION ADOPTED AGAINST PLASTIC POLLUTION GLOBALLY

Nanoplastics can accumulate in organisms, potentially leading to toxic effects and disrupting ecosystems (Shi et al., 2024). Their impact helps preserve biodiversity and ecological balance. Nanoplastics can enter the food chain, affecting human health through the consumption of contaminated food and water (Jiang et al., 2020). Investigating their impact on ecosystems informs strategies to mitigate health risks. NPs influence soil properties and interactions between particles, water, chemicals, and microorganisms (Verma et al., 2023). Understanding the impacts of NPs and policy decisions regarding waste management, pollution prevention, and environmental protection (Joksimovic et al., 2022) is critical in crafting effective regulations to mitigate nanoplastic pollution (Figure 17.3).

The Microbead-Free Waters Act (2005) in the United States banned microbeads in personal care products, promoted biodegradable alternatives, enhanced plastic recycling, reduced energy use, and supported wastewater treatment and bioremediation research (Wu et al., 2017). California, Hawaii, and New York implemented statewide bans on single-use plastic bags. California enacted the legislation in 2014, Hawaii followed suit with non-biodegradable plastic bags in populous counties, and New York passed a similar ban in 2019.

The United Kingdom aimed for all plastic packaging to be recyclable, reusable, or compostable by 2025, aligning with goals for zero avoidable waste by 2050 and eliminating avoidable plastic waste by 2042, to leave a better environment for future generations. The UK government's 2019 manifesto commitment, now part of the Environment Bill 2021–2022, aimed to ban the export of plastic waste to non-OECD (Organization for Economic Co-operation and Development) countries, ensuring responsible management of plastic waste domestically (Smith L., 2020). China's regulations under the Law on the Prevention and Control of Environmental Pollution by Solid Wastes (LPCEPSW) aimed to control plastic pollution, banning dumping in water bodies and promoting a circular economy (Yang et al., 2021). Malaysia's 2018 roadmap aimed to achieve zero single-use plastic usage by 2030, addressing pollution and promoting alternatives through research and development, aligning with long-term targets to combat plastic pollution. Australia's Recycling and Waste Reduction Bill of 2020 integrated the Product Stewardship Act of 2011, outlining a waste management and recycling framework. The country banned the export of plastic, paper, and glass waste, assuming full responsibility for its disposal. Canada proposed an order to include plastic products in Schedule 1 of the Canadian Environmental Protection Act of 1999 (CEPA), marking a regulatory measure to manage plastic products effectively. France

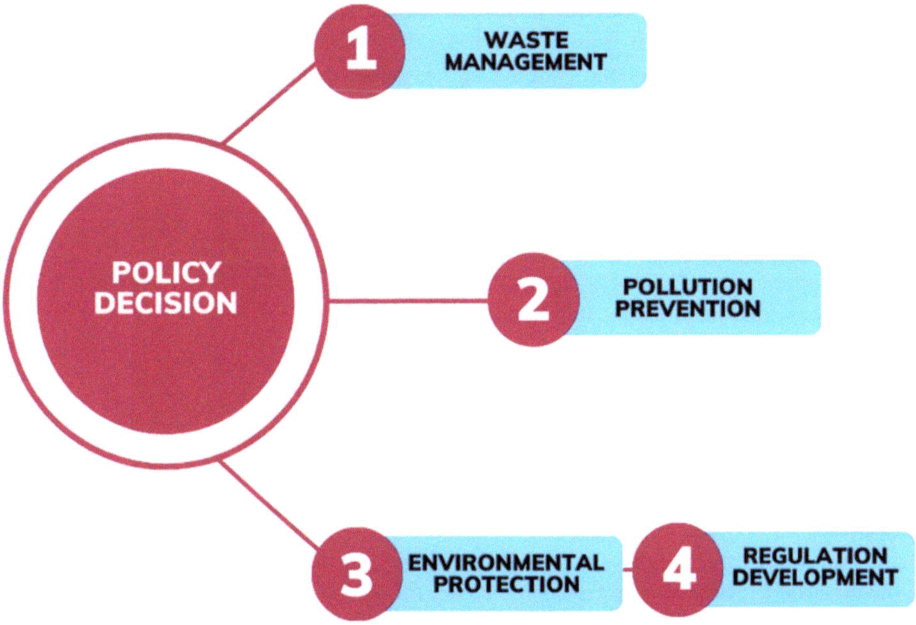

FIGURE 17.3 Policy decisions are made to control different aspects of the environment in addition to waste management and prevention of pollution.

regulated packaging and plastic waste through the French Environmental Code (FEC) amended by the Circular Economy Law (Law No. 2020-105). This legislation imposed obligations on plastic waste, influencing both production and consumption practices (Cloché-Dubois et al., 2021). The Korean government enacted the Resource Circulation Act (RCA) in 2016 to promote a circular economy and resource efficiency in waste management. This led to the creation of the Plastic Waste Control Plan (PWCP) in 2018, focused on comprehensive management of plastic waste (Shin et al., 2020). The Swedish parliament implemented a tax on plastic bags, varying by size, to combat plastic pollution. It was enforced from March 1, 2020, and full taxation came into effect on April 1, 2020. Existing policies aimed at mitigating plastic pollution suffer from delays in development and insufficient implementation, exacerbated by increasing plastic production and disposal rates. They often overlook certain types of plastics, notably microplastics, and lack effective measurement tools to gauge their impact on various fronts such as aquatic life, food safety, human health, and alternative materials. Inadequate attention is given to research and development in these critical areas, hindering comprehensive solutions. Looking ahead, there is a pressing need to aggressively expand and harmonize policies both domestically and internationally. This expansion should integrate robust, measurable tools to evaluate policy effectiveness and environmental repercussions comprehensively, encompassing all stages of a plastic's lifecycle and its diverse impacts. Moreover, forthcoming policies should prioritize research and development endeavors, with a focus on understanding the implications for aquatic ecosystems, human health, and sustainable alternatives to plastics. By incorporating these measures, policies can better confront the multifaceted challenges posed by plastic pollution, thereby reducing its adverse effects and fostering a global shift toward more sustainable practices. These actions are crucial for preserving environmental health and human well-being amid the escalating threat of plastic pollution.

17.5 CONCLUSIONS

Estimates suggest that there are about 5 trillion plastic particles floating in the sea. Plastic pollution is present worldwide in the aquatic environment. A significant amount of this plastic waste originates from anthropogenic activities and enters the marine environment mostly via rivers, industrial and urban effluents, and runoff from nearby fields. Therefore, to fully understand the extent and causes of plastic pollution in aquatic systems, and association with other contaminating substances which can potentially stick to their surfaces and get transported over long distances, more research is warranted. This will help to develop effective mitigation strategies that can reduce the environmental risks associated with plastic pollution.

ACKNOWLEDGEMENTS

The authors appreciate the School of Agricultural Science, Sharda University, Noida, Uttar Pradesh, India, which provided facilities for this chapter.

REFERENCES

Amobonye, A., Bhagwat, P., Raveendran, S., Singh, S. and Pillai, S., 2021. Environmental impacts of microplastics and nanoplastics: a current overview. *Frontiers in Microbiology, 12*, p.3728.

Astner, A.F., Gillmore, A.B., Yu, Y., Flury, M., DeBruyn, J.M., Schaeffer, S.M. and Hayes, D.G., 2023. Formation, behavior, properties and impact of micro-and nanoplastics on agricultural soil ecosystems (A Review). *NanoImpact,* p.100474.

Barría, C., Brandts, I., Tort, L., Oliveira, M. and Teles, M., 2020. Effect of nanoplastics on fish health and performance: a review. *Marine Pollution Bulletin, 151*, p.110791.

Bashir, I., Lone, F.A., Bhat, R.A., Mir, S.A., Dar, Z.A. and Dar, S.A., 2020. Concerns and threats of contamination on aquatic ecosystems. *In* Khalid Rehman Hakeem, Rouf Ahmad Bhat, Humaira Qadri (Eds.), *Bioremediation and Biotechnology: Sustainable Approaches to Pollution Degradation* 27, (pp. 1–26) Springer Nature Switzerland AG.

Bhat, M.A., Gedik, K. and Gaga, E.O., 2023. Atmospheric micro (nano) plastics: future growing concerns for human health. *Air Quality, Atmosphere & Health*, *16*(2), pp.233–262.

Bulannga, R.B. and Schmidt, S., 2024. Micro-and nanoplastics in freshwater ecosystems—interaction with and impact upon bacterivorous ciliates. *Frontiers in Earth Science*, *12*, p.1349865.

Catarino, A.I., Patsiou, D., Summers, S., Everaert, G., Henry, T.B. and Gutierrez, T., 2023. Challenges and recommendations in experimentation and risk assessment of nanoplastics in aquatic organisms. *TrAC Trends in Analytical Chemistry*, p.117262.

Chen, M., Chen, F., Li, Z., Haider, M.R., Wei, J., Chen, G., Wang, W. and Wang, J., 2023. Environmental risk assessment of microplastics and nanoplastics generated from biodegradable plastics in marine ecosystem. *TrAC Trends in Analytical Chemistry*, p.117381.

Cloché-Dubois C., Plisson A., Bourdon A., Peraudeau C., 2021. CMS Expert Guide to plastics and packaging laws: A snapshot of legal developments Plastics and packaging laws in France. C. Law. Tax. Future. 6:1–9. (https://cms.law/en/int/expert-guides/plastics-and-packaging-laws/france).

Cunningham, B.E., Sharpe, E.E., Brander, S.M., Landis, W.G. and Harper, S.L., 2023. Critical gaps in nanoplastics research and their connection to risk assessment. *Frontiers in Toxicology*, *5*, p.1154538.

de Sousa, F.D.B., 2021. The role of plastic concerning the sustainable development goals: the literature point of view. *Cleaner and Responsible Consumption*, *3*, p.100020.

Facciolà, A., Visalli, G., Pruiti Ciarello, M. and Di Pietro, A., 2021. Newly emerging airborne pollutants: current knowledge of health impact of micro and nanoplastics. *International Journal of Environmental Research and Public Health*, *18*(6), p.2997.

Gao, G., Li, G., Liu, M., Liu, J., Ma, S., Li, D., Liang, X., Wu, M. and Li, Z., 2024. Microbial carbon metabolic activity and bacterial cross-profile network in paddy soils of different fertility. *Applied Soil Ecology*, 195, p.105233.

Garcia-Muñoz, P., Robert, D., Ruppert, A.M. and Keller, N., 2023. Microplastics (MPs) and nanoplastics (NPs): introduction. In *Current Developments in Biotechnology and Bioengineering* (pp. 1–32). Elsevier.

Geremia, E., Muscari Tomajoli, M.T., Murano, C., Petito, A. and Fasciolo, G., 2023. The impact of micro-and nanoplastics on aquatic organisms: mechanisms of oxidative stress and implications for human health—A review. *Environments*, *10*(9), p.161.

González-Pleiter, M., Tamayo-Belda, M., Pulido-Reyes, G., Amariei, G., Leganés, F., Rosal, R. and Fernández-Piñas, F., 2019. Secondary nanoplastics released from a biodegradable microplastic severely impact freshwater environments. *Environmental Science: Nano*, *6*(5), pp.1382–1392.

https://www.iucn.org/resources/issues-brief/marine-plastic-pollution

https://www.un.org/en/observances/zero-waste-day

Iqbal, S., Xu, J., Khan, S., Arif, M.S., Yasmeen, T., Nadir, S. and Schaefer, D.A., 2021. Deciphering microplastic ecotoxicology: impacts on crops and soil ecosystem functions. *Circular Agricultural Systems*, 1(1), pp.1–7.

Jiang, B., Kauffman, A.E., Li, L., McFee, W., Cai, B., Weinstein, J., Lead, J.R., Chatterjee, S., Scott, G.I. and Xiao, S., 2020. Health impacts of environmental contamination of micro-and nanoplastics: a review. *Environmental Health and Preventive Medicine*, *25*, pp.1–15.

Jiang, M., Wang, B., Ye, R., Yu, N., Xie, Z., Hua, Y., Zhou, R., Tian, B. and Dai, S., 2022. Evidence and impacts of nanoplastic accumulation on crop grains. *Advanced Science*, *9*(33), p.2202336.

Joksimovic, N., Selakovic, D., Jovicic, N., Jankovic, N., Pradeepkumar, P., Eftekhari, A. and Rosic, G., 2022. Nanoplastics as an invisible threat to humans and the environment. *Journal of Nanomaterials*, *2022*, pp.1–15.

Kibria, M.G., Masuk, N.I., Safayet, R., Nguyen, H.Q. and Mourshed, M., 2023. Plastic waste: challenges and opportunities to mitigate pollution and effective management. *International Journal of Environmental Research*, *17*(1), p.20.

Kögel, T., Bjorøy, Ø., Toto, B., Bienfait, A.M. and Sanden, M., 2020. Micro-and nanoplastic toxicity on aquatic life: determining factors. *Science of the Total Environment*, *709*, p.136050.

Kukkola, A., Krause, S., Lynch, I., Smith, G.H.S. and Nel, H., 2021. Nano and microplastic interactions with freshwater biota–current knowledge, challenges and future solutions. *Environment International*, *152*, p.106504.

kwun Omang, D.I., John, G.E., Inah, S.A. and Bisong, J.O., 2021. Public health implication of solid waste generated by households in Bekwarra Local Government area. *African Health Sciences*, *21*(3), pp.1467–1473.

Lai, H., Liu, X. and Qu, M., 2022. Nanoplastics and human health: hazard identification and biointerface. *Nanomaterials*, *12*(8), p.1298.

Liu, S., Junaid, M., Liao, H., Liu, X., Wu, Y. and Wang, J., 2022. Eco-corona formation and associated ecotoxicological impacts of nanoplastics in the environment. *Science of the Total Environment*, *836*, p.155703.

Luo, D., Chu, X., Wu, Y., Wang, Z., Liao, Z., Ji, X., Ju, J., Yang, B., Chen, Z., Dahlgren, R. and Zhang, M., 2024. Micro-and nano-plastics in the atmosphere: a review of occurrence, properties and human health risks. *Journal of Hazardous Materials*, p.133412.

Maddela, N.R., Ramakrishnan, B., Kadiyala, T., Venkateswarlu, K. and Megharaj, M., 2023. Do microplastics and nanoplastics pose risks to biota in agricultural ecosystems?. *Soil Systems*, *7*(1), p.19.

Moeck, C., Davies, G., Krause, S. and Schneidewind, U., 2023. Microplastics and nanoplastics in agriculture—A potential source of soil and groundwater contamination?. *Grundwasser*, *28*(1), pp.23–35.

Mofijur, M., Ahmed, S.F., Rahman, S.A., Siddiki, S.Y.A., Islam, A.S., Shahabuddin, M., Ong, H.C., Mahlia, T.I., Djavanroodi, F. and Show, P.L., 2021. Source, distribution and emerging threat of micro-and nano-plastics to marine organism and human health: socio-economic impact and management strategies. *Environmental Research*, *195*, p.110857.

Musa, I.O., Auta, H.S., Ilyasu, U.S., Aransiola, S.A., Makun, H.A., Adabara, N.U., Abioye, O.P., Aziz, A., Jayanthi, B., Maddela, N.R. and Prasad, R., 2024. Micro-and nanoplastics in environment: degradation, detection, and ecological impact. *International Journal of Environmental Research*, *18*(1), p.1.

Noventa, S., Boyles, M.S., Seifert, A., Belluco, S., Jiménez, A.S., Johnston, H.J., Tran, L., Fernandes, T.F., Mughini-Gras, L., Orsini, M. and Corami, F., 2021. Paradigms to assess the human health risks of nano- and microplastics. *Microplastics and Nanoplastics*, *1*(1), pp.1–27.

Okeke, E.S., Chukwudozie, K.I., Addey, C.I., Okoro, J.O., Ezeorba, T.P.C., Atakpa, E.O., Okoye, C.O. and Nwuche, C.O., 2023. Micro and nanoplastics ravaging our agroecosystem: a review of occurrence, fate, ecological impacts, detection, remediation, and prospects. *Heliyon*, *9*, p.e13296.

Omidoyin, K.C. and Jho, E.H., 2023. Effect of microplastics on soil microbial community and microbial degradation of microplastics in soil: a review. *Environmental Engineering Research,* 28(6).

Pérez-Reverón, R., Álvarez-Méndez, S.J., González-Sálamo, J., Socas-Hernández, C., Díaz-Peña, F.J., Hernández-Sánchez, C. and Hernández-Borges, J., 2023. Nanoplastics in the soil environment: Analytical methods, occurrence, fate and ecological implications. *Environmental Pollution,* 317, p.120788.

Pérez-Reverón, R., Álvarez-Méndez, S.J., González-Sálamo, J., Socas-Hernández, C., Díaz-Peña, F.J., Hernández-Sánchez, C. and Hernández-Borges, J., 2023. Nanoplastics in the soil environment: Analytical methods, occurrence, fate and ecological implications. *Environmental Pollution,* 317, p.120788.

Rodrigues, A.C.B., de Jesus, G.P., Waked, D., Gomes, G.L., Silva, T.M., Yariwake, V.Y., da Silva, M.P., Magaldi, A.J. and Veras, M.M., 2022. Scientific evidence about the risks of micro and nanoplastics (MNPLs) to human health and their exposure routes through the environment. *Toxics*, *10*(6), p.308.

Saidon, N.B., Szabó, R., Budai, P. and Lehel, J., 2024. Trophic transfer and biomagnification potential of environmental contaminants (heavy metals) in aquatic ecosystems. *Environmental Pollution*, *340*, p.122815.

Sala, S. and Castellani, V., 2019. The consumer footprint: monitoring sustainable development goal 12 with process-based life cycle assessment. *Journal of Cleaner Production*, *240*, p.118050.

Schmaltz, E., Melvin, E.C., Diana, Z., Gunady, E.F., Rittschof, D., Somarelli, J.A., Virdin, J. and Dunphy-Daly, M.M., 2020. Plastic pollution solutions: emerging technologies to prevent and collect marine plastic pollution. *Environment International*, *144*, p.106067.

Shi, C., Liu, Z., Yu, B., Zhang, Y., Yang, H., Han, Y., Wang, B., Liu, Z. and Zhang, H., 2024. Emergence of nanoplastics in the aquatic environment and possible impacts on aquatic organisms. *Science of the Total Environment*, 906, p.167404.

Shin, S.K., Um, N., Kim, Y.J., Cho, N.H. and Jeon, T.W., 2020. New policy framework with plastic waste control plan for effective plastic waste management. *Sustainability*, *12*(15), p.6049.

Smith, L., 2020. Plastic waste. House of Commons Briefing Paper, (08515).

Soares, J., Miguel, I., Venâncio, C., Lopes, I. and Oliveira, M., 2020. Perspectives on micro (Nano) plastics in the marine environment: biological and societal considerations. *Water*, *12*(11), p.3208.

Somashekharappa, K.K. and Rajendrachari, S., 2022. Sustainable development information management of carbon nanomaterial-based sensors. In *Carbon Nanomaterials-Based Sensors* (pp. 3–12). Elsevier.

Strungaru, S.A., Jijie, R., Nicoara, M., Plavan, G. and Faggio, C., 2019. Micro-(nano) plastics in freshwater ecosystems: abundance, toxicological impact and quantification methodology. *TrAC Trends in Analytical Chemistry*, *110*, pp.116–128.

Tekman, M.B., Walther, B., Peter, C., Gutow, L. and Bergmann, M., 2022. *Impacts of Plastic Pollution in the Oceans on Marine Species, Biodiversity and Ecosystems.* p 1–221. WWF Germany, Berlin. Doi: 10.5281/zenodo.5898684 .

Tong, H., Zhong, X., Duan, Z., Yi, X., Cheng, F., Xu, W. and Yang, X., 2022. Micro-and nanoplastics released from biodegradable and conventional plastics during degradation: formation, aging factors, and toxicity. *Science of The Total Environment*, *833*, p.155275.

Trevisan, R., Ranasinghe, P., Jayasundara, N. and Di Giulio, R.T., 2022. Nanoplastics in aquatic environments: impacts on aquatic species and interactions with environmental factors and pollutants. *Toxics*, *10*(6), p.326.

Tuuri, E.M. and Leterme, S.C., 2023. How plastic debris and associated chemicals impact the marine food web: a review. *Environmental Pollution*, *321*, p.121156.

Verma, K.K., Song, X.P., Xu, L., Huang, H.R., Seth, C.S. and Li, Y.R., 2023. Nano-microplastic and agro-ecosystems: a mini-review. *Frontiers in Plant Science*, *14*, p.1283852.

Wang, L., Wu, W.M., Bolan, N.S., Tsang, D.C., Li, Y., Qin, M. and Hou, D., 2021. Environmental fate, toxicity and risk management strategies of nanoplastics in the environment: current status and future perspectives. *Journal of Hazardous Materials*, *401*, p.123415.

Wu, W.M., Yang, J. and Criddle, C.S., 2017. Microplastics pollution and reduction strategies. *Frontiers of Environmental Science & Engineering*, *11*, pp.1–4.

Yan, H., Cordier, M. and Uehara, T., 2024. Future projections of global plastic pollution: scenario analyses and policy implications. *Sustainability*, *16*(2), p.643.

Yang, L., Zhang, Y., Kang, S., Wang, Z. and Wu, C., 2021. Microplastics in soil: a review on methods, occurrence, sources, and potential risk. *Science of the Total Environment*, *780*, p.146546.

Yee, M.S.L., Hii, L.W., Looi, C.K., Lim, W.M., Wong, S.F., Kok, Y.Y., Tan, B.K., Wong, C.Y. and Leong, C.O., 2021. Impact of microplastics and nanoplastics on human health. *Nanomaterials*, *11*(2), p.496.

Zhang, P., Liu, Y., Zhang, L., Xu, M., Gao, L. and Zhao, B., 2022. The interaction of micro/nanoplastics and the environment: effects of ecological corona on the toxicity to aquatic organisms. *Ecotoxicology and Environmental Safety*, *243*, p.113997.

Zhou, P., Wang, L., Gao, J., Jiang, Y., Adeel, M. and Hou, D., 2023. Nanoplastic–plant interaction and implications for soil health. *Soil Use and Management*, *39*(1), pp.13–42.

Part VI

Challenges and Future Opportunities in Aquatic Ecosystem Monitoring

18 Challenges and Future Implications in Monitoring and Assessment of Aquatic Ecosystems

*Smrutisikha Mohanty, Prem C. Pandey**
Department of Life Sciences, School of Natural Sciences, Shiv Nadar Institution of Eminence, Greater Noida, Uttar Pradesh, India

Prashant K. Srivastava
Institute of Environment and Sustainable Development, Banaras Hindu University, Varanasi, Uttar Pradesh, India

Sanjeev Kumar Srivastava
School of Science, Technology & Engineering, University of the Sunshine Coast, Queensland, Australia

18.1 INTRODUCTION

Aquatic ecosystems are well recognized for providing a substrate where biotic and abiotic elements of an ecosystem interact (Lamberti et al., 2010). There are several classifications of aquatic ecosystems, e.g., freshwater and marine ecosystems (Moss, 2009), and freshwater, brackish water, and marine ecosystems. Each subgroup of the aquatic ecosystem, whether occupying a large or small proportion, performs crucial functions (NRC, 1992). The marine ecosystem, the largest category of aquatic ecosystem, covers almost 70% of the Earth's surface (El-Regal et al., 2023). Oceans, estuaries, coral reefs, and coastal ecosystems are several categories included in marine ecosystems (McLean et al., 2001). Freshwater ecosystems account for less than one percent of the total Earth's surface. This ecosystem can be further categorized as lotic, lentic, and wetland ecosystems (Reddy et al., 2018). The key factors characterizing aquatic ecosystems are hydromorphological, biological, hydrological, chemical, and physicochemical parameters (Moldoveanu et al., 2017; Stefanidis et al., 2019). The state of aquatic ecosystems can be characterized using a combination of qualitative and quantitative parameters to derive metrics for each of these categories (Davies et al., 2006). Most of the biodiversity hotspots of the world are located in coral reefs and wetland ecosystems mainly because of their capability to support an enormous variety of life forms, genetic diversity, biological balance, and the continuous possibility of discovering new life forms (Berde et al., 2022; Chivian, E., 2002). Apart from supporting diverse life forms, oceans and other aquatic environments play a key role in regulating global climate. The marine ecosystem operates as a large carbon sink by continuously absorbing substantial amounts of CO_2 from the atmosphere (Heinze et al., 2015; El-regal & Satheesh., 2023). The microscopic phytoplanktons are well recognized for photosynthesizing carbon dioxide to produce oxygen. Wetlands also act as climate change buffers by storing carbon

* Corresponding author.

DOI: 10.1201/9781003354000-24

in their rich soil and plant life (Moomaw et al., 2018). These wetlands present a common linkage to a few Sustainable Development Goals (SDGs), either directly or indirectly (Mohanty et al., 2024a). This chapter emphasizes the interconnection between wetlands and SDGs, particularly SDG6 (Clean Water and Sanitation), SDG12 (Responsible Consumption and Production), SDG13 (Climate Action), SDG14 (Life below Water), and SDG15 (Life on Land). In addition to several ecological roles, aquatic ecosystems provide valuable economic services such as fisheries, tourism, raw materials, transportation routes, and hydropower energy. Fisheries are directly impacted by the health of these ecosystems, serving millions of people with protein as a source of food (Yusoff et al., 2021; Dugan et al., 2010). These ecosystems are of cultural and spiritual importance and offer monetary value to different communities. Overall, this chapter focuses on the methods of monitoring using traditional and advanced technologies, with their limitations and advantages, along with knowledge of different models and plugin tools/extensions in licensed or open-source software, ArcGIS, and QGIS. Finally, the chapter provides future scope and implications in aquatic ecosystem monitoring.

18.2 MONITORING OF AQUATIC ECOSYSTEMS

Given the vastness and diversity of aquatic ecosystems, their monitoring is a complex task that requires continuous observation across spatial and temporal scales. For example, studies on aquatic ecosystems are conducted at local, regional, continental, and global scales where seasonal as well as annual changes are monitored. The diverse monitoring methods include traditional field data collection for molecular-level assessment and characterization of physicochemical and biological parameters. The physical states of these ecosystems can be monitored using a variety of remote sensing (RS) methods, especially for regional or global-level monitoring. Traditional monitoring of aquatic ecosystems includes Hydromorphological, Physical–Chemical, and Biological Monitoring with field measurements and in situ data collection. Hydromorphological monitoring includes the assessment of the physical habitat and geomorphology of aquatic ecosystems, mainly riverine ecosystems. This assessment and monitoring is crucial for the ecological integrity of lotic ecosystems such as rivers and streams. Certain activities pertaining to surveying and monitoring can directly or indirectly affect riverine systems and their surrounding areas, such as inter-basin water management, land drainage, channel alignment and realignment, flood control, and water navigation.

18.2.1 Molecular-Level Monitoring: Environmental DNA (eDNA)

Environmental DNA (eDNA) is one of the evolving and transformative technologies in biology, chemistry, and bioinformatics that can be used to monitor the current status and anticipate future trends of aquatic ecosystems under stress, as well as monitor species (Huang et al.,2022; Altenburger et al., 2019). It is a novel method for biomonitoring and ecosystem monitoring. eDNA is a type of organismal DNA that is found in the environment and is derived from cells that have been released into either terrestrial or aquatic habitats such as lakes, rivers, and streams as well as into the ground (Thomsen & Willerslev, 2015). To find rare and cryptic species as well as to detect invasive and native species early on, it is crucial to sample and monitor it using molecular approaches (Darling & Blum, 2007). Species identification from DNA sequences is frequently simpler and more accurate than identification by observation of external morphology as eDNA monitoring reduces the field survey time and has little impact on the ecosystem (Armstrong & Ball, 2005). Another advantage is the ability to detect many aquatic species from a single water sample (Rees et al., 2014). eDNA involves a series of steps including identification of DNA in the environment, sample collection, DNA extraction from samples, DNA amplification, DNA sequencing, and species identification using the DNA (Ruppert et al., 2019; Beng & Corlett, 2020). All living organisms shed their cells, tissues, and waste, and leave their DNA, the exclusive molecule that retains their genetic blueprints in the ecosystem (Sultan, 2015). Following sample collection from lakes, rivers, or sediment, DNA and RNA can be isolated and analyzed in a lab. eDNA analysis can obtain high levels of taxonomic specificity without the requirement for time-consuming, labor-intensive organism collections and identifications (Sivakamavalli, 2022). Furthermore, a single DNA or RNA molecule can be detected by employing these techniques (Ficetola et al., 2008).

18.2.2 Traditional In Situ or Lab-Based Experiments

In situ or lab-based analysis depends upon the collection of samples from aquatic ecosystems, and data is recorded either on the spot using probe sensors or carried to the lab for analysis using chemical methods or smart instruments. Physical and chemical characteristics of water can be determined by analyzing samples collected from the water bodies during field surveys. The physical characteristics or parameters, which are analyzed using traditional methods, include temperature, conductivity, clarity and transparency of water bodies using SDD (Secchi disk depth methods), turbidity, and total suspended solids (TSS) (Batina & Krtalić, 2023; Gholizadeh et al., 2016, Ritchie et al., 2003). The chemical parameters of the water are mostly determined or analyzed using chemical-based analysis or smart water analyzers to assess different parameters, including dissolved oxygen, biochemical oxygen demand, chemical oxygen demand, nutrients (phosphates, nitrates), pH, and other elements, including lead, zinc, copper, iron, free iodine, and free bromine (Onojake et al., 2023; Yaroshenko et al., 2020). The changes in physical and chemical parameters of water are mainly due to anthropogenic activities, leading to contamination and pollution, and increased concentration of chemical pollutants or heavy metals in water bodies, such as insecticides, paints, nutrients (nitrogen, phosphate), and organic pollutants (Weldeslassie et al., 2018; Madhav et al., 2020) (Akhtar et al., 2021). Most of the time, addition of nutrients lead to increased nutrients level, converting lakes or waterbodies from oligotrophic to mesotrophic and again mesotrophic to eutrophic condition (Qin et al., 2013). The procedure includes probe tests in the field or collection of samples for analysis in the laboratory. The collection of samples is done using standardized procedures for accurate outcomes (Shelton, 1994). However, even after very carefully adopting standard procedures or protocols, the physical and chemical parameters of water exhibit significant spatiotemporal variations meaning that measurements can vary greatly throughout a day, month, season, or year, depending on several factors (Skalska et al., 2020; Maleki et al., 2024). Water samples collected from the lake could provide information on the health of the ecosystems, and the Trophic Status Index (TSI) measured using field data collected from the sample sites using boat and can be derived from Satellite images (Landsat) which provides the health status of Chilika lake, if the lake is oligotrophic or eutrophic, depending upon the TSI values of TSITN (total Phosphate), TSITN (Total Nitrogen), TSISDD (Sechhi Disk Depth-SDD), $TSI_{Chl\text{-}a}$ (Chlorophyll-a concentration) and TSItotal. Remote Sensing images are utilised to predict different water parameters such as turbidity, pH, Silica, Dissolved oxygen, conductivity, bathymetry, alkalinity, total hardness, and these can be corelated to the field measured values for assessing health and lake conditions like sedimentation or clear water.

Therefore, traditional ways to study aquatic ecosystems mostly consist of designing experiments with good sampling strategies and constructing baselines and transects for collecting data in the field (Madsen & Wersal, 2018). One of the challenges of using such methods is the accessibility of key locations because of remoteness and unsafe environments. This often leads to data collection with biases, e.g., having more sampling points in the vicinity of accessibility points such as roads, tracks, bridges, etc. Datasets mostly come from topographic maps, toposheets, and coarse/high-resolution remote sensing images. Drainage patterns and stream orders used to be calculated from topographic maps (Singh et al., 2014). There are several merits along with associated demerits with traditional methods of measurement in the field for aquatic ecosystems.

18.2.3 Regional and Global-Level Monitoring Using Geospatial Technology

Remote sensing provides water quality data with a high spatial and temporal resolution for thousands of lakes at a time. It supports the evaluation of environmental problems and potential health risks through the analysis of changes in water quality and the detection of harmful algal blooms

(Khan et al., 2021; Klemas, 2012). Remote sensing is widely used to monitor natural resources including water resources, vegetation resources, biological species, habitat and species distribution, and landscape ecosystems (Pandey & Sharma, 2021a; Pandey & Sharma, 2021b; Nagendra et al., 2013). With the continuous emergence of higher (high resolution), faster (short revisit period), and finer (high spectral resolution) datafrom satellites, drones, and terrestrial devices, remote sensing data are used for the classification of regions based on different algorithms and techniques to separate land use and land cover categories (Pandey et al., 2021). Remote sensing data provides the opportunity to quickly and effectively reveal the intricate connections between water environment elements, mine varied hydrological, environmental, and ecological knowledge, and achieve fine-grained, omnidirectional, high-temporal, and multilevel simulations of water ecosystems (Chen et al., 2022). Mishra et al. (2024) provide a comprehensive review of the spectral indices used for different research themes and applications. The use and incorporation of different combinations of spectral channels could benefit communities with the identification of vegetation, soil, and water regions. There are several indices used in aquatic ecosystems research to delineate water surfaces, water regions, occurrences of water areas, vegetation, and so on, known generally as the water index. This is just like the vegetation indices, the normalized difference vegetation index (NDVI) and modified normalized difference vegetation index (MNDVI). The water index, also known as the normalized difference water index (NNDWI), provides a straightforward and efficient approach for water extraction, typically derived from analyzing multiple bands to distinguish water from various land covers. There are numerous spectral indices for water-snow-vegetation extraction using a combination of spectral channels (Mishra et al., 2024). Several of them, such as the NDWI, MNDWI, and the world-view water index (WV-WI), as mentioned in Table 18.1, introduced the tasseled cap humidity (TCW) index for the first time. After this, numerous water indices emerged and gained widespread adoption and applicability in water extraction research. For instance, McFeeters (1996) introduced the normalized difference water index (NDWI) in 1996, utilizing green and near-infrared (NIR) bands. Similarly, the modified NDWI (MNDWI) was devised by substituting the green and NIR bands in the NDWI with red and medium infrared (MIR) bands, effectively addressing the limitation of NDWI in distinguishing water bodies from built-up areas. This modification has proven successful in identifying wetlands and water bodies with varying turbidity levels (Xu,

TABLE 18.1
Most Widely Used Spectral Indices for Water Identification and Related Environment*

Spectral index	Expression/combination of channels	Applications or use	References
NDWI	(Red – NIR)/(RED + NIR)	Used for mapping analysis of surface water bodies	McFeeters, 1996; Gao, 1996
MNDWI	(Green – SWIR)/(Green + SWIR)	Enhancing and extracting water from a background that is dominated by built-up land areas	Xu, 2007
WV-WI	(Coastal – NIR2)/(Coastal + NIR2)	To highlight areas of standing water greater than one pixel in size	Wolf, 2012
TCWI	= b1*Blue + b2*Green +b3*Red + b4*NIR + b5*SWIR1 + b6*SWIR2 where b1, b2, b3, b4, b5, and b6 are coefficients specific to the TCW index, determined through calibration or empirical analysis	To assess humidity levels in land areas by converting satellite imagery data into a set of standardized brightness, greenness, and wetness values	Christ, 1985

*There are several spectral indices that are used for vegetation, water, snow, and urban applications, see Mishra et al. 2024.

2006). Apart from these, Mohanty et al., (2024b) utilised combination of spectral indices to assess the sensitivity of indices with sensitivity models (taylor Plot and Lek Profile) towards chlorophyll prediction delivering best indices and also implemented machine learning techniques for aquatic vegetation classification and discrimination.

18.3 IMPORTANCE OF REMOTE SENSING IN AQUATIC ECOSYSTEM MONITORING

The incorporation of remote sensing techniques in the monitoring and assessment of aquatic ecosystems is imperative due to their reliability and wide coverage, as evident from their high scope. RS plays a pivotal role in aquatic ecosystem monitoring due to its ability to provide comprehensive and timely information over large spatial extents within a specified time period, which can be highlighted through a few main features: wide spatial coverage, temporal monitoring, spectral assessment, and high spatial resolution for natural resources monitoring and mapping (Pandey & Sharma, 2021; Glasgow et al., 2004; Adjovu et al., 2023; Giardino et al., 2019). These dimensions provide robust monitoring and assessment of aquatic ecosystems with ease, speed, and reliability in a timely manner with wider coverage (Borja et al., 2016) and reflect the capability of RS to cover vast regions, including rivers, wetlands, lakes, estuaries, and ocean, which researchers may otherwise struggle to cover comprehensively by traditional methods that are tedious, expensive, and time consuming. Geospatial techniques provide real-time monitoring and simultaneous observation of vast water bodies, thereby offering valuable insights into their condition and dynamic nature (Sogno et al., 2022). Moreover, geospatial technology with space-borne platforms has the potential to acquire data at regular intervals, thus enabling regular and updated monitoring of aquatic ecosystems (Srivastava et al., 2022). Temporal assessment is pivotal for detecting seasonal variations in water parameters, short-term events such as algal blooms, and long-term trends of events such as the impact of climate on water bodies (Wells et al., 2015; Yang et al., 2019).

Spectral dimensions have been employed to understand the vegetation types and turbidity conditions using reflectance parameters while portraying pigment concentrations and the level of dissolved suspended particles, respectively. This approach provides the discrimination of vegetation types and clear and turbid water conditions (Rowan & Kalacska, 2021; Lee et al., 2011). Thus, spectral diversity from multispectral and hyperspectral data allows for the detection and quantification of various water quality parameters, such as chlorophyll-a (Chl-a) concentration, turbidity, dissolved organic matter, nutrient levels, and pigment concentrations (Gholizadeh et al., 2016; Chang et al., 2015). Narrow spectral channels offer even greater spectral resolution, enabling the discrimination of subtle changes in aquatic features, especially phytoplankton, and can be used for monitoring invasive weed species and their habitats (Dierssen et al., 2021; Hestir et al., 2008). Remote sensing technology can be employed to monitor and assess other parameters such as salinity, pH, temperature, sedimentation, habitat mapping, and harmful algal blooms (HABs), which are critical for aquatic ecosystem health (Chang et al., 2015; Kavanaugh et al., 2021). Understanding these parameters is crucial for habitat suitability of diverse organisms, identification of sensitive regions of ecological importance, and prevention of further pollution or damage due to anthropogenic disturbances (Zacharias & Gregr, 2005; Geist, 2011). Remote sensing also plays a crucial role in monitoring the impacts of climate change on aquatic ecosystems, mainly glaciers and ocean systems (Yang et al., 2013; Cui et al., 2019). It provides information and assessment about the retreat and melting of glaciers, snow cap melting, sea level rise, sea surface temperature variation, and ocean acidification (Raup et al., 2015; Bhushan & Sharma, 2022). The level of information provided by the outcomes and products of remote sensing helps in understanding these changes and provides a platform for predicting future scenarios to adopt strategies to mitigate the adverse effects on aquatic biodiversity and ecosystem services.

For example, Mandlburger (2022) demonstrated a critical review of the applications of active and passive optical methods in hydrography for shallow, medium, and deep water using different optical methods. The analysis was demonstrated according to the depth categories (i.e., 0 to 50 m, 50 to 500 m, and >500 m) and the appropriate hydrographic techniques used for their monitoring. These techniques incorporate optical remote sensing, including drones, airborne remote sensing, SONAR, and satellite gravimetry. According to the definition issued by the International Hydrographic Organization (IHO):

> Hydrography is the branch of applied sciences that deals with the measurement and description of the physical features of oceans, seas, coastal areas, lakes and rivers, as well as with the prediction of their change over time, for the primary purpose of safety of navigation and in support of all other marine activities, including economic development, security and defense, scientific research, and environmental protection.
>
> **(IHO, 2022)**

Several applications of hydrography are mentioned later in Tablular form, where datasets are provided by GEE (cloud computing).

Therefore, with the above discussion, it becomes evident that remote sensing data, when integrated with other sources of information (e.g., field observations, modeling), offers significant contributions to decision-making and the management of aquatic ecosystems (Rose et al., 2015; Pettorelli et al., 2018) (C.C. de Araujo Barbosa et al., 2015). This integration aids policymakers, government agencies, resource managers, and stakeholders in pinpointing crucial conservation areas, devising strategies for restoration projects, and evaluating the efficacy of management actions (Lim & Do, 2023; Ervin et al., 2010). In conclusion, remote sensing is indispensable for aquatic ecosystem monitoring, offering powerful tools and techniques for understanding the complex dynamics of aquatic ecosystems and their environments, assessing their health and resilience, and guiding sustainable management practices to ensure their long-term conservation.

18.4 CHALLENGES: CONVENTIONAL AND TECHNOLOGICAL

The aquatic ecosystems offer a number of challenges for scientific investigations because of their shape and complex interactions with the surrounding environment comprising land use types, as well as weather and climate (Soininen et al., 2015). For example, river systems are largely ribbon shaped and while studying a given location one has to consider upstream conditions (Kleinhans et al., 2013). The recent advancements in geospatial data collection and analytics can address most of the challenges associated with methods for investigating such ecosystems, due to temporal data availability and large area coverage (Bishop et al., 2012; Palmer et al., 2015).

There are distinct factors which are studied and analyzed to understand and assess aquatic ecosystem health, such as physical habitats, water quality, sedimentation, hydrophytes, phytoplanktons, zooplanktons, periphyton, and macrophytes (Siddiqui et al., 2023; Majumder et al., 2015; Davies & Jackson, 2006). There are several challenges associated with water quality monitoring, aquatic vegetation monitoring, and mapping. These factors pose hurdles in accurate mapping and providing optimal outcomes and these challenges must be addressed to ensure accurate aquatic ecosystem monitoring. First of all, the cost and monetary aspect pose challenges in terms of high-quality sensors, sampling devices, labor costs, travel costs, and data loggers for continuous data collection which require significant monetary investment (Calderwood et al., 2020). This involves infrastructure setup for testing the samples, data integration software, and storage of solutions for analysis, creating a significant financial burden (Slaný et al., 2020). Data may be required for short-term monitoring, long-term monitoring, off-site monitoring, or real-time monitoring, depending on the several technology-related challenges across the research. These are related to the availability, accuracy, robustness, reliability, scalability, interoperability, and security of data for monitoring purposes. Different types of sensors are required for freshwater, saline water, and brackish water monitoring to provide reliability in different conditions and varying weather along with changing water chemistry (Younos & Heyer, 2015; Hopmans et al., 2021). Scaling is one of the biggest challenges for monitoring large lakes, wetlands, and rivers due to topography, rock types, and confluence of streams or pouring of water from another source, which may result in different water chemistry (Ramachandra & Solanki, 2007). Therefore, the use of geospatial technology and employing remote sensing data in a GIS environment could enhance the spatial coverage while monitoring and analysis could be done simultaneously.

Aquatic field sampling is tiring, tedious, and requires a large number of samples from different locations covering all parts of the target water bodies at different time periods, which is quite expensive too. In other words, multiple samples are required for the accurate analysis of water quality in aquatic ecosystems, which are essential for the statistical analysis and trends in water quality due to high variability in the data at spatiotemporal scales (Bierman et al., 2011; Shi et al., 2017). Spatial analysis and trend measurements require collecting samples from various locations covering the aquatic environment. Therefore, spatial and temporal scales pose major challenges in aquatic ecosystem monitoring. Moreover, once statistical analysis is performed, it is difficult to determine the actual cause or source of degradation of water quality or contents, in the case of a lotic aquatic ecosystem, which degrade water quality or increase nutrients (Akhtar et al., 2021).

Phytoplankton monitoring and assessment also face several challenges due to their variable spatial distribution over time. Moreover, the transient nature of phytoplankton is also one of the hurdles in accurate monitoring (Limates et al., 2016). If research is conducted using any indicators for the presence of phytoplankton, their concentration or availability is of a variable nature and often changes with the season (Platt et al., 2009). As their life spans are short, long-term monitoring cannot be done and their seasonal outcomes are available for any monitoring and assessment purposes. These variations are caused by changing water quality over seasons, substrate changes, or seasonal succession. Habitat represents the interaction of living organisms with the structural environment, which directly affects the aquatic ecosystem (Soukup et al., 2022). These physical habitats are associated with the riparian zones, watershed regions, and channels. Mostly these are assessed using a combination of field data and remote sensing datasets (Rusnák et al., 2022). These challenges highlight the complexities involved in using remote sensing technology for monitoring various aquatic ecosystems and underscore the importance of addressing these challenges through advanced sensor technologies, data processing techniques, and interdisciplinary collaborations (Carey et al., 2019; Shaikh & Birajdar, 2024). Predicting future implications in aquatic ecosystems involves considering a range of factors, including climate change, pollution, overfishing, habitat destruction, invasive species, and human activities. Challenges with remote sensing in aquatic ecosystems monitoring involve several factors including atmospheric interference, signal attenuation, signal penetration in water at different levels, and spectral and spatial resolution for each type, as mentioned in Table 18.2.

18.5 CURRENT SCENARIO, FUTURE SCOPE, AND IMPLICATIONS

Future scope of RS in aquatic ecosystem monitoring involve the use of Cloud Computing and Emerging Sensors/Platforms to Monitor natural resources including Waterbodies mapping and Changes assessment in Intensive Time Series (Sharma et al., 2021). The recent advent in cloud computing platforms provided by Google is one of the best platforms for aquatic ecosystem monitoring, used for all research applications (Zhao et al., 2021). This is popularly known as The Google Earth Engine (GEE) platform, which provides a collection of satellite imageries, server-side data processing, and algorithms for analysis and monitoring purposes (Tamiminia et al., 2020). GEE contains a collection of large datasets, with different spatial resolutions and temporal dimensions, such as Landsat, Sentinel series, Hyperion, MODIS-derived products, and so on (Amani et al., 2020; Tamiminia et al., 2020). The Landsat and Sentinel data catalogs offer good coverage of the river and its basin due to coarse spatial resolution and assist in local and regional scale monitoring (Uereyen & Kuenzer, 2019). Though Sentinel data provides a high ground resolution of 10 m, Landsat provides multi-temporal historic coverage at a good 30 m resolution, which is utilized for temporal

TABLE 18.2
Challenges in Aquatic Ecosystem Monitoring Using Remote Sensing Technologies

	Remote sensing challenges				
Aquatic ecosystems affected	**Atmospheric interference**	**Signal attenuation and penetration**	**Spectral complexity**	**Spatial resolution and scales**	**Validation and ground truthing**
Wetlands	Dense vegetation and water vapour can cause scattering and absorption of radiation, which can affect image quality.	Signals may be attenuated by water absorption, variation in depth may affect the limit on depth penetration, and there is limited ability to detect bottom characteristics.	Diverse vegetation and water quality variation composition may result in complex spectral signatures, which requires advanced machine learning and AI tools and classification algorithms.	Wetlands occur at various scales, from small marshes to large swamp complexes, requiring remote sensing data with appropriate spatial and temporal resolution to capture fine-scale features and temporal dynamics.	Wetlands require validation and ground truthing to ensure accuracy, which can be challenging due to the inaccessible nature of many wetland areas and the dynamic nature of wetland ecosystems.
Riverine ecosystems	Riverine ecosystems are ribbon-like systems with irregular meandering patterns. This often leads to the requirements of multiple image scenes with varying atmospheric interference. River systems often have narrow channels and dense vegetation in the riparian zones along the banks, leading to shadowing effects and difficulties in image interpretation.	Rivers and streams with turbid or murky waters may impede signal penetration, making it challenging to observe underwater topography and habitat features.	Riverine ecosystems exhibit variability in water quality, sediment load, and aquatic vegetation, leading to spectral ambiguity and difficulty in distinguishing between different features.	Riverine ecosystems span different spatial scales, from small streams to large river systems, necessitating remote sensing data at multiple resolutions to accommodate variations in habitat characteristics and ecological processes.	Similar to wetlands, riverine ecosystems also require validation and ground truthing. Given the large coverage of these particular ecosystems, ground truthing for riverine ecosystems involves field data collection along riverbanks and waterways, which can be logistically difficult and time-consuming, especially in remote or inaccessible areas.

(Continued)

TABLE 18.2 (CONTINUED)
Challenges in Aquatic Ecosystem Monitoring Using Remote Sensing Technologies

	Remote sensing challenges				
Aquatic ecosystems affected	**Atmospheric interference**	**Signal attenuation and penetration**	**Spectral complexity**	**Spatial resolution and scales**	**Validation and ground truthing**
Lakes	Lakes may experience atmospheric distortion, particularly in larger bodies of water, affecting the accuracy of remote sensing data collection.	Lakes with varying water clarity may exhibit differences in signal attenuation, affecting the depth and quality of imagery captured by remote sensors.	Lakes may have varying levels of algae, sediments, and dissolved organic matter, resulting in spectral confusion and challenges in differentiating between water constituents.	Lakes vary in size and morphology, from small ponds to large freshwater bodies, requiring remote sensing data at suitable scales to capture detailed information about shoreline dynamics, bathymetry, and water quality parameters.	Ground truthing efforts for lake monitoring may involve boat-based surveys and water sampling to validate remote sensing data, which can be constrained by weather conditions and accessibility issues.
Estuaries	Influence of tidal dynamics and sediment transport on water clarity and bottom reflectance, affecting the interpretation of remote sensing data.	Fresh water, saline water and sometimes, brackish water properties cause fluctuations and variations.	Mixing of freshwater and saltwater in estuarine environments leads to variable water properties and spectral signatures.	Challenges in delineating estuarine boundaries and distinguishing between terrestrial and aquatic features.	Water level variation is a major limiting factor. Seasonal variations in water levels and vegetation cover influence the accuracy of habitat mapping and land-water classification.
Ocean	Large area coverage and influence of sea surface water interaction with the atmosphere may cause disturbances in radiation and low-quality images.	Influence of atmospheric conditions, such as aerosols and water vapor, on satellite imagery acquired over marine environments.	A limited spectral resolution of satellite sensors for distinguishing between different ocean features and biogeochemical properties.	The large spatial extent and remote location of open ocean areas require specialized satellite sensors and data processing techniques.	Large area and collection of ground truthing is tedious, rigorous, and dangerous in open ocean and affected by thermal and saline gradients, which fluctuate over time.

change assessment at local and global levels (Mahdianpari et al., 2020). Several water applications could be harnessed from the GEE and associated datasets, such as visualization of water bodies, changes in their extents or pathways, and monitoring over a period of time (Pu et al., 2021). For this purpose, spectral indices are incorporated from bands of multispectral datasets. There are several script codes that can be applied to datasets in GEE to assess and monitor aquatic ecosystems, mainly wetlands, rivers, glaciers, lakes, and oceans (Singh et al., 2020; Wang et al., 2022). The techniques most utilized are water dynamic change assessment, threshold-based water extraction, machine learning-based water extraction, and spectral analysis based on spectral indices-based water extraction (Bijeesh & Narasimhamurthy, 2020; Wu et al., 2024). Several studies demonstrated the utilization of Google Earth Engine to develop an operational cyanobacteria bloom monitoring across inland water (Jia et al., 2019), including spatiotemporal patterns of cyanobacteria blooms and interannual coverage of cyanobacteria blooms. Chlorophyll-a estimation using RS has been used to assess the health and response of aquatic ecosystems, as well as coastal ecosystems, to changing climatic conditions. Zhao et al. (2024) demonstrated the assessment and spatiotemporal monitoring of Chl-a concentrations using Sentinel-2 images in GEE and developed a Chl-a retrieval algorithm based on 13 optical water types (OWTs) with optimal performance and higher accuracy. Therefore, GEE is especially suitable for long-term and large-scale spatiotemporal series of RS applications; its multi-petabyte geospatial data and high-performance computation capacity provide a new approach to understanding the Earth (Jia et al., 2019). Conventional remote sensing algorithms and methods, which were initially conducted with small-scale data, have the potential to be embedded and integrated in GEE, and applied in a variety of fields and scenarios (Zhou et al., 2020).

Modern ways to study aquatic ecosystems nowadays include sophisticated instruments and large complex datasets such as the availability of high-resolution digital elevation models, new bathymetry data (LiDAR, SONAR, and physics-based models applied on multispectral images), aerial survey data from drones, and smart sensors installed in situ for continuous monitoring (Wölfl et al., 2019; Vélez-Nicolás et al., 2021).

LiDAR is nowadays used to survey and map underwater details in aquatic environments (Purkins et al., 2013). Underwater or bathimetric LiDAR is used for exploration as it can provide high-resolution 3D maps of the underwater environment with high accuracy (Pittman et al., 2013) for bathymetry and depth data, bottom structure (Klemas, 2011), and underwater landscapes with clarity for better understanding of changes over time (Szafarczyk & Toś, 2022). It can be used to perform underwater stream surveys and inspections for changes or monitoring in real time. Thus LiDAR is the future of underwater surveillance with precision mapping, enhanced exploration capabilities, and the ability to assist in efficient surveys across aquatic ecosystems such as marine macroalgal communities (Zavalas et al., 2014), submerged vegetation, reefs (Collins et al., 2020, Collins et al., 2018), or bottom landscape of rivers and marine environments (Filisetti et al., 2018).

Apart from advanced technologies in sensors and instruments, several advanced algorithms and models are being developed to understand and simulate flood-prone zones as well as spatial hydrology algorithms to derive watershed boundaries and drainage patterns (Kumar et al., 2023; Daniel et al., 2011, Omran et al., 2016). These models are also used as plugins and are available in GIS software such as ArcGIS, SAGA GIS, etc., e.g. drainage density, total drop, catchment shape, and a range of other catchment characteristics (see Table 18.4). There are several data and models used in the aquatic environment for different applications, such as ANUDEM, improved DEM/DTM, better runoff models, accurate characterization of land use with remote sensing images, Global Navigation Satellite System (GNSS) receiver for better field data collection, flood models, and hydrologic and hydraulic models. Aquatic ecosystems and its environment can be effectively monitoring using remote sensing data in GIS environment (Mohanty et al., 2024). There are several lists of tools and functionalities (extensions/plugins) available in SAGA GIS and other geospatial-related software for assessing hydrology, catchment characteristics, and related research themes enabling comprehensive estimation and management of water resources and aquatic ecosystems (as mentioned in Table 18.3).

TABLE 18.3
Lists of Tools and Functionalities (Extensions/Plugins) Available in Different Software

Software	Extensions/plugins	Features for aquatic research themes	References
ArcGIS	Spatial Analyst Extension	ArcGIS Spatial Analyst provides a suite of tools for terrain analysis, including extraction of catchment areas, flow direction, flow accumulation, and stream network delineation.	Merwade, 2012; Chinnayakanahalli et al., 2006; Wu & Lane, 2017; Bajjali, 2023
	Hydrology Toolset	This toolset includes tools for calculating drainage density, stream order, stream slope, and other hydrological parameters.	Palaka & Sankar, 2016; Gupta, 2016
	3D Analyst Extension	3D Analyst can be used for visualizing and analyzing terrain characteristics, such as elevation profiles and slope analysis.	Conde et al., 2014; Carreño-Conde et al., 2012
	Arc Hydro Tools	A set of tools specifically designed for hydrological analysis and modeling, including watershed delineation, stream segmentation, and flow accumulation.	Shamsi, 2008; Kraemer & Panda, 2009; Djokic et al., 2011; Chinnayakanahalli et al., 2006
	GeoAnalytics Tools	These tools offer scalable analysis capabilities for processing large datasets, including spatial aggregation and summarization.	Xie et al., 2019; Jern et al., 2008; Jolma et al., 2008;
	Remote Sensing Tools	ArcGIS integrates with remote sensing data for land use and land cover classification, allowing for the characterization of catchment areas using satellite imagery. Surface temperature of water bodies.	Mustaquim, 2024; Sameen & Kubaisy, 2014
	Morphometric Toolbox	A new technique in basin morphometric analysis using ArcGIS.	Beg, 2015; Beg, 2020
SAGA GIS	Terrain Analysis Tools	SAGA GIS provides a comprehensive set of tools for terrain analysis, including hydrological flow routing, catchment delineation, and slope stability analysis.	Olaya & Conrad, 2009; Chen et al., 2010; Köthe & Bock, 2006
	Hydrology Tools	Similar to ArcGIS, SAGA GIS offers tools for calculating flow accumulation, flow direction, stream network extraction, and other hydrological parameters.	Chen et al., 2010; Passy & Théry, 2018; Fárek et al., 2014
	Remote Sensing Modules	SAGA GIS includes modules for processing remote sensing data, such as image classification, vegetation indices calculation, and land cover change detection.	
	Flood Risk Assessment Tools	SAGA GIS can be used for flood risk assessment and modeling, including inundation mapping and floodplain analysis.	Feloni et al., 2020; Aksoy et al., 2016; Aleksova et al., 2023
	Geostatistics Tools	SAGA GIS provides tools for geostatistical analysis, including interpolation methods for generating improved DEMs or DTMs.	Khal et al., 2018; Malamos & Koutsoyiannis, 2018; Hengl, 2009; Passy & Théry, 2018; Conrad et al., 2015

(Continued)

biodiversity, sustainable growth, and health maintenance of these fragile environments. Through systematic monitoring efforts using either traditional or advanced technologies, researchers can gather valuable data on water quality, biodiversity, and ecosystem dynamics for detailed analysis, such as spatiotemporal assessment and change analysis and its trends over time. Rapid detection of HABs using remote sensing helps in implementing timely management strategies to mitigate their impacts. The timely outcomes or generated products are essential for making well-planned policies and management decisions, implementing effective conservation measures, and mitigating the impacts of human activities on aquatic habitats. By investing in ongoing monitoring programs, emerging techniques, and incorporating newly advanced technologies and methodologies such as artificial intelligence, machine learning, and geospatial technology, we can better understand the nature of aquatic ecosystems and will be involved in the sustainable use, conservation, and protection for our future generations. Achieving SDGs linked with delicate aquatic ecosystems are facing major challenges such as pollution, chemical discharge, unsustainable practices (overharvesting, habitat destruction), while climate change exacerbates these challenges and issues posing a threat to efficient and effective management and conservation efforts. Finally, the healthier aquatic ecosystem is impossible without adequate governance, enough financial resources, sufficient datasets on ecosystem along with global actions to facilitate smooth monitoring and sustainable practices of aquatic ecosystems across the world.

ACKNOWLEDGEMENTS

The authors thank the Shiv Nadar Institution of Eminence (Deemed to be University) for continuous help and support. Dr. Pandey is grateful to Science and Engineering Research Board (SERB), Department of Science and Technology, Government of India for financial support SERB-sponsored Project- SRG/02021/00129

REFERENCES

Adjovu, G. E., Stephen, H., James, D., & Ahmad, S. (2023). Overview of the application of remote sensing in effective monitoring of water quality parameters. *Remote Sensing*, *15*(7), 1938.

Akhtar, N., Syakir Ishak, M. I., Bhawani, S. A., & Umar, K. (2021). Various natural and anthropogenic factors responsible for water quality degradation: A review. *Water*, *13*(19), 2660.

Aksoy, H., Kirca, V. S. O., Burgan, H. I., & Kellecioglu, D. (2016). Hydrological and hydraulic models for determination of flood-prone and flood inundation areas. *Proceedings of the International Association of Hydrological Sciences*, *373*, 137–141.

Aleksova, B., Lukić, T., Milevski, I., Spalević, V., & Marković, S. B. (2023). Modelling water erosion and mass movements (wet) by using GIS-based multi-hazard susceptibility assessment approaches: A case study—Kratovska Reka Catchment (North Macedonia). *Atmosphere*, *14*(7), 1139.

Altenburger, R., Brack, W., Burgess, R. M., Busch, W., Escher, B. I., Focks, A., ... Krauss, M. (2019). Future water quality monitoring: Improving the balance between exposure and toxicity assessments of real-world pollutant mixtures. *Environmental Sciences Europe*, *31*(1), 1–17.

Amani, M., Ghorbanian, A., Ahmadi, S. A., Kakooei, M., Moghimi, A., Mirmazloumi, S. M., ... & Brisco, B. (2020). Google earth engine cloud computing platform for remote sensing big data applications: A comprehensive review. *IEEE Journal of Selected Topics in Applied Earth Observations and Remote Sensing,* 13, 5326–5350.

Armstrong, K. F., & Ball, S. L. (2005). DNA barcodes for biosecurity: Invasive species identification. *Philosophical Transactions of the Royal Society B: Biological Sciences*, *360*(1462), 1813–1823.

Bajjali, W. (2023). Watershed delineation. In *ArcGIS Pro and ArcGIS Online: Applications in Water and Environmental Sciences* (pp. 243–258). Cham: Springer International Publishing.

Batina, A., & Krtalić, A. (2023). A review of remote sensing applications for determining lake water quality. Preprints, 2023090489. https://doi.org/10.20944/preprints202309.0489.v1

Beg, A. A. F. (2015). Morphometric toolbox: A new technique in basin morphometric analysis using ArcGIS. *Global Journal of Earth Science and Engineering*, *2*(2), 21–30.

Beg, A. A. F. (2020, February). Spatial analysis of the geomorphic evolution of Tigris River basin using developed ArcGIS-Morphometric toolbox. In *IOP Conference Series: Materials Science and Engineering* (Vol. 737, No. 1, p. 012235). IOP Publishing.

Beng, K. C., & Corlett, R. T. (2020). Applications of environmental DNA (eDNA) in ecology and conservation: Opportunities, challenges and prospects. *Biodiversity and Conservation*, *29*(7), 2089–2121.

Berde, V. B., Chari, P. V. B., & Berde, C. V. (2022). Wetland and biodiversity hotspot conservation. In *Research Anthology on Ecosystem Conservation and Preserving Biodiversity* (pp. 775–784). IGI Global.

Bhushan, B., & Sharma, A. (2022). Sea-level rise due to climate change. In *Flood Handbook* (pp. 265–284). CRC Press.

Bishop, M. P., James, L. A., Shroder Jr, J. F., & Walsh, S. J. (2012). Geospatial technologies and digital geomorphological mapping: Concepts, issues and research. *Geomorphology,* 137(1), 5–26.

Bijeesh, T. V., & Narasimhamurthy, K. N. (2020). Surface water detection and delineation using remote sensing images: A review of methods and algorithms. *Sustainable Water Resources Management,* 6(4), 68.

Bishop, M. P., James, L. A., Shroder Jr, J. F., & Walsh, S. J. (2012). Geospatial technologies and digital geomorphological mapping: Concepts, issues and research. *Geomorphology,* 137(1), 5–26.

Borja, A., Elliott, M., Andersen, J. H., Berg, T., Carstensen, J., Halpern, B. S., ... Rodriguez-Ezpeleta, N. (2016). Overview of integrative assessment of marine systems: The ecosystem approach in practice. *Frontiers in Marine Science*, *3*, 20.

Calderwood, A. J., Pauloo, R. A., Yoder, A. M., & Fogg, G. E. (2020). Low-cost, open source wireless sensor network for real-time, scalable groundwater monitoring. *Water*, *12*(4), 1066.

Carey, C. C., Ward, N. K., Farrell, K. J., Lofton, M. E., Krinos, A. I., McClure, R. P., ... & Arzberger, P. (2019). Enhancing collaboration between ecologists and computer scientists: lessons learned and recommendations forward. *Ecosphere,* 10(5), e02753.

Carreño-Conde, F., García-Martínez, S., Lillo-Ramos, J., & Fernández-Martínez, R. (2012). 3D Detrital aquifer modelling for water resources management of the Regional Park of the Lower Courses of Manzanares and Jarama Rivers (Madrid, Spain). In *Management of Water Resources in Protected Areas* (pp. 161–167). Springer Berlin Heidelberg.

Chang, N. B., Imen, S., & Vannah, B. (2015). Remote sensing for monitoring surface water quality status and ecosystem state in relation to the nutrient cycle: A 40-year perspective. *Critical Reviews in Environmental Science and Technology*, *45*(2), 101–166.

Chen, D., Shams, S., Carmona-Moreno, C., & Leone, A. (2010). Assessment of open source GIS software for water resources management in developing countries. *Journal of Hydro-environment Research*, *4*(3), 253–264.

Chen, J., Chen, S., Fu, R., Li, D., Jiang, H., Wang, C., ... Hicks, B. J. (2022). Remote sensing big data for water environment monitoring: Current status, challenges, and future prospects. *Earth's Future*, *10*(2), e2021EF002289.

Chinnayakanahalli, K., Kroeber, C., Hill, R. A., Tarboton, D. G., Olson, J. R., & Hawkins, C. P. (2006). The multi-watershed delineation tool: GIS software in support of regional watershed analyses. *Utah State University*, Logan.

Chivian, E. (2002). *Biodiversity: Its importance to human health.* Center for Health and the Global Environment, Harvard Medical School, Cambridge, MA, 23.

Cloete, N. A., Malekian, R., & Nair, L. (2016). Design of smart sensors for real-time water quality monitoring. *IEEE access,* 4, 3975–3990.

Collin, A., Ramambason, C., Pastol, Y., Casella, E., Rovere, A., Thiault, L., ... Davies, N. (2018). Very high resolution mapping of coral reef state using airborne bathymetric LiDAR surface-intensity and drone imagery. *International Journal of Remote Sensing*, *39*(17), 5676–5688.

Collings, S., Martin, T. J., Hernandez, E., Edwards, S., Filisetti, A., Catt, G., ... Embry, C. (2020). Findings from a combined subsea LiDAR and multibeam survey at Kingston reef, Western Australia. *Remote Sensing*, *12*(15), 2443.

Conde, F. C., Martínez, S. G., Ramos, J. L., Martínez, R. F., & Mabeth-Montoya Colonia, A. (2014). Building a 3D geomodel for water resources management: Case study in the Regional Park of the lower courses of Manzanares and Jarama Rivers (Madrid, Spain). *Environmental Earth Sciences*, *71*, 61–66.

Conrad, O., Bechtel, B., Bock, M., Dietrich, H., Fischer, E., Gerlitz, L., ... Böhner, J. (2015). System for automated geoscientific analyses (SAGA) v. 2.1. 4. *Geoscientific Model Development*, *8*(7), 1991–2007.

Cui, X., Guo, X., Wang, Y., Wang, X., Zhu, W., Shi, J., ... Gao, X. (2019). Application of remote sensing to water environmental processes under a changing climate. *Journal of Hydrology*, *574*, 892902.

Cummings, J. A., & Smedstad, O. M. (2013). Variational data assimilation for the global ocean. In *Data Assimilation for Atmospheric, Oceanic and Hydrologic Applications* vol II, chapter 13, 303–343.

Maidment, D. R., & Djokic, D. (Eds.). (2000). *Hydrologic and Hydraulic Modeling Support: With Geographic Information Systems.* ESRI, Inc.

Majumder, S., Dhua, R. P., Kar, S., Mishra, T., Mahapatra, S.S., Shit, S., & Patra, A. (2015). Zooplankton diversity influenced by hydro biological parameters in some ponds of south eastern part of Bankura town of WB, India. *International Journal of Advanced Research,* 3(5), 354–368.

Malamos, N., & Koutsoyiannis, D. (2018). Field survey and modelling of irrigation water quality indices in a Mediterranean island catchment: A comparison between spatial interpolation methods. *Hydrological Sciences Journal, 63*(10), 1447–1467.

Maleki, P., Rahman, P., Jafaryan, H., Salmanmahiny, A., Ghorbani, R., Gholizadeh, M., & Harsij, M. (2024). Spatial and temporal distribution of physicochemical parameters of water in Gorgan Bay (the southeast of the Caspian Sea-Iran): Fundamentals for the monitoring of the Caspian Sea coastal areas. *Marine Pollution Bulletin, 201,* 116236.

Mandlburger, G. (2022). A review of active and passive optical methods in hydrography. *The International Hydrographic Review,* (28), 8–52.

Mars Brisbin, M., Brunner, O. D., Grossmann, M. M., &Mitarai, S. (2020). Paired high-throughput, in situ imaging and high-throughput sequencing illuminate acantharian abundance and vertical distribution. *Limnology and Oceanography,* 65(12), 2953–2965.

Martin, N., St-Onge, B., & Waaub, J. (2019). 13 geographic tools for decision making in watershed management. In *Spatial Multicriteria Decision Making and Analysis: A Geographic Information Sciences Approach* (pp. 309–334), Jean-Claude Thill (Ed.), Routledge.

McFeeters, S. K. (1996). The use of the Normalized Difference Water Index (NDWI) in the delineation of open water features. *International journal of remote sensing,* 17(7), 1425–1432.

McLean, R. F., Tsyban, A., Burkett, V., Codignotto, J. O., Forbes, D. L., Mimura, N., ... Ittekkot, V. (2001). Coastal zones and marine ecosystems. *Climate Change,* 343–379.

Menon, G. S., Ramesh, M. V., & Divya, P. (2017, October). A low cost wireless sensor network for water quality monitoring in natural water bodies. In *2017 IEEE Global Humanitarian Technology Conference (GHTC)* (pp. 1–8). IEEE.

Merwade, V. (2012). Watershed and stream network delineation using ArcHydro tools. *School of Civil Engineering, Purdue University,* 1–7.

Mishra, K., Siddiqui, A., Kumar, V., Pandey, K. and Garg, R.D., 2024. Examining effect of super-resolution on AVIRIS-NG data: A precursor to generation of large-scale urban material and natural cover maps. *Advances in Space Research,* 73(2), pp.1573–1596.

Mizaikoff, B. (2003). Infrared optical sensors for water quality monitoring. *Water Science and Technology, 47*(2), 35–42.

Moeller, M. L. (2021). *Low-light Camera field evaluation for underwater nighttime operations (Doctoral dissertation, Monterey, CA; Naval Postgraduate School).*

Mohany, S., Pandey, P. C., Pandey, M., Srivastava, P. K., & Dwivedi, C. S. (2024a). Wetlands contribution and linkage to support SDGs, its indicators and targets- A critical review. *Sustainable Development.* https://onlinelibrary.wiley.com/doi/full/10.1002/sd.2948

Mohanty, S., Pandey, P. C., Singh, P., Dugesar, V., & Srivastava, P. K. (2024b). Vegetation discrimination based on chlorophyll prediction in Marshy wetland using Unmanned Aerial Vehicles. *Aquatic Conservation: Marine and Freshwater Ecosystems, 34*(5), e4170.

Mohanty, S., Wasim, M., Pandey, P. C., & Srivastava, P. K. (2024c). Exploring aquatic environments through geographical information science: A comprehensive review and applications. Petropoulos, GP & Chalkias, C (Eds.) *Geographical Information Science,* 231–251. John Wiley & Sons.

Moldoveanu, M., Sarbu, D., Galie, A. C., Bojan, V., & Mezdroiu, I. (2017). Assessment of river water quality using an integrated hydro-morphological, physicochemical and biological approach. *Advances in Environmental Sciences, 9*(2), 158–172.

Moomaw, W. R., Chmura, G. L., Davies, G. T., Finlayson, C. M., Middleton, B. A., Natali, S. M., ... Sutton-Grier, A. E. (2018). Wetlands in a changing climate: Science, policy and management. *Wetlands, 38*(2), 183–205.

Moss, B. R. (2009). *Ecology of Fresh Waters: Man and Medium, Past to Future.* John Wiley & Sons.

Mueller, N., Lewis, A., Roberts, D., Ring, S., Melrose, R., Sixsmith, J., ... Ip, A. (2016). Water observations from space: Mapping surface water from 25 years of Landsat imagery across Australia. *Remote Sensing of Environment, 174,* 341–352.

Mustaquim, S. M. (2024). Utilizing remote sensing data and ArcGIS for advanced computational analysis in land surface temperature modeling and land use property characterization. *World Journal of Advanced Research and Reviews, 21*(1), 1496–1507.

Nagendra, H., Lucas, R., Honrado, J. P., Jongman, R. H., Tarantino, C., Adamo, M., & Mairota, P. (2013). Remote sensing for conservation monitoring: Assessing protected areas, habitat extent, habitat condition, species diversity, and threats. *Ecological Indicators*, *33*, 45–59.

NASA. (2024). Earth science missions: Upcoming. https://science.nasa.gov/earth-science/missions/earth-missions/earth-missions-future Accessed March 2024.

National Research Council, Division on Earth, Life Studies, Commission on Geosciences, Committee on Restoration of Aquatic Ecosystems, & Public Policy. (1992). *Restoration of Aquatic Ecosystems: Science, Technology, and Public Policy.* National Academies Press.

Olaya, V., & Conrad, O. (2009). Geomorphometry in SAGA. *Developments in Soil Science*, *33*, 293–308.

Omran, A., Dietrich, S., Abouelmagd, A., & Michael, M. (2016). New ArcGIS tools developed for stream network extraction and basin delineations using Python and java script. *Computers & Geosciences*, *94*, 140–149.

Onojake, M. C., Obi, C., & Mahmoud, A. E. D. (2023). Water quality monitoring using sensors and models. In *Artificial Intelligence and Modeling for Water Sustainability* (pp. 97–127). CRC Press.

Palaka, R., & Sankar, G. J. (2016). Study of watershed characteristics using Google Elevation Service. *Geospatial World.*

Palmer, S. C., Kutser, T., & Hunter, P. D. (2015). Remote sensing of inland waters: Challenges, progress and future directions. *Remote sensing of Environment,* 157, 1–8.

Pandey, P. C., Koutsias, N., Petropoulos, G. P., Srivastava, P. K., & Ben Dor, E. (2021). Land use/land cover in view of earth observation: Data sources, input dimensions, and classifiers—A review of the state of the art. *Geocarto International, 36*(9), 957–988.

Pandey, P. C., & Sharma, L. K. (2021a). Introduction to natural resource monitoring using remote sensing technology. Pandey PC., & Sharma LK., (Eds.), *Advances in Remote Sensing for Natural Resource Monitoring*, 1–8, Wiley Blackwell.

Pandey, P. C., & Sharma, L. K. (Eds.). (2021b). *Advances in remote sensing for natural resource monitoring.* John Wiley & Sons.

Park, J., Kim, K. T., & Lee, W. H. (2020). Recent advances in information and communications technology (ICT) and sensor technology for monitoring water quality. *Water, 12*(2), 510.

Passy, P., & Théry, S. (2018). The use of SAGA GIS modules in QGIS. *QGIS and Generic Tools*, *1*, 107–149.

Pekel, J.-F., Cottam, A., Gorelick, N., Belward, A.-S. (2016). High-resolution mapping of global surface water and its long-term changes. *Nature, 540*, 418–422. https://doi.org/10.1038/nature20584

Pettorelli, N., Schulte to Bühne, H., Tulloch, A., Dubois, G., Macinnis-Ng, C., Queirós, A. M., ... Nicholson, E. (2018). Satellite remote sensing of ecosystem functions: Opportunities, challenges and way forward. *Remote Sensing in Ecology and Conservation*, *4*(2), 71–93.

Pipitone, C., Maltese, A., Dardanelli, G., Lo Brutto, M., & La Loggia, G. (2018). Monitoring water surface and level of a reservoir using different remote sensing approaches and comparison with dam displacements evaluated via GNSS. *Remote Sensing, 10*(1), 71.

Pittman, S. J., Costa, B., & Wedding, L. M. (2013). LiDAR applications. In *Coral Reef Remote Sensing: A Guide for Mapping, Monitoring and Management* (pp. 145–174). Springer Netherlands.

Pu, G., Quackenbush, L. J., & Stehman, S. V. (2021). Using google earth engine to assess temporal and spatial changes in river geomorphology and riparian vegetation. *JAWRA Journal of the American Water Resources Association,* 57(5), 789–806.

Pu, G., Quackenbush, L. J., & Stehman, S. V. (2021). Using google earth engine to assess temporal and spatial changes in river geomorphology and riparian vegetation. *JAWRA Journal of the American Water Resources Association,* 57(5), 789–806.

Purkis, S. J., & Brock, J. C. (2013). LiDAR overview. In *Coral Reef Remote Sensing: A Guide for Mapping, Monitoring and Management* (pp. 115–143). *Dordrecht:Springer Netherlands.*

Qin, B., Gao, G., Zhu, G., Zhang, Y., Song, Y., Tang, X., ... Deng, J. (2013). Lake eutrophication and its ecosystem response. *Chinese Science Bulletin*, *58*, 961–970.

Ramachandra, T. V., & Solanki, M. (2007). Ecological assessment of lentic water bodies of Bangalore. *The Ministry of Science and Technology*, *25*(96), 2.

Raup, B. H., Andreassen, L. M., Bolch, T., & Bevan, S. (2015). Remote sensing of glaciers. In Bruce H. Raup, Liss M. Andreassen, Tobias Bolch, Suzanne Bevan (Eds.). *Remote Sensing of the Cryosphere* (pp. 123–156).

Reddy, M. T., Sivaraj, N., Kamala, V., Pandravada, S. R., Sunil, N., & Dikshit, N. (2018). Classification, characterization and comparison of aquatic ecosystems in the landscape of Adilabad District, Telangana, Deccan Region, India. *Open Access Library Journal*, *5*(4), 1–49.

Rees, H. C., Maddison, B. C., Middleditch, D. J., Patmore, J. R., & Gough, K. C. (2014). The detection of aquatic animal species using environmental DNA–a review of eDNA as a survey tool in ecology. *Journal of Applied Ecology, 51*(5), 1450–1459.

Ritchie, J. C., Zimba, P. V., & Everitt, J. H. (2003). Remote sensing techniques to assess water quality. *Photogrammetric engineering & remote sensing, 69*(6), 695–704.

Rose, R. A., Byler, D., Eastman, J. R., Fleishman, E., Geller, G., Goetz, S., ... Wilson, C. (2015). Ten ways remote sensing can contribute to conservation. *Conservation Biology, 29*(2), 350–359.

Rowan, G. S., & Kalacska, M. (2021). A review of remote sensing of submerged aquatic vegetation for non-specialists. *Remote Sensing, 13*(4), 623.

Ruppert, K. M., Kline, R. J., & Rahman, M. S. (2019). Past, present, and future perspectives of environmental DNA (eDNA) metabarcoding: A systematic review in methods, monitoring, and applications of global eDNA. *Global Ecology and Conservation, 17*, e00547.

Salameh, E., Frappart, F., Almar, R., Baptista, P., Heygster,G., Lubac, B., ... & Laignel, B. (2019). Monitoring beach topography and nearshore bathymetry using spaceborne remote sensing: A review. *Remote Sensing, 11*(19), 2212.

Salameh, E., Frappart, F., Almar, R., Baptista, P., Heygster,G., Lubac, B., ... & Laignel, B. (2019). Monitoring beach topography and nearshore bathymetry using spaceborne remote sensing: A review. *Remote Sensing, 11*(19), 2212.

Sameen, M. I., & Al Kubaisy, M. A. (2014). Automatic surface temperature mapping in arcgis using landsat-8 tirs and envi tools, case study: Al Habbaniyah Lake. *Journal of Environmental & Earth Sciences, 4*(12), 12–17.

Shaikh, M., & Birajdar, F. (2024). Advancements in remote sensing and GIS for sustainable groundwater monitoring: applications, challenges, and future directions. *International Journal of Research in Engineering, Science and Management, 7*(3), 16–24.

Shamsi, U. M. S. (2008). Arc hydro: A framework for integrating GIS and hydrology. *Journal of Water Management Modeling.* R228–11. doi: 10.14796/JWMM.R228-11.

Sharma, L. K., Gupta, R., & Pandey, P. C. (2021). Future aspects and potential of the remote sensing technology to meet the natural resource needs. *Advances in remote sensing for natural resource monitoring*, 445-464. Wiley Blackwell.

Shelton, L. R. (1994). Field guide for collecting and processing stream-water samples for the National Water-Quality Assessment Program (Vol. 94, No. 455). US Geological Survey.

Siddiqui, A. J., Jahan, S., Adnan, M., Ashraf, S. A., & Singh, R. (2023). Macrophytes and Their Role in Wetland Ecosystems. In *Aquatic Macrophytes: Ecology, Functions and Services* (pp. 119–138). Singapore: Springer Nature Singapore.

Siddiqui, A. J., Jahan, S., Adnan, M., Ashraf, S. A., & Singh, R. (2023). Macrophytes and Their Role in Wetland Ecosystems. In *Aquatic Macrophytes: Ecology, Functions and Services* (pp. 119–138). Singapore: Springer Nature Singapore.

Singh, G., Reynolds, C., Byrne, M., & Rosman, B. (2020). A remote sensing method to monitor water, aquatic vegetation, and invasive water hyacinth at national extents. *Remote Sensing, 12*(24), 4021.

Singh, P., Gupta, A., & Singh, M. (2014). Hydrological inferences from watershed analysis for water resource management using remote sensing and GIS techniques. *The Egyptian Journal of Remote Sensing and Space Science, 17*(2), 111–121.

Sivakamavalli, J. (2022). Environment biomonitoring with eDNA—A new perspective to identify biodiversity. In *New Paradigms in Environmental Biomonitoring Using Plants* (pp. 109–164). Elsevier.

Skalska, K., Ockelford, A., Ebdon, J. E., & Cundy, A. B. (2020). Riverine microplastics: Behaviour, spatio-temporal variability, and recommendations for standardised sampling and monitoring. *Journal of Water Process Engineering, 38*, 101600.

Slaný, V., Lučanský, A., Koudelka, P., Mareček, J., Krčálová, E., & Martínek, R. (2020). An integrated iot architecture for smart metering using next generation sensor for water management based on lorawan technology: A pilot study. *Sensors, 20*(17), 4712.

Sogno, P., Klein, I., & Kuenzer, C. (2022). Remote sensing of surface water dynamics in the context of global change—a review. *Remote Sensing, 14*(10), 2475.

Soininen, J., Bartels, P. I. A., Heino, J., Luoto, M., &Hillebrand, H. (2015). Toward more integrated ecosystem research in aquatic and terrestrial environments. *BioScience, 65*(2), 174–182.

Soukup, P. R., Näslund, J., Höjesjö, J., & Boukal, D. S.(2022). From individuals to communities: Habitat complexity affects all levels of organization in aquatic environments. *Wiley Interdisciplinary Reviews: Water, 9*(1), e1575.

Srivastava, A., Singhal, A., & Jha, P. K. (2022). Geospatial technology for sustainable management of water resources. In *Ecological Significance of River Ecosystems* (pp. 105–132). Elsevier.

Stefanidis, K., Papaioannou, G., Markogianni, V., & Dimitriou, E. (2019). Water quality and hydromorphological variability in Greek rivers: A nationwide assessment with implications for management. *Water, 11*(8), 1680.

Sultan, S. E. (2015). *Organism and Environment: Ecological Development, Niche Construction, and Adaptation*. Oxford University Press.

Szafarczyk, A., & Toś, C. (2022). The use of green laser in LiDAR bathymetry: State of the art and recent advancements. *Sensors, 23*(1), 292.

Tabibi, S., & Francis, O. (2020). Can GNSS-R detect abrupt water level changes?. *Remote Sensing, 12*(21), 3614.

Tamiminia, H., Salehi, B., Mahdianpari, M., Quackenbush, L., Adeli, S., & Brisco, B. (2020). Google Earth Engine for geo-big data applications: A meta-analysis and systematic review. *ISPRS journal of photogrammetry and remote sensing, 164,* 152–170.

Tellman, B., Sullivan, J. A., Kuhn, C., Kettner, A. J., Doyle, C. S., Brakenridge, G. R., Erickson, T., & Slayback, D. A. (2021). Satellites observe increasing proportion of population exposed to floods. *Nature*. https://doi.org/10.1038/s41586-021-03695-w

Thomsen, P. F., & Willerslev, E. (2015). Environmental DNA–An emerging tool in conservation for monitoring past and present biodiversity. *Biological Conservation, 183*, 4–18.

Vélez-Nicolás, M., García-López, S., Barbero, L., Ruiz-Ortiz, V., & Sánchez-Bellón, Á. (2021). Applications of unmanned aerial systems (UASs) in hydrology: A review. *Remote Sensing, 13*(7), 1359.

Vélez-Nicolás, M., García-López, S., Barbero, L., Ruiz-Ortiz, V., & Sánchez-Bellón, Á. (2021). Applications of unmanned aerial systems (UASs) in hydrology: A review. *Remote Sensing, 13*(7), 1359.

Wang, H., Kong, X., Luo, J., Li, P., Xie, T., Yi, X., ... & Xiao, J. (2022). A Novel Approach for Monitoring the Ecoenvironment of Alpine Wetlands using Big Geospatial Data and Cloud Computing. *Advances in Meteorology,* 2022(1), 7451173.

Weldeslassie, T., Naz, H., Singh, B., & Oves, M. (2018). Chemical contaminants for soil, air and aquatic ecosystem. *Modern Age Environmental Problems and their Remediation*, (pp. 1–22).

Wells, M. L., Trainer, V. L., Smayda, T. J., Karlson, B. S., Trick, C. G., Kudela, R. M., ... Cochlan, W. P. (2015). Harmful algal blooms and climate change: Learning from the past and present to forecast the future. *Harmful Algae, 49*, 68–93.

White, A. M., Gardner, W. P., Borsa, A. A., Argus, D. F., & Martens, H. R. (2022). A review of GNSS/GPS in hydrogeodesy: Hydrologic loading applications and their implications for water resource research. *Water Resources Research, 58*(7), e2022WR032078.

Winkler, S., Rieger, L., Saracevic, E., Pressl, A., & Gruber, G. (2004). Application of ion-sensitive sensors in water quality monitoring. *Water Science and Technology, 50*(11), 105–114.

Wölfl, A. C., Snaith, H., Amirebrahimi, S., Devey, C. W., Dorschel, B., Ferrini, V., ... & Wigley, R. (2019). Seafloor mapping–the challenge of a truly global ocean bathymetry. *Frontiers in Marine Science, 6,* 283.

Wolny, J. L., Tomlinson, M. C., Schollaert Uz, S., Egerton,T. A., McKay, J. R., Meredith, A., ... & Stumpf, R. P. (2020). Current and future remote sensing of harmful algal blooms in the Chesapeake Bay to support the shellfish industry. *Frontiers in Marine Science, 7,* 337.

Wu, Q., & Lane, C. R. (2017). Delineating wetland catchments and modeling hydrologic connectivity using lidar data and aerial imagery. *Hydrology and Earth System Sciences, 21*(7), 3579–3595.

Wu, Y., Ma, J., Zhang, W., Sun, L., Liu, Y., Liu, B., ...& Chen, Z. (2024). Rapid evaluation of drought tolerance of winter wheat cultivars under water-deficit conditions using multi-criteria comprehensive evaluation based on UAV multispectral and thermal images and automatic noise removal. *Computers and Electronics in Agriculture, 218,* 108679.

Xie, C., Li, M., Wang, H., & Dong, J. (2019). A survey on visual analysis of ocean data. *Visual Informatics, 3*(3), 113–128.

Xu, H. (2006). Modification of normalised difference water index (NDWI) to enhance open water features in remotely sensed imagery. *International journal of remote sensing, 27*(14), 3025–3033.

Yang, J., Gong, P., Fu, R., Zhang, M., Chen, J., Liang, S., ... Dickinson, R. (2013). The role of satellite remote sensing in climate change studies. *Nature Climate Change, 3*(10), 875–883.

Yang, J., Holbach, A., Wilhelms, A., Qin, Y., Zheng, B., Zou, H., ... Norra, S. (2019). Highly time-resolved analysis of seasonal water dynamics and algal kinetics based on in-situ multi-sensor-system monitoring data in Lake Taihu, China. *Science of the Total Environment, 660*, 329–339.

Yang, Q., Van Niel, T. G., McVicar, T. R., Hutchinson, M. F., & Li, L. (2005). Developing a digital elevation model using ANUDEM for the Coarse Sandy Hilly Catchments of the Loess Plateau, China. *CSIRO Land and Water Technical Report*, *7*(5), 1–74.

Yaroshenko, I., Kirsanov, D., Marjanovic, M., Lieberzeit, P. A., Korostynska, O., Mason, A., ... Legin, A. (2020). Real-time water quality monitoring with chemical sensors. *Sensors*, *20*(12), 3432.

Younos, T., & Heyer, C. J. (2015). Advances in water sensor technologies and real-time water monitoring. In Tamim Younos, Tammy E. Parece (Eds.) *Advances in Watershed Science and Assessment* (pp. 171–203), Springer.

Yusoff, F. M., Abdullah, A. F., Aris, A. Z., & Umi, W. A. D. (2021). Impacts of COVID-19 on the aquatic environment and implications on aquatic food production. *Sustainability*, *13*(20), 11281.

Zacharias, M. A., & Gregr, E. J. (2005). Sensitivity and vulnerability in marine environments: An approach to identifying vulnerable marine areas. *Conservation Biology*, *19*(1), 86–97.

Zavalas, R., Ierodiaconou, D., Ryan, D., Rattray, A., & Monk, J. (2014). Habitat classification of temperate marine macroalgal communities using bathymetric LiDAR. *Remote Sensing*, *6*(3), 2154–2175.

Zhao, D., Huang, J., Li, Z., Yu, G., & Shen, H. (2024). Dynamic monitoring and analysis of chlorophyll-a concentrations in global lakes using Sentinel-2 images in Google Earth Engine. *Science of the Total Environment*, *912*, 169152.

Zhao, L., & Yin, Y. (2012, May). The new multi-spectral water quality analyzer based on MSP430. In *2012 International Conference on Systems and Informatics (ICSAI2012)* (pp. 320–325). IEEE.

Zhou, B., Okin, G. S., & Zhang, J. (2020). Leveraging Google Earth Engine (GEE) and machine learning algorithms to incorporate in situ measurement from different times for rangelands monitoring. *Remote Sensing of Environment, 236,* 111521.

Zhou, B., Okin, G. S., & Zhang, J. (2020). Leveraging Google Earth Engine (GEE) and machine learning algorithms to incorporate in situ measurement from different times for rangelands monitoring. *Remote Sensing of Environment, 236,* 111521.

Index